Michel Capderou

Satellites
Orbits and Missions

Translated from French by Stephen Lyle

Springer

Michel Capderou
L.M.D.
École Polytechnique
91128 Palaiseau
capderou@lmd.polytechnique.fr

Translator
Stephen Lyle
Andebu
09240 Alzen
s.lyle@wanadoo.fr

Ouvrage publié avec le concours du Ministère français chargé de la culture - Centre national du livre
Publication supported by the French Ministry of Culture - National Book Centre

ISBN : 2-287-21317-1
Additional material to this book can be downloaded from http://extras.springer.com

Imprimé en France
Springer-Verlag France est membre du groupe Springer Science + Business Media

SPIN: 10994245

Cover design : Nadia OUDDANE

Foreword

In all, the European Space Agency (ESA) has developed 60 spacecrafts over the last few decades. Some of these, such as the ERS satellites and Envisat, are dedicated to monitoring the Earth and providing vital data on the health status of our planet. Other spacecraft have helped to improve the accuracy of weather forecasting. Galileo, the joint ESA/EU satellite navigation programme, demonstrates the political dimension of space as service-provider for the benefit of European citizens. While these missions involve utilitarian space activities, others are devoted to either exploring the solar system, including the Sun, or achieving a better understanding of the Universe and the cosmic beginnings.

The orbits of these satellites cover a wide range (Sun-synchronous, geostationary, highly eccentric, at Lagrange points, etc.) so that a complete novice may be astonished when tackling the space technology field. The main credit of Michel Capderou's book is to take the reader (whether it be a student, an engineer or a research scientist) progressively from the basic Kepler laws to the most complex equations of space mechanics. His educational concern has led him to propose many examples and graphical illustrations from ESA, but also from the American, Russian, Indian, Japanese or even Chinese space agencies. These programmes provide scientific insights and moreover appear to fascinate the general public, in particular the younger generation. Those wishing to understand the orbital mechanisms behind these programmes will find the explanations they seek in this book.

***Jean-Jacques Dordain*, ESA Director General**

Paris, France
December 2004

Foreword to the French Edition

For more than forty years now, space science has revolutionised a whole range of fields including telecommunications, data transmission, meteorology, geodesy, precise positioning on the Earth's surface, and Earth observation for both civilian and military purposes. Within science itself, it has added a new dimension to astronomy and astrophysics, to the study of the Solar System and planets, and to our understanding of the mechanisms underlying the behaviour of the various subsystems making up the Earth environment, i.e., the atmosphere, the oceans, the biosphere, the continental surfaces, and the cryosphere. Space is now opening up to biology and medecine, and is used as a laboratory for fundamental physics.

These advances are of course due to the invention and development of ever more sophisticated methods of communication and observation, making either active or passive use of the whole range of the electromagnetic spectrum, from the shortest wavelengths associated with X and gamma rays, to microwaves and centimetre waves. However, they also owe much to the capability developed by research teams and engineers in the world's space agencies to position satellites and space probes with greater and greater accuracy on orbits selected to fulfill the commercial, operational or scientific aims of today's missions. Indeed, an Earth-observation system, no matter how advanced it may be, will never yield the results required of it unless it is placed on an appropriate orbit and with suitable orientation, so that it can deliver the predefined spatial and temporal resolutions. Likewise, a communications or positioning system, based for example on a satellite constellation, will not achieve its objectives unless the orbits are correctly chosen to guarantee full temporal coverage. Finally, space probes and large space observatories in different ranges of the electromagnetic spectrum will only lead us to a better understanding of the origin and evolution of our Universe and the Solar System if they are adequately placed to observe the relevant objects.

With this in mind, Michel Capderou's book makes an essential contribution to our understanding of the problems raised by the precise determination of satellite and space probe orbits. His approach, based on physics and mechanics, reminds us that the underlying principles of celestial mechanics based on the theory of gravitation were brought to light by the great names of science, from Tycho Brahe and Copernicus to Newton, from Kepler and Galileo to Laplace, from Lagrange to Poincaré and Einstein. But this does not mean that orbital calculations are any the simpler, and they appear in their full complexity as soon as the relevant problem involves the mutual attractions of several interacting bodies, or requires some consideration of perturbations to the system, due for example to spatial and temporal anomalies in the Earth's gravitational field. But it is to Michel Capderou's credit that he manages to initiate us in the mathematical and physical techniques of orbit calculations whilst nevertheless bringing out the full scope of the problem. Basing his presentation on concrete examples which refer to real missions under development or already launched, we are brought to a better understanding of the relative advantages of the different types of satellite orbit and positioning currently used. The completeness of this treatise guarantees its success as a textbook, just as the accompanying software, developed and perfected by the author himself, will help the reader to build up the kind of three-dimensional vision that is a prerequisite to space observation, so that he or she may then be able to design the orbit and mission appropriate to the problem at hand. Michel Capderou's book is an essential element for any student, engineer or research scientist involved in space science.

Paris, France
July 2002

***Gérard Mégie*, President CNRS**
Professor, Pierre & Marie Curie University

Preface

Space mechanics is one thing, but space missions are many and varied. Indeed, they are involved in the detection of forest fires and cyclones, television relay, positioning on the globe to within a metre, measurement of sea level, and the search for water on Mars, to mention but a few. Since 1957, several thousand satellites have been launched and 1 200 satellite names appear in the index to this book.

It is in this spirit of duality that the book was written: on the one side, the 'eternal' laws of mechanics, and on the other, the rationale behind each type of orbit, given the temporal constraints laid down by the nature of its mission. The text is divided into four parts.

In the first part (Chaps. 1–3), which deals with the motion of the satellite, it is difficult to be innovative with the fundamental laws of dynamics and the theory of perturbations. We have introduced a great many figures in the hope of making this part appear less theoretical.

In the second part (Chaps. 4–7), which is devoted to the orbit and its ground track, we begin to present new tools that we have developed to provide a clearer presentation of these key notions, such as the use of the constant of Sun-synchronicity k_{h}, or the frequency κ and the index Φ for the study of recurrence. The theory is abundantly illustrated by examples relating to a wide range of satellite trajectories, be they circular or elliptical, low or high, or characterised in some other way.

In the next part (Chaps. 8–9), which treats the questions of spatial and temporal sampling, we begin to examine the instruments carried aboard in such a way as to describe what they will see from this vantage point. We study the geometry of the satellite–target and satellite–target–Sun configurations. This part is particularly concerned with Earth-observation satellites.

The results of these three parts are all taken up again in the last part of the book, comprising Chaps. 10 and 11, where we consider a satellite around Mars or some other celestial body.

Ixion, our orbitography and sampling software, provides the backbone of this book. It originated in preparations for the ScaRaB project, a radiometer designed by the LMD. Two models were carried aboard the Russian satellites Meteor-3-07 in 1994 and Resurs-O1-4 in 1998. The software was originally devised to study the angular properties of radiation detected by the instrument.

Having confronted the reality of ScaRaB's pixels, Ixion was extended to all Earth-observation satellites, then to all satellites, whatever their orbit, and even to satellites revolving around attractive bodies other than the Earth. Other features have been developed, such as ways of displaying cycles, precession or recurrence, or the detailed study of sighting conditions in relation to solar illumination. All these ideas are discussed in the book. The software Ixion has since been used for preliminary studies pertaining to observation strategy, in which the relevant target phenomena are tied in with the question of orbital characteristics. The latest applications concern the French–Indian project known as Megha-Tropiques, the US–French mission Calipso and the Mambo instrument for the French mission Premier to Mars.

In the Program part of the adjoined CD, the reader will find the program IxionPC, which is the non-graphing part of Ixion. It can be used interactively to study the orbit of any satellite and establish sampling tables.

The mathematical cartography software Atlas that we have created has been coupled with Ixion to produce graphical representations of orbits and ground tracks. We hope the reader will find the charts we have presented there not just useful for a better understanding of what is at issue, but also pleasant to the eye. Perhaps they will go some way towards banishing the typically grey world of cartographic representations encountered in this field. Apart from the maps appearing in the book itself, the reader will find hundreds of others on the CD (Graphs part). With the exception of a few photographs (indicated by the symbol ©), all the illustrations in this book are original.

Since the main subjects here are space mechanics and the geometry of observation, this book makes no attempt to cover the many technological questions relating to satellites. There is not a word on launch vehicles or ways of getting satellites into orbit, nor on the problem of attitude control or even the instruments carried aboard, except regarding their sighting geometry, which may more or less directly affect the mission.

Throughout the book, our main concern has been to move forward in a gradual way, demonstrating the formulas used and generously interspersing the text with examples and illustrations. We have emphasised the more subtle points, such as different definitions of period and conditions for Sun-synchronicity, to cite but two examples, and we have given some attention to questions which may rouse the reader's curiosity, such as the Ellipso Borealis orbits or the orbits of satellites located at Lagrange points. We have also added a few notes on the history of astronomy, or the more recent history of astronautics and the conquest of space, together with several comments of an etymological nature.

So let me wish you a good trip into space!

Acknowledgements

This work was brought to fruition at the Laboratoire de Météorologie Dynamique (LMD) and I thank the successive directors, Claude Basdevant and Hervé Le Treut, for their support. The LMD is a mixed research unit of the Centre National de la Recherche Scientifique (CNRS), the Ecole Normale Supérieure, the Ecole Polytechnique and the Pierre & Marie Curie University.

I am grateful to all my colleagues at the university and in the CNRS for their comments, criticisms, help, and careful readings, but especially to François Barlier, Christophe Boitel, André Capderou, Olivier Chomette, François Forget, Jacques Lefrère, Patrick Rabéranto, Karim Ramage, Rémy Roca, Jean Roux, Carsten Standfuss, and Pierre Thomas. I am also grateful to Nathalie Huilleret and Nicolas Puech of Springer. *Et je n'oublie pas* Stephen Lyle.

Palaiseau, France
September 2002
November 2004

Michel Capderou

Contents

1 Keplerian Motion

1.1 The Satellite and its Motion

Satellites move discreetly across the heavens, on their tireless rounds. As we scan the sky for shooting stars at nightfall on a summer evening, a satellite may come into view and cross the firmament in ten minutes or so. But it is soon forgotten, like a train which arrives on time. The passing satellite causes no stir.

When the mass media are interested in a satellite, beware! There is disaster in the air. In March 2001, over a period of one week, the Russian space station Mir was the subject of everyone's conversation. Was the sky to fall upon us? Three fatal injections were administered by ground control engineers and the station burnt up in the atmosphere. Its glowing embers fell to Earth in the Pacific Ocean, more or less where they were intended to.

To the uninitiated but curious, this raised several questions. Why destroy the space station? Could it not have prolonged its course, even indefinitely? Was this not perpetual motion? And why was the final splashdown only given approximately, when astronomers are renowned for their accuracy?

When a satellite re-enters the Earth's atmosphere (a crucial stage, as one might imagine, for passengers aboard the American Space Shuttle), determination of the trajectory involves highly complex empirical models. On the other hand, apart from this re-entry stage, a satellite's motion can be explained by the simple formalism invented by Kepler and Newton. This motion, based on Newton's second law, applies to every single artificial satellite launched since 1957, provided that it did not burn up in the atmosphere, together with all space probes ever sent to the Moon or other planets, just as it also applies to natural satellites (e.g., the Moon around the Earth or Titan around Saturn), or the planets and comets that revolve about the Sun. This is Keplerian motion.

In this chapter, we shall thus consider Keplerian motion, first from a general standpoint, using the fundamental equations, then applying it to a body in periodic motion. One might ask whether it is possible to explain every aspect of a satellite trajectory in terms of Keplerian motion. We shall see in Chap. 3 that, although Kepler's laws account for many things, indeed for the main features of the motion, they cannot explain the details. There are certain small differences which, as time goes by, can develop into a large discrepancy. In astronomy, time is of the essence: a period of rotation or revolution of a planet or a satellite can be measured over many years, giving highly accurate results. Approximations must be brought into line.

We shall show that a satellite's motion will accord perfectly with Kepler's laws precisely in the situation where everything is perfect for that satellite. That is, the satellite must be a point object in orbit around a spherical planet devoid of any defect, and the two bodies, satellite and planet, must be alone in the universe. It is because this perfect situation does not exist in the real world that one must make some slight modifications to Keplerian motion, thereby generating a realistic motion that accounts for the various small interactions. This is the so-called perturbed motion.

But let us begin with Keplerian motion. Using the basic laws of kinematics and Newton's second law, we shall deduce the various relations which define and characterise Keplerian motion. The main subject of this study – the satellite – is of macroscopic dimensions and it moves slowly relative to the speed of light. We therefore remain within the framework of of classical mechanics and the Galilean approximation, except for one or two cases that will be clearly distinguished, where relativistic effects will be observed. The frame of reference will thus be assumed to be Galilean (see Chap. 2).

The area of mechanics discussed here is known as celestial mechanics or space mechanics. The term 'celestial' is generally reserved for the study of celestial bodies making up the Solar System. Since satellites and probes have been launched, that is, since the conquest of space, the term 'space mechanics' has been used to refer to our many artificial moons.

1.2 General Acceleration

1.2.1 Velocity and Acceleration

We consider a material point S in space, referred to an origin O and three fixed directions. The position vector, velocity and acceleration of the point S are denoted by

$$\boldsymbol{r} = \boldsymbol{OS} \; , \quad \dot{\boldsymbol{r}} = \frac{\mathrm{d}\boldsymbol{r}}{\mathrm{d}t} \; , \quad \ddot{\boldsymbol{r}} = \frac{\mathrm{d}^2\boldsymbol{r}}{\mathrm{d}t^2} \; .$$

Consider the plane containing the position and velocity vectors. An orthonormal frame $(O; \boldsymbol{i}, \boldsymbol{j})$ is defined in this plane, together with a polar coordinate basis $(\boldsymbol{e}_r, \boldsymbol{e}_\theta)$. Adjoining the unit vector along the Oz axis, we obtain the right-handed system:

$$\boldsymbol{k} = \boldsymbol{i} \wedge \boldsymbol{j} = \boldsymbol{e}_r \wedge \boldsymbol{e}_\theta \; .$$

Let r be the length of the position vector, i.e., $r = \|\boldsymbol{r}\|$, and θ the angle between the $\boldsymbol{i}$ axis and the position vector, viz., $\theta = (\boldsymbol{i}, \boldsymbol{r})$. The unit vectors

$\boldsymbol{e}_r$ and $\boldsymbol{e}_\theta$ are defined by $\boldsymbol{e}_r = \boldsymbol{r}/r$ and $(\boldsymbol{e}_r, \boldsymbol{e}_\theta) = \pi/2$. The angular speed and acceleration of the motion are denoted by $\dot\theta$ and $\ddot\theta$.

In this plane coordinate system, the velocity and acceleration of the point S are obtained by successive derivatives of the expression for $\boldsymbol{OS}$. This gives

$$\boldsymbol{r} = r\boldsymbol{e}_r \ , \tag{1.1}$$

$$\dot{\boldsymbol{r}} = \dot r\boldsymbol{e}_r + r\dot\theta\boldsymbol{e}_\theta \ , \tag{1.2}$$

$$\ddot{\boldsymbol{r}} = (\ddot r - r\dot\theta^2)\boldsymbol{e}_r + (r\ddot\theta + 2\dot r\dot\theta)\boldsymbol{e}_\theta \ . \tag{1.3}$$

Note the relations (for nonzero r)

$$\boldsymbol{r}\cdot\dot{\boldsymbol{r}} = r\dot r \ , \tag{1.4}$$

$$r\ddot\theta + 2\dot r\dot\theta = \frac{1}{r}\frac{\mathrm{d}}{\mathrm{d}t}(r^2\dot\theta) \ . \tag{1.5}$$

1.2.2 Angular Momentum

The angular momentum per unit mass is defined by

$$\boldsymbol{C} = \boldsymbol{r} \wedge \dot{\boldsymbol{r}} \ . \tag{1.6}$$

Differentiating with respect to time, we obtain

$$\frac{\mathrm{d}\boldsymbol{C}}{\mathrm{d}t} = \dot{\boldsymbol{r}} \wedge \dot{\boldsymbol{r}} + \boldsymbol{r} \wedge \ddot{\boldsymbol{r}} = \boldsymbol{r} \wedge \ddot{\boldsymbol{r}} \ . \tag{1.7}$$

Moreover, using the definition and the relations (1.1) and (1.2), we obtain the following expression for $\boldsymbol{C}$:

$$\boldsymbol{C} = r^2\dot\theta\boldsymbol{k} \ . \tag{1.8}$$

1.3 Central Acceleration

1.3.1 Definition and Properties

Consider a material point S in space. Its motion is said to undergo central acceleration if there is some fixed point O such that, at each moment of time, the vector $\boldsymbol{OS}$ and the acceleration vector are collinear.

The motion of the point has central acceleration if and only if

$$\boldsymbol{r} \wedge \ddot{\boldsymbol{r}} = \boldsymbol{0} \ . \tag{1.9}$$

Inserting this defining relation in (1.7), we obtain

$$\frac{\mathrm{d}\boldsymbol{C}}{\mathrm{d}t} = \mathbf{0} , \tag{1.10}$$

which shows that, for this type of motion, the angular momentum is constant in time:

$$\boldsymbol{C} = \boldsymbol{r} \wedge \dot{\boldsymbol{r}} = \text{const.} \tag{1.11}$$

If this constant is zero, the motion is in a straight line, since $\boldsymbol{r}$ and $\dot{\boldsymbol{r}}$ are then collinear.

If the constant vector on the right-hand side of (1.11) is not zero, and this is the general case that we shall consider hereafter, the motion of the point is contained in the plane orthogonal to the constant vector. Let $\mathcal{P}$ denote this plane.

The quantity C calculated from (1.8) is thus constant:

$$C = r^2\dot{\theta} , \quad \text{with} \quad \boldsymbol{C} = C\boldsymbol{k} . \tag{1.12}$$

Note that $\dot{\theta}$ cannot change sign during the motion.

1.3.2 Characterising the Motion

Acceleration

Setting

$$\ddot{\boldsymbol{r}} = \gamma \boldsymbol{e}_r , \quad \text{with} \quad \gamma = f(r) , \tag{1.13}$$

under the assumption that the acceleration is central, it follows that γ is the signed magnitude of the acceleration vector, with a plus sign when it lies parallel to the radial vector $\boldsymbol{e}_r$. Using the value for $\ddot{\boldsymbol{r}}$ calculated in (1.3), we obtain the following two relations, one for each component:

$$\ddot{r} - r\dot{\theta}^2 = \gamma , \tag{1.14}$$

$$\frac{1}{r}\frac{\mathrm{d}}{\mathrm{d}t}(r^2\dot{\theta}) = 0 . \tag{1.15}$$

The last relation shows once again that C is constant.

Areal Law

Let $\mathcal{A}$ be (the magnitude of) the area swept out by the vector $\boldsymbol{r}$. The area element is the area of the triangle with base $r\mathrm{d}\theta$ and height r. Hence, $\mathrm{d}\mathcal{A} = (r/2)r\mathrm{d}\theta$ and the areal speed (area swept out per unit time) is thus

$$\frac{\mathrm{d}\mathcal{A}}{\mathrm{d}t} = \frac{1}{2}r^2\dot{\theta} = \frac{1}{2}C . \tag{1.16}$$

This is the areal law. It tells us that the area swept out is proportional to the time. Alternatively, equal areas are swept out in equal times.

Binet's Equations

Binet's equations give the velocity and acceleration as a function of the angle θ and the auxiliary variable u defined by

$$u = \frac{1}{r} .$$

To find these relations, we eliminate the time t using (1.12), whence

$$\dot{\theta} = \frac{C}{r^2} = Cu^2 ,$$

noting further that

$$r\dot{\theta} = Cu , \quad r\dot{\theta}^2 = C^2u^3 .$$

Using the chain rule, we now have

$$\dot{r} = \frac{\mathrm{d}r}{\mathrm{d}u}\frac{\mathrm{d}u}{\mathrm{d}\theta}\frac{\mathrm{d}\theta}{\mathrm{d}t} = -\frac{\dot{\theta}}{u^2}\frac{\mathrm{d}u}{\mathrm{d}\theta} = -C\frac{\mathrm{d}u}{\mathrm{d}\theta}$$

and

$$\ddot{r} = \frac{\mathrm{d}\dot{r}}{\mathrm{d}t} = \left(\frac{\mathrm{d}\dot{r}}{\mathrm{d}\theta}\right)\dot{\theta} = \dot{\theta}\frac{\mathrm{d}}{\mathrm{d}\theta}\left(-C\frac{\mathrm{d}u}{\mathrm{d}\theta}\right) = -C^2u^2\frac{\mathrm{d}^2u}{\mathrm{d}\theta^2} .$$

Then from (1.2) and (1.3), we obtain the velocity and acceleration vectors relative to the basis $(\boldsymbol{e}_r, \boldsymbol{e}_\theta)$:

$$\dot{\boldsymbol{r}} = C\left(-\frac{\mathrm{d}u}{\mathrm{d}\theta}\boldsymbol{e}_r + u\boldsymbol{e}_\theta\right) ,$$

$$\ddot{\boldsymbol{r}} = -C^2u^2\left(\frac{\mathrm{d}^2u}{\mathrm{d}\theta^2} + u\right)\boldsymbol{e}_r .$$

Setting

$$V = \|\dot{\boldsymbol{r}}\| ,$$

we obtain the two Binet equations from the last two equations. One refers to the velocity via its magnitude V and the other refers to the acceleration via the quantity γ:

$$V^2 = C^2\left[\left(\frac{\mathrm{d}u}{\mathrm{d}\theta}\right)^2 + u^2\right] , \tag{1.17}$$

$$\gamma = -C^2u^2\left(\frac{\mathrm{d}^2u}{\mathrm{d}\theta^2} + u\right) . \tag{1.18}$$

1.4 Newtonian Acceleration

1.4.1 Equation for the Trajectory

A Newtonian acceleration is a central acceleration proportional to r^{-2}. We also speak of a Coulombic acceleration.[1] The expression for the acceleration given in (1.13) thus becomes

$$\gamma = \varepsilon \frac{\mu}{r^2} , \tag{1.19}$$

where $\varepsilon = -1, 0, +1$ and μ is a positive constant.

The quantity ε, which takes three discrete values, allows us to deal simultaneously with the attractive case ($\varepsilon = -1$), the repulsive case ($\varepsilon = +1$), and the situation with no acceleration ($\varepsilon = 0$). In terms of the auxiliary variable u, we now have

$$\gamma = \varepsilon \mu u^2 . \tag{1.20}$$

This reveals the usefulness of the Binet equation for the acceleration (1.18) in the case of a Newtonian acceleration: comparing with (1.20), a factor of u^2 cancels out. This leads to the equation

$$\frac{\mathrm{d}^2 u}{\mathrm{d}\theta^2} + u = -\varepsilon \frac{\mu}{C^2} . \tag{1.21}$$

This is a second order linear differential equation with constant coefficients and the right-hand side is constant. It is easy to solve. The solution is the sum of the general solution to the homogeneous equation (with zero on the right-hand side), bringing in two constants of integration, and a particular solution:

$$u = A\cos(\theta - \theta_0) - \varepsilon \frac{\mu}{C^2} ,$$

where A and θ_0 are the two constants of integration.

The expression for r is

$$r = \frac{p}{-\varepsilon + e\cos(\theta - \theta_0)} , \tag{1.22}$$

where

$$p = \frac{C^2}{\mu} \tag{1.23}$$

[1] The term 'Coulombic' is usually reserved for electrostatic phenomena, where forces may be attractive or repulsive, whilst the word 'Newtonian' refers to gravitational phenomena, where forces are exclusively attractive. In this chapter, we shall use the term 'Newtonian', since we are concerned here with gravity. However, we shall indicate the repulsive case in what follows.

and

$$e = Ap\ . \tag{1.24}$$

This is the equation for a conic section in polar coordinates with one focus at the origin O. p is the focal parameter and e is the eccentricity. The quantities e and θ_0 are determined by the initial conditions. In order to study the relationship between these quantities and the initial conditions, that is, to investigate the main features of the trajectory, we use the Binet equation for the speed (1.17).

1.4.2 Types of Trajectory

Eccentricity

Returning to the variable u, we now write

$$u = \frac{-\varepsilon + e\cos(\theta - \theta_0)}{p}\ . \tag{1.25}$$

Differentiating this relation,

$$\frac{\mathrm{d}u}{\mathrm{d}\theta} = -\frac{e}{p}\sin(\theta - \theta_0)\ .$$

Substituting into (1.17), we have

$$e^2 - 2\varepsilon e\cos(\theta - \theta_0) + \varepsilon^2 = \frac{V^2}{C^2}p^2\ .$$

Equation (1.25) implies $e\cos(\theta - \theta_0) = up + \varepsilon$ and since $p = C^2/\mu$, we now have

$$e^2 - 2\varepsilon\left(u\frac{C^2}{\mu} + \varepsilon\right) + \varepsilon^2 = \frac{V^2C^2}{\mu^2}\ .$$

The eccentricity is therefore given by

$$e^2 = \varepsilon^2 + \frac{C^2}{\mu^2}\left(V^2 + 2\varepsilon\mu\frac{1}{r}\right)\ . \tag{1.26}$$

Since e, C and μ are constants, this equation shows the constancy of the quantity $\mathcal{K}$ defined by

$$\mathcal{K} = V^2 + 2\varepsilon\frac{\mu}{r} = \text{const.} \tag{1.27}$$

At the end of the chapter, using the relation (1.83), we shall see that this quantity corresponds to the mechanical energy. For motion under a Newtonian acceleration, r and V vary in such a way that the relation (1.27) is always satisfied.

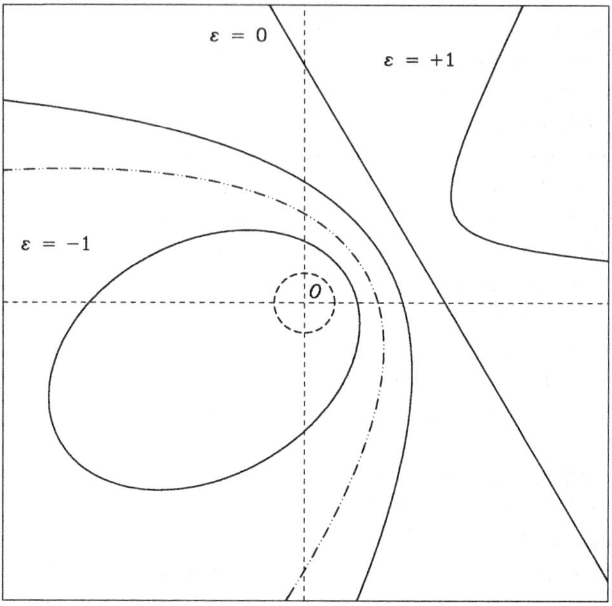

Figure 1.1. Trajectories in the various possible cases. $\varepsilon = +1$: hyperbola – convex branch. $\varepsilon = 0$: straight line. $\varepsilon = -1$: hyperbola – concave branch, parabola, ellipse, circle, in that order as we approach the origin O. In the example shown, all conic sections have O as focus and the same axis of symmetry ($\theta_0 = \pi/6$)

Different Cases

We shall now investigate the quantities r and e defined by (1.22) and (1.26) for the three possible values of ε (see Fig. 1.1).

- Repulsive Case $\varepsilon = +1$: The equation for the trajectory is

$$r = \frac{p}{-1 + e\cos(\theta - \theta_0)} \,. \tag{1.28}$$

The eccentricity is

$$e^2 = 1 + \frac{C^2}{\mu^2}\left(V^2 + 2\frac{\mu}{r}\right) \tag{1.29}$$

and we find that $e > 1$ since the quantity $\mathcal{K}$ is in this case always positive. The denominator in (1.28) may vanish, and r can thus be infinite. The trajectory is a hyperbola with focus at O. This is the branch which is convex with respect to the origin O, since the force is repulsive. Recall that this cannot happen for gravitational interactions.

- Attractive Case $\varepsilon = -1$: The equation for the trajectory is

$$r = \frac{p}{1 + e\cos(\theta - \theta_0)} . \tag{1.30}$$

The eccentricity is

$$e^2 = 1 + \frac{C^2}{\mu^2}\left(V^2 - 2\frac{\mu}{r}\right) . \tag{1.31}$$

Depending on the value of $\mathcal{K}$, e may be greater than or less than unity. We see that the quantity

$$V_\mathrm{e} = \sqrt{2\frac{\mu}{r}} \tag{1.32}$$

plays a specific role in this demarcation.

- If $\mathcal{K} > 0$, i.e., $V > V_\mathrm{e}$, then $e > 1$. We have a hyperbola with focus at O. This time we have the branch which is concave with respect to the origin O, since the force is attractive.
- If $\mathcal{K} = 0$, i.e., $V = V_\mathrm{e}$, then $e = 1$. We have a parabola with focus at O.
- If $\mathcal{K} < 0$, i.e., $V < V_\mathrm{e}$, then $e < 1$. We now have an ellipse with one focus at O. In this case, there is a condition for the equation (1.31): the right-hand side cannot be negative. We must therefore have

  $$V^2 - 2\frac{\mu}{r} \geqslant -\frac{\mu^2}{C^2} \quad \Longrightarrow \quad V^2 \geqslant 2\frac{\mu}{r} - \frac{\mu}{p} ,$$

  whence the condition

  $$V \geqslant V_\mathrm{s} ,$$

  where

  $$V_\mathrm{s} = \sqrt{2\frac{\mu}{r} - \frac{\mu}{p}} . \tag{1.33}$$

  V_s is the orbital insertion speed for putting a satellite into orbit at the distance r. We shall return to this case shortly. We shall also consider the special case of an ellipse with zero eccentricity, which is in fact a circle with centre at O.

It is easy to understand the significance of the speed V_e defined above. When $V \geqslant V_\mathrm{e}$, the point S describing a parabola or a branch of a hyperbola can go to infinity. On the other hand, when $V < V_\mathrm{e}$, the point S remains forever within a finite distance of O and the motion is periodic. This speed (which depends on r) is therefore known as the escape velocity at the distance r.

If the motion is to be periodic, the speed V must therefore satisfy the conditions

$$V_\mathrm{s} \leqslant V < V_\mathrm{e} . \tag{1.34}$$

- No Acceleration $\varepsilon = 0$: The equation for the trajectory is

$$r = \frac{p}{e\cos(\theta - \theta_0)} \,. \tag{1.35}$$

The product $r\cos(\theta - \theta_0)$ is constant. This is the equation for a straight line in polar coordinates. The trajectory of a point that is subjected to no forces is indeed a straight line, covered at constant speed.
In this case $\mathcal{K} = V^2$, implying that V is constant and $e = CV/\mu$. Substituting for p, we obtain

$$r = \frac{C}{V}\frac{1}{\cos(\theta - \theta_0)} \,.$$

The magnitude of the angular momentum defined by (1.6) is now

$$\|\boldsymbol{r} \wedge \dot{\boldsymbol{r}}\| = rV\cos(\theta - \theta_0) = C \,.$$

Note. To end this discussion, note that the nature of the trajectory does not depend on the orientation of the velocity vector, but only on its magnitude V at a point at distance r from the origin.

1.5 Trajectory and Period for Keplerian Motion

1.5.1 Definition of Keplerian Motion

Keplerian motion is the motion of a point mass in a central force field going as $1/r^2$. We shall see in Chap. 2 that this corresponds to a gravitational field due to another mass, assumed motionless. We consider only these two masses and make no attempt to include perturbations due to other bodies. In this chapter, we shall study the motion of this material point S and it will suffice to consider that it is subject to a Newtonian acceleration, or, in terms of forces, that it undergoes a central force of the form $1/r^2$ in a Galilean frame of reference.

Since the aim here is to study satellite trajectories, we shall hereafter consider only periodic trajectories, that is, elliptical trajectories with $\mathcal{K} < 0$ or $e < 1$.

1.5.2 Periodic Trajectories

Elliptical Trajectories

In the case of an elliptical trajectory, given the eccentricity in (1.31), we can write

$$1 - e^2 = -\frac{C^2}{\mu^2}\mathcal{K} \,, \quad \text{with} \quad \mathcal{K} < 0 \,.$$

We also know that an ellipse is defined by its semi-major axis a and its eccentricity e (see Sect. 1.10 for a review). The parameter p is related to these two quantities by $p = a(1 - e^2)$. Hence,

$$1 - e^2 = \frac{p}{a} = \frac{C^2}{\mu a} . \tag{1.36}$$

These relations imply

$$\mathcal{K} = V^2 - \frac{2\mu}{r} = -\frac{\mu}{a} . \tag{1.37}$$

We may now deduce the expression for the speed V as a function of r:

$$V^2 = \mu \left(\frac{2}{r} - \frac{1}{a} \right) . \tag{1.38}$$

We can check that this is always positive since, in an ellipse, we have $r < 2a$.

To sum up, the equation for the elliptical trajectory can be written in polar coordinates in the form

$$r = r(\theta) = \frac{C^2}{\mu} \frac{1}{1 + e \cos(\theta - \theta_0)} , \tag{1.39}$$

where

$$e^2 = 1 - \frac{C^2}{\mu a} , \tag{1.40}$$

or alternatively,

$$r = r(\theta) = \frac{a(1 - e^2)}{1 + e \cos(\theta - \theta_0)} . \tag{1.41}$$

During the periodic motion of the point S, the distance r goes through a minimum and a maximum, denoted respectively by r_p and r_a :[2]

$$r_\mathrm{p} = r(\theta = \theta_0) = a(1 - e) , \tag{1.42}$$

$$r_\mathrm{a} = r(\theta = \theta_0 + \pi) = a(1 + e) . \tag{1.43}$$

The sum of these two lengths is equal to the major axis of the ellipse, i.e.,

[2] The subscripts p and a stand for the perigee and the apogee, respectively, for motion around the Earth (ἡ γῆ, ῆς), or perihelion and aphelion for motion around the Sun (ὁ ἥλιος, ου). More generally, when the gravitational source is not specified, we speak of the periastron and apoastron, or pericenter and apocenter. The prefixes 'peri' and 'apo' come from the adverbs (περί), meaning 'above' and (ἀπό), meaning 'far away'.

$$r_\mathrm{p} + r_\mathrm{a} = 2a \ .$$

Moreover, referring to (1.38), we see that the speed V goes through a maximum V_p for $r = r_\mathrm{p}$ and a minimum V_a for $r = r_\mathrm{a}$, with respective values

$$V_\mathrm{p} = \sqrt{\frac{\mu}{a}}\sqrt{\frac{1+e}{1-e}} = \sqrt{\frac{\mu}{p}}(1+e) \ , \tag{1.44}$$

$$V_\mathrm{a} = \sqrt{\frac{\mu}{a}}\sqrt{\frac{1-e}{1+e}} = \sqrt{\frac{\mu}{p}}(1-e) \ . \tag{1.45}$$

It follows that

$$r_\mathrm{p} V_\mathrm{p} = r_\mathrm{a} V_\mathrm{a} = \sqrt{\frac{\mu}{p}} a(1-e^2) = \sqrt{\mu p} = C \ ,$$

leading us to the angular momentum C, since, for these two extremal points on the ellipse, the velocity and the radial vector are orthogonal.

Note also that

$$\frac{V_\mathrm{p}}{V_\mathrm{a}} = \frac{r_\mathrm{a}}{r_\mathrm{p}} = \frac{1+e}{1-e} \ , \tag{1.46}$$

$$e = \frac{r_\mathrm{a} - r_\mathrm{p}}{r_\mathrm{a} + r_\mathrm{p}} = 1 - \frac{r_\mathrm{p}}{a} = \frac{r_\mathrm{a}}{a} - 1 \ . \tag{1.47}$$

Special Case of Circular Trajectory

A circle is an ellipse with zero eccentricity, i.e., $e = 0$. Equation (1.40) gives

$$\frac{C^2}{\mu} = a \ .$$

Substituting in (1.30), we obtain the expected relation for a circle, viz.,

$$r = \frac{C^2}{\mu} = a = p \ .$$

The speeds V_e and V_s defined by (1.32) and (1.33) become in this case

$$V_l = \sqrt{2\frac{\mu}{a}} \ , \quad V_\mathrm{s} = \sqrt{\frac{\mu}{a}} = \frac{V_l}{\sqrt{2}} \ .$$

Moreover, the relation involving the constant $\mathcal{K}$ gives

$$\mathcal{K} = V^2 - \frac{2\mu}{a} = -\frac{\mu}{a} \ ,$$

whence

$$V = \sqrt{\frac{\mu}{a}} = \frac{C}{a} . \tag{1.48}$$

This shows that the magnitude V of the velocity is constant and equal to V_s. The motion is uniform.

Considering the value of γ given by (1.19), it can also be checked that we do retrieve the usual value for the acceleration in the case of a uniform circular motion:

$$\gamma = -\frac{\mu}{a^2} = -\frac{V^2}{a} .$$

1.5.3 Period and Angular Speed

Period

The period is the time T taken by the point S to describe the whole ellipse. Integrating (1.16) over one period, we obtain

$$\mathcal{A} = \frac{1}{2} C T .$$

In this case, $\mathcal{A}$ represents the area of the ellipse, i.e., $\mathcal{A} = \pi ab$, where b is the semi-minor axis of the ellipse. Recall further that $b^2 = pa$. Hence,

$$\mathcal{A} = \pi a \sqrt{pa} = \pi C \sqrt{\frac{a^3}{\mu}} .$$

The period is therefore

$$T = 2\pi \sqrt{\frac{a^3}{\mu}} . \tag{1.49}$$

This is called the period of revolution, the orbital period, or the Keplerian period of the motion.[3] Note that, for an attractive body μ, the Keplerian period depends only on the semi-major axis a, and not the eccentricity e.

Mean Motion

The corresponding angular speed n, called the mean motion, is defined as

[3] As will be seen in Chap. 3, for a perturbed Keplerian motion, one can define several periods relative to the actual motion, such as the nodal (or draconitic) period and the anomalistic period.

$$n = \frac{2\pi}{T} = \sqrt{\frac{\mu}{a^3}} \,. \tag{1.50}$$

The mean motion is the angular speed of a fictitious point in uniform circular movement at radius a and with the same period as a point in Keplerian motion on an orbit with semi-major axis a.

The daily frequency ν of a satellite measures the number of revolutions the satellite covers in one day (24 hours). This auxiliary quantity is often used to present results. We have

$$\nu = \frac{86\,400}{T(\text{sec})} = \frac{1\,440}{T(\text{min})} \,. \tag{1.51}$$

The quantity ν is known as the daily orbital frequency or daily frequency of revolution.

1.6 Position Along Orbit: The Three Anomalies

The formulas established above give us the trajectory of the material point S in polar coordinates, i.e., the relation between r and θ. These relations were obtained in a way which did not involve time, since the starting point was the Binet equations, set up by eliminating the time. To find an expression for the time t as a function of the polar coordinates, that is, to establish the position of the point S at any time t, we must return to the constant in the areal law for motion with a central acceleration, specified in (1.12). To obtain t, we integrate this relation:

$$\mathrm{d}t = \frac{1}{C} r^2 \mathrm{d}\theta \,. \tag{1.52}$$

We shall examine several approaches based on the integration of $r^2\mathrm{d}\theta$.

- In the first method, we eliminate r, expressing it as a function of θ, i.e., $r = f(\theta)$. We then obtain t from

$$t = \frac{1}{C} \int [f(\theta)]^2 \, \mathrm{d}\theta \,.$$

- In the second approach, we eliminate θ, expressing $\mathrm{d}\theta$ in the form $\mathrm{d}\theta = g(r)\mathrm{d}r$. We obtain t from

$$t = \frac{1}{C} \int r^2 g(r) \mathrm{d}r \,.$$

- The third and final method is based on the geometrical properties of the ellipse.

1.6.1 $t = t(\theta)$ and the True Anomaly v

In this first method, the starting point is (1.52). We replace r by its expression as a function of θ, as given by (1.39):

$$\mathrm{d}t = \frac{1}{C} \frac{p^2}{[1 + e\cos(\theta - \theta_0)]^2} \mathrm{d}\theta$$

The minimal value of r, namely $r = r_\mathrm{p}$ is obtained when $\theta = \theta_0$ [see (1.42)], and it is convenient to measure angles from this origin. We therefore make the change of variable

$$v = \theta - \theta_0 . \tag{1.53}$$

The angle v is called the true anomaly.[4]

We now calculate $t = t(v)$ from

$$t = \frac{p^2}{C} \int \frac{\mathrm{d}v}{(1 + e\cos v)^2} .$$

This type of function integrates as follows:

$$\mathcal{I} = \int \frac{\mathrm{d}v}{(1 + e\cos v)^2} = -\frac{e\sin v}{(1 - e^2)(1 + e\cos v)} + \frac{1}{(1 - e^2)} \int \frac{\mathrm{d}v}{1 + e\cos v} ,$$

$$\int \frac{\mathrm{d}v}{1 + e\cos v} = \frac{2}{\sqrt{1 - e^2}} \arctan\left(\sqrt{\frac{1 - e}{1 + e}} \tan\frac{v}{2}\right) .$$

Using (1.36), we have

$$\frac{p^2}{C} = \sqrt{\frac{a^3}{\mu}}(1 - e^2)^{3/2} = \frac{(1 - e^2)^{3/2}}{n} ,$$

with the expression for the mean motion n given by (1.50).

[4] Kepler invented the term for this angle, from *anomalia, æ* in Latin. It originally comes from the Greek word (ἡ ἀνωμαλία, ας), which means 'irregularity' (prefix ἀν privative, adjective ὁμαλός, 'self-similar', 'regular'). The idea behind this was to express the irregular behaviour of the angle in time (since the motion does not appear to be circular and regular). Kepler first used the term to indicate the position of Mars with respect to the Sun and he defined several anomalies. Among these were the three described in this chapter: the true anomaly (*anomalia coæquata vera*), the eccentric anomaly (*anomalia eccentri*), and the mean anomaly (*anomalia media*). In his work *Astronomia Nova*, apart from the true anomaly, Kepler used the 'artificial' anomaly (*anomalia coæquata fictitia*) and four other anomalies (*anomalia circularis & elliptica, anomalia distantaria, anomalia scrupularia*).

Taking the time origin as $t = t_\text{p}$ for $r = r_\text{p}$ and $v = 0$, we now obtain

$$t - t_\text{p} = \frac{(1-e^2)^{3/2}}{n}\mathcal{I}\,,$$

which gives the time as a function of θ via v:

$$n(t - t_\text{p}) = 2\arctan\left(\sqrt{\frac{1-e}{1+e}}\tan\frac{v}{2}\right) - \frac{e\sqrt{1-e^2}\sin v}{1+e\cos v}\,. \tag{1.54}$$

1.6.2 $t = t(r)$ and the Eccentric Anomaly E

In the second method, instead of expressing $\mathrm{d}\theta$ directly as a function of r, we adopt a neighbouring approach, starting from (1.31). This relates V and r to the parameters of the ellipse and allows us to write

$$V^2 = \frac{2\mu}{r} - \frac{\mu^2}{C^2}(1-e^2)\,.$$

From the vector form for the velocity (1.2), we have

$$V^2 = \dot{r}^2 + r^2\dot{\theta}^2 = \dot{r}^2 + \frac{C^2}{r^2}\,,$$

where θ has been eliminated using the areal law (1.16). This leads to

$$\dot{r}^2 = -\frac{\mu^2}{C^2}(1-e^2) + \frac{2\mu}{r} - \frac{C^2}{r^2}\,,$$

and the differential equation

$$\mathrm{d}t = \frac{r\mathrm{d}r}{\sqrt{-\mu^2(1-e^2)r^2/C^2 + 2\mu r - C^2}}\,.$$

Replacing C^2 by $\mu a(1-e^2)$, we obtain the following simplifications:

$$\begin{aligned}
-\frac{\mu^2}{C^2}(1-e^2)r^2 + 2\mu r - C^2 &= -\frac{\mu}{a}r^2 + 2\mu r - \mu a(1-e^2)\\
&= \frac{\mu}{a}\left(-r^2 + 2ar - a^2 + a^2e^2\right)\\
&= \frac{\mu}{a}\left[a^2e^2 - (r-a)^2\right]\,,
\end{aligned}$$

whereupon

$$\mathrm{d}t = \sqrt{\frac{a}{\mu}}\frac{r\mathrm{d}r}{\sqrt{a^2e^2-(r-a)^2}}\,. \tag{1.55}$$

To integrate this equation, it is convenient to introduce the auxiliary angle variable E such that $a - r = ae\cos E$, which can also be defined by

$$\cos E = \frac{1}{e}\left(1 - \frac{r}{a}\right) . \tag{1.56}$$

This angle is called the eccentric anomaly.[5] It is illustrated in Fig. 1.2. Below, we describe its geometric meaning with respect to the ellipse. Note that $E = 0$ for $r = r_{\mathrm{p}}$ and $E = \pi$ for $r = r_{\mathrm{a}}$ [see (1.42) and (1.43)].

We now change variables,

$$r = a(1 - e\cos E) . \tag{1.57}$$

so that $\mathrm{d}r = ae\sin E\,\mathrm{d}E$. As a function of E, (1.55) becomes

$$\mathrm{d}t = \sqrt{\frac{a}{\mu}}a(1 - e\cos E)\mathrm{d}E .$$

The integration is carried out taking the time origin at $t = t_{\mathrm{p}}$ for $r = r_{\mathrm{p}}$ and $E = 0$:

$$t - t_{\mathrm{p}} = \sqrt{\frac{a^3}{\mu}}(E - e\sin E) .$$

Bringing in the mean motion n, we obtain

$$n(t - t_{\mathrm{p}}) = E - e\sin E . \tag{1.58}$$

This is known as Kepler's equation. We have thus expressed t as a function of r via the variable E.

1.6.3 Geometric Approach to Kepler's Equation

In the third approach, we use the areal law and the fact that the ellipse is an affine transformation of the principal circle with expansion $\sqrt{1 - e^2}$, axis $\boldsymbol{Ox}$ and direction $\boldsymbol{Oy}$ (see Sect. 1.10).

Integrating the areal law (1.16) from t_{p} to t, we obtain

$$t - t_{\mathrm{p}} = \frac{2}{C}\mathcal{A} ,$$

where the quantity $\mathcal{A}$ is the area swept out between these two times, i.e., in the notation of Fig. 1.2, the area of the curvilinear triangle OPS. Let $\mathcal{A}'$ be the area of the curvilinear triangle OPT, where T is the point giving S under the affine transformation. We thus have

[5] Eccentric means 'off-centre'. The centre in question here is not the centre of the circle or the ellipse, but the focus of the ellipse, which is the centre of attraction.

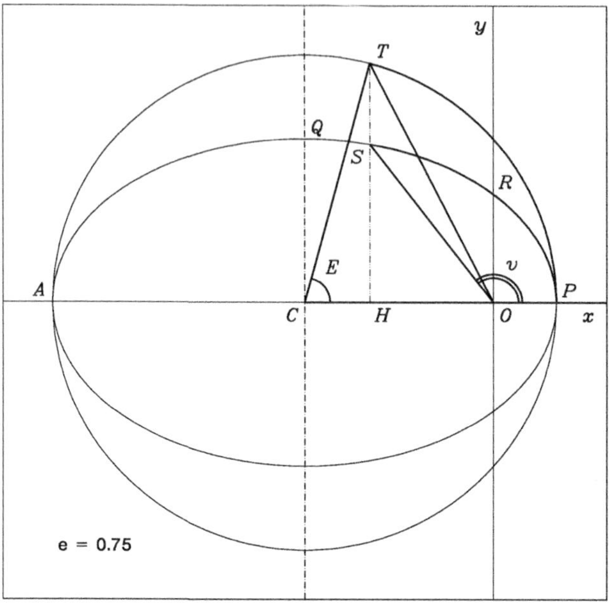

Figure 1.2. Elliptical trajectory. Ellipse and principal circle, indicating the notation for points and angles used to establish Kepler's equation geometrically

$$\mathcal{A} = \sqrt{1-e^2}\mathcal{A}' \ .$$

Replacing C by $\sqrt{\mu a(1-e^2)}$, we now have

$$t - t_{\rm p} = \frac{2}{\sqrt{\mu a}}\mathcal{A}' \ .$$

The area $\mathcal{A}'$ is given by

$$\begin{aligned}\mathcal{A}' &= \text{sector } CPT - \text{triangle } COT \\ &= \text{sector } \{\text{angle } E\} - \frac{1}{2}CO \times HT \\ &= \frac{1}{2}a^2E - \frac{1}{2}(ae)(a\sin E) = \frac{1}{2}a^2(E - e\sin E) \ .\end{aligned}$$

This geometrical method yields Kepler's equation very quickly, with the expression for $\mathcal{A}'$ and introducing the mean motion:

$$n(t - t_{\rm p}) = E - e\sin E \ . \qquad (1.59)$$

1.6.4 Relating the Anomalies: Mean Anomaly M

Relation Between the Anomalies v and E

To establish the relations between v and E (see Fig. 1.2), the ellipse is equipped with a frame $(\boldsymbol{Ox}, \boldsymbol{Oy})$. Let O be the principal focus, the centre of the Newtonian field, and C the centre of the ellipse. The major axis is AP, where P is the point on the ellipse closest to O and A the point furthest away. We choose the axis $\boldsymbol{Ox}$ along OP and the axis $\boldsymbol{Oy}$ at 90° to it in the anticlockwise direction. Let R be the intersection of $\boldsymbol{Oy}$ with the ellipse, and Q the intersection of the line parallel to $\boldsymbol{Oy}$ through C with the ellipse. We then have the following correspondence:

$$\begin{aligned} a &= CP\ , & b &= CQ\ , \\ p &= OR\ , & ae &= CO\ , \\ r_{\mathrm{p}} &= OP\ , & r_{\mathrm{a}} &= OA\ . \end{aligned}$$

We also draw the circle with centre C and radius CP which contains the ellipse and is tangent to it at P and A. This is the principal circle. Let S be an arbitrary point on the ellipse and H its projection onto $\boldsymbol{Ox}$.

The true anomaly can be defined immediately as the polar angle

$$v = (\boldsymbol{Ox}, \boldsymbol{OS})\ .$$

The eccentric anomaly is obtained geometrically from its definition as

$$E = (\boldsymbol{Cx}, \boldsymbol{CT})\ ,$$

where the point T is the intersection of the straight line through H and S with the principal circle. Indeed, according to the relation (1.56), we have $\cos E = (a-r)/ae$. Transforming $(a-r)$ in such a way as to bring in v, i.e.,

$$a - r = a(1-e^2) - r + ae^2 = r(1 + e\cos v) - r + ae^2 = e(r\cos v + ae)\ ,$$

we find

$$\cos E = \frac{r\cos v + ae}{a} = \frac{CH}{CT}\ ,$$

using the notation from Fig. 1.2. The angle E is indeed the angle C of the right-angled triangle HCT.

To obtain the relation between the angles v and E, we write down the coordinates of the point S,

$$x = r\cos v = a(\cos E - e)\ , \tag{1.60}$$

$$y = r\sin v = a\sqrt{1-e^2}\sin E\ , \tag{1.61}$$

so that

$$\sqrt{x^2+y^2} = r = a(1 - e\cos E)\ ,$$

and deduce the relations between the true and eccentric anomalies. Note that v and E change sign together.

For v as a function of E:

$$\cos v = \frac{\cos E - e}{1 - e\cos E}\,, \qquad \sin v = \frac{\sqrt{1-e^2}\sin E}{1 - e\cos E}\,,$$

$$\tan\frac{v}{2} = \sqrt{\frac{1+e}{1-e}}\tan\frac{E}{2}\,. \tag{1.62}$$

For E as a function of v:

$$\cos E = \frac{\cos v + e}{1 + e\cos v}\,, \qquad \sin E = \frac{\sqrt{1-e^2}\sin v}{1 + e\cos v}\,,$$

$$\tan\frac{E}{2} = \sqrt{\frac{1-e}{1+e}}\tan\frac{v}{2}\,. \tag{1.63}$$

We can also express the difference $(v - E)$ as a function of v or E:

$$\tan\frac{v-E}{2} = \frac{\beta\sin E}{1-\beta\cos E} = \frac{\beta\sin v}{1+\beta\cos v}\,, \tag{1.64}$$

where

$$\beta = \frac{e}{1+\sqrt{1-e^2}}\,. \tag{1.65}$$

Using these relations, the expression (1.54) for t as a function of v gives the expression (1.58) for t as a function of E. The integration leading to (1.54) serves no purpose if we use the trigonometric relations between v and E.

Definition of the Mean Anomaly M

The product of n with a time gives a dimensionless quantity, representing an angle in this case. The mean anomaly is simply defined as the angle M, product of n and t, with the usual initial conditions: $t = t_p$ for $r = r_p$ and $M = 0$:

$$n(t - t_p) = M\,. \tag{1.66}$$

The mean anomaly can be viewed as the angle determining the position of a fictitious point in uniform circular motion with angular speed n (mean motion).

Differential Relations Between Anomalies

From the definition of M, we have a fundamental relation between the mean anomaly and the mean motion, namely,

$$\frac{\mathrm{d}M}{\mathrm{d}t} = n \, . \tag{1.67}$$

The relation between $\mathrm{d}M$ and $\mathrm{d}E$ follows from (1.58) and (1.66):

$$M = E - e \sin E \, , \tag{1.68}$$

$$n = \frac{\mathrm{d}M}{\mathrm{d}t} = (1 - e \cos E) \frac{\mathrm{d}E}{\mathrm{d}t} \, . \tag{1.69}$$

The relation between $\mathrm{d}M$ and $\mathrm{d}v$ is derived from the areal law. Equation (1.16) yields

$$\frac{\mathrm{d}\theta}{\mathrm{d}t} = \frac{C}{r^2} = \frac{na^2\sqrt{1-e^2}}{r^2} \, ,$$

and since $\mathrm{d}\theta$ and $\mathrm{d}v$ represent the same quantity, we have

$$n = \frac{\mathrm{d}M}{\mathrm{d}t} = \frac{r^2}{a^2\sqrt{1-e^2}} \frac{\mathrm{d}v}{\mathrm{d}t} \, . \tag{1.70}$$

1.6.5 Kepler's Problem

We have just seen how to express the time in terms of the three anomalies, in an analytical way. The converse problem consists in expressing the true anomaly in terms of time. This is called Kepler's problem. It has no analytical solution.

The three anomalies v, E and M play quite different roles. $v(t)$ allows one to identify the position of the body in its orbit and gives the radial vector r via (1.30). $M(t)$ is another way of representing time, whilst $E(t)$ is only really used to solve Kepler's problem.

Solution of Kepler's Problem

At a given time t, defining the value $M = M(t)$ of the mean anomaly, Kepler's equation becomes

$$E - e \sin E = M \, . \tag{1.71}$$

Since Kepler, dozens of methods have been put forward to solve this equation. We seek to obtain E as a function of M, so that we may subsequently obtain v.

If the eccentricity is not too close to unity ($e < 0.99$), the fastest method is Newton's method, which involves approximating a curve at a point by its tangent there (first order Taylor expansion). We then proceed by iteration. Consider the point $[x_n, f(x_n)]$ on the curve. Drawing the tangent at this point, we find the point $[x_{n+1}, f(x_{n+1}) = 0]$ where it intersects the x axis.

Let $f(x)$ be the function

$$f(x) = x - e \sin x - M \ .$$

The solution we seek, $x = E$, will be such that $f(E) = 0$. Differentiating f gives

$$f'(x) = 1 - e \cos x \ ,$$

so that we may express x_{n+1} as a function of x_n by

$$x_{n+1} = x_n - \frac{f(x_n)}{f'(x_n)} = x_n - \frac{x_n - e \sin x_n - M}{1 - e \cos x_n} \ . \tag{1.72}$$

Starting with the estimate $x_0 = M$, this method generally gives the solution to an accuracy of around 10^{-3} degrees within two or three iterations. With the solution obtained for E, we can express v using (1.62), i.e.,

$$v = 2 \arctan \left(\sqrt{\frac{1+e}{1-e}} \tan \frac{E}{2} \right) \ . \tag{1.73}$$

Example 1.1. *Calculate the true anomaly for the planet Mars when the mean anomaly is $M = 98.679°$.*

The eccentricity of the orbit of Mars around the Sun is $e = 0.09340$. Angles are given in degrees, but should be converted to radians for the purposes of calculation. Set $x_0 = M$. With the iteration (1.72), we obtain successively:

$$x_0 = 98.679 \ , \quad x_1 = 103.896 \ , \quad x_2 = 103.875 \ , \quad x_3 = 103.875 \ .$$

To the required accuracy, two iterations suffice. We obtain v from (1.73):

$$E = x_3 = 103.875 \quad \Longrightarrow \quad v = 109.020 \ .$$

1.7 Representation of Anomalies

1.7.1 Summary of Anomalies

Let us sum up the results of the last two sections. If we express the time t, represented by M, as a function of polar coordinates θ and r, themselves represented by v and E, respectively, we obtain the analytical relations $M(v)$ and $M(E)$:

$$
\begin{array}{ll}
v \longmapsto M = M(v) & \text{equation (1.54)}\,, \\
E \longmapsto M = M(E) & \text{equation (1.58)}\,, \\
v = v(E) \longleftrightarrow E = E(v) & \text{equations (1.62), (1.63)}\,.
\end{array}
$$

If we express the polar angle θ, represented by v, as a function of time t, itself represented by M, we must go through E, solving Kepler's problem by iteration:

$$
\begin{array}{ll}
M \longmapsto E = E(M) & \text{iteration (1.72)}\,, \\
E \longmapsto v = v[E(M)] = v(M) & \text{equation (1.73)}\,.
\end{array}
$$

When we study a particular trajectory, we generally consider a sequence of times t with a given separation, and at each time $t_i, t_{i+1}, \ldots$, we solve Kepler's problem to obtain the true anomaly.

1.7.2 Representation of the Anomalies $v(M)$ and $E(M)$

During a period, up to a factor of n, the mean anomaly M thus represents the time elapsed since the passage at point P. We can plot graphs giving the evolution of v and E as a function of time, i.e., the functions $v(M)$ and $E(M)$. On the graphs shown in Fig. 1.3, M varies over one period, from $M = -\pi$ (S at A) to $M = +\pi$ (S at A), passing through $M = 0$ (S at P).

The upper graph of Fig. 1.3 shows the function $v(M)$ for various values of the eccentricity between $e = 0.0$ and $e = 0.9$, at intervals of 0.1. The two angles are equal when $e = 0.0$ (circular trajectory), whilst the deviation increases with e. When S is close to the periastron P ($M = 0, v = 0$), ever bigger variations in v correspond to small variations in M (i.e., the time) as e increases. In contrast, when S is close to the apoastron A ($|M| = \pi, |v| = \pi$), large variations in M correspond to small variations in v. This is, of course, an illustration of the areal law.

The lower graph of Fig. 1.3 shows the function $E(M)$ for various values of the eccentricity between $e = 0.0$ and $e = 0.9$, at intervals of 0.1. As for the last function, the two angles are equal for $e = 0.0$ and the deviation increases with e, although in a less marked way than for $v(M)$. We have once again an illustration of the areal law.

1.7.3 Equation of Centre

In Keplerian motion, it is useful to compare the true and mean anomalies. In astronomy, one defines the equation of centre, denoted by E_C, as the difference between these two anomalies. This quantity[6] is an angle:

[6] The term 'equation' taken from algebra was defined in its modern sense by Descartes in 1637. Prior to this, the word came from astronomy and was specified and used by Kepler (*æquatio, nis,* in Latin) as 'that variable quantity, determined by calculation, which must be added or subtracted from the mean motion to obtain the true motion'. This is how one should understand the name 'equation of centre', but also 'equation of time', to be described later.

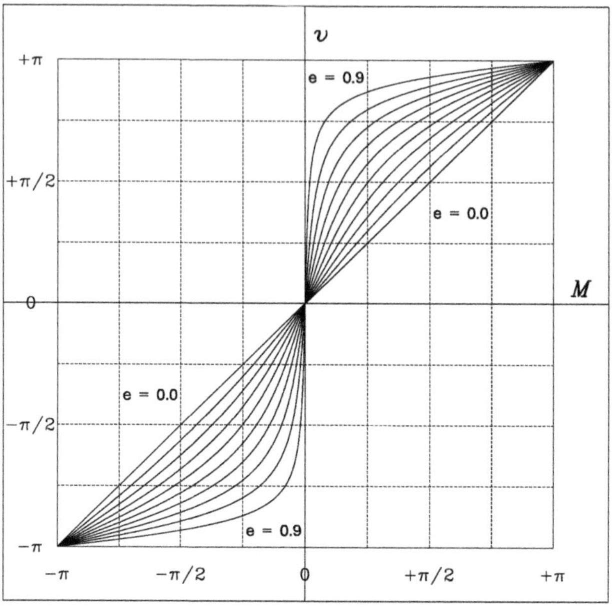

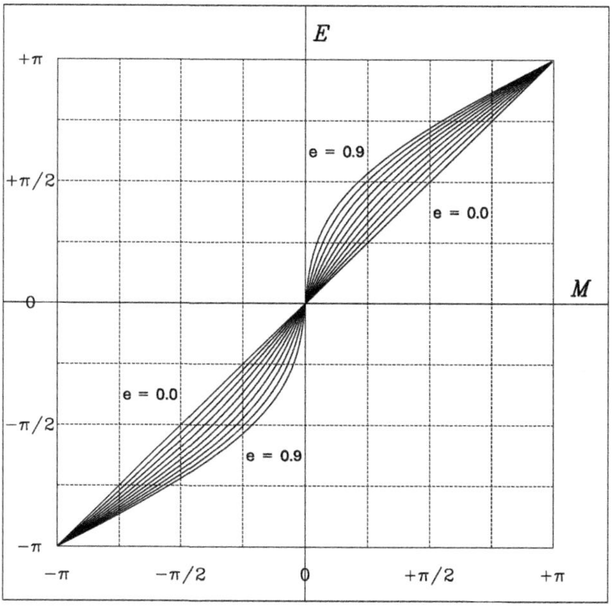

Figure 1.3. *Top*: Variation $v(M)$ of the true anomaly v as a function of the mean anomaly M over a period for ten values of the eccentricity between $e = 0.0$ and $e = 0.9$, at intervals of $e = 0.1$. Angles in radians. *Bottom*: Variation $E(M)$ of the eccentric anomaly E as in the upper graph

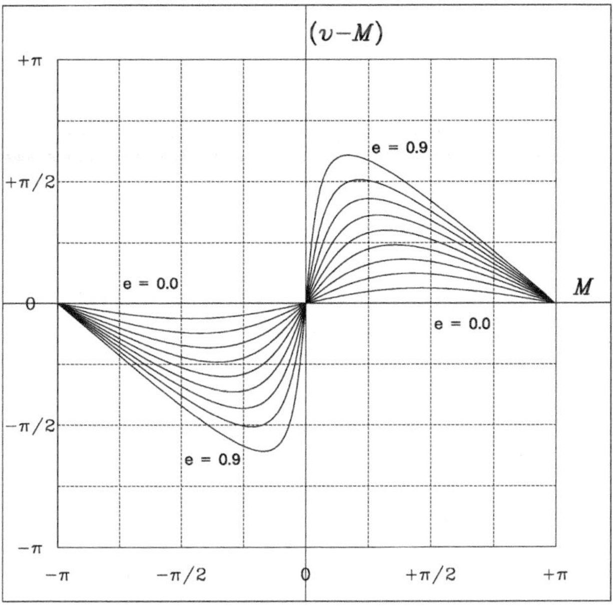

Figure 1.4. Equation of centre. $(v - M)(M)$: variation of the difference between the true anomaly v and the mean anomaly M over a period for ten values of the eccentricity between $e = 0.0$ and $e = 0.9$, at intervals of $e = 0.1$. Angles in radians

$$E_{\mathrm{C}} = v - M \ . \tag{1.74}$$

We shall need this when studying the apparent motion of the Sun around the Earth or Mars.

In the general case, we can study the position of the extrema of E_{C}, given by $\mathrm{d}E_{\mathrm{C}} = 0$, i.e.,

$$\frac{\mathrm{d}v}{\mathrm{d}t} = \frac{\mathrm{d}M}{\mathrm{d}t} \ .$$

Using (1.70), this corresponds to the value r_{m} of r such that

$$r_{\mathrm{m}}^2 = a^2\sqrt{1-e^2} \ , \quad \text{or } r_{\mathrm{m}} = \sqrt{ab} \ .$$

Considering the value of $r = r(v, a, e)$ given by (1.41), we have

$$r_{\mathrm{m}} = \frac{a(1-e^2)}{1+e\cos v_{\mathrm{m}}} = a(1-e^2)^{1/4} \ ,$$

where v_{m} is the value of the true anomaly corresponding to the extremum. We deduce that

$$\cos v_{\mathrm{m}} = \frac{1}{e}\left[(1-e^2)^{3/4} - 1\right] \ . \tag{1.75}$$

The graph of the function $(v - M)(M)$ is shown in Fig. 1.4, which should be compared with the upper graph in Fig. 1.3a. The values of v corresponding to the maximum and minimum of $(v - M)$ are symmetric with respect to the origin (the periastron). In Chap. 2, we shall return to the equation of centre in the case of low eccentricities.

1.8 First Integrals of the Motion

1.8.1 Conservation Laws

Starting from the expression for the Newtonian acceleration, we have obtained the equation of motion by two integrations, to go from $\ddot{\boldsymbol{r}}$ to $\boldsymbol{r}$. Following this rather detailed solution, we shall now give a brief presentation of a more synthetic method. The motivation for this is that it brings out the quantities remaining constant throughout the motion. These values are obtained with just one integration, to go from $\ddot{\boldsymbol{r}}$ to $\dot{\boldsymbol{r}}$. This is why they are referred to as first integrals of the motion. Starting from the equation of motion in the case of a Newtonian acceleration, viz.,

$$\ddot{\boldsymbol{r}} = -\frac{\mu}{r^2}\boldsymbol{e}_r \,, \tag{1.76}$$

we obtain the conservation of energy, angular momentum and Laplace vector (which gives the equation of motion).

- Conservation of energy. We take the scalar product of both sides of (1.76) with the velocity vector $\dot{\boldsymbol{r}}$, using the relation (1.4):

$$\ddot{\boldsymbol{r}}\cdot\dot{\boldsymbol{r}} = \frac{1}{2}\frac{\mathrm{d}}{\mathrm{d}t}\left(\dot{\boldsymbol{r}}^2\right) = \frac{1}{2}\frac{\mathrm{d}}{\mathrm{d}t}\left(\dot{r}^2\right) ,$$

$$-\frac{\mu}{r^2}\boldsymbol{e}_r\cdot\dot{\boldsymbol{r}} = -\frac{\mu}{r^2}\dot{r} = \frac{\mathrm{d}}{\mathrm{d}t}\left(\frac{\mu}{r}\right) .$$

Equation (1.76) thus gives

$$\frac{\mathrm{d}}{\mathrm{d}t}\left(\frac{1}{2}\dot{r}^2 - \frac{\mu}{r}\right) = 0 \,, \tag{1.77}$$

whence

$$\frac{1}{2}\mathcal{K} = \frac{1}{2}\dot{r}^2 - \frac{\mu}{r} = \text{const.} \tag{1.78}$$

This expresses conservation of energy. We can recover (1.27) by setting $V^2 = \dot{r}^2$ in the case $\varepsilon = -1$.

- Conservation of angular momentum. We take the vector product of each side of (1.76) with the radial vector $\boldsymbol{r}$. This yields

$$\boldsymbol{r} \wedge \ddot{\boldsymbol{r}} = -\frac{\mu}{r^2}\boldsymbol{r} \wedge \boldsymbol{e}_r = \boldsymbol{0} \ .$$

Using the derivative (1.7) of the definition (1.6), we obtain

$$\boldsymbol{C} = \boldsymbol{r} \wedge \dot{\boldsymbol{r}} = \text{const.} \ , \tag{1.79}$$

which expresses conservation of angular momentum. The motion is restricted to a plane. We recover the relation (1.11).

- Conservation of the Laplace vector. Equation of motion. We consider the vector product of the acceleration and the angular momentum:

$$\ddot{\boldsymbol{r}} \wedge \boldsymbol{C} = -\frac{\mu}{r^2}\boldsymbol{e}_r \wedge r^2\dot{\theta}\boldsymbol{k} = \mu\dot{\theta}\boldsymbol{e}_\theta = \mu\frac{\mathrm{d}\boldsymbol{e}_r}{\mathrm{d}t} \ .$$

Now, since $\boldsymbol{C}$ is constant, we have

$$\frac{\mathrm{d}}{\mathrm{d}t}(\dot{\boldsymbol{r}} \wedge \boldsymbol{C}) = \ddot{\boldsymbol{r}} \wedge \boldsymbol{C} \ ,$$

whence

$$\frac{\mathrm{d}}{\mathrm{d}t}\left(\dot{\boldsymbol{r}} \wedge \boldsymbol{C}\right) = \frac{\mathrm{d}}{\mathrm{d}t}\left(\mu\boldsymbol{e}_r\right) \ . \tag{1.80}$$

Considering this relation, we define the vector $\boldsymbol{\Lambda}$, known as the Laplace vector (or the Laplace–Runge–Lenz vector), which has the property of being a constant vector:

$$\boldsymbol{\Lambda} = \frac{\dot{\boldsymbol{r}} \wedge \boldsymbol{C}}{\mu} - \boldsymbol{e}_r = \text{const.} \tag{1.81}$$

This vector $\boldsymbol{\Lambda}$ is perpendicular to $\boldsymbol{C}$, since $\boldsymbol{\Lambda} \wedge \boldsymbol{C} = \boldsymbol{0}$. It thus lies in the plane of motion. In order to evaluate $\boldsymbol{\Lambda}$, we calculate $\dot{\boldsymbol{r}} \wedge \boldsymbol{C}$:

$$\dot{\boldsymbol{r}} \wedge \boldsymbol{C} = \left(\dot{r}\boldsymbol{e}_r + r\dot{\theta}\boldsymbol{e}_\theta\right) \wedge r^2\dot{\theta}\boldsymbol{k} = r^3\dot{\theta}^2\boldsymbol{e}_r - r^2\dot{r}\dot{\theta}\boldsymbol{e}_\theta \ .$$

If we project $\boldsymbol{\Lambda}$ onto $\boldsymbol{e}_r$ and use the value of C given by (1.8), we have

$$\boldsymbol{\Lambda} \cdot \boldsymbol{e}_r = \frac{1}{\mu}r^3\dot{\theta}^2 - 1 = \frac{1}{\mu}\frac{C^2}{r} - 1 \ .$$

We thus obtain the expression for r:

$$r = \frac{C^2}{\mu}\frac{1}{1 + \boldsymbol{\Lambda} \cdot \boldsymbol{e}_r} \ . \tag{1.82}$$

Let v be the angle between the radial vector and the fixed vector $\boldsymbol{\Lambda}$. Setting $\boldsymbol{\Lambda} \cdot \boldsymbol{e}_r = ||\boldsymbol{\Lambda}|| \cos v$ and $p = C^2/\mu$, we find that the trajectory is an ellipse. Comparing formulas, we see that $||\boldsymbol{\Lambda}||$ corresponds to the eccentricity. The distance r goes through a minimum when the vectors $\boldsymbol{\Lambda}$ and $\boldsymbol{e}_r$ are collinear (with $v = 0$): the vector $\boldsymbol{\Lambda}$ goes through the periastron and v therefore represents the true anomaly as defined earlier. Once we know the Laplace vector $\boldsymbol{\Lambda}$, we automatically have the eccentricity and the direction of the periastron.

Note. The conservation of these quantities is expressed by Noether's theorem.[7]

1.8.2 Note on Energy

When we study the Keplerian motion of a satellite, its mass never enters our considerations.[8] This is why we have always spoken of acceleration rather than force. Everything we have said so far can be recast. For example, one could speak of a Newtonian force applied to a material point S of mass m. This would lead us to introduce the standard definition of energy. In the case of a Newtonian attraction, the force is

$$\boldsymbol{F} = -m\frac{\mu}{r^2}\boldsymbol{e}_r \ .$$

This corresponds to a potential energy $\mathcal{U}$ (recalling that $\boldsymbol{F} = -\boldsymbol{\nabla}\mathcal{U}$ with the convention that $\mathcal{U}$ vanishes for infinite r),

$$\mathcal{U} = -m\frac{\mu}{r} \ .$$

The point S moving at speed V has kinetic energy $\mathcal{T}$ given by

$$\mathcal{T} = \frac{1}{2}mV^2 \ .$$

The mechanical energy $\mathcal{E}$ is thus

---

[7] Emmy Noether (1882–1935) was a German mathematician, considered as the founder of modern algebra (inventor of rings and ideals). Noether's theorem (1918) says that a conservation law is a consequence of the invariance of a physical law under a continuous transformation with one parameter. (This is proven using the Lagrangian formalism for the equations of classical mechanics.) As far as we are concerned here, conservation of momentum (resp. angular momentum) results from the invariance of the laws of physics under translation (resp. rotation) due to the homogeneity (resp. isotropy) of space, whilst conservation of energy results from invariance under time translations due to the uniformity of the flow of time.

[8] In the study of perturbed motion, the mass of the body is relevant in certain specific instances, such as the study of air resistance in the upper atmosphere or radiation pressure.

$$\mathcal{E} = \mathcal{T} + \mathcal{U} = \frac{1}{2} m \left(V^2 - \frac{2\mu}{r} \right) . \tag{1.83}$$

The angular momentum $\boldsymbol{L}$ is by definition

$$\boldsymbol{L} = \boldsymbol{r} \wedge m\dot{\boldsymbol{r}} . \tag{1.84}$$

The quantities considered earlier are thus equivalent to those related to the energy:

$$\mathcal{K} = \frac{2\mathcal{E}}{m} , \quad C = \frac{L}{m} .$$

Equation (1.27) which establishes that $\mathcal{K} =$ const. is thus equivalent to

$$\mathcal{E} = \frac{1}{2} m \mathcal{K} = \text{const.} , \tag{1.85}$$

which expresses the conservation of mechanical energy $\mathcal{E}$.

For periodic motion, (1.37) yields

$$\mathcal{E} = -\frac{1}{2} m \frac{\mu}{a} , \tag{1.86}$$

which is negative (this arises due to the convention for $\mathcal{U}$). We see that a is related to $\mathcal{E}/m$, which shows that the period depends only on the mechanical energy per unit mass of the material point under consideration.

1.9 Historical Note on Universal Attraction

1.9.1 Kepler's Laws

Using observations made by Tycho Brahe,[9] Kepler[10] explained the motion of the planets in the Solar System by the following three propositions (Kepler's laws):

---

[9] The Danish astronomer Tycho Brahe (1546–1601) spent twenty years in his observatory in Uraniborg ('city of the sky'), in Denmark, making very accurate astronomical measurements. He was the first to take into account the refraction of light. The accuracy of his observations was $1'$ (1 arcmin, or 1/60 of a degree), whereas his contemporaries were not doing better than $10'$.
He measured the motion of the planet Mars, observing ten oppositions. His model of the universe was a compromise between Ptolemy's geocentric model and the heliocentric model of Copernicus. In 1597, he left for Bohemia where he worked with Kepler to set up the astronomical tables known as *Tabulae Rudolphinae.*

[10] The German astronomer Johannes Kepler (1571–1630) published the first two laws in 1609, in *Astronomia Nova (Αι τωλογητ ος seu Physica Cœlestis* and the third in 1619, in *Harmonices Mundi.*
It would be wrong to think that the laws appear in a totally clear manner in these writings, as they would in today's scientific papers. The mathematical ter-

1. Law of ellipses. (a) The trajectory of each planet lies in a plane and (b) it is an ellipse in which one focus is the Sun.
2. Areal law. The area swept out by the radial vector is proportional to the time it takes to sweep them out.
3. Harmonic law. The square of the period of revolution is proportional to the cube of the length of the major axis.

These correspond to our earlier equations in the following way:

$$\begin{aligned} &\text{Law 1a} \Longleftrightarrow \text{equation (1.10)} ,\\ &\text{Law 1b} \Longleftrightarrow \text{equation (1.39)} ,\\ &\text{Law 2} \Longleftrightarrow \text{equation (1.16)} ,\\ &\text{Law 3} \Longleftrightarrow \text{equation (1.49)} . \end{aligned}$$

Note that laws 1a and 2 apply in the case of central accelerations, and laws 1b and 3 in the case of central accelerations with $1/r^2$ dependence, i.e., Newtonian accelerations.

---

minology was heavy and the explanations hard to follow. The first of these two books is almost exclusively devoted to describing the orbit of Mars (whence the subtitle *Tradita comentariis de motibus stellæ Martis ex observationibus G. V. Tychonis Brahe*). The second law appears at the beginning of this work and the first at the end. Naturally, they concern only the elliptical trajectory of the planet Mars. The Greek word attached to the title *Astronomia Nova* is the substantive arising from the verb (αἰτιολογέω), 'to seek the causes'.
Although these works seem difficult to follow nowadays, they nevertheless attest to the extraordinary discoveries made by their author. To demonstrate the eccentricity of the orbit of Mars or the Earth required a great level of trust in Tycho Brahe's observations, made with the naked eye, and a considerable degree of mathematical ability. Many other moral qualities were also involved. Courage and self-confidence were essential to take such a revolutionary theory to its logical conclusion in the face of universal opposition, in a climate of family problems and widespread religious hostility, as the wars of religion tore Europe apart. Perseverance was another quality we may safely attribute: in *Astronomia Nova*, following fifteen pages of close calculations, Kepler tells us that he had to repeat them seventy times in order to arrive at the result!
Kepler was deeply convinced that the cosmic and hence divine order had to be perfect, and had great difficulty renouncing the perfection of the circular orbit in favour of the ellipse, blemished as it was by the failings of the real world. Throughout his approach to science, Kepler was guided by this search for divine harmony. In this last work, this led to the musical harmony of the planets and the geometrical harmony of the regular polyhedra he fitted around the planetary orbits. In this context, Kepler wrote: "There are six planets because there are five regular polyhedra. I cannot begin to express my wonder before this discovery."

1.9.2 Newton and the Law of Universal Attraction

Whilst Kepler was the founder of celestial mechanics, it was Galileo[11] who laid the foundations for terrestrial mechanics, at the same period.

But it required the genius of Newton[12] to synthesise these two aspects of the same phenomenon into a single idea: the universal law of attraction. Today, we show that an attraction of type $1/r^2$ between the Sun and the planets leads to an elliptical trajectory for the planets. But historically, Newton had to piece together the rudiments of infinitesimal calculus (called the method of fluxions) and, starting from the elliptical trajectories discovered by Kepler, deduce that the forces at play had to have the $1/r^2$ dependence, whereupon he could put forward his law of universal attraction.

Let us rewrite Newton's demonstration using the notation of modern mathematics. We begin, as he did, with the principle that the forces were central. We consider a material point, whose position is defined by a radial vector $\boldsymbol{r}$ describing an ellipse. We can then write

$$u = \frac{1}{r} = \frac{1 + e\cos\theta}{a(1-e^2)} ,$$

from which we deduce

[11] Galileo Galilei (1564–1642) was an Italian physicist and astronomer. Shortly after the invention of the refracting telescope, he began to use this instrument to observe the sky. In 1610, he discovered four moons in orbit around Jupiter and it was this observation that persuaded him that the Earth and the other planets were in orbit around the Sun. The discovery of the crescent of Venus (impossible in a geocentric system), confirmed this idea and he communicated it to Kepler. Galileo wrote down all his astronomical discoveries in *Sidereus Nuncius.* He propounded the principle of inertia, which corresponds to Newton's first law, in 1638, in *Discorsi intorno a due nuove scienze.* (This was not written in Latin, but in the local vernacular, *volgare*, of Italy or Tuscany – this too was revolutionary.)

[12] Isaac Newton (1642–1727) was an English mathematician, physicist and astronomer. In 1687, he proclaimed the three laws of mechanics in his *Philosophæ Naturalis Principia Mathematica*:

1. the law of inertia,
2. the fundamental law of dynamics – in a Galilean frame of reference, the force is equal to the product of the (inertial) mass and the acceleration,
3. the law of action and reaction.

It can be shown that (1) is a special case of (2) and that (3) follows from (2). The fundamental law (2) (usually referred to as Newton's second law) was not originally expressed in this way by Newton.
Newton's work dominated eighteenth century mathematics (analysis, solution of equations, etc.) and physics, and in particular, optics (with the publication of his book *Opticks*).

$$\frac{\mathrm{d}u}{\mathrm{d}\theta} = -\frac{e\sin\theta}{a(1-e^2)}\,, \qquad \frac{\mathrm{d}^2u}{\mathrm{d}\theta^2} = -\frac{e\cos\theta}{a(1-e^2)}\,.$$

Substituting this into Binet's equation (1.18) (recall that this refers to central forces), we obtain

$$\begin{aligned}\ddot{\boldsymbol{r}} &= -C^2u^2\left[-\frac{e\cos\theta}{a(1-e^2)} + \frac{1+e\cos\theta}{a(1-e^2)}\right]\boldsymbol{e}_r \\ &= -C^2u^2\frac{1}{a(1-e^2)}\boldsymbol{e}_r \\ &= -\frac{C^2}{a(1-e^2)}\frac{1}{r^2}\boldsymbol{e}_r\,.\end{aligned}$$

Now C is given by $\pi ab = CT/2$ (this is the areal law, which also refers to central forces), i.e.,

$$C = \frac{2\pi}{T}a^2\sqrt{1-e^2}\,.$$

We can therefore deduce that

$$\ddot{\boldsymbol{r}} = -\frac{4\pi^2}{T^2}a^3\frac{1}{r^2}\boldsymbol{e}_r\,,$$

whereby we obtain

$$\ddot{\boldsymbol{r}} = -\frac{\mu}{r^2}\boldsymbol{e}_r = -\frac{\mu}{r^3}\boldsymbol{r} \quad \text{with} \quad \mu = \frac{4\pi^2}{T^2}a^3\,.$$

Tycho Brahe's observations as interpreted by Kepler showed that this quantity μ remains constant for all planets gravitating around the Sun. Newton thus deduced that it could be written in the form

$$\mu = \mathcal{G}M_{\mathrm{S}}\,,$$

where M_{S} is the mass of the Sun and $\mathcal{G}$ is a universal constant.

Finally, the attractive force exerted by a body of mass M on a body of mass m (and conversely) could be written

$$\boldsymbol{F} = m\ddot{\boldsymbol{r}} = -\mathcal{G}\frac{Mm}{r^2}\boldsymbol{e}_r\,.$$

This is indeed the law of universal attraction between two bodies of masses M and m, where $\mathcal{G}$ is the gravitational constant.

1.10 Appendix: Geometry of the Ellipse

1.10.1 History, Definition, and Properties

The Greek mathematicians (notably Apollonios of Perga, ca. 200 BC) studied the geometric figures obtained by intersecting a cone of revolution with a plane. They called these conic sections[13] and classified them into various types (ellipse, parabola, hyperbola).[14]

We shall be concerned here with the ellipse, which is given the following definition today:

(a) The ellipse is the locus of points in the plane such that the sum of their distances to two fixed points (F and F' in this plane, called foci) is equal to some given value.

The etymological definition of the ellipse is:

(b) The section of a cone of revolution by a plane is an ellipse if the plane through the apex, parallel to the intersecting plane, lies entirely outside the cone.

Property (b) is proven from the definition (a), along with a great many other properties, showing the tremendous wealth of this family of geometrical objects. Let us mention the main properties:

(c) A light ray emerging from a focus[15] goes through the other focus after reflecting from the ellipse.
(d) The orthogonal projection of a circle on a plane is an ellipse.
(e) The ellipse is obtained from a circle by an affine transformation[16] of expansion $\sqrt{1-e^2}$, where e is the eccentricity of the ellipse.

---

[13] The word 'conic' is borrowed from the late Greek adjective *kônikos*, which itself comes from *kônos*, (ὁ κῶνος, ου), meaning 'pine cone'.

[14] Apollonios of Perga used the following three terms, all borrowed from the language used by the Pythagoreans:

- ellipse: (ἡ ἔλλειψις, εως) means 'lack' ('omission of a word'), thereby reminding us that this figure is an imperfect circle,
- parabola: (ἡ παραβολή, ῆς), action of 'throwing next to', whence 'comparing',
- hyperbola: (ἡ ὑπερβολή, ῆς), action of 'throwing over', whence 'exceeding'.

The last two words are built around the Greek verb (βάλλει ν), 'to throw'. It is interesting to note that, taken in this order, the three nouns 'lack', 'equality' (comparison) and 'excess' associated with the three conics correspond to the relation that the eccentricity bears to unity. The names of the conic sections were introduced by Kepler, Desargues and Descartes in the first half of the seventeenth century, first in Latin, then in the modern European languages.

[15] It is this property that justifies the term 'focus'.

[16] An affine transformation with axis D, direction δ and expansion k ($k \neq 0$) is the point transformation which maps any point M in the plane to the point M' constructed as follows: the straight line MM' parallel to δ meets D at H, where

$$\boldsymbol{HM'} = k(\boldsymbol{HM}) .$$

1.10.2 Cartesian and Polar Coordinates

We define a Cartesian coordinate system $(O; x, y)$. Consider a circle with radius a and centre at O. The ellipse is the affine transformation of the circle with expansion $k = b/a$, axis $\boldsymbol{Ox}$ and direction $\boldsymbol{Oy}$. In this case, a and b are the lengths of the semi-major and semi-minor axes, respectively, as shown in Fig. 1.5. The equation for the ellipse in Cartsian coordinates is thus

$$\frac{x^2}{a^2} + \frac{y^2}{b^2} = 1 \,. \tag{1.87}$$

The foci F and F' of the ellipse are symmetrical with respect to O. We set

$$OF = c \,,$$

and let M be an arbitrary point on the ellipse. Hence, from the definition of the ellipse,

$$MF + MF' = 2a \,.$$

When the point M is at B, we have $MF = a$, and we may write

$$a^2 = b^2 + c^2 \,.$$

The eccentricity is defined by

$$c = ae \,.$$

We then obtain the relation

$$b^2 = a^2(1 - e^2) \,, \tag{1.88}$$

and the expansion k of the affine transformation mentioned above is equal to $\sqrt{1 - e^2}$.

We take F as the origin of the polar coordinate system (r, θ). Let M be an arbitrary point on the ellipse and H its projection onto $\boldsymbol{Ox}$. Polar coordinates are defined by

$$r = \|\boldsymbol{FM}\| \,, \quad \theta = (\boldsymbol{Fx}, \boldsymbol{FM}) \,.$$

The relation between Cartesian and polar coordinates is thus

$$\begin{cases} x = OH = ae + r\cos\theta \,, \\ y = HM = r\sin\theta \,. \end{cases}$$

The equation for the ellipse becomes

$$(1 - e^2)(a^2e^2 + 2are\cos\theta + r^2\cos^2\theta) + r^2\sin^2\theta = (1 - e^2)a^2 \,.$$

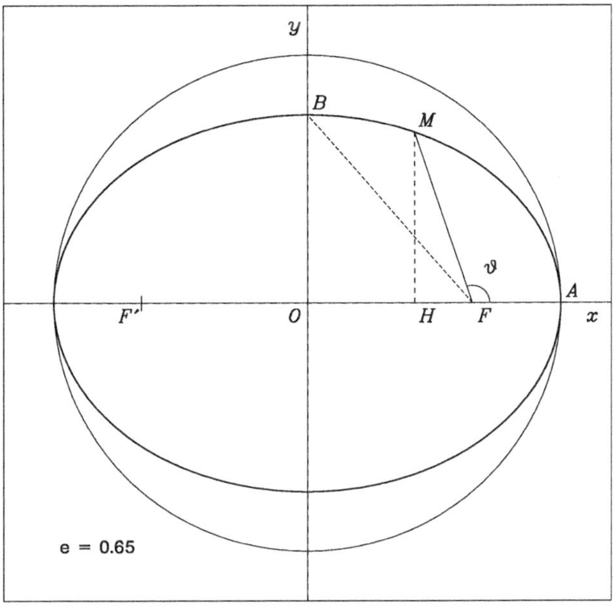

Figure 1.5. Ellipse and principal circle, using the notation of the Cartesian coordinates (axes Ox, Oy) and polar coordinates ($r = FM$, angle θ)

With some calculation,

$$r^2 - (1-e^2)(1-e^2)a^2 + (1-e^2)2are\cos\theta - e^2r^2\cos^2\theta = 0 \,,$$

$$r^2 = \left[a(1-e^2) - re\cos\theta\right]^2 \,,$$

$$r(1+e\cos\theta) = a(1-e^2) \,,$$

whereupon the equation for the ellipse in polar coordinates becomes

$$r = \frac{p}{1+e\cos\theta} \,. \tag{1.89}$$

We have introduced the focal parameter p

$$p = a(1-e^2) \,.$$

We can express p and e in terms of a and b:

$$p = \frac{b^2}{a} \,, \quad e^2 = \frac{a^2-b^2}{a^2} \,,$$

or conversely,

$$a = \frac{p}{1-e^2} \,, \quad b = \frac{p}{\sqrt{1-e^2}} \,.$$

Low-Eccentricity Orbits

We thus see that, for an ellipse, the distance from the centre to the focus is proportional to e and, for orbits with low eccentricity, the difference between the semi-major axis and the semi-minor axis is proportional to e^2:

$$\frac{c}{a} = e \,, \qquad \frac{a-b}{a} \simeq \frac{1}{2}e^2 \,. \tag{1.90}$$

The elliptical orbit with eccentricity 1/10 can be considered as a circle of radius R but with a focus situated at a distance $R/10$ from the centre.[17]

Angle of Eccentricity

The eccentricity can be expressed via the auxiliary angle ψ, known as the angle of eccentricity, defined by

$$e = \sin\psi \,. \tag{1.91}$$

It is easy to see (with $\sin\psi = ae/a$, for example) the geometrical meaning of this angle introduced by Gauss. For most relations involving e, we obtain very elegant expressions in terms of ψ. Some examples are:

- For the semi-minor axis,

$$b = a\cos\psi \,.$$

- For the formulas giving the maximum and minimum speeds (1.44) and (1.45), respectively,

$$V_{\mathrm{p}} = \sqrt{\frac{\mu}{a}}\tan\left(\frac{\pi}{4}+\frac{\psi}{2}\right) \,, \qquad V_{\mathrm{a}} = \sqrt{\frac{\mu}{a}}\tan\left(\frac{\pi}{4}-\frac{\psi}{2}\right) \,.$$

- For the relation (1.64) between the anomalies, which uses the auxiliary variable β defined by (1.65), we have

$$\beta = \tan\frac{\psi}{2} \,.$$

[17] In the case of Mars, $e = 9.3\times10^{-2}$, $e^2/2 = 4.4\times10^{-3}$. This orbit can therefore be represented as a circle of radius 1 m ($a = 1\,000$ mm, $b = 996$ mm), but with focus 9.3 cm from the centre.
This was how Kepler formulated, in 1600, what was to become his first law: the orbit of Mars is circular and the Sun is not at its centre. Kepler used the idea of the ellipse from 1603: *Itaque plane hoc est ... Planetæ orbita non est figura circula, sed ovalis (elliptica).* Later he wrote: "I began by assuming that the planetary orbits were perfect circles. This mistake cost me all the more time in that it was upheld by the authority of all the philosophers and seemed altogether plausible from a metaphysical standpoint."

1.10.3 Eccentricity and Flattening

When one focus plays a privileged role, as in the study of motion under the action of a central force, we use the eccentricity and polar coordinates. When the ellipse is treated as a flattened circle, under affine transformation, it is the centre of the circle (and the ellipse) that plays a privileged role. We then use the flattening and Cartesian coordinates.

The flattening f is defined by

$$f = \frac{a-b}{a} . \tag{1.92}$$

The following relations hold between e and f:

$$\frac{b}{a} = \sqrt{1-e^2} = 1 - f ,$$

$$f = 1 - \sqrt{1-e^2} ,$$

$$e^2 = 1 - (1-f)^2 .$$

We shall now describe two examples.

Example 1.2. *Calculate the polar radius of the Earth, treating it as an ellipsoid of revolution with flattening $f = 1/298.257$.*

Let R_{e} be the equatorial radius and R_{p} the polar radius (i.e., with the standard notation for the ellipse, $R_{\mathrm{e}} = a$ and $R_{\mathrm{p}} = b$). We take $R_{\mathrm{e}} = R = 6378.137$ km. For the polar radius, we thus have

$$R_{\mathrm{p}} = R(1-f) = 6\,356.752 \text{ km} ,$$

and for the difference between the radii,

$$R_{\mathrm{e}} - R_{\mathrm{p}} = Rf = 21.384 \text{ km} .$$

For the eccentricity, we find $e = 0.0818$, but this idea is not used in the case of the terrestrial ellipsoid for the reasons discussed above. Note that $e^2 \approx 2f \approx 1/150$.

Example 1.3. *Calculate the eccentricity of the orbit of an artificial satellite in orbit around the Earth whose altitudes at perigee and apogee are 500 km and 40 000 km, respectively (the so-called Molniya orbit). Take the value $R = 6\,400$ km.*

Let h_{p} and h_{a} be the altitudes at perigee and apogee, respectively. Using the distances r_{p} and r_{a} defined by (1.42) and (1.43), we have

$$r_{\mathrm{p}} = R + h_{\mathrm{p}} , \quad r_{\mathrm{a}} = R + h_{\mathrm{a}} ,$$

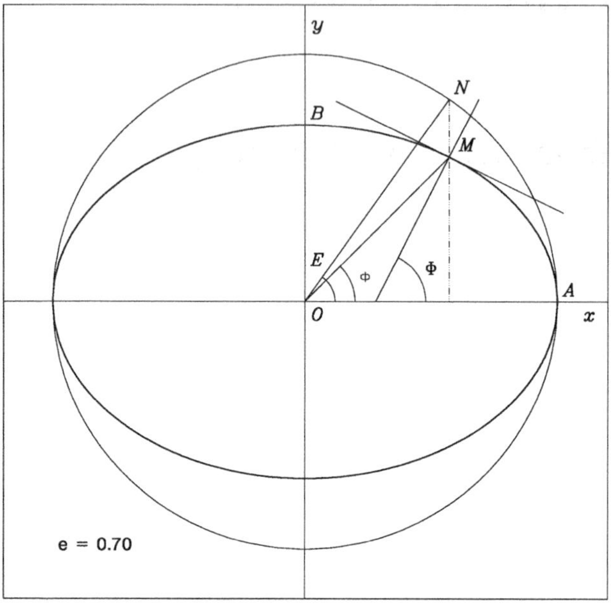

Figure 1.6. Different latitudes for a point M on the ellipse (representing the terrestrial ellipsoid): geocentric latitude ϕ (angle AOM), geodetic latitude Φ, and reduced latitude E

$$a = R + \frac{h_\mathrm{a} + h_\mathrm{p}}{2} \, ,$$

which imply, with (1.47), that the eccentricity is given by

$$e = \frac{h_\mathrm{a} - h_\mathrm{p}}{2R + h_\mathrm{a} + h_\mathrm{p}} \, .$$

In the case of the Molniya orbit, we have

$$e = \frac{39\,500}{12\,800 + 40\,500} \approx 0.75 \, .$$

Moreover, (1.46) gives $V_\mathrm{a}/V_\mathrm{p} = (1-e)/(1+e) = 0.25/1.75 = 1/7$. This type of orbit gives the satellite a speed seven times slower at the apogee than at the perigee.

1.10.4 Radius of the Ellipse

To study the terrestrial ellipsoid, we consider a polar coordinate system with origin at the centre of the Earth. This is a compromise between the two coordinate systems discussed above. Consider a plane through the axis of revolution of the Earth (the axis joining the poles). Let M be a point in this plane on the surface of the Earth, hence on the ellipse, as shown in Fig. 1.6 (where the flattening of the ellipse has of course been exaggerated).

We set

$$\rho = \|\boldsymbol{OM}\| , \quad \phi = (\boldsymbol{Ox}, \boldsymbol{OM}) .$$

ρ is called the radius of the ellipse. For the terrestrial ellipsoid, the angle ϕ represents the latitude. To be more precise, we call this the geocentric latitude to distinguish it from the geodetic latitude, which is denoted by Φ in the figure.[18]

Now consider the point N on the principal circle, of radius a, for which M is the affine transform. We define the angle $E = (\boldsymbol{Ox}, \boldsymbol{ON})$, known as the reduced latitude, or parametric latitude, which is nothing other than the eccentric anomaly defined earlier.

Later, when studying satellite altitudes, we shall need to know ρ as a function of ϕ. To this end, we adopt Cartesian coordinates x and y,

$$x = \rho \cos \phi , \quad y = \rho \sin \phi ,$$

inserting them into the definition (1.87) of the ellipse. We thus obtain

$$\rho^2 \left(\frac{\cos^2 \phi}{a^2} + \frac{\sin^2 \phi}{b^2} \right) = 1 ,$$

which gives ρ as a function of the geocentric latitude ϕ:

$$\rho(\phi) = \frac{a}{\sqrt{\cos^2 \phi + \dfrac{\sin^2 \phi}{(1-f)^2}}} , \tag{1.93}$$

with f the flattening.

Note that

$$\rho(0) = \rho(\pi) = a ,$$

18 The geodetic latitude Φ is defined by the angle between the normal to the tangent to the ellipse at M and the Ox axis. (We also define the astronomical latitude by the angle between the vertical, given by the direction of a plumbline at a given location, and the equatorial plane.) One can then prove the following relation between Φ and ϕ:

$$\tan \Phi = \frac{\tan \phi}{(1-f)^2} .$$

For small f ($f \ll 1$), this formula simplifies to $\Phi \approx \phi + f \sin(2\phi)$. Such an approximation is of course justified for the the terrestrial ellipsoid, and the difference $(\Phi - \phi)$ is therefore always very small. It reaches its maximum for $\phi = \pm 45°$ where it has the value $\Phi - \phi = f$ (rad) $= 0.19°$. Consequently, in this book, we shall only consider the latitude ϕ, which we shall call the geographical latitude.

$$\rho(\pi/2) = \rho(3\pi/2) = b \ .$$

When f is small ($f \ll 1$), we can expand (1.93) to first order in f to obtain the simplified relation

$$\rho(\phi) \approx a(1 - f \sin^2 \phi) \ .$$

2 Satellite in Keplerian Orbit

From now on, we shall be mainly concerned with the periodic motion of a body, the artificial satellite,[1] in the gravitational field of the Earth.

2.1 Gravitational Field

2.1.1 Universal Attraction

The law of gravity, or universal attraction, propounded by Newton says that two point bodies A and B with masses M and m, respectively, attract each other with a force proportional to each of their masses and inversely proportional to the square of the distance separating them:

$$\boldsymbol{f}_{A \to B} = -\boldsymbol{f}_{B \to A} = -\mathcal{G}\frac{Mm}{r^2}\boldsymbol{e_r} \; , \tag{2.1}$$

where

[1] In Latin, *satelles, satellitis* was a bodyguard, soldier, assistant, or accomplice. The origins of the word are obscure. Some claim an Etruscan origin. The word *satelite* appears in French around 1265 to denote an armed man who carries out the orders of a commander, then in the form *satellite*, around 1500, to refer to a man depending in some way on another, or accompanying another. It was Kepler, in 1611, who gave it the modern meaning of 'satellite' in the Latin term *satelles*, which he used to refer to the four satellites of Jupiter, recently discovered by Galileo with his refracting telescope. He wrote: *De quattuor Jovis satellibus erronibus*, that is, 'Concerning the four wandering companions of Jupiter'. The term 'artificial satellite' appeared around 1950.
In many languages, 'satellite' is expressed by a word coming directly from the Latin term modernised by Kepler, as in the Latin and Anglo-Saxon languages. In others, it is the word for 'Moon' which is used, as in Arabic (*qamar ṣanā'i*, meaning 'artifical moon').
However, certain languages have kept to the first idea of *satelles*. In modern Greek, the satellite is still a bodyguard, since it is called *doryphoros*, ὁ δορυφόρος, ου, 'armed with a spear', built up from τὸ δόρυ, ατος, 'spear' and the suffix φορός, 'which carries'. In Russian, *sputnik* is the travel companion (*put*, 'way'). In Chinese, the satellite is called *wei xing*, 'guardian star', a word written with the two ideograms *wei*, 'guard' and *xing*, 'star'. The same form is found in Japanese.

$$\boldsymbol{f}_{A\to B} \text{ is the force exerted by } A \text{ on } B$$

and

$$\boldsymbol{AB} = \boldsymbol{r} = r\boldsymbol{e_r} \ .$$

The value of $\mathcal{G}$, the gravitational constant, is not used directly when studying trajectories (see Table 3.7). Instead, one uses the specific gravitational constant μ, which is the product of $\mathcal{G}$ and the mass of the relevant attracting body (e.g., the geocentric or heliocentric gravitational constant).

The relation (2.1) is symmetric. In order to distinguish the role of one of the two bodies, we may express the fact that body A, for example, creates a gravitational field to which body B is subjected. This field $\boldsymbol{g}$ is such that

$$\boldsymbol{f}_{A\to B} = m\boldsymbol{g} \ ,$$

or, with $\mu = \mathcal{G}M$,

$$\boldsymbol{g} = -\frac{\mu}{r^2}\boldsymbol{e_r} = -\frac{\mu}{r^3}\boldsymbol{r} \ . \tag{2.2}$$

2.1.2 Gauss' Theorem

When studying the motion of the Earth with respect to the Sun, the two bodies can be treated as pointlike compared with the vast distance that separates them. But this is not the case for a satellite gravitating around the Earth at an altitude of just 800 km, or when we consider the weight of an object resting on the Earth's surface. To face this difficulty, we apply Gauss' theorem. There are many ways to prove this theorem. We shall use a method based on the idea of solid angle.

Consider a closed surface S enclosing a volume V. We can thus define an inside and an outside. Consider a surface element $\mathrm{d}S$ and a unit normal vector $\boldsymbol{n}$ pointing from the inside towards the outside. The flux of an arbitrary vector field $\boldsymbol{g}$ through the surface element is by definition

$$\mathrm{d}\Phi = \boldsymbol{g}\cdot\mathrm{d}\boldsymbol{S} \ , \quad \text{with} \quad \mathrm{d}\boldsymbol{S} = \boldsymbol{n}\mathrm{d}S \ .$$

The flux of $\boldsymbol{g}$ out through the whole of the surface S is

$$\Phi = \oint_S \boldsymbol{g}\cdot\mathrm{d}\boldsymbol{S} \ ,$$

where the integral is taken over the whole of the closed surface S.

Consider a surface S enclosing a distribution of masses: each point A_i is attributed a mass M_i. The field created by each mass M_i at a point B is

$$\boldsymbol{g}_i = -\mathcal{G}M_i\frac{\boldsymbol{A}_i\boldsymbol{B}}{\|\boldsymbol{A}_i\boldsymbol{B}\|^3} \ .$$

As the point P runs over the surface S, the flux out of S is

$$\Phi = \oint_S \boldsymbol{g}\cdot\mathrm{d}\boldsymbol{S} = \mathcal{G}\oint_S \sum_i M_i \frac{\boldsymbol{A_iP}}{\|\boldsymbol{A_iP}\|^3}\cdot\mathrm{d}\boldsymbol{S} = \mathcal{G}\sum_i M_i \oint_S \frac{\boldsymbol{n}\cdot(\boldsymbol{A_iP})}{\|\boldsymbol{A_iP}\|^3}\mathrm{d}S\ .$$

Now,

$$\frac{\boldsymbol{n}\cdot(\boldsymbol{A_iP})}{\|\boldsymbol{A_iP}\|^3}\mathrm{d}S = \frac{\mathrm{d}S\cos\alpha_i}{\|\boldsymbol{A_iP}\|^2} = \frac{\mathrm{d}\Sigma}{\|\boldsymbol{A_iP}\|^2} = \mathrm{d}\Omega_i\ ,$$

where α_i is the angle between the normal and $\boldsymbol{A_iP}$, and $\mathrm{d}\Sigma$ is the projection of $\mathrm{d}S$ on the plane perpendicular to $\boldsymbol{A_iP}$. The quantity $\mathrm{d}\Omega_i$ is then the element of solid angle, represented by the elementary cone with apex A_i standing on the surface element $\mathrm{d}S$ (or $\mathrm{d}\Sigma$, which amounts to the same thing).

The integration over $\mathrm{d}\Omega_i$ is independent of the surface S. We thus take a sphere of centre A_i and radius R. This gives

$$\Omega_i = \oint \mathrm{d}\Omega_i = \oint \frac{\mathrm{d}\Sigma}{R^2} = \frac{1}{R^2}\oint \mathrm{d}\Sigma = \frac{4\pi R^2}{R^2} = 4\pi\ .$$

On the other hand, an external mass, i.e., with A_i outside S, produces a field whose flux through S is zero. Indeed, a cone with apex A_i standing on a surface element $\mathrm{d}S$ determines two opposing flux elements, whose total contribution cancels. ($\mathrm{d}\Phi$ is a scalar whose sign depends on the scalar product.)

Finally, letting $M_{\mathrm{int}} = \sum_{\mathrm{int}} M_i$ be the sum of the masses contained within the surface S, the flux out of S is

$$\Phi = -4\pi\mathcal{G}\sum_{\mathrm{int}} M_i\ ,$$

and Gauss' theorem is

$$\oint_S \boldsymbol{g}\cdot\mathrm{d}\boldsymbol{S} = -4\pi\mathcal{G}M_{\mathrm{int}}\ . \tag{2.3}$$

For a continuous mass distribution with density ρ at a given point, M_{int} is

$$M_{\mathrm{int}} = \int_V \rho(\boldsymbol{r})\mathrm{d}V\ .$$

2.1.3 Calculating the Field by Gauss' Theorem

If the density depends only on r (the magnitude of $\boldsymbol{r}$), i.e., if the mass distribution is spherically symmetric, the field it produces will also have spherical symmetry:

$$g(r) = \|g(r)\| \frac{r}{r} .$$

Its flux is then easy to calculate. For S, we choose the surface of a sphere of radius r containing all the mass M_{int}. From the symmetry, the field g must be orthogonal to S at every point. Remembering that r is constant over the whole surface S, we obtain

$$\Phi = \oint_S g \cdot \mathrm{d}S = \oint_S \|g(r)\| \frac{r}{r} \cdot n \mathrm{d}S = \|g(r)\| \oint_S \mathrm{d}S = 4\pi \|g(r)\| r^2 .$$

Applying Gauss' theorem (2.3), we find that

$$4\pi \|g(r)\| r^2 = -4\pi \mathcal{G} M_{\text{int}} ,$$

and hence the expression for the gravitational field g:

$$g(r) = -\mathcal{G} M_{\text{int}} \frac{r}{r^3} . \tag{2.4}$$

This shows that the field is the same as would be produced by a point mass M_{int} located at the centre of the spherical distribution.

2.1.4 Gravitational Field of the Earth

In this chapter, we treat the Earth as a sphere whose density is a function of distance from the centre O. If we consider a point outside or on the surface of the Earth, the field at distance r from O is

$$g(r) = -\mu \frac{r}{r^3} , \quad \text{with} \quad \mu = \mathcal{G} M , \tag{2.5}$$

where M is the total mass of the Earth. In this case, μ is called the geocentric gravitational constant. Under such conditions, a satellite in orbit around the Earth can be considered to follow Keplerian motion: a point mass moving in a $1/r^2$ field created by a motionless point mass (the mass of the Earth). This is precisely what we studied in Chap. 1. Under the conditions (1.34) on the speed, we have a periodic motion whose trajectory is a plane ellipse.

In the real world (as we shall see in the next chapter), the motion of a satellite is close to Keplerian, but slightly different nonetheless. There are two main reasons for this:

- the Earth is not perfectly spherical and the mass distribution does not depend on r alone,
- the gravitational fields of the Sun, the Moon and the planets in the Solar System are superposed on the Earth's own gravitational field.

The motion of a satellite is thus perturbed by these various effects. Satellite trajectories are not therefore ideal Keplerian orbits, but orbits arising from them in ways we shall now examine.

2.2 N-Body and 2-Body Problems

2.2.1 N-Body Problem

We consider N bodies in space, each producing a gravitational field. If we wish to know the motion of one of these bodies, we must take into account the fact that it is attracted to the other $(N-1)$ bodies. This is called the N-body problem. It has no exact analytical solution in the general case, if N is greater than or equal to 3. The 3-body problem was studied by Euler, Lagrange, Laplace, and Poincaré. The motion of the Moon clearly falls into the category of 3-body problems (involving the Moon, the Earth and the Sun).

Poincaré showed that a system of 3 or more bodies is chaotic: it evolves in a way that depends critically on the initial conditions, to such an extent that those conditions would have to be determined to greater accuracy than is actually possible. For the planets in orbit around the Sun, the problem is simplified by considering the masses of the planets to be very small compared with the mass of the Sun. (This is justified: the mass of Jupiter is only one thousandth of the mass of the Sun.) The N-body problem is transformed into $(N-1)$ two-body problems. Each planet is treated as subject to the gravitational field of the Sun, taken as motionless, whilst the attractions due to the other planets are treated as perturbations.[2] The same can be done for artificial satellites around the Earth: the action of the Sun and Moon are treated as perturbations.

2.2.2 2-Body Problem

Consider two bodies A_1 and A_2, with masses m_1 and m_2, respectively, moving in a Galilean reference frame $(O_0; x, y, z)$ (the notion of Galilean frame is defined below). The system is isolated in the sense that each body feels only the attractive force of the other. Taking an arbitrary origin O, Newton's second law takes the form

$$m_1 \ddot{\boldsymbol{OA}}_1 = -\mathcal{G} m_1 m_2 \frac{\boldsymbol{A_1A_2}}{\|\boldsymbol{A_1A_2}\|^3} ,$$

$$m_2 \ddot{\boldsymbol{OA}}_2 = -\mathcal{G} m_1 m_2 \frac{\boldsymbol{A_2A_1}}{\|\boldsymbol{A_2A_1}\|^3} ,$$

[2] The first use of this approach was for a prediction of the return of comet Halley by Clairaut, Lalande and Reine Lepaute. The comet (with period 76 years) was expected in 1758, but these astronomers successfully announced its return for March 1759, with a delay of 200 days due to the influence of Jupiter and Saturn, having accurately calculated their perturbing effect. See also the note on Le Verrier.

which yields the relation

$$m_1 \ddot{\boldsymbol{OA}}_1 + m_2 \ddot{\boldsymbol{OA}}_2 = \boldsymbol{0} \,. \tag{2.6}$$

The barycenter C of two points A_1 and A_2 (also called the centre of mass) is defined by

$$m_1 \boldsymbol{OA}_1 + m_2 \boldsymbol{OA}_2 = (m_1 + m_2)\boldsymbol{OC} \,.$$

We thus have

$$\ddot{\boldsymbol{OC}} = \boldsymbol{0} \quad \Longrightarrow \quad \boldsymbol{OC} = \boldsymbol{v}_0 t + \boldsymbol{u}_0 \,,$$

where the vectors $\boldsymbol{v}_0$ and $\boldsymbol{u}_0$ are constant.

Since C is in uniform motion with respect to $(O_0; x, y, z)$, this shows that the frame $(C; x, y, z)$ is Galilean. Using (2.6) and the expressions for the accelerations, we obtain

$$\ddot{\boldsymbol{A_1A}}_2 = -\mathcal{G}(m_1 + m_2)\frac{\boldsymbol{A}_1\boldsymbol{A}_2}{\|\boldsymbol{A}_1\boldsymbol{A}_2\|^3} \,. \tag{2.7}$$

Taking the barycenter C of these two points as the origin, we thus have

$$\boldsymbol{r}_1 = \boldsymbol{CA}_1 \,, \quad \boldsymbol{r}_2 = \boldsymbol{CA}_2 \,, \quad \boldsymbol{r}_{12} = \boldsymbol{r}_2 - \boldsymbol{r}_1 \,,$$

and from the definition of the barycenter,

$$m_1 \boldsymbol{r}_1 + m_2 \boldsymbol{r}_2 = \boldsymbol{0} \,.$$

Equation (2.7) gives the equation of motion

$$\ddot{\boldsymbol{r}}_{12} = -\mathcal{G}(m_1 + m_2)\frac{\boldsymbol{r}_{12}}{\|\boldsymbol{r}_{12}\|^3} \,, \tag{2.8}$$

as observed with respect to a Galilean frame of reference. The motion of the points A_1 and A_2 can then be deduced from the motion of $\boldsymbol{r}_{12}$ via the relations

$$\boldsymbol{r}_1 = -\frac{m_2}{m_1 + m_2}\boldsymbol{r}_{12} \,, \quad \boldsymbol{r}_2 = +\frac{m_1}{m_1 + m_2}\boldsymbol{r}_{12} \,.$$

As an example, for the motion of the Moon (A_2) around the Earth (A_1), we have $m_1 = 81 m_2$: the motion is studied relative to the barycenter of the Earth–Moon system.

For an artificial satellite (A_2) in orbit around the Earth (A_1), m_2 is negligible in comparison with m_1, and we have

$$\boldsymbol{r}_1 = \boldsymbol{0} \,, \quad \boldsymbol{r}_2 = \boldsymbol{r}_{12} \,.$$

Briefly, the motion of an artificial satellite around the Earth will be treated as follows:

- To begin with, it will be considered as a two-body problem in which one body (the satellite) has negligible mass compared with the other. It will therefore be in Keplerian motion and the orbit it follows will be called the Keplerian orbit.
- In a second step, this motion will be considered to be perturbed. The real orbit which results will be called the perturbed orbit.

2.3 Orbital Elements

2.3.1 Defining the Frame of Reference

The fundamental principle of mechanics known as the Galilean principle of relativity says that the properties of space and time are the same, and the laws of mechanics are identical, in all inertial frames of reference.

An inertial frame, or Galilean frame, is any frame of reference at rest with respect to an absolute frame of reference called the Copernican frame.[3] A representation of this Copernican frame can be obtained as follows: it is a frame with origin at the Sun and axes in three perpendicular directions fixed relative to the distant stars. The Sun and stars are considered to be fixed. A frame centred on the Earth and with axes parallel to the axes of a Copernican frame is not Galilean, because the Earth is in motion around the Sun. However, in most cases, this frame can be considered as inertial.[4]

---

[3] Nicolaj Kopernik (1473–1543), known in English as Nicolaus Copernicus and in Latin as *Nicolaus Copernicus Torinensis*, was a Polish astronomer. By the time of Copernicus, the inadequacies of the Ptolemaic system had become clear to several astronomers, including those at the University of Padua. However, Copernicus was the first, in his work *De Revolutionibus orbium cœlestium, Libri VI*, to reject the geocentric system in favour of the heliocentric system. In this model, the Sun occupies a central position and the planets all move on concentric spheres. Fearing the reactions of contemporary theologians, he delayed publication of his book, which came out in the year he died. (According to the story, it came out on the very day of his death: Copernicus laid his hand on the printed copy of *De Revolutionibus* and expired.)

[4] The orbital motion (hence, accelerating) of the terrestrial frame explains the phenomenon known in astronomy as aberration: the apparent position of a star in the sky describes an ellipse, flattened to varying degrees depending on the position of the star with respect to the plane of the ecliptic. The semi-major axis of this ellipse subtends $20''.5$, a value which depends only on the ratio of the speed of the Earth in its orbit to the speed of light.
Aberration is superposed in a quite independent manner on the phenomenon of astronomical parallax. This is the displacement of the apparent position of a star due to the change in position of the Earth in its orbit during the year, and hence the change in position of the point of observation. Parallax decreases in proportion to the distance of the star, and is relatively unimportant. Indeed, it is less than 1 arcsec for the nearest star.

Consider a satellite in periodic motion around the Earth. Let us define the frame $(O; x, y, z)$. The origin O is the centre of the Earth, which is taken to be a sphere Σ. The axis $\boldsymbol{Oz}$ is the axis joining the poles, oriented from the south to the north. The plane xOy is the equatorial plane of the Earth, denoted $\mathcal{E}$, which cuts the terrestrial sphere at the equator. The axis $\boldsymbol{Ox}$ is chosen arbitrarily to point towards a distant star. The axis $\boldsymbol{Oy}$ is deduced from the other two axes in such a way as to obtain a right-handed orthonormal frame. The frame associated with this coordinate system is considered to be Galilean and will be denoted by $\mathfrak{R}$.

The motion of the satellite is Keplerian, i.e., it occurs on a Keplerian orbit. In $\mathfrak{R}$, the trajectory is a conic section, in this case an ellipse, with one focus at the centre of attraction O, and lying in a plane $\mathcal{P}$, the orbital plane.

In this Galilean frame, the orbital plane $\mathcal{P}$ is fixed. Let OZ denote the straight line perpendicular to $\mathcal{P}$ at O. The intersection of the planes $\mathcal{P}$ and $\mathcal{E}$ is a straight line through O, called the line of nodes.

2.3.2 Specifying a Point on an Orbit

In order to specify a point in Keplerian motion in space, the first step is to identify the orbit, and then the point on the orbit. We thus define successively:

(a) the location of the orbital plane in this frame,
(b) the position of the elliptical orbit in this plane,
(c) the characteristics of the ellipse,
(d) the position of the moving point (i.e., the satellite) on the orbit.

We shall find that six parameters are necessary and sufficient to determine the position of the satellite in $\mathfrak{R}$. Let us now go through each of these points.

Points (a) and (b): Locating the orbital plane in the coordinate system, and the orbit in the plane. The orbit is considered as a solid. When a solid has a fixed point in a frame $\mathfrak{R}$, its position relative to $\mathfrak{R}$ is determined by three parameters corresponding to the three degrees of freedom. We choose the Euler angles, defined classically as the angle of precession ψ, the angle of nutation θ, and the angle of proper rotation[5] χ.

We shall return to the details of the decomposition of a rotation into elementary rotations when we come to study the ground track in Chap. 5. In the present context, to specify the plane $\mathcal{P}$ of the ellipse, we consider the intersection of $\mathcal{P}$ with the sphere Σ, which gives the circle $\mathcal{T}$, as shown in Fig. 2.1. $\mathcal{T}$ is called the ground track of the orbit.

The projection of P, the perigee, on the ground track is P_0 (the intersection of $\boldsymbol{OP}$ with Σ). The two points of intersection of the ground track $\mathcal{T}$

[5] The three Euler angles are traditionally denoted by ϕ, θ and ψ, respectively, in the literature. We have broken with this tradition by using χ, to avoid confusion with the latitude, which already uses the symbol ϕ.

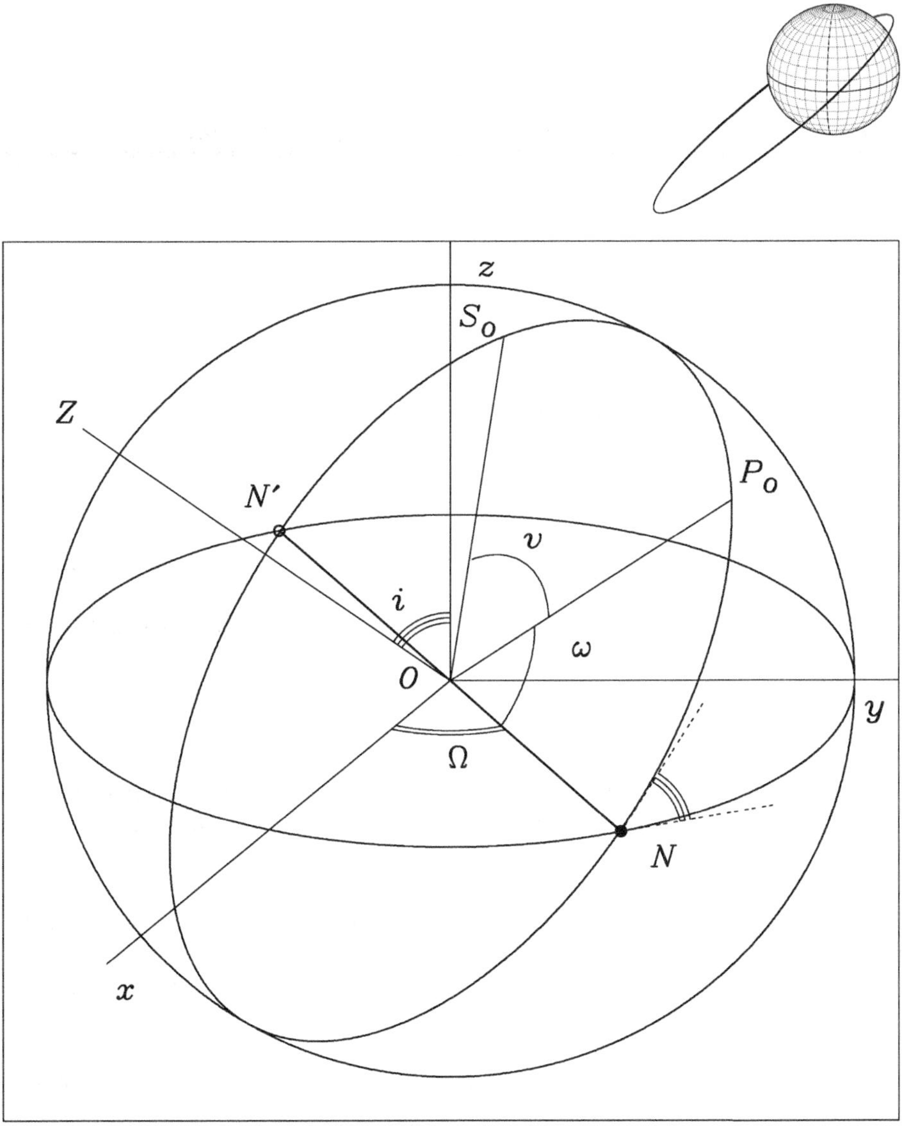

Figure 2.1. Ground track and orbital elements. The point S_0 is the projection of the point S (satellite) and the point P_0 is the projection of the perigee P onto the ground track. The point N is the projection of the ascending node and N' is the projection of the descending node. The equatorial plane of the Earth is (xOy,N, N'), normal to Oz. Orbital plane: (P_0, S_0, N, N'), normal to OZ. Three of the orbital elements are the Euler angles: the longitude of the ascending node (Ω), the inclination (i), and the argument of the perigee (ω). The fourth, here the true anomaly (v), specifies the point on the ellipse. The two other parameters (a and e) serve to define the shape of the ellipse and are not shown here. *Upper image*: Orbit viewed in a Galilean frame

with the equator are denoted N and N'. N is the projection of the ascending node (where the moving point S passes from the southern hemisphere into the northern hemisphere) onto the ground track, and N' is the projection of the descending node (where the moving point S passes from the northern hemisphere into the southern hemisphere) onto the ground track. In the following, we shall simply refer to N and N' as the ascending and descending nodes, respectively.

For a circular trajectory, P, and hence P_0, are not defined. If the orbit is equatorial ($\mathcal{P}$ and $\mathcal{E}$ coincide), points N and N' are not defined. We shall return to these cases below.

Using standard astronomical notation, the three Euler angles are:

- Angle of precession Ω, called here the right ascension of the ascending node or the longitude of the ascending node:

$$\Omega = (\boldsymbol{Ox}, \boldsymbol{ON}) \ .$$

- Angle of nutation i, called here the inclination. This is the dihedral angle $i = (\mathcal{E}, \mathcal{P})$ between the equatorial and orbital planes, viz.,

$$i = (\boldsymbol{Oz}, \boldsymbol{OZ}) \ .$$

- Angle of proper rotation ω, called here the argument of the perigee:

$$\omega = (\boldsymbol{ON}, \boldsymbol{OP}_0) \ .$$

The angles Ω and i achieve (a), that is, they locate $\mathcal{P}$ in $\Re$, whilst the angle ω achieves (b), that is, it specifies the position of the orbit in $\mathcal{P}$.

Point (c): Characterising the ellipse. Since its axes are specified ($\boldsymbol{OP}$ orients the ellipse), the ellipse is characterised by two parameters:

- Length of the semi-major axis and eccentricity,

$$a \quad \text{and} \quad e \ ,$$

  respectively.

Point (d): Locating the satellite in its orbit. A point on the ellipse is perfectly determined by an angle. We choose from the three anomalies discussed in Chap. 1, i.e., v, E or M. If S_0 is the ground position of the satellite S (the intersection of $\boldsymbol{OS}$ with Σ), the true anomaly v is defined by

$$v = (\boldsymbol{OP}_0, \boldsymbol{OS}_0) \ ,$$

and the mean anomaly M by

$$M = n(t - t_\mathrm{p}) \ ,$$

where n is the mean motion and t_p the time of passage at perigee (when S is at P, or S_0 at P_0), as given in (1.66).

We can also use the angle α, called the argument of the latitude or the position on orbit:

$$\alpha = \omega + v , \tag{2.9}$$

which specifies the position of S relative to the ascending node.

2.3.3 Keplerian Elements

The parameters discussed above define the orbit and the position of the satellite on the orbit. These parameters constitute the six orbital elements, also known as the Keplerian elements. They are generally organised in the following order:

$$a, \ e, \ i, \ \Omega, \ \omega, \ M .$$

The parameter a has dimensions of length, whilst the five others (e and the four angles) are dimensionless.

We should ask why there are 6 parameters. Here are two equivalent reasons:

- Three points define the position of a solid in space. Once the point O is fixed, 6 parameters (2 times 3 position coordinates) define the two other points.
- The position of a point (3 position coordinates) and its velocity (3 velocity components) at a given time can provide the initial conditions required to integrate the equations of motion, thereby defining the position of a point on its trajectory.

In Sect. 5.7, we shall examine the orbital elements which allow one in practice to locate a satellite with great accuracy. We shall describe the relation between these elements and the Keplerian elements discussed here.

2.3.4 Adapted Orbital Elements

In certain cases (mathematically rather special but often encountered in practice), the Keplerian elements do not provide a precise enough system of parameters. They are then combined in such a way as to provide better adapted parameters. We shall now list the combinations used to get around the difficulties. No attempt has been made to give more than a simple inventory. These methods are used for a highly refined description of satellite motion (accurate positioning and stationkeeping), which lie beyond the scope of the present book.

- **Near-Circular Orbits.** For very low eccentricities, the perigee is extremely difficult to establish and small fluctuations in e can lead to large variations in ω. To specify the position of the satellite in its orbit, it is more convenient to take the origin at the ascending node, with $\omega + M$, than at the perigee, with M. The Keplerian elements are then replaced by the following adapted parameters:

$$a,\ e\cos\omega,\ e\sin\omega,\ i,\ \Omega,\ \omega + M\ .$$

- **Near-Equatorial Orbits.** If the orbit lies almost in the equatorial plane, the ascending node is ill-defined. There may even be discontinuities in Ω when the plane of the orbit crosses the equatorial plane: the ascending node becomes the descending node. In this case, it is preferable to use the following parameters:

$$a,\ e,\ \omega + \Omega,\ 2\sin\frac{i}{2}\cos\Omega,\ \ 2\sin\frac{i}{2}\sin\Omega,\ M\ .$$

- **Near-Circular Near-Equatorial Orbits.** If the conditions of the last two cases are brought together simultaneously (as in the case of geostationary satellites, as we shall see below), the following parameters are preferred:

$$\begin{gathered} a,\ e\cos(\omega+\Omega),\ e\sin(\omega+\Omega), \\ 2\sin\frac{i}{2}\cos\Omega,\ 2\sin\frac{i}{2}\sin\Omega,\ \omega+\Omega+M\ . \end{gathered}$$

2.4 Near-Circular Orbits

Let us now reconsider near-circular orbits. We adopt a different standpoint to the one taken above, where we discussed specially adapted parameters. The aim here is not to specify the position of the satellite with high accuracy, but rather to assess the error involved in treating a near-circular orbit as actually being circular.

2.4.1 Low-Eccentricity Orbits

For low-eccentricity orbits, the equation of centre defined by (1.74) can be used to assess the deviation from uniform circular motion. Expanding to first order, we can obtain simplified relations between v and E, then between E and M, using (1.64) and (1.68):

$$\begin{cases} v - E \approx e\left(1 + \dfrac{1}{2}e\cos E\right)\sin E\ , \\ E - M = e\sin E\ . \end{cases} \tag{2.10}$$

Neglecting e^2 in comparison with e and replacing E by M in the argument of the sine function, we obtain

$$\begin{cases} v - E \approx e \sin M \; , \\ E - M \approx e \sin M \; , \end{cases} \tag{2.11}$$

which gives for the equation of centre E_C,

$$E_C = v - M \approx 2e \sin M \; . \tag{2.12}$$

The maximum of $(v - M)$, denoted $(v - M)_m$ or E_{C_m}, is thus equal to

$$E_{C_m} = (v - M)_m = 2e \; . \tag{2.13}$$

Using (1.75) and truncating at first order, we obtain[6]

$$v_m \approx \frac{\pi}{2} + \frac{3}{4}e \; . \tag{2.14}$$

We thus obtain the corresponding value of M_m, viz.,

$$M_m = v_m - 2e \approx \frac{\pi}{2} - \frac{5}{4}e \; . \tag{2.15}$$

The values at the minimum are obtained by symmetry.

2.4.2 Near-Circular Orbits

Representing an elliptical orbit by a circular one amounts to identifying the anomalies v, E and M. We can therefore evaluate the error involved by using the quantity $(v - M)$. We shall now give an example of this procedure.

Example 2.1. *Find the error involved in calculating the various quantities associated with the orbit of the satellite SPOT-4 under the assumption that its orbit is circular.*

The eccentricity of the orbit of SPOT-4 is $e = 1.14 \times 10^{-3}$, as it is for all the SPOT satellites. We thus have $(v - M)_m = 2.28 \times 10^{-3}$ rad, corresponding on the ground track to a maximal error Δx of

$$\Delta x = 2.28 \times 10^{-3} R \approx 14 \text{ km} \, ,$$

where R is the radius of the Earth. For the time of passage, given that the period of this satellite is $T = 101$ min, the maximal error is Δt given by

[6] Recall that the Taylor expansion of arcsin is

$$\arcsin x = x + \frac{1}{2}\frac{x^3}{3} + \frac{1 \cdot 3}{2 \cdot 4}\frac{x^5}{5} + \cdots \, ,$$

and $\arccos x = \pi/2 - \arcsin x$.

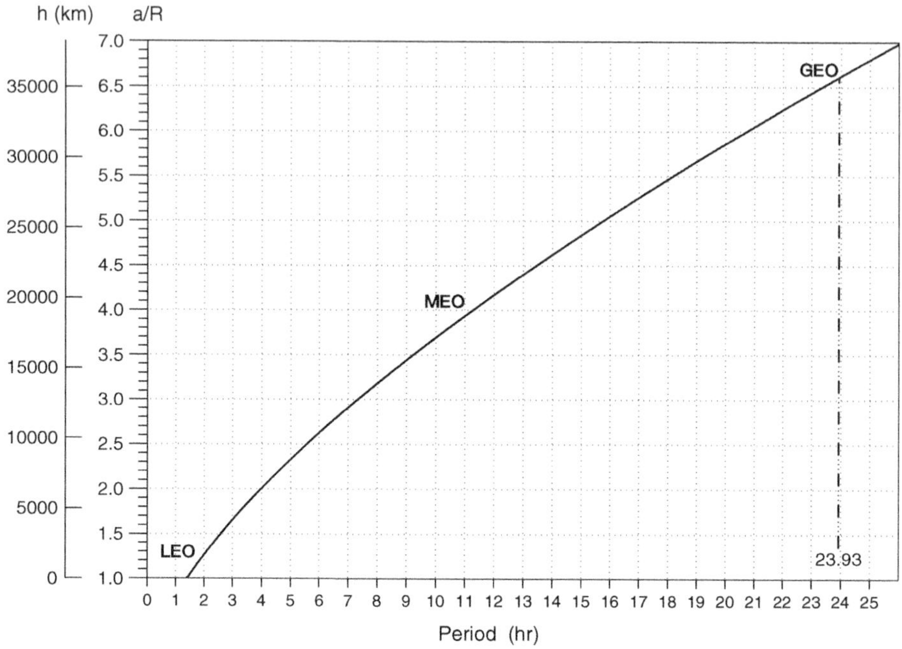

Figure 2.2. Keplerian period of a satellite as a function of altitude. Abbreviations LEO, MEO and GEO are explained in Chap. 5

$$\Delta t = 2.28 \times 10^{-3} \frac{T}{2\pi} \approx 2 \text{ s} .$$

In the context of precision orbitography, an error of 14 km is unacceptable. On the other hand, for a time sampling study, an error of this order, corresponding to 2 s, is quite negligible. Note also with regard to the orbit that, when applying the relations (1.90), we may consider the trajectory of this satellite as a circle (the difference between a and b is just 4 m for a radius of 7 200 km), with centre located 8 km from the centre of the Earth.

To end this section, it should be noted that the errors (between a near-circular orbit and an orbit treated as circular) do not accumulate. After each round trip, the errors are reset to zero: whether the orbit is highly elliptical or circular, the period of a Keplerian motion depends only on the value of the semi-major axis.

This shows that the accuracy required depends on the type of study one intends to make. In precise orbitography studies, an ellipse, even a near-circular one, remains an ellipse, whereas in sampling studies, these circular orbits lead to non-cumulative errors of a few seconds at the time of passage, and these can be ignored.

2.4.3 Reduced Orbital Elements

Near-circular orbits ($e \sim 0.001$) are generally called circular orbits. In this case, the position of the satellite can be defined by four parameters, which we refer to as the reduced parameters. The plane of the orbit is defined by the inclination i. The semi-major axis a, which is the radius of the orbital circle, can be written in the form

$$a = R + h \, ,$$

where R is the equatorial radius of the Earth and h is the altitude. As the perigee P is not defined, we have seen that the satellite position can be specified by the angle α, known as the nodal elongation, defined by

$$\alpha = (\boldsymbol{ON}, \boldsymbol{OS}_0) = \omega + M \, .$$

This is equivalent to the position on orbit, an angle defined above by (2.9).

In circular motion, the three anomalies coincide. This is a uniform motion and we may set

$$\alpha = n(t - t_{\mathrm{NA}}) \, ,$$

where t_{NA} is the time of passage of the ascending node ($t = t_{\mathrm{NA}}$ when S_0 is at N).

When studying satellites in circular orbit, we thus use the following four orbital elements:

$$h, \; i, \; \Omega, \; t_{\mathrm{NA}} \, .$$

However, it is clear that these reduced parameters can only be used to designate orbits in the case when they are considered to be circular. In calculations regarding orbits of arbitrary type, the six orbital elements must be used.

2.5 Keplerian Period

Whatever the eccentricity of an elliptical orbit may be, the Keplerian period is given by (1.49). We denote this by T_0 to distinguish it from periods defined for perturbed motion:

$$T_0 = 2\pi \sqrt{\frac{a^3}{\mu}} \, . \tag{2.16}$$

It is convenient to define the period $T_{0(h=0)}$ of a (fictitious) satellite in circular orbit at ground level on a spherical Earth of radius R:

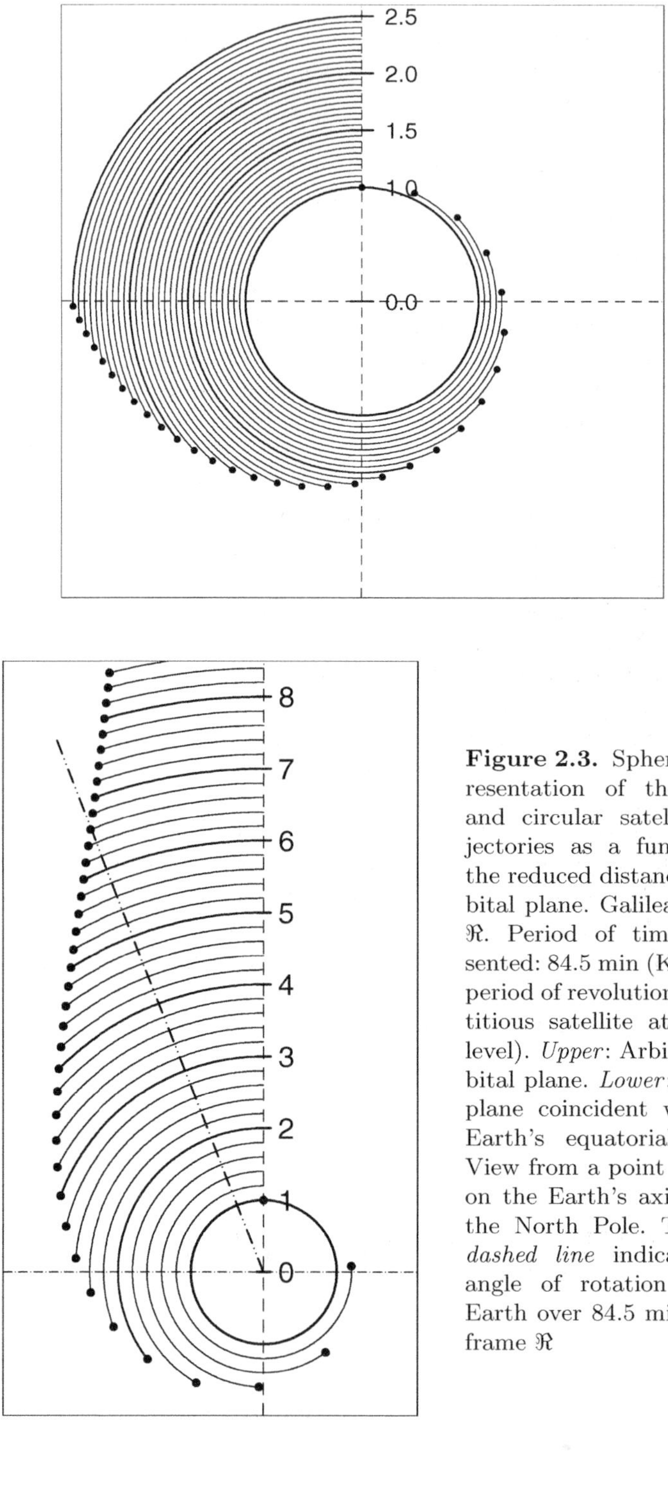

Figure 2.3. Spherical representation of the Earth and circular satellite trajectories as a function of the reduced distance η. Orbital plane. Galilean frame $\mathfrak{R}$. Period of time represented: 84.5 min (Keplerian period of revolution of a fictitious satellite at ground level). *Upper*: Arbitrary orbital plane. *Lower*: Orbital plane coincident with the Earth's equatorial plane. View from a point situated on the Earth's axis, above the North Pole. The *dot-dashed line* indicates the angle of rotation of the Earth over 84.5 min in the frame $\mathfrak{R}$

$$T_{0(h=0)} = 2\pi\sqrt{\frac{R^3}{\mu}} \,. \tag{2.17}$$

We also define the reduced (dimensionless) distance of the satellite from the centre of the Earth, denoted throughout this book by η:

$$\eta = \frac{a}{R} \,. \tag{2.18}$$

It is now easy to obtain T_0 in terms of a, η or h:

$$\frac{T_0(a)}{T_{0(h=0)}} = \left(\frac{a}{R}\right)^{3/2} , \tag{2.19}$$

$$T_0(\eta) = \eta^{3/2} T_{0(h=0)} \,, \quad T_0(h) = \left(1 + \frac{h}{R}\right)^{3/2} T_{0(h=0)} \,. \tag{2.20}$$

Numerical calculation of the period of the satellite orbiting at ground level gives

$$T_{0(h=0)} = 5069.34 \text{ s} = 84.4891 \text{ min} \,. \tag{2.21}$$

Figure 2.2 can be used as a quick way to evaluate the period as a function of altitude. Several satellite appellations appear on the figure (LEO, MEO, GEO). These will be explained later.

For small h compared with R (the case of low altitudes, i.e., LEO satellites), we have

$$T_0(h) \approx \left(1 + \frac{3}{2}\frac{h}{R}\right) T_{0(h=0)} \,.$$

Circular Motion

The upper graph in Fig. 2.3 shows the trajectory of a satellite with circular orbit (in its orbital plane) for various altitudes and over the same period of time, chosen here to be equal to $T_{0(h=0)}$, i.e., 84.5 min. The frame is the Galilean frame $\mathfrak{R}$. The satellite is moving in a direct or prograde orbit. The trajectory begins on the vertical axis, on which the values of $\eta = a/R$ are indicated, and ends at a black dot. For $\eta = 1.0$, the trajectory makes a complete round trip, for $\eta = 1.6$, just half a revolution, and for $\eta = 2.5$, one quarter of a revolution (since $2^{2/3} \approx 1.6$ and $4^{2/3} \approx 2.5$).

Consider now the lower graph in Fig. 2.3, representing the Earth's equatorial plane. On this graph, which extends the last one, we have also marked with a dot-dashed line the angle through which the Earth has rotated in 84.5 min relative to $\mathfrak{R}$. For $\eta < 6.6$, the satellite moves around faster than the Earth, otherwise more slowly. For $\eta = 6.6$, an equatorial satellite revolves at the same angular speed as the Earth in the frame $\mathfrak{R}$. In a terrestrial frame $\mathfrak{R}_\mathrm{T}$ fixed with respect to the Earth, this satellite will appear to be motionless. It is said to be geostationary. We shall return to this case in Chap. 4.

Synodic Period

The idea of synodic period will often be used in this book, whether to speak of planets or satellites. Consider two bodies in the same Galilean frame, moving uniformly with angular speeds (mean motions) n and n_1. The motion of the first relative to the second is a (relative) motion of angular speed n' given by

$$n' = n - n_1 \ . \tag{2.22}$$

In terms of periods, this means that

$$\frac{1}{T'} = \frac{1}{T} - \frac{1}{T_1} \ , \tag{2.23}$$

where T' is the synodic period. A negative value of the period T' indicates that the motion has period $|T'|$, but in the opposite direction.

3 Satellite in Perturbed Orbit

Up to now, we have been discussing the Keplerian motion of a satellite around the Earth: the satellite is pointlike and subject to the gravitational attraction of a pointlike Earth. We have seen that the Earth's attraction reduces to the attraction due to a point mass, provided that the Earth can be treated as spherical, with a spherically symmetric mass distribution. But when we observe the motion of a satellite with sufficient accuracy, we find that it does not follow exactly this kind of Keplerian motion. The difference is only very slight over one revolution, but increases steadily as time goes by until it becomes easily discernible after a few days. One might say that the Keplerian orbit is gradually distorted.

The difference between the true motion and the ideal Keplerian motion results from two considerations, as we have already mentioned:

- The Earth is not exactly spherical and the mass distribution is not exactly spherically symmetrical.
- The satellite feels other forces apart from the Earth's attraction: attractive forces due to other heavenly bodies and forces that can be globally categorised as frictional.

3.1 Perturbing Forces

3.1.1 Order of Magnitude of Perturbing Forces

The forces felt by a satellite in geocentric orbit are examined below as a function of the altitude of the satellite and summed up in Fig. 3.1. From a physical standpoint, it is useful to divide these forces into two categories, depending on whether or not they are conservative.

Conservative forces are ones that can be derived from a potential, e.g., gravitational forces such as the gravitational attraction of the Earth and attraction to other heavenly bodies. Non-conservative forces are forces that cannot be derived from a potential, i.e., dissipative forces involving energy loss. Apart from atmospheric drag, which falls off very rapidly with increasing altitude, the relevant forces here are due to radiation pressure.

It is important to assess the orders of magnitude of the various forces. For example, for a satellite at an altitude of 800 km in a near-circular orbit, taking the central attraction to be unity, the other attractive effects have the following much lower values:

- 10^{-3} for the perturbation due to the flattening of the Earth,
- 10^{-6} for perturbations due to other irregularities of the geoid,
- 10^{-7} for the attraction of the Moon,
- 10^{-8} for the attraction of the Sun.

The other forces (conservative and non-conservative) generally never exceed about 10^{-8}.

These are orders of magnitude, to which we shall return later. However, we can already see that all these forces (except the main one) can be treated as perturbations. They are not dealt with together in one global treatment, but one by one, as quantities that remain small relative to the main force, which is the Newtonian attraction.

The set of forces mentioned above can be written in terms of the accelerations

$$\boldsymbol{\gamma} = \boldsymbol{\gamma}_{\mathrm{CCC}} + \boldsymbol{\gamma}_{\mathrm{CCN}} + \sum_i \boldsymbol{\gamma}_{\mathrm{C}i} + \sum_i \boldsymbol{\gamma}_{\mathrm{D}i} \,, \tag{3.1}$$

using the suffixes given in Table 3.3. The leading term is $\boldsymbol{\gamma}_{\mathrm{CCC}}$, compared with which all the others are very small. Naturally, we have the equivalence

$$\boldsymbol{\gamma}_{\mathrm{CCC}} = \boldsymbol{g}(r) = -\frac{\mu}{r^2}\boldsymbol{e_r} \,, \tag{3.2}$$

where the vector field $\boldsymbol{g}(r)$ represents the Newtonian gravitational field of the Earth, previously defined in (2.5).

3.1.2 Potentials

To begin with, let us consider only those terms due to conservative forces, so that we may use the idea of potential. We know that the potential U associated with a vector field $\boldsymbol{\gamma}$ is obtained from

$$\boldsymbol{\gamma} = \boldsymbol{\nabla} U \,.$$

Then by the linearity of the gradient operator, we obtain

$$U_{\mathrm{C}} = U_{\mathrm{CCC}} + U_{\mathrm{CCN}} + \sum_i U_{\mathrm{C}i} \,. \tag{3.3}$$

Using the expression for $\boldsymbol{\gamma}_{\mathrm{CCC}}$, the leading term U_{CCC} is found immediately to be

$$U_{\mathrm{CCC}} = \frac{\mu}{r} = \frac{\mathcal{G}M}{r} \,. \tag{3.4}$$

To find the form of U_{CCN}, the main perturbative term, we study the terrestrial potential, known as the geopotential,

$$U_{\mathrm{CC}} = U_{\mathrm{CCC}} + U_{\mathrm{CCN}} \ . \tag{3.5}$$

Note. The definition $\boldsymbol{\gamma} = \boldsymbol{\nabla} U$ used here is the usual one in astronomy. In mechanics, one generally adopts the convention $\boldsymbol{\gamma} = -\boldsymbol{\nabla} U$.

3.2 Geopotential

3.2.1 Potential Element

The temporal variation in the terrestrial mass distribution (due to land and ocean tides and phenomena linked to internal geophysical processes) and the variation in the direction of the Earth's axis of rotation (motion of the poles) are not taken into account here. We only consider the averaged effect of these phenomena over a given period and calculate the static geopotential produced by a fixed mass distribution.

Let O be the centre of the Earth and $(O; x, y, z)$ a coordinate system based on the Galilean frame defined in Fig. 2.1, where $\boldsymbol{Oz}$ is the axis of the poles and (xOy) the equatorial plane. Let S be a point outside the Earth (the satellite). Its position is specified by the three spherical coordinates r, λ, ϕ. If S_1 is the projection of S onto the equatorial plane, we have

$$r = \|\boldsymbol{OS}\| \ ,$$

$$\lambda = (\boldsymbol{Ox}, \boldsymbol{OS}_1) \ ,$$

$$\phi = \frac{\pi}{2} - (\boldsymbol{Oz}, \boldsymbol{OS}) = (\boldsymbol{OS}_1, \boldsymbol{OS}) \ .$$

The angles λ and ϕ give the longitude and latitude of the point S, which correspond to the geographical longitude and latitude of the projection S_0 of S on the Earth's surface. (The geographical latitude coincides with the geocentric latitude here.)

Let T be a point inside the Earth. Its position is also specified by the three spherical coordinates ρ, α, β, where ρ is the magnitude of $\boldsymbol{OT}$, α the longitude, and β the latitude. We now have the standard relations giving components in Cartesian coordinates:

$$\frac{\boldsymbol{OS}}{r} = \begin{pmatrix} \cos\phi \, \cos\lambda \\ \cos\phi \, \sin\lambda \\ \sin\phi \end{pmatrix} , \qquad \frac{\boldsymbol{OT}}{\rho} = \begin{pmatrix} \cos\beta \, \cos\alpha \\ \cos\beta \, \sin\alpha \\ \sin\beta \end{pmatrix} . \tag{3.6}$$

Let θ be the angle between the two radial vectors, viz.,

$$\theta = (\boldsymbol{OS}, \boldsymbol{OT}) \ ,$$

so that the distance D between the two points S and T is

$$D^2 = \|\boldsymbol{TS}\|^2 = r^2 - 2r\rho\cos\theta + \rho^2 \ ,$$

$$D(T,S) = D = r\left[1 - 2\frac{\rho}{r}\cos\theta + \left(\frac{\rho}{r}\right)^2\right]^{1/2} .$$

The scalar product $\boldsymbol{OS} \cdot \boldsymbol{OT}$ yields

$$\cos\theta = \sin\phi\,\sin\beta + \cos\phi\,\cos\beta\,\cos(\lambda - \alpha) \ .$$

The potential element is given by

$$\mathrm{d}U = \frac{\mathcal{G}\mathrm{d}M}{\|\boldsymbol{TS}\|} \ .$$

3.2.2 Obtaining the Potential by Integration

The potential U_{CC} which we are seeking here is obtained by summing all the potential elements produced by the mass elements making up the mass distribution. The mass element $\mathrm{d}M$ is associated with the point T which ranges over the whole of the Earth:

$$U_{\mathrm{CC}} = U(S) = \int_{\mathrm{Earth}} \mathrm{d}U = \mathcal{G}\int_{T\in\mathrm{Earth}} \frac{\mathrm{d}M(T)}{D(T,S)} \ . \tag{3.7}$$

The expression for D arising in the calculation of the potential is given as a function of θ by

$$\frac{1}{D} = \frac{1}{r}\frac{1}{\sqrt{1 - 2\frac{\rho}{r}\cos\theta + \left(\frac{\rho}{r}\right)^2}} \ . \tag{3.8}$$

This expression can be expanded in terms of Legendre polynomials (see Sect. 3.15). The expansion converges if $\rho/r < 1$. The calculation is thus valid if S remains strictly outside the sphere containing all the mass elements. We may then write

$$\frac{1}{D} = \frac{1}{r}\sum_{l=0}^{\infty}\left(\frac{\rho}{r}\right)^l P_l(\cos\theta) \ , \tag{3.9}$$

where P_l is the lth Legendre polynomial (or Legendre polynomial of degree l). Replacing $\cos\theta$ by its value in terms of spherical coordinates, the angles $\lambda, \phi, \alpha, \beta$, or more precisely, $\phi, \beta, (\lambda - \alpha)$, we now use the Legendre addition formula

$$P_l(\cos\theta) = P_l(\sin\phi)P_l(\sin\beta) \\ +2\sum_{m=1}^{l}\frac{(l-m)!}{(l+m)!}P_{lm}(\sin\phi)P_{lm}(\sin\beta)\cos m(\lambda-\alpha)\ ,$$

where P_{lm} are the associated Legendre functions. We thereby obtain $1/D$ in terms of the six spherical coordinates. Substituting this expression into (3.9) and then into (3.7), we obtain

$$\begin{aligned} U(r,\lambda,\phi) &= \mathcal{G}\int_\rho\int_\alpha\int_\beta \frac{\mathrm{d}M(\rho,\alpha,\beta)}{D(r,\lambda,\phi,\rho,\alpha,\beta)} \\ &= \mathcal{G}\frac{1}{r}\int_{\rho=0}^{R}\int_{\alpha=0}^{2\pi}\int_{\beta=-\pi/2}^{\pi/2}\sum_{l=0}^{\infty}\left(\frac{\rho}{r}\right)^l\Bigg[P_l(\sin\phi)P_l(\sin\beta) \\ &+2\sum_{m=1}^{l}\frac{(l-m)!}{(l+m)!}P_{lm}(\sin\phi)\cos(m\lambda)P_{lm}(\sin\beta)\cos(m\alpha) \\ &+2\sum_{m=1}^{l}\frac{(l-m)!}{(l+m)!}P_{lm}(\sin\phi)\sin(m\lambda)P_{lm}(\sin\beta)\sin(m\alpha)\Bigg]\,\mathrm{d}M\ . \end{aligned}$$

Finally, we obtain the expression for U in terms of the associated Legendre functions P_{lm} and the coefficients C_{lm} and S_{lm}:

$$U(r,\lambda,\phi) = \frac{\mu}{r}\sum_{l=0}^{\infty}\left(\frac{R}{r}\right)^l\left\{\sum_{m=0}^{l}\Big[C_{lm}\cos(m\lambda)+S_{lm}\sin(m\lambda)\Big]P_{lm}(\sin\phi)\right\}\ , \tag{3.10}$$

with $\mu = \mathcal{G}M$ and M the mass of the Earth given by

$$M = \int_{\rho=0}^{R}\int_{\alpha=0}^{2\pi}\int_{\beta=-\pi/2}^{\pi/2}\mathrm{d}M(\rho,\alpha,\beta)\ ,$$

and C_{lm}, S_{lm} the harmonic coefficients of the geopotential of degree l and order m.

In the expression (3.10), the terms for $m = 0$ refer to the Legendre polynomial P_l, and the sum from $m = 1$ to $m = l$ refers to the associated Legendre functions P_{lm}. The coefficients C_{lm} and S_{lm} are obtained by identifying the two formulas for U. There are two cases, depending on whether m is zero or not:

- Harmonic coefficients for $m = 0$, C_{l0} and S_{l0}:

$$C_{l0} = \frac{1}{MR^l}\int_{\rho=0}^{R}\int_{\alpha=0}^{2\pi}\int_{\beta=-\pi/2}^{\pi/2}\rho^l P_l(\sin\beta)\mathrm{d}M(\rho,\alpha,\beta)\ , \tag{3.11}$$

$$S_{l0} = 0\ . \tag{3.12}$$

The coefficient S_{l0} can be chosen arbitrarily since it arises as a prefactor of a zero term.

- Harmonic coefficients for $m \neq 0$, C_{lm} and S_{lm}:

$$C_{lm} = \frac{2}{MR^l}\frac{(l-m)!}{(l+m)!}\int_\rho\int_\alpha\int_\beta \rho^l P_{lm}(\sin\beta)\cos(m\alpha)\,\mathrm{d}M\ , \tag{3.13}$$

$$S_{lm} = \frac{2}{MR^l}\frac{(l-m)!}{(l+m)!}\int_\rho\int_\alpha\int_\beta \rho^l P_{lm}(\sin\beta)\sin(m\alpha)\,\mathrm{d}M\ . \tag{3.14}$$

3.2.3 Spherical Harmonics

The potential U has been given as a linear combination of spherical functions F_{lm} and G_{lm} defined by

$$F_{lm}(\lambda,\phi) = \cos(m\lambda)P_{lm}(\sin\phi)\ ,$$
$$G_{lm}(\lambda,\phi) = \sin(m\lambda)P_{lm}(\sin\phi)\ .$$

These can be considered as the real and imaginary parts of functions H_{lm}, called spherical harmonics:

$$H_{lm}(\lambda,\phi) = \mathrm{e}^{\mathrm{i}m\lambda}P_{lm}(\sin\phi)\ .$$

These functions have many mathematical properties (such as orthogonality) and there exists an extensive literature. In the present context, they can be used to give a graphical decomposition of the geopotential.

One can gain an idea of the way the spherical functions vary by plotting the points on the sphere where they vanish. To do so, the spherical harmonics are divided into three groups: the zonal harmonics, the sectorial harmonics, and the tesseral harmonics.

- Zonal harmonics. These are obtained when $m = 0$. In this case,

$$F_{l0} = P_{l0}(\sin\phi) = P_l(\sin\phi)\ ,$$
$$G_{l0} = 0\ ,$$
$$H_{l0} = P_l(\sin\phi)\ .$$

  Hence, $H_{l0}(\lambda,\phi) = H_{l0}(\phi)$ depends only on the latitude. Zonal harmonics have axial symmetry about the axis through the poles. In particular, they take into account the flattening of the Earth.
- Sectorial harmonics. These are obtained when $m = l$. In this case,

$$P_{lm}(\sin\phi) = P_{ll}(\sin\phi) = \frac{(2l)!}{2^l l!}(\cos^2\phi)^{l/2}\ .$$

  This function of ϕ is never zero, except at the poles. Hence, H_{ll} is only zero for certain values of λ. The sectorial harmonics only vanish on the meridians and one generally gives a picture of the sphere that looks like an orange separated into segments that meet at the poles.
- Tesseral harmonics. These are obtained in all other cases. The zeros produce a kind of spherical chessboard pattern, marked out by the meridians and parallels.

3.2.4 Second Degree Expansion of the Potential

Theoretical Calculation of Coefficients

If we expand the potential U given by (3.10) up to second degree, we obtain

$$\begin{aligned} U(r,\lambda,\phi) = \frac{\mu}{r}\Bigg\{ & C_{00}P_0(\sin\phi) \\ & + \left(\frac{R}{r}\right)\Big[C_{10}P_1(\sin\phi) + (C_{11}\cos\lambda + S_{11}\sin\lambda)P_{11}(\sin\phi)\Big] \\ & + \left(\frac{R}{r}\right)^2\Big[C_{20}P_2(\sin\phi) + (C_{21}\cos\lambda + S_{21}\sin\lambda)P_{21}(\sin\phi) \\ & + (C_{22}\cos 2\lambda + S_{22}\sin 2\lambda)P_{22}(\sin\phi)\Big]\Bigg\} . \end{aligned} \tag{3.15}$$

Recall now the values of the first few Legendre polynomials and functions for the argument $(\sin\beta)$:

$$P_0(\sin\beta) = 1 , \quad P_1(\sin\beta) = \sin\beta , \quad P_2(\sin\beta) = \frac{1}{2}(3\sin^2\beta - 1) ,$$

$$P_{11}(\sin\beta) = \cos\beta , \quad P_{21}(\sin\beta) = 3\sin\beta\cos\beta , \quad P_{22}(\sin\beta) = 3\cos^2\beta .$$

We can now calculate the harmonic coefficients C_{lm} and S_{lm} from the four formulas (3.11) to (3.14), using the spherical coordinates of the interior point T defined by (3.6). The coordinates of the centre of gravity of the Earth are (x_G, y_G, z_G) and the components of the Earth's inertia tensor are I_x (for I_{x^2}), I_{xy}, and so on. The results are displayed in Table 3.1.

As an example, the calculation of C_{20} goes as follows:

$$\begin{aligned} C_{20} &= \frac{1}{MR^2}\int_\rho\int_\alpha\int_\beta \rho^2\frac{3\sin^2\beta - 1}{2}\mathrm{d}M(\rho,\alpha,\beta) \\ &= \frac{1}{2MR^2}\int\left[3z^2 - (x^2+y^2+z^2)\right]\mathrm{d}M \\ &= \frac{1}{2MR^2}\int\left[(x^2+z^2) + (y^2+z^2) - 2(x^2+y^2)\right]\mathrm{d}M \\ &= \frac{1}{2MR^2}(I_x + I_y - 2I_z) . \end{aligned}$$

Case of the Terrestrial Ellipsoid

In the case of a solid Earth, the origin of the coordinate system for decomposing the geopotential is taken at the centre of the Earth. We then have $x_G = y_G = z_G = 0$, which implies

Table 3.1. Harmonic coefficients C_{lm} and S_{lm} of the geopotential of degree l and order m, up to $l = 2$, $m = 2$

$C_{00} = 1$		
$C_{10} = \dfrac{z_G}{R}$	$C_{11} = \dfrac{x_G}{R}$	$S_{11} = \dfrac{y_G}{R}$
$C_{20} = \dfrac{1}{2MR^2}(I_x + I_y - 2I_z)$	$C_{21} = \dfrac{1}{MR^2} I_{xz}$	$S_{21} = \dfrac{1}{MR^2} I_{yz}$
	$C_{22} = \dfrac{1}{4MR^2}(I_x - I_y)$	$S_{22} = \dfrac{1}{2MR^2} I_{xy}$

$$C_{10} = 0 \,, \quad C_{11} = 0 \,, \quad S_{11} = 0 \,.$$

If we consider that the axis $\boldsymbol{Oz}$ passes through the centre of inertia, we have $I_{xz} = I_{yz} = 0$, which implies

$$C_{21} = 0 \,, \quad S_{21} = 0 \,.$$

The most significant inhomogeneity in the terrestrial mass distribution is due to the flattening at the poles. As a first approximation, we may assume that the Earth is an ellipsoid of revolution with axis $\boldsymbol{Oz}$. In this case, the axial symmetry implies that $I_{xy} = 0$ and $I_x = I_y$, which in turn implies that

$$C_{22} = 0 \,, \quad S_{22} = 0 \,. \tag{3.16}$$

The flattening at the poles is expressed by the fact that $I_z > I_x$. Hence,

$$C_{20} = \frac{1}{MR^2}(I_x - I_z) \,, \quad C_{20} < 0 \,. \tag{3.17}$$

When we expand the geopotential to second order with the above assumptions, the only nonzero term (apart from the leading term $C_{00} = 1$) is thus the term C_{20} (which is negative). It is customary to introduce the coefficients J_l defined by

$$J_l = -C_{l0} \,. \tag{3.18}$$

The geopotential is then

$$U(r, \lambda, \phi) = U(r, \phi) = \frac{\mu}{r}\left[1 - \left(\frac{R}{r}\right)^2 J_2 \frac{3\sin^2\phi - 1}{2}\right] \,, \tag{3.19}$$

with

$$J_2 = 1.0826 \times 10^{-3} \,.$$

Table 3.2. Harmonic coefficients J_n for the geopotential, up to $n = 10$. Values taken from the GRIM5-C1 model

$J_n = -C_{n0}$	Value (dimensionless)
J_0	1
J_1	0
J_2	$+1\,082.626\,220\,70 \times 10^{-6}$
J_3	$-2.536\,150\,69 \times 10^{-6}$
J_4	$-1.619\,363\,55 \times 10^{-6}$
J_5	$-0.223\,101\,38 \times 10^{-6}$
J_6	$+0.540\,289\,52 \times 10^{-6}$
J_7	$-0.360\,260\,16 \times 10^{-6}$
J_8	$-0.207\,767\,04 \times 10^{-6}$
J_9	$-0.114\,567\,39 \times 10^{-6}$
J_{10}	$-0.233\,800\,81 \times 10^{-6}$

This term is dimensionless, like all the coefficients C_{lm} and S_{lm}. The value of the coefficient J_2 was known for a long time from geodetic considerations (see the historical note in Sect. 3.10), and then to very high accuracy by studying the trajectories of the first artificial satellites.

3.2.5 Expanding the Potential to Higher Degrees

For degrees higher than 2, the potential can be written

$$U(r,\lambda,\phi) = \frac{\mu}{r}\left\{1 - \sum_{l=1}^{\infty}\left(\frac{R}{r}\right)^l J_l P_l(\sin\phi) \right. \qquad (3.20)$$

$$\left. + \sum_{l=1}^{\infty}\sum_{m=1}^{l}\left(\frac{R}{r}\right)^l \Big[C_{lm}\cos(m\lambda) + S_{lm}\sin(m\lambda)\Big] P_{lm}(\sin\phi)\right\},$$

with the above notation. In the part between curly brackets, there are three groups of terms. The first comprises only the number 1, representing the central potential. The second, with J_l and P_l, constitutes the contribution of the zonal harmonics, whilst the third, involving C_{lm}, S_{lm} and P_{lm}, gives the contribution of the sectorial and tesseral harmonics.

For the Earth as it really is (dropping the ellipsoid approximation), the numerical values of J_l (J_l generally denoted by J_n) are given in Table 3.2 (see also Table 3.6). Hence, for the geoid, the coefficients C_{10} (or J_1), C_{11} and S_{11} are zero, whilst the coefficients C_{21} and S_{21} ($\sim 10^{-9}$), C_{22} and S_{22} ($\sim 10^{-6}$) are not zero. Concerning orders of magnitude, the term J_2 is seen to be about 10^3 times smaller than the leading term, but 10^3 times bigger than the following coefficients.

To sum up, considering the expansion of the potential given by (3.20), we observe that:

- The term of order 0 is the leading term, causing the Keplerian motion, in which the Earth is considered to be spherical and made up of homogeneous layers.
- The term of degree 1, which would correspond to a shift in the centre of mass of the Earth away from the geometrical centre, is made to vanish by choice of the coordinate origin.
- The term of degree 2 corresponds to the flattening of the Earth when the latter is considered as an ellipsoid of revolution.
- The terms of degree 3 and higher cater for deviations between the geoid and the ellipsoid.

3.3 Perturbations and Altitude of a Satellite

Let us now investigate the whole range of perturbative forces (accelerations) affecting a satellite S at distance r from the centre O of the Earth. We may also express them in terms of h, the altitude of the satellite, given by $h = r - R$, where R is the equatorial radius of the Earth. Figure 3.1 shows the value of the acceleration γ as a function of r on a log–log scale (both abscissa and ordinate are logarithmic scales).

The figure shows typical altitudes of three types of satellite (we shall discuss these types in more detail later):

- $h = 1\,000$ km for satellites in low orbit (LEO),
- $h = 20\,000$ km for positioning satellites (MEO),
- $h = 36\,000$ km for geostationary satellites (GEO).

The figure shows the sensitivity of the different types of satellite to the various perturbing forces depending on their altitudes. The forces, divided into conservative and non-conservative (or dissipative) forces, are summarised in Table 3.3.

3.3.1 Conservative Forces

The gravitational forces acting on the satellite come from the following sources:

(a) Attraction by the Earth

The central acceleration, denoted here by γ_{CCC}, is given by (2.5) or (3.2):

$$\gamma_{\mathrm{CCC}}(r) = g(r) = \frac{\mu}{r^2} \,. \tag{3.21}$$

It is thus represented by a straight line of gradient $p = -2$ in a graph with log–log scale. The value at the origin ($r = R$ or $h = 0$) is

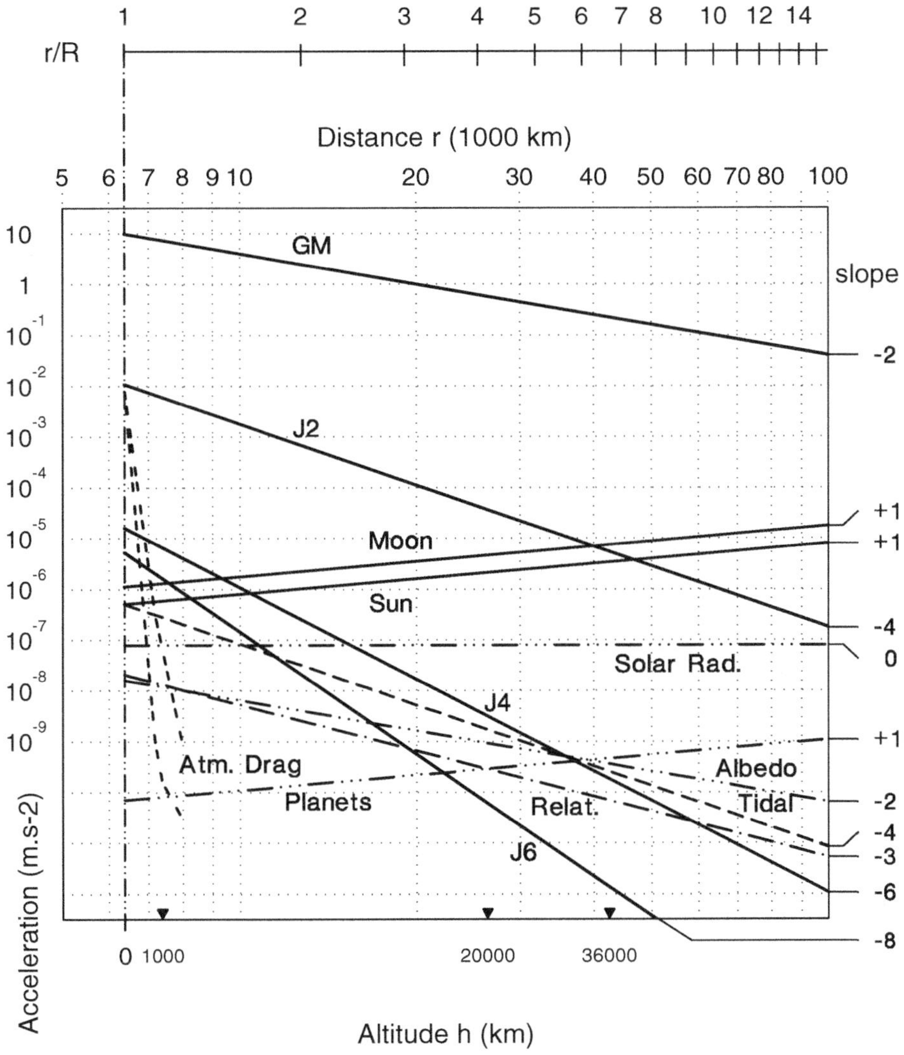

Figure 3.1. Central acceleration and perturbative accelerations as a function of the distance r of the satellite from the centre of the Earth, shown on a log–log scale. Over the ranges considered, the curves can be approximated as straight lines with gradients as noted. The altitudes of the three types of satellite have also been indicated

$$\gamma_{\mathrm{CCC}}(R) = g(R) = g_0 = \frac{\mu}{R^2} , \tag{3.22}$$

with numerical value

$$g_0 = 9.80 \mathrm{~m\,s^{-2}} .$$

Concerning the J_2 term, the potential $U_{\mathrm{CCN}.J2}$ given by (3.19) goes as r^{-3}. The corresponding acceleration $\gamma_{\mathrm{CCN}.J2}$ therefore goes as r^{-4}, whence the gradient $p = -4$ of the straight line representing it on the log–log graph. As value at the origin ($h = 0$), we have taken

$$\gamma_{\mathrm{CCN}.J2}(R) = g_0 J_2 ,$$

which is an average value over the latitudes frequented by the satellite. The acceleration $\gamma_{\mathrm{CCN}.JN}$ for the other terms J_n, $n > 2$, leads to an even steeper slope. We shall see below that only those terms J_n with even n have any long-term (secular) influence on the satellite. Equation (3.20) shows that the potential goes as $r^{-(n+1)}$, and the acceleration therefore as $r^{-(n+2)}$. The slope is $p = -6$ for J_4, $p = -8$ for J_6, and so on. At the origin, the numerical values are

$$g_0 J_2 = 1.1 \times 10^{-2} \mathrm{~m\,s^{-2}} ,$$

$$g_0 J_4 = 1.6 \times 10^{-5} \mathrm{~m\,s^{-2}} , \quad g_0 J_6 = 5.3 \times 10^{-6} \mathrm{~m\,s^{-2}} .$$

where we have, of course, taken the absolute values of the terms J_n.

(b) Attraction by the Sun and Moon

We consider an attracting body (the Sun or Moon) and calculate its action on the satellite. As the Earth also feels an attraction from the same source, we must calculate the differential attraction felt by the satellite in a coordinate system fixed relative to the centre of the Earth.

The differential attraction due to the Sun produces the acceleration γ_{CS} (calculated in Sect. 3.13). Equation (3.96) yields

$$\gamma_{\mathrm{CS}} = 2\frac{\mu_{\mathrm{S}}}{a_{\mathrm{S}}^3} r , \tag{3.23}$$

where μ_{S} is the heliocentric gravitational constant and a_{S} is the Earth–Sun distance (the semi-major axis of the Earth's orbit around the Sun).

For the differential attraction due to the Moon, the acceleration γ_{CL} can be calculated to a first approximation by a similar argument:

$$\gamma_{\mathrm{CL}} = 2\frac{\mu_{\mathrm{L}}}{a_{\mathrm{L}}^3} r , \tag{3.24}$$

where μ_L is the lunar gravitational constant and a_L the Earth–Moon distance (the semi-major axis of the geocentric lunar orbit).

For the range of values represented here, γ_{CS} and γ_{CL} are proportional to r, so that their variation has gradient $p = +1$. At the origin ($h = 0$), numerical values are

$$\gamma_{CS}(R) = 2\frac{\mu_S}{a_S^3}R = 5 \times 10^{-7}\ \mathrm{m\,s^{-2}}\ , \quad \gamma_{CL}(R) = 2\frac{\mu_L}{a_L^3}R = 1.1 \times 10^{-6}\ \mathrm{m\,s^{-2}}\ .$$

(c) Attraction by Other Planets

This differential attraction causes a very small acceleration of the satellite, denoted γ_{CP}, also of slope $p = +1$. For each planet, the order of magnitude is given by

$$\gamma_{CP} = 2\frac{\mu_P}{a_P^3}r\ , \tag{3.25}$$

where μ_P is the gravitational attraction of the planet and a_P the Earth–planet distance. Depending on the configuration of the planets, the maximal perturbative accelerations are

$$\gamma_{CP} \sim 10^{-10}\ \mathrm{m\,s^{-2}}\quad \text{due to Venus}\ , \quad \gamma_{CP} \sim 10^{-11}\ \mathrm{m\,s^{-2}}\quad \text{due to Jupiter}\ .$$

(d) Tidal Effects

The ocean tides are caused by the perturbing effects of the Sun and Moon. This idea was first put forward by Newton, then Bernoulli, and the theory was completed by Laplace and Kelvin. The phenomenon is familiar to us and easy to observe. It is less well known that this same perturbation also affects the Earth's crust: twice a day, the Earth's solid envelope rises and falls with an amplitude of around one decimetre.

Both ocean and land tides involve friction, so this phenomenon is not conservative. Indeed, this effect explains the gradual reduction of the Earth's rotation. However, the effect of the tides on the satellite can be found using the fact that the relevant stresses derive from a potential. It can be shown that the interaction potential goes as r^{-3}, giving rise to an acceleration that goes as r^{-4}. For $r = R$, we have $\gamma_{CT} \sim 5 \times 10^{-7}\ \mathrm{m\,s^{-2}}$. The effect of the ocean tides is roughly one tenth of the effect of the land tides.

(e) Relativistic Effects

The speed of a satellite V never exceeds a few kilometres per second. This is very small compared with the speed of light c and a relativistic treatment is generally unnecessary. However, since the TOPEX/Poseidon mission, relativistic effects have been taken into account for altimetry and positioning

(GPS-type) satellites. It can be shown that the correction amounts to considering a so-called relativistic acceleration, whose leading term is γ_{CR} given by

$$\gamma_{\mathrm{CR}} = \frac{\mu}{r^2} \frac{3V^2}{c^2} \,. \tag{3.26}$$

For a circular orbit, with (1.48), we obtain

$$\gamma_{\mathrm{CR}} = \frac{3\mu^2}{c^2} \frac{1}{r^3} \,, \tag{3.27}$$

giving a gradient of $p = -3$.

At the origin ($h = 0$), we have

$$\gamma_{\mathrm{CR}}(R) = g_0 \frac{3\mu}{c^2 R} = 1.6 \times 10^{-8} \ \mathrm{m\,s^{-2}} \,.$$

The main consequence of this is a secular effect on the argument of the perigee: the perigee of the orbit moves around more quickly than classical calculations would suggest. Indeed, the effect is known as the advance of the perigee or the precession of the perigee. For planets in orbit around the Sun, we speak of the advance or the precession of the perihelion, a phenomenon first explained by Albert Einstein.[1]

[1] Albert Einstein (1879–1955) was a German physicist. In 1905, he published his first papers on the special theory of relativity. By 1915, he had established the foundations for the general theory of relativity. In this, Euclidean space is replaced by a Riemannian space, which is said to be curved by the masses distributed within it. The first confirmation of general relativity was obtained by Einstein himself when he explained the advance of the perihelion of Mercury.
The orbit of Mercury is an ellipse with $e = 0.2056$. Newtonian mechanics gives a value of 531″ per century for the motion of the perigee due to the perturbing effects of the other planets. However, very precise measurements give the value 574″ per century. This is a discrepancy of 43″ per century, which remained unexplained until the advent of general relativity, when Einstein was able to produce precisely this value. In order to understand the accuracies involved here, it is interesting to compare with the Earth's precession constant of 5 026″ per tropical century.
In 1889, F. Tisserand ended his monumental work *Traité de Mécanique Céleste* with the statement that the advance of the perihelion of Mercury was the greatest outstanding mystery of astronomy. He refuted all the current explanations. For example, Le Verrier had tried to explain it by the presence of a hypothetical planet situated between Mercury and the Sun, Newcomb by the non-spherical aspect of the Sun, and Hall by an attractive gravitational force going as $r^{-2.000000151}$. Tisserand concluded: "We may consider that the attraction between two celestial bodies can only be transmitted over a distance by the intermediary effect of some kind of medium. But for the moment we know nothing of these transmission modes."

Table 3.3. Forces felt by a satellite

Symbol			Type of force
C			Conservative forces
	CC		• Earth's attraction
		CCC	• ○ Central term $\mu = \mathcal{G}M$
		CCN	• ○ Terms other than CCC
	CL		• Lunar attraction
	CS		• Solar attraction
	CP		• Planetary attraction
	CT		• Tidal effects (land and ocean)
	CR		• Relativistic effects
D			Dissipative forces
	DF		• Atmospheric friction
	DP		• Solar radiation pressure
	DA		• Albedo effect

Calculations of this apsidal precession give a variation $\Delta_1\omega$ in the argument of the perigee ω for each revolution, where

$$\Delta_1\omega = \frac{6\pi\mu}{a(1-e^2)c^2} \,. \tag{3.28}$$

The subscript 1 indicates that the value corresponds to one revolution, whilst a subscript Y is used for the value obtained over one year. This value is very small (a fraction of an arcmin per year), whether the satellite is in near-circular or eccentric orbit:

- for SPOT-4 ($a = 7.2 \times 10^6$ m), $\Delta_Y\omega = 1.16 \times 10^{-8}$ rad $= 12''.4$,
- for TOPEX/Poseidon ($a = 7.7 \times 10^6$ m), $\Delta_Y\omega = 1.09 \times 10^{-8}$ rad $= 10''.5$,
- for Molniya ($a = 2.6 \times 10^7$ m, $e = 0.75$), $\Delta_Y\omega = 7.33 \times 10^{-9}$ rad $= 1''.1$.

The advance of the perihelion can also be observed in the orbits of other planets, although the effect diminishes as the planet is further from the Sun (because it moves more slowly relative to a Copernican frame). Values are $8''63$ per century for Venus, $3''84$ per century for the Earth, and $1''35$ per century for Mars. For the Moon in its motion around the Earth, the value is $2''$ per century. All these values can be calculated using (3.28).
Returning to Einstein and his contribution to astronomy, the second confirmation of general relativity was obtained during an eclipse of the Sun on 29 May 1919: observations of stars close to the occulted disk of the Sun confirmed that light rays were deviated and gave results corresponding to the values calculated from the theory. (These observations were mainly the work of the astronomer A. Eddington on Principe, an island in the Gulf of Guinea.)

3.3.2 Non-Conservative Forces

Non-gravitational perturbing forces are independent of the satellite mass m. The corresponding accelerations thus go as $1/m$. Along with the forces discussed below, one must include forces applied to the satellite (usually by gas jets) when modifying its trajectory. These are the non-conservative forces which allow one to guide the satellite.

(a) Atmospheric Drag

For satellites in low orbit ($h < 800$ km), friction with molecules of residual atmospheric gases can be quite significant. This effect is difficult to model theoretically. Indeed, it depends on the state of the upper atmosphere, which in turn depends on various factors, such as the level of solar activity. It also depends on the shape of the satellite,[2] and in particular, the shape of its solar panels. There exist a great many highly sophisticated models.

Atmospheric effects become very weak above $h = 1\,000$ km and fall off very quickly with altitude. Hence, they are very slight for TOPEX/Poseidon at an altitude of 1 300 km, and almost imperceptible for LAGEOS at 6 000 km.

(b) Radiation Pressure

Solar radiation pressure goes as a_{S}^{-2}, and is therefore independent of r (since $r \ll a_{\mathrm{S}}$). The actual consequences of this radiation pressure on the satellite depend on its shape, covering materials, and configuration.[3] Naturally, there are no consequences at all when the satellite passes into the Earth's shadow with respect to the Sun. The corresponding perturbative acceleration is evaluated to be

$$\gamma_{\mathrm{DP}}(r) = \gamma_{\mathrm{DP}} = \text{const.} \sim 10^{-7}\ \mathrm{m\,s^{-2}}\ .$$

Another effect is the albedo effect, wherein radiation scatters from the Earth and creates a radiation pressure in the visible (short wavelengths) and infrared (long wavelengths). This depends on the region overflown and the altitude. These effects can be considered to go as r^{-2}.

[2] In order to minimise these effects, the passive (i.e., without power supply) satellite Starlette, is a sphere covered with 60 laser reflectors. Its core is made of uranium (density 18.7) and it has a total mass of 47 kg and a diameter of 48 cm.

[3] This perturbative effect was quite large for the first balloon-type satellites such as Echo-1, launched in 1960, Echo-2 in 1964, and PAGEOS in 1966. These very light satellites, comprising an aluminised mylar envelope just 13 μm thick, blown up after the launch, had a diameter of 30 to 40 m.

3.4 Perturbative Methods

3.4.1 Perturbed Equation of Motion

In this chapter, we have been concerned so far with perturbations which represent the difference between the Newtonian potential and the true potential. We shall now see how this difference of potential can lead to a deviation in the satellite motion.

In order to give the satellite position at an arbitrary time, i.e., to establish its equation of motion, there are two methods available: one involves integrating the equation of motion using Cartesian coordinates at the point, and the other uses the orbital elements. The latter allows one to express the modifications of the motion in a clear and pictorial way, so that one might say whether the eccentricity increases, or the orbit begins to precess, and so on.

(a) Keplerian Motion

For Keplerian motion, the two approaches function in the following way. We consider the equation of motion

$$\ddot{\boldsymbol{r}} = \boldsymbol{\nabla} U \,, \quad \text{where} \quad U = \frac{\mu}{r} \,. \tag{3.29}$$

- Using Cartesian coordinates: We integrate (3.29) for suitable initial conditions. The system of equations can be written

$$\begin{cases} \ddot{\boldsymbol{r}} = -\mu \dfrac{\boldsymbol{r}}{r^3} \,, \\ \boldsymbol{r}(t=0) = \boldsymbol{r}_0 \,, \quad \dot{\boldsymbol{r}}(t=0) = \dot{\boldsymbol{r}}_0 \,. \end{cases} \tag{3.30}$$

- Using the orbital elements: We recall from Chap. 2 that the Keplerian motion of a satellite is defined by six orbital parameters, the so-called Keplerian orbital elements. We show that there exists a bijective map between the six parameters and the six Cartesian components (in the Galilean frame $\Re$) of the vectors $\boldsymbol{r}$ and $\dot{\boldsymbol{r}}$:

$$\Big\{x(t), y(t), z(t), \dot{x}(t), \dot{y}(t), \dot{z}(t)\Big\} \stackrel{\mathcal{T}}{\longmapsto} \Big\{a(t), e(t), i(t), \Omega(t), \omega(t), M(t)\Big\} \,.$$

In the Keplerian motion, the 5 parameters fixing the position of the orbit, viz., a, e, i, Ω, ω, remain constant, whilst M varies linearly with time, i.e., $M = n(t - t_{\mathrm{p}})$, where t_{p} is the time of the passage at perigee. The dynamical equation (3.30) is equivalent to 6 relations:

$$\begin{cases} \dot{a} = 0 \,, \quad \dot{e} = 0 \,, \quad \dot{i} = 0 \,, \\ \dot{\Omega} = 0 \,, \quad \dot{\omega} = 0 \,, \quad \dot{M} - n = 0 \,, \end{cases} \tag{3.31}$$

where the mean motion is given by $n = \sqrt{\mu/a^3}$.

(b) Perturbed Motion

In the case where the satellite is not subject to a Newtonian acceleration but the sum of a Newtonian acceleration and a perturbation, we shall assume that, since the perturbation is relatively weak, the trajectory remains close to the conic section trajectory. The equation of motion is then

$$\ddot{\boldsymbol{r}} = \boldsymbol{\nabla} U_0 + \boldsymbol{\gamma}_{\mathrm{P}} , \qquad U_0 = \frac{\mu}{r} , \tag{3.32}$$

where γ_{P} is the perturbative acceleration, which remains small compared with the leading term:

$$\gamma_{\mathrm{P}} \ll \frac{\mu}{r^2} .$$

We consider here only perturbative acceleration fields that can be derived from a potential. In this case,

$$\boldsymbol{\gamma}_{\mathrm{P}} = \boldsymbol{\nabla} \mathcal{R} ,$$

where $\mathcal{R}$ is the perturbing potential.

The satellite is thus subject to the potential U given by

$$U = U_0 + \mathcal{R} . \tag{3.33}$$

- Using Cartesian coordinates: The system of equations is in this case

$$\begin{cases} \ddot{\boldsymbol{r}} = -\mu \dfrac{\boldsymbol{r}}{r^3} + \boldsymbol{\gamma}_{\mathrm{P}} , \\ \boldsymbol{r}(t=0) = \boldsymbol{r}_0 , \quad \dot{\boldsymbol{r}}(t=0) = \dot{\boldsymbol{r}}_0 . \end{cases} \tag{3.34}$$

  This system is very difficult to integrate because it involves second order integrations. Although it is not used to find analytical solutions, it is used for purely numerical methods.
- Using the orbital elements: Examination of the transformation $\mathcal{T}$ shows that (3.34) is equivalent to the 6 relations

$$\begin{cases} \dot{a} = g_1 , \quad \dot{e} = g_2 , \quad \dot{i} = g_3 , \\ \dot{\Omega} = g_4 , \quad \dot{\omega} = g_5 , \quad \dot{M} - n_0 = g_6 , \end{cases} \tag{3.35}$$

  where $n_0 = \sqrt{\mu/a_0^3}$ and a_0 is, to begin with, the value of a without perturbation. The terms g_i are small.

We then apply an iterative method. The parameters, which now vary in time, are known as osculating orbital elements. They correspond to the parameters of the Keplerian orbit that the satellite would follow if the perturbations suddenly ceased to act. These osculating elements[4] provide a better way of describing the deformation of the orbit than the values of the position and velocity. As an example, this method allows one to establish the critical inclination of the orbit, as we shall see later.

The perturbation method consists in solving the six equations above, known as Lagrange's equations.[5]

3.4.2 Basic Principles

The actual motion is obtained by calculating the small variations around the first integrals of the unperturbed motion. For a perturbative acceleration field deriving from a potential, (3.32) and (3.33) give the equation of motion as

$$\ddot{\boldsymbol{r}} = \boldsymbol{\nabla} U \;, \quad \text{where} \quad U = U_0 + \mathcal{R} \;. \tag{3.36}$$

Transcribing this for each Cartesian component of $\boldsymbol{r}$, we have

$$\frac{\mathrm{d}^2 x}{\mathrm{d}t^2} = \frac{\partial U}{\partial x} \;, \quad \frac{\mathrm{d}^2 y}{\mathrm{d}t^2} = \frac{\partial U}{\partial y} \;, \quad \frac{\mathrm{d}^2 z}{\mathrm{d}t^2} = \frac{\partial U}{\partial z} \;.$$

Let us write down the results for x. The first of the above equations yields,

---

[4] The word 'osculating' does not mean the same thing here as in geometry. In the purely geometrical context, two curves are said to be osculating if their two centres of curvature coincide for some point of contact between them. In the present context, in the investigation of trajectories in space, the osculating ellipse defined by the osculating orbital elements is tangent to the actual trajectory, since the velocity vector is the same, but it does not have exactly the same radius of curvature, since the accelerations are different. The term was originally invented by geometers in 1752 and then slightly deflected from its geometrical meaning by the needs of astronomy. The etymology of the word attests to this corruption. Indeed, it comes from the Latin *osculatio*, the noun from the verb *osculare* meaning 'to kiss'. This in turn derives from *osculum*, 'little mouth', a diminutive form of *os, oris*, which is the standard term for 'mouth'. The idea which therefore underpins this term is therefore one of extended and continued contact.

[5] Joseph Louis de Lagrange (1736–1813) was a French mathematician. He applied his analytical theories to the motion of the Moon and the periodic variation of the major axes of the planets, published in *Théorie de la libration de la Lune et autres phénomènes qui dépendent de la figure non sphérique de cette planète* (1763). He also invented the idea of gravitational potential in 1772. All these ideas were brought together in his magnum opus *Mécanique analytique* (1788). He invented the theory of perturbations to study the motion of the heavenly bodies, published in *Sur la théorie des variations des éléments des planètes* (1808). He also continued Euler's work, devising a final version of the method known as variation of constants, published in *Sur la théorie générale de la variation des constantes arbitraires dans tous les problèmes de mécanique* (1810). His name is still associated with the equations and mathematical tools used in these theories.

$$\dot{x} = \frac{dx}{dt}, \tag{3.37}$$

$$\frac{d\dot{x}}{dt} = \frac{\partial U}{\partial x}. \tag{3.38}$$

Bringing in the 6 orbital elements (or any 6 conveniently chosen variables), we obtain $\dot{x}$ as a sum of 6 terms, and hence $\ddot{x}$ as a sum of 36 terms.

The perturbation method solves the differential equations by the so-called method of variation of constants. The choice of certain variables, called canonical variables, allows us to obtain Lagrange's equations in a very simple form, the canonical form, as the relations (3.43). Several sets of 6 variables offer this possibility, such as the Delaunay variables, the Poincaré variables, or the Whittaker variables.

More will be said about the Delaunay variables later, but for the moment, let us not enter into the details of these powerful mathematical theories, developed principally to calculate planetary orbits, to which we now attach the names of Euler,[6] Lagrange, Laplace,[7] Gauss and Poincaré.[8]

[6] Leonhard Euler (1707–1783) was a Swiss mathematician. In astronomy, he studied the mutual perturbations of Saturn and Jupiter in *Theoria motuum planetarium et cometarum* (1744), the precession of the equinoxes (1749), the restricted three-body problem and the motion of the Moon (1772). In mathematics, his vast contribution includes mechanics (the Euler angles), analysis, trigonometry (the notation e for the exponential, $\mathrm{i} = \sqrt{-1}$, and π were his and he obtained the famous result $\mathrm{e}^{\mathrm{i}\pi} = -1$), algebra, and geometry.

[7] Pierre Simon de Laplace (1749–1827) was a French mathematician, astronomer and physicist. Not only did he seek to understand mathematically how universal attraction could be compatible with the observed stability of the distances between the stars, but noting that the 43 observable bodies in the Solar System all rotate in the same direction (the planets, moons and even the Sun itself), and taking the cue from observations of the nebulas made with Herschel, he devised a first cosmogonic system in *Exposition du système du Monde* (first edition in 1796 and fifth edition in 1824). According to this view, the Solar System and all other objects in the universe were produced by the condensation of a primordial nebula. During cooling, the rotation of this nebula would have generated a succession of rings in the same plane (the ecliptic), and these would have given birth to the planets and their moons, with the central nucleus becoming the Sun. The emperor Napoleon, in wonder before this work, raised the inevitable question: "And what about God in all this?", to which he replied: "Sire, I did not need that hypothesis." In the two centuries from Kepler to Laplace, the relationship between astronomers and the divine order had changed considerably!
In *Mécanique céleste* (first edition in 1798 and sixth edition in 1825), he returned to all the theories elaborated since Newton. He established fundamental results in mathematics (harmonic functions, differential equations, probabilities) and in physics (electromagnetism, thermodynamics).

[8] Henri Poincaré (1854–1912) was a French mathematician. He studied changes of variables which conserve the canonical form of the equations of mechanics (in the Jacobi formulation). He also achieved new results with regard to the three-body

3.5 Perturbative Method: Solution

3.5.1 Lagrange's Equations

The perturbative method is explained in many standard textbooks on celestial mechanics. The main developments involving calculations with Lagrange brackets can be found in [51]. Here we shall simply state the results.

We obtain six equations giving the time derivative of the six orbital elements as a function of the perturbing potential $\mathcal{R}$ and the orbital elements. The results are given in Table 3.4. This system of equations constitutes the full set of Lagrange's equations, which correspond, by successive equivalences, to the original equation given in (3.32), i.e., the equation of motion. This can be written in matrix form using the matrix $\mathcal{L}$, which has two important properties: it is antisymmetric and depends on only three of the parameters, viz., a, e and i, i.e.,

$$\mathcal{L} = \mathcal{L}(a, e, i) \; .$$

To conclude this section, we shall show that, when the perturbative acceleration field derives from a potential, the satellite motion is defined by Lagrange's equations. We check that, if $\mathcal{R} = 0$, we recover the solution of the two-body problem, i.e., Kepler's solution: a, e, i, Ω, ω are constants and the mean anomaly is given by $M = n(t - t_\mathrm{p})$.

3.5.2 Metric and Angular Orbital Elements

The matrix $\mathcal{L}$ clearly brings out the separation of the orbital elements into two groups: on the one hand, a, e, i, and on the other, Ω, ω, M. Indeed we see that, in Lagrange's equations, the time derivatives of a, e, i involve only the partial derivatives of $\mathcal{R}$ with respect to Ω, ω, M, and conversely. This can be expressed in a global manner as follows:

$$\left\{ \frac{\mathrm{d}a}{\mathrm{d}t}, \frac{\mathrm{d}e}{\mathrm{d}t}, \frac{\mathrm{d}i}{\mathrm{d}t} \right\} = f_1 \left(a, e, i; \frac{\partial \mathcal{R}}{\partial \Omega}, \frac{\partial \mathcal{R}}{\partial \omega}, \frac{\partial \mathcal{R}}{\partial M} \right) \; ,$$

$$\left\{ \frac{\mathrm{d}\Omega}{\mathrm{d}t}, \frac{\mathrm{d}\omega}{\mathrm{d}t}, \frac{\mathrm{d}M}{\mathrm{d}t} \right\} = f_2 \left(a, e, i; \frac{\partial \mathcal{R}}{\partial a}, \frac{\partial \mathcal{R}}{\partial e}, \frac{\partial \mathcal{R}}{\partial i} \right) \; ,$$

where f_1 and f_2 are functions of (a, e, i) and the partial derivatives mentioned.

---

problem in *Sur le problème à trois corps et les équations de la dynamique* (1889). He developed this work in *Les méthodes nouvelles de la mécanique céleste* (1899) and explained the stability of the Solar System.

Table 3.4. Lagrange's equations for the six orbital elements as a function of the perturbing potential $\mathcal{R}$

$$\frac{\mathrm{d}a}{\mathrm{d}t} = \frac{1}{na}\left(2\frac{\partial \mathcal{R}}{\partial M}\right)$$

$$\frac{\mathrm{d}e}{\mathrm{d}t} = \frac{1}{na^2}\frac{1-e^2}{e}\left(-\frac{1}{\sqrt{1-e^2}}\frac{\partial \mathcal{R}}{\partial \omega} + \frac{\partial \mathcal{R}}{\partial M}\right)$$

$$\frac{\mathrm{d}i}{\mathrm{d}t} = \frac{1}{na^2\sqrt{1-e^2}\sin i}\left(-\frac{\partial \mathcal{R}}{\partial \Omega} + \cos i\frac{\partial \mathcal{R}}{\partial \omega}\right)$$

$$\frac{\mathrm{d}\Omega}{\mathrm{d}t} = \frac{1}{na^2\sqrt{1-e^2}\sin i}\left(\frac{\partial \mathcal{R}}{\partial i}\right)$$

$$\frac{\mathrm{d}\omega}{\mathrm{d}t} = \frac{1}{na^2\sqrt{1-e^2}}\left(\frac{1-e^2}{e}\frac{\partial \mathcal{R}}{\partial e} - \frac{\cos i}{\sin i}\frac{\partial \mathcal{R}}{\partial i}\right)$$

$$\frac{\mathrm{d}M}{\mathrm{d}t} - n = \frac{1}{na^2}\left(-2a\frac{\partial \mathcal{R}}{\partial a} - \frac{1-e^2}{e}\frac{\partial \mathcal{R}}{\partial e}\right)$$

In matrix form:

$$\begin{bmatrix} \frac{\mathrm{d}a}{\mathrm{d}t} \\ \frac{\mathrm{d}e}{\mathrm{d}t} \\ \frac{\mathrm{d}i}{\mathrm{d}t} \\ \frac{\mathrm{d}\Omega}{\mathrm{d}t} \\ \frac{\mathrm{d}\omega}{\mathrm{d}t} \\ \frac{\mathrm{d}M}{\mathrm{d}t} - n \end{bmatrix} = \mathcal{L} \begin{bmatrix} \frac{\partial \mathcal{R}}{\partial a} \\ \frac{\partial \mathcal{R}}{\partial e} \\ \frac{\partial \mathcal{R}}{\partial i} \\ \frac{\partial \mathcal{R}}{\partial \Omega} \\ \frac{\partial \mathcal{R}}{\partial \omega} \\ \frac{\partial \mathcal{R}}{\partial M} \end{bmatrix}$$

$$\mathcal{L} = \frac{1}{na^2}\begin{bmatrix} 0 & 0 & 0 & 0 & 0 & +2a \\ 0 & 0 & 0 & 0 & -\tau\overline{\tau} & +\tau \\ 0 & 0 & 0 & -\sigma\overline{\tau} & +\sigma\overline{\tau}\cos i & 0 \\ 0 & 0 & +\sigma\overline{\tau} & 0 & 0 & 0 \\ 0 & +\tau\overline{\tau} & -\sigma\overline{\tau}\cos i & 0 & 0 & 0 \\ -2a & -\tau & 0 & 0 & 0 & 0 \end{bmatrix}$$

where $\sigma = \frac{1}{\sin i}$, $\tau = \frac{1-e^2}{e}$, $\overline{\tau} = \frac{1}{\sqrt{1-e^2}}$.

The parameters (a, e, i) are called metric orbital elements,[9] or more briefly, metric elements. If p_i denote the metric elements and q_i the angular elements, the last two relations can be written in the form

$$\dot{p}_i = f_1\left(p_i; \frac{\partial \mathcal{R}}{\partial q_i}\right) , \quad \dot{q}_i = f_2\left(p_i; \frac{\partial \mathcal{R}}{\partial p_i}\right) . \tag{3.39}$$

Note. The quantity $na^2\sqrt{1-e^2}$ often arises in the denominator of the expressions in the system of equations presented in Table 3.4. Returning to the Keplerian orbit relations for the osculating elements, (1.23) becomes

$$C^2 = \mu p = \mu a(1-e^2) = n^2 a^4 (1-e^2) ,$$

and we thus see that the quantity in question is the magnitude of the angular momentum of the satellite, denoted by C:

$$C = na^2\sqrt{1-e^2} = nab . \tag{3.40}$$

3.5.3 Delaunay Elements

The form of the matrix $\mathcal{L}$ given in Table 3.4, together with the symmetries and similarities in the Lagrange brackets, suggest making a change of variables to obtain an even simpler formulation of the results and to group the elements into two homogeneous sets. These elements are known as the Delaunay variables or Delaunay elements.[10] These variables are written as follows, clearly separating the elements L, G, H, which have the dimensions of angular momentum per unit mass (action variables), and the associated elements l, g, h, which are dimensionless (angle variables):

---

[9] This term serves to distinguish the two groups of orbital elements. However, the term 'metric' is open to discussion. If it refers to the notion of length, only the quantity a is actually a length. If it is intended to contrast with the notion of angle, the inclination is actually an angle. Here, when the word 'metric' is attributed to e and i, it indicates that these two elements behave mathematically like a. The Delaunay variables discussed in the next section avoid this ambiguity: variables in the same group have the same dimensions.

[10] Charles Delaunay (1816–1872) was a French astronomer and author of many works, including a very detailed study of the motion of the Moon in 1860, in which he produced a perturbation expansion with 1967 terms. He detected a slight disagreement between his predictions and observational results. Le Verrier claimed that the error lay in Delaunay's formulas, but the latter replied that the disagreements were the result of unknown causes. In 1865, Delaunay put forward the hypothesis that the discrepancy was due to a very gradual slowing down of the Earth's rotation, caused by friction due to the tides. This is indeed the accepted theory today. Delaunay's method is no longer used to study the motion of the Moon, but it is still current practice for the moons of other planets. His ideas were taken up by Von Zeipel, then Brouwer in 1959, to study the motion of artificial satellites.

$$\begin{cases} L = \sqrt{\mu a} \,, \quad G = L\sqrt{1-e^2} \,, \quad H = G\cos i \,, \\ l = M \,, \quad g = \omega \,, \quad h = \Omega \,. \end{cases} \tag{3.41}$$

Lagrange's equations can now be written in a very simple form, called the Delaunay equations:

$$\begin{cases} \dfrac{\mathrm{d}L}{\mathrm{d}t} = \dfrac{\partial F}{\partial l} \,, \quad \dfrac{\mathrm{d}G}{\mathrm{d}t} = \dfrac{\partial F}{\partial g} \,, \quad \dfrac{\mathrm{d}H}{\mathrm{d}t} = \dfrac{\partial F}{\partial h} \,, \\ \dfrac{\mathrm{d}l}{\mathrm{d}t} = -\dfrac{\partial F}{\partial L} \,, \quad \dfrac{\mathrm{d}g}{\mathrm{d}t} = -\dfrac{\partial F}{\partial G} \,, \quad \dfrac{\mathrm{d}h}{\mathrm{d}t} = -\dfrac{\partial F}{\partial H} \,, \end{cases} \tag{3.42}$$

where

$$F = \frac{\mu^2}{2L^2} + \mathcal{R} \,.$$

In canonical form, the equations are even simpler than in (3.39):

$$\dot{p}_i = \frac{\partial F}{\partial q_i} \,, \quad \dot{q}_i = -\frac{\partial F}{\partial p_i} \,, \tag{3.43}$$

where the p_i are the action variables and the q_i are the associated angle variables.

According to (3.40), G is the angular momentum C and H is its projection on the axis of the poles $\boldsymbol{Oz}$:

$$H = C\cos i = C_z \,. \tag{3.44}$$

We shall not use the Delaunay variables in what follows (except for H a little later). They are given here as an example of a homogeneous notation, in the sense of physical dimensions, leading to canonical equations. The Hamiltonian method due to Von Zeipel and Brouwer involves integrating Lagrange's equations using the Delaunay variables.

3.5.4 Poorly Defined Parameters

There are two situations, already discussed in Chap. 2, which can raise difficulties for the definition of certain parameters: the case $e = 0$ and the case $i = 0$:

- The eccentricity e occurs in the denominator of the expressions for $\dot{e}, \dot{\omega}$ and $\dot{M}$ in Lagrange's equations (see Table 3.4). If e is zero, these quantities are not defined. This is to be expected since, as we have seen, the perigee is not then defined, and this means that ω and M cannot be either.
- The inclination i occurs in the denominator of $\dot{i}, \dot{\Omega}$ and $\dot{\omega}$. If i is zero, these quantities are not defined. This is also to be expected since, as the ascending node is not defined, Ω and ω cannot be either.

In these two cases, one solution is to drop the 'standard' orbital elements and replace them by others, obtained by suitably chosen combinations of these. However, for the present purposes, this is not necessary. We shall see later that, when the perturbing potential is limited to the J_{2n} term of the geopotential, the function $\mathcal{R}$ is such as to remove the indeterminacy when $e = 0$ or $i = 0$. For example, with regard to the indeterminacy that arises in $\dot{\Omega}$ when $i = 0$, we shall see that $\mathcal{R}$ is a function of $(\sin^2 i)$. Its derivative with respect to i yields a term in $\sin i \cos i$ which cancels the term in $\sin i$ in the denominator of $\dot{\Omega}$. In this case, the quantity $\dot{\Omega}$ is perfectly well defined for $i = 0$.

With this type of perturbing potential, we shall find that all the angular speeds $\dot{\Omega}, \dot{\omega}, \dot{M}$ are well defined for $e = 0$ or $i = 0$ (or $e = 0$ and $i = 0$), even if the zero point of the angles is not.

3.5.5 Perturbative Accelerations not Derived from a Potential

When the perturbing acceleration γ_{P} does not derive from a potential, we use a coordinate system moving with the orbit and decompose the vector relative to this. This gives a system of equations called Gauss' equations. We shall not discuss this method further, for it only concerns non-conservative forces such as atmospheric drag, radiation pressure, and so on, which we have seen to be very small compared with the gravitational forces.

3.6 Perturbative Method: Results for the Geopotential up to J_2

In order to achieve our aim, that is, to find the time derivatives of the six orbital elements, we must apply Lagrange's equations. We must therefore feed in the value of the perturbing potential $\mathcal{R}$. We shall consider here the perturbation due to the geopotential. We shall not tackle the other gravitational potentials, such as the lunisolar attraction potential, treated here as negligible.

As far as the geopotential is concerned, we shall proceed in steps, initially considering the first zonal harmonic (see Sect. 3.6.1), corresponding to the term in J_2, then all the zonal harmonics up to the term in J_n (see Sect. 3.7.1), and finally the general case, including all the zonal, sectorial and tesseral harmonics (see Sect. 3.7.2).

3.6.1 Expression for Perturbative Potential up to J_2

We expand the potential to second degree. Using (3.19) and putting $U = U_0 + \mathcal{R}$, we find the following value for $\mathcal{R}$:

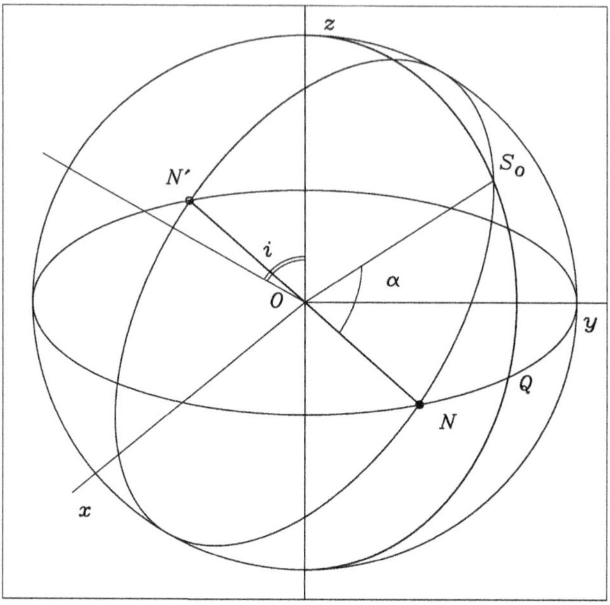

Figure 3.2. Subsatellite point S_0, specified by the position on orbit α and the geographical latitude

$$\mathcal{R} = -\frac{\mu R^2}{r^3} J_2 \frac{3\sin^2\phi - 1}{2} . \tag{3.45}$$

To integrate this over a period, we first find r and ϕ.

Expressions for r and ϕ

The distance r can be expressed as a function of a, e, and v, as we saw in Chap. 1. Equation (1.41) gives

$$r = \frac{a(1-e^2)}{1+e\cos v} . \tag{3.46}$$

To find an expression for the angle ϕ, the latitude of the satellite, we consider the following points, as shown in Fig. 3.2: the projection N of the ascending node on the Earth's surface, the projection S_0 of the satellite, and the point Q where the meridian through S_0 intersects the equator (i.e., the intersection of the half-plane S_0Oz with the equatorial circle). In the spherical triangle NS_0Q, with a right angle at Q, the angle N is the dihedral angle $(\mathcal{E}, \mathcal{P})$, that is, the inclination i. The known sides of the triangle (arcs of a great circle) are

$$\widehat{QS_0} = \phi \,, \quad \widehat{NS_0} = \alpha \,.$$

Along with (ST VIII), the sine rule (see Sect. 3.16) gives

$$\frac{1}{\sin\alpha} = \frac{\sin i}{\sin\phi} .$$

Now, with $\alpha = \omega + v$ (since α is the position on orbit), this leads to

$$\sin\phi = \sin i \, \sin(\omega + v) . \tag{3.47}$$

Hence, for the second Legendre polynomial,

$$P_2(\sin\phi) = \frac{3\sin^2\phi - 1}{2} = \frac{3}{2}\frac{\sin^2 i}{2}\left[1 - \cos 2(\omega + v)\right] - \frac{1}{2} .$$

According to what was said above, we can write $\mathcal{R}$ in the form

$$\mathcal{R} = -\frac{3}{2}\frac{\mu R^2}{a^3}\left(\frac{1 + e\cos v}{1 - e^2}\right)^3 J_2 \left\{\frac{\sin^2 i}{2}\left[1 - \cos 2(\omega + v)\right] - \frac{1}{3}\right\} , \tag{3.48}$$

which shows that $\mathcal{R}$ is a function of the constant quantities (a, e, i, ω are considered to be constant over one revolution) and v. It thus varies periodically, with the same period $T = 2\pi/n$ as the Keplerian motion. We note that $\mathcal{R}$ does not depend on Ω. This is to be expected insofar as the only perturbation we are considering arises from the replacement of the sphere by an ellipsoid of revolution. In this situation, the terrestrial longitude is irrelevant. The position of the ascending node, and hence the value of Ω, is of no consequence, and we have $\partial\mathcal{R}/\partial\Omega = 0$.

Integrating $\mathcal{R}$ over a Period

We now calculate $\langle\mathcal{R}\rangle$, the average value of $\mathcal{R}$ over one period, by integrating with respect to time, i.e., with respect to the mean anomaly M. We first express $\mathrm{d}M$ as a function of $\mathrm{d}v$. This relation is given by (1.70), which we obtained in Chap. 1 by applying the areal law:

$$\mathrm{d}M = \frac{r^2}{a^2\sqrt{1 - e^2}}\mathrm{d}v .$$

Using the variables r, ϕ and v, and as the bounds of integration are the same for v and M, we obtain

$$2\pi\langle\mathcal{R}\rangle = -\frac{\mu R^2 J_2}{2a^3(1 - e^2)^{3/2}}\int_0^{2\pi}\frac{3\sin^2\phi - 1}{r}\mathrm{d}v .$$

Using (3.46) and (3.47), this yields

$$2\pi\langle\mathcal{R}\rangle = \int_0^T \mathcal{R}\mathrm{d}t = \int_0^{2\pi}\mathcal{R}\mathrm{d}M = -\frac{\mu R^2 J_2}{2a^3(1 - e^2)^{3/2}}\mathcal{I} ,$$

where

$$\mathcal{I} = \int_0^{2\pi} \left\{ \frac{3\sin^2 i}{2} [1 - \cos 2(v+\omega)] - 1 \right\} (1 + e\cos v) \mathrm{d}v \ .$$

Expanding the terms in this expression, we see that integration of the terms periodic in v over the interval $[0, 2\pi]$ will give zero. The only nonzero contribution comes from the constant terms, independent of v, which yields

$$\int_0^{2\pi} \left(\frac{3\sin^2 i}{2} - 1 \right) \mathrm{d}v = 2\pi \left(\frac{3\sin^2 i}{2} - 1 \right) \ .$$

The average value of the potential is thus

$$\langle \mathcal{R} \rangle = -\frac{3}{2} \frac{\mu R^2}{a^3(1-e^2)^{3/2}} J_2 \left(\frac{1}{2}\sin^2 i - \frac{1}{3} \right) \ . \tag{3.49}$$

Periodic and Secular Variations

The integration of $\mathcal{R}$ over a period shows that we can decompose this quantity into two parts:

$$\mathcal{R}(v) = \mathcal{R}_{\mathrm{s}} + \mathcal{R}_{\mathrm{p}}(v) \ , \tag{3.50}$$

where the constant $\mathcal{R}_{\mathrm{s}} = \langle \mathcal{R} \rangle$ represents the average value and the periodic $\mathcal{R}_{\mathrm{p}}$ averages to zero over one period. The only part of the potential having a long-term effect (over times longer than the period T) is therefore $\mathcal{R}_{\mathrm{s}}$. These slow variations, proportional to the time, are called secular variations (in contrast to $\mathcal{R}_{\mathrm{p}}$, which causes only periodic effects).

The secular variations of the elements are obtained by differentiating this part $\mathcal{R}_{\mathrm{s}}$ of the perturbative potential $\mathcal{R}$. Equation (3.49) shows that $\mathcal{R}_{\mathrm{s}} = \langle \mathcal{R} \rangle$ can be expressed entirely in terms of the metric elements:

$$\mathcal{R}_{\mathrm{s}} = \mathcal{R}_{\mathrm{s}}(a, e, i) \ .$$

3.6.2 Variation of the Orbital Elements

Calculating the Variation of the Orbital Elements

We can at last apply Lagrange's equations. Referring to the six equations in Table (3.4), we replace the perturbing potential $\mathcal{R}$ by its secular part $\mathcal{R}_{\mathrm{s}}$ and replace μ by n^2a^3. The time derivatives $\mathrm{d}a/\mathrm{d}t$, $\mathrm{d}e/\mathrm{d}t$, and $\mathrm{d}i/\mathrm{d}t$ of the metric elements obtained by differentiating with respect to the angle elements are therefore all zero. Consequently, the parameters a, e, i remain constant in time.

On the other hand, we see that the time derivatives of the angle elements $\mathrm{d}\Omega/\mathrm{d}t$, $\mathrm{d}\omega/\mathrm{d}t$, and $\mathrm{d}M/\mathrm{d}t$, obtained by differentiating with respect to the metric elements are nonzero. The partial derivatives of $\mathcal{R}_s$ expressed in (3.49) yield

$$\frac{\partial \mathcal{R}_s}{\partial a} = -\frac{3}{a}\mathcal{R}_s \,, \qquad \frac{\partial \mathcal{R}_s}{\partial e} = -\frac{3e}{1-e^2}\mathcal{R}_s \,, \qquad \frac{\partial \mathcal{R}_s}{\partial i} = \frac{\sin i \, \cos i}{\frac{1}{2}\sin^2 i - \frac{1}{3}}\mathcal{R}_s \,.$$

The final result is:

$$\dot{a} = 0 \,, \tag{3.51}$$

$$\dot{e} = 0 \,, \tag{3.52}$$

$$\dot{i} = 0 \,, \tag{3.53}$$

$$\dot{\Omega} = -\frac{3}{2(1-e^2)^2} n J_2 \left(\frac{R}{a}\right)^2 \cos i \,, \tag{3.54}$$

$$\dot{\omega} = \frac{3}{4(1-e^2)^2} n J_2 \left(\frac{R}{a}\right)^2 \left(5\cos^2 i - 1\right) \,, \tag{3.55}$$

$$\dot{M} - n = \Delta n = \frac{3}{4(1-e^2)^{3/2}} n J_2 \left(\frac{R}{a}\right)^2 \left(3\cos^2 i - 1\right) \,. \tag{3.56}$$

To begin with, we may take $n = n_0$.

As a function of time t, and starting from the origin $t = 0$, the orbital elements are thus

$$a(t) = a_0 \,, \quad e(t) = e_0 \,, \quad i(t) = i_0 \,,$$

$$\Omega(t) = \Omega_0 + \dot{\Omega} t \,, \quad \omega(t) = \omega_0 + \dot{\omega} t \,, \quad M(t) = M_0 + nt + (\Delta n)t \,.$$

This system of equations represents the solution of the equation of motion (3.32) as a function of the orbital elements.

Remarks Concerning the Variation of the Orbital Elements

To sum up, if we compare the actual trajectory of the satellite (perturbed by the action of the J_2 term of the geopotential) with the Keplerian trajectory, we observe the following points:

1. The semi-major axis a of the orbit remains constant.
2. The eccentricity e of the orbit remains constant.

3. The inclination i of the orbit with respect to the equatorial plane is constant.
4. The orbital plane rotates uniformly about the polar axis with a constant angular speed $\dot{\Omega}$. This motion is known as precession of the orbit or nodal precession.[11] When we speak of precession without further specification, we are generally referring to this motion. According to (3.54), it can occur either in the prograde or the retrograde direction, depending on the inclination of the satellite:

$$\text{prograde} \quad \dot{\Omega} \geqslant 0 \quad \Longleftrightarrow \quad \cos i \leqslant 0 \quad \Longleftrightarrow \quad i^\circ \notin \mathcal{D}_1 \,,$$

$$\text{retrograde} \quad \dot{\Omega} \leqslant 0 \quad \Longleftrightarrow \quad \cos i \geqslant 0 \quad \Longleftrightarrow \quad i^\circ \in \mathcal{D}_1 \,,$$

where $\mathcal{D}_1 = [0.00, 90.00]$ in degrees of arc.
5. The perigee, and hence the whole orbit, rotate uniformly in the plane of the orbit with constant angular speed $\dot{\omega}$. This motion is called apsidal precession.[12] According to (3.55), it can occur in the prograde or retrograde direction depending on the inclination of the satellite:

$$\text{prograde} \quad \dot{\omega} \geqslant 0 \quad \Longleftrightarrow \quad \sin^2 i \leqslant \frac{4}{5} \quad \Longleftrightarrow \quad i^\circ \notin \mathcal{D}_2 \,,$$

$$\text{retrograde} \quad \dot{\omega} \leqslant 0 \quad \Longleftrightarrow \quad \sin^2 i \geqslant \frac{4}{5} \quad \Longleftrightarrow \quad i^\circ \in \mathcal{D}_2 \,,$$

where $\mathcal{D}_2 = [63.43, 116.57]$ in degrees of arc. We define the critical inclination by

$$i_{\mathrm{C}} = \arcsin\left(\frac{2}{\sqrt{5}}\right) \,. \tag{3.57}$$

For the two values, $i = i_{\mathrm{C}} = 63.43°$ and $i = 180° - i_{\mathrm{C}} = 116.57°$, the rate of apsidal precession $\dot{\omega}$ is zero. The value of the critical inclination is independent of a and e. It is important for certain satellites in highly elliptical orbit, because one seeks to avoid apsidal precession: as we shall see, in the case of Molniya-type communications satellites, the position of the apogee needs to be fixed on the orbit.
When $\dot{\omega}$ is calculated using an expansion going beyond the J_2 term, the value obtained for i_{C} depends very slightly on a and e. It differs by a few hundredths of a degree from the value given by (3.57).

---

11 The word 'nodal' means that the motion concerns the line of nodes, i.e., the line joining the ascending and descending nodes. This straight line is the intersection of the equatorial plane with the orbital plane.

12 The term 'apsidal' means that the motion concerns the line of apsides, i.e., the line joining the perigee and the apogee. This line segment represents the major axis of the ellipse. The perigee and apogee are the two apsides. The word comes from the Greek (ἡ αψίς, ῖδος), meaning 'arch'.

6. The (true) mean motion of the satellite is not the same as it would have been if there were no flattening. According to (3.56), it may be faster or slower depending on the inclination of the satellite:

$$\text{faster} \quad \Delta n \geqslant 0 \quad \Longleftrightarrow \quad \sin^2 i \leqslant \frac{2}{3} \quad \Longleftrightarrow \quad i^\circ \notin \mathcal{D}_3 \,,$$

$$\text{slower} \quad \Delta n \leqslant 0 \quad \Longleftrightarrow \quad \sin^2 i \geqslant \frac{2}{3} \quad \Longleftrightarrow \quad i^\circ \in \mathcal{D}_3 \,,$$

where $\mathcal{D}_3 = [54.74, 125.26]$ in degrees of arc. A little later, we shall give definitions of the various periods of the motion.

Example calculations are given in the following chapters. In Chap. 4, Figs. 4.1 and 4.2 show the quantities $\dot{\Omega}$, $\dot{\omega}$ and $\Delta n/n$ as a function of the inclination.

Note. The signs of the three quantities $\dot{\Omega}$, $\dot{\omega}$ and $\dot{M} - n$ relating to the angle elements depend only on the inclination i.

3.7 Perturbative Method: Results for General Case

3.7.1 Geopotential up to J_n

Remark Concerning Zonal Harmonics

There is an interesting point when we assume that the Earth has an axial symmetry, as we shall below. In this case, the geopotential involves only zonal harmonics (i.e., the expansion involves only the J_n terms and not the sectorial or tesseral harmonics), whose contribution to the full potential is by far the greatest, as we have seen. The perturbative acceleration γ_{P} defined by (3.34) then lies in the plane containing the satellite and the polar axis. It can thus be decomposed as

$$\gamma_{\mathrm{P}} = \gamma_1 \boldsymbol{e}_r + \gamma_2 \boldsymbol{e}_z \,, \tag{3.58}$$

where the unit vector along $\boldsymbol{OS}$ is denoted by $\boldsymbol{e}_r$ and the unit vector along the polar axis $\boldsymbol{Oz}$ by $\boldsymbol{e}_z$. Using (1.7), the time derivative of the angular momentum $\boldsymbol{C}$ is

$$\begin{aligned} \frac{\mathrm{d}\boldsymbol{C}}{\mathrm{d}t} &= \boldsymbol{r} \wedge \ddot{\boldsymbol{r}} = r\boldsymbol{e}_r \wedge \left[\left(-\frac{\mu}{r^2} + \gamma_1 \right) \boldsymbol{e}_r + \gamma_2 \boldsymbol{e}_z \right] \\ &= r\gamma_2 \left(\boldsymbol{e}_r \wedge \boldsymbol{e}_z \right) \,, \end{aligned}$$

which is perpendicular to $\boldsymbol{e}_z$. If C_z is the projection of $\boldsymbol{C}$ onto the polar axis, we thus have

$$\frac{\mathrm{d}C_z}{\mathrm{d}t} = \frac{\mathrm{d}\boldsymbol{C}}{\mathrm{d}t} \cdot \boldsymbol{e}_z = 0 \,.$$

We saw earlier, in (3.44), that C_z corresponds to the Delaunay variable H, which gives here

$$\frac{\mathrm{d}H}{\mathrm{d}t} = \frac{\mathrm{d}C_z}{\mathrm{d}t} = 0 \ .$$

We deduce the following property:

$$H = \sqrt{\mu a(1-e^2)}\cos i = \sqrt{\mu p}\cos i = \text{const.} \tag{3.59}$$

This relation is a very general feature of orbits perturbed by zonal terms with the same axis.

As H is the metric element associated with the angle element $h = \Omega$, it follows from (3.42) that

$$\frac{\mathrm{d}H}{\mathrm{d}t} = \frac{\partial F}{\partial \Omega} = 0 \ ,$$

which shows that the function F does not depend on Ω. We may then recover the result noted earlier: if the geopotential does not involve the longitude, that is, if it only involves the zonal harmonics, then $\mathcal{R}$ (or F) is independent of Ω.

We also note the following formula, obtained by differentiating the last:

$$\frac{1}{2a}\mathrm{d}a = \frac{e}{1-e^2}\mathrm{d}e + \tan i \, \mathrm{d}i \ . \tag{3.60}$$

Calculating the Perturbative Potential $\mathcal{R}$

We now consider the Earth as a body with axial symmetry. The potential U involves only the terms J_n. The full formula (3.20) reduces to

$$U(r,\phi) = \frac{\mu}{r}\left[1 - \sum_{n=2}^{\infty}\left(\frac{R}{r}\right)^n J_n P_n(\sin\phi)\right] \ . \tag{3.61}$$

The perturbative potential $\mathcal{R}$ is obtained as the difference between U and U_0. This then replaces the value given by (3.45).

Let us begin our calculation of $\mathcal{R}$ by expanding up to third degree, thereby bringing in the Legendre polynomials $P_2(\sin\phi)$ and $P_3(\sin\phi)$. We obtain

$$\mathcal{R} = -\frac{\mu}{r}\left[J_2\left(\frac{R}{r}\right)^2 \frac{3\sin^2\phi - 1}{2} + J_3\left(\frac{R}{r}\right)^3 \frac{5\sin^2\phi - 3}{2}\sin\phi\right] \ . \tag{3.62}$$

Written in simplified form (with $\mathcal{R}_n$ for degree n),

$$\mathcal{R} = \mathcal{R}_2 + \mathcal{R}_3 \ ,$$

where $\mathcal{R}_2$ corresponds to the sum $\mathcal{R}_{\rm s} + \mathcal{R}_{\rm p}$ given in (3.50).

As before for $\mathcal{R} = \mathcal{R}_2$, we calculate the average value $\langle \mathcal{R}_3 \rangle$ of $\mathcal{R}_3$ over a period, to evaluate the part which contributes to the secular variation:

$$2\pi \langle \mathcal{R}_3 \rangle = -\frac{\mu R^3 J_3}{2a^2(1-e^2)^{1/2}} \int_0^{2\pi} \frac{5\sin^2\phi - 3}{r^2} \sin\phi \, \mathrm{d}v \ .$$

With the help of (3.46) and (3.47), this gives

$$2\pi \langle \mathcal{R}_3 \rangle = \int_0^T \mathcal{R}_3 \mathrm{d}t = \int_0^{2\pi} \mathcal{R}_3 \mathrm{d}M = -\frac{\mu R^3 J_3}{2a^4(1-e^2)^{5/2}} \mathcal{J} \ ,$$

where

$$\mathcal{J} = \int_0^{2\pi} \left\{ \frac{5\sin^2 i}{2} \left[1 - \cos 2(v+\omega)\right] - 3 \right\} (1 + e\cos v) \sin i \, \sin(v+\omega) \mathrm{d}v \ .$$

Over one period, only the element v varies. Since the term $\sin(v+\omega)$ appears as a factor in the above expression, the integral $\mathcal{J}$ over the interval $[0, 2\pi]$ is zero. This amounts to saying that $\langle \mathcal{R}_3 \rangle = 0$, because $\sin\phi$ is a factor in $\mathcal{R}_3$, and this because $x = \sin\phi$ is a factor in the Legendre polynomial $P_3(x)$.

Referring to Sect. 3.15, we see that the variable x is a factor in the expressions for all the polynomials of odd order. We deduce that the terms J_n with n odd make no contribution to the secular variation of the orbital elements.

Periodic and Secular Variations

From (3.61), we obtain $\mathcal{R}$ as a function of five orbital parameters (since Ω does not occur):

$$\mathcal{R} = \mathcal{R}(a, e, i, \omega, M) \ .$$

The potential $\mathcal{R}$ decomposes into a sum of terms $\mathcal{R}_n$, one for each term J_n. We consider Lagrange's equations as given in Table 3.4. It can be shown that:

- If a term $\mathcal{R}_n$ depends on neither M, nor ω, it causes a variation in the relevant orbital element which is proportional to the time, i.e., a secular variation.
- If a term $\mathcal{R}_n$ depends on ω, but not M, there will be a variation whose period will be of the order of the apsidal precession, i.e., a long-period periodic variation.
- If a term $\mathcal{R}_n$ depends on both ω and M, there will be a variation whose period will be of the order of the Keplerian period, i.e., a short-period periodic variation.

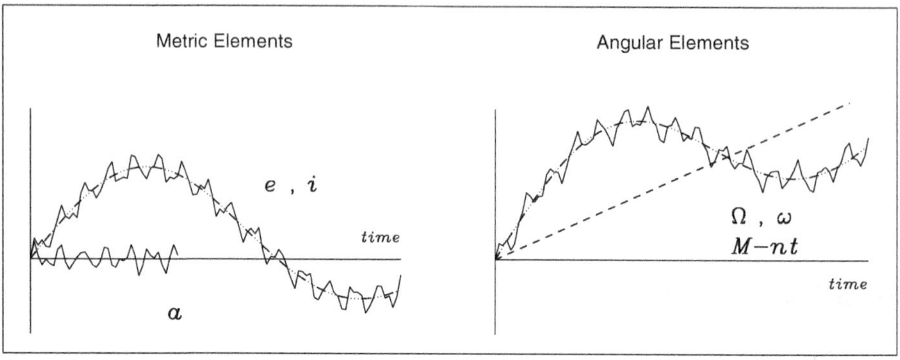

Figure 3.3. Schematic representation of the time variation of the orbital elements. Short-period variations are shown with a *continuous curve* and long-period variations with a *dot-dashed curve*. Secular variations are shown with a *dashed line*

To sum up, periodic variations are divided into short-period variations with period less than the orbital period T, corresponding to the period of M, and long-period variations occurring over several tens of days ($\sim 1\,000T$), corresponding to the apsidal period of ω. The latter are mainly due to the influence of the J_3 term, which affects e, i and the angle elements in particular. There is no long-period perturbation of the semi-major axis a.[13] This theorem is known as the invariability of the major axis and it applies to many types of motion in astronomy.

Secular variations are variations proportional to time. Naturally, it is this kind of variation that deviates the satellite from its Keplerian orbit. We have seen that the J_2 term causes secular deviations in the three angle elements, whilst having no effect on the metric elements. The other zonal harmonics, although only the even ones, i.e., $J_4, J_6, \ldots, J_{2n}, \ldots$, also contribute to secular deviations [see (4.12), (4.13) and (4.14) in Table 4.1]. Figure 3.3 shows in a schematic way all the periodic and secular variations of the orbital elements.

Example 3.1. *Calculate the effect of perturbations when the geopotential is expanded up to the term in J_3. Long-period variations.*

The potential $\mathcal{R}$ is given by (3.62). We have just seen that it can be written in the form

$$\mathcal{R} = (\mathcal{R}_\mathrm{s} + \mathcal{R}_\mathrm{p}) + \mathcal{R}_3 \; .$$

The term $\mathcal{R}_\mathrm{s}(a, e, i)$ is given by (3.49). The term $\mathcal{R}_\mathrm{p}(a, e, i, \omega, M)$ leads to short-period variations which we shall not discuss here. The term $\mathcal{R}_3(a, e, i, \omega)$ can be written in the form

[13] Laplace showed in 1776 that perturbations of the semi-major axis had no secular terms to first order. In 1809, Poisson proved that there were no secular terms to second order either.

$$\mathcal{R}_3 = \frac{3}{2}\frac{\mu R^3 e}{a^4(1-e^2)^{5/2}} J_3 \left(1 - \frac{5}{4}\sin^2 i\right)\sin\omega\ . \tag{3.63}$$

The term $\mathcal{R}_3(a,e,i,\omega)$ leads to long-period variations.

Since $\partial\mathcal{R}_3/\partial M = 0$ and $\partial\mathcal{R}_3/\partial\Omega = 0$, the equations for the metric elements (the first three Lagrange equations in Table 3.4) become

$$\frac{\mathrm{d}a}{\mathrm{d}t} = 0\ , \qquad \frac{\mathrm{d}e}{\mathrm{d}t} = -\frac{1}{na^2}\frac{\sqrt{1-e^2}}{e}\frac{\partial\mathcal{R}_3}{\partial\omega}\ , \qquad \frac{\mathrm{d}i}{\mathrm{d}t} = \frac{1}{na^2}\frac{1}{\sqrt{1-e^2}}\frac{\cos i}{\sin i}\frac{\partial\mathcal{R}_3}{\partial\omega}\ .$$

This implies

$$\frac{\mathrm{d}e}{\mathrm{d}i} = -\frac{1-e^2}{e}\tan i\ , \tag{3.64}$$

which is equivalent to (3.60) when a is constant (which is almost always the case for the kinds of motion occurring in astronomy). To calculate the variation arising in i as a result of $\mathcal{R}_3$, we start with $\mathrm{d}i/\mathrm{d}t$ above, differentiate $\mathcal{R}_3$ with respect to ω, and replace μ by n^2a^3. This gives

$$\frac{\mathrm{d}i}{\mathrm{d}t} = \frac{3e}{2(1-e^2)^3} nJ_3 \left(\frac{R}{a}\right)^3 \cos i \left(1 - \frac{5}{4}\sin^2 i\right)\cos\omega\ . \tag{3.65}$$

To find the long-period variation in i due to the zonal harmonic J_3, we integrate (3.65) with respect to ω, using the relation $(\mathrm{d}i/\mathrm{d}t) = (\mathrm{d}i/\mathrm{d}\omega)/(\mathrm{d}\omega/\mathrm{d}t)$ and the value calculated for $\dot\omega$ from (3.55), to give finally

$$\frac{\mathrm{d}i}{\mathrm{d}\omega} = \frac{1}{\dot\omega}\frac{\mathrm{d}i}{\mathrm{d}t} = \frac{1}{2}\frac{e}{1-e^2}\frac{J_3}{J_2}\left(\frac{R}{a}\right)\cos i\cos\omega\ .$$

The required quantity is thus

$$\Delta_{\mathrm{LP}_3} i = \frac{1}{2}\frac{e}{1-e^2}\frac{J_3}{J_2}\left(\frac{R}{a}\right)\cos i\sin\omega\ , \tag{3.66}$$

where the subscript LP_3 indicates that this is a long-period effect due to the zonal harmonic J_3. To obtain the corresponding variation e, we use (3.64), taking $(\Delta_{\mathrm{LP}_3} e/\Delta_{\mathrm{LP}_3} i)$ for $\mathrm{d}e/\mathrm{d}i$:

$$\Delta_{\mathrm{LP}_3} e = -\frac{1}{2}\frac{J_3}{J_2}\left(\frac{R}{a}\right)\sin i\sin\omega\ . \tag{3.67}$$

Conversely, one can determine the value of J_3 by measuring the values of $\Delta_{\mathrm{LP}_3} i$ and $\Delta_{\mathrm{LP}_3} e$.

Variation of Orbital Parameters over One Revolution

The above results can be illustrated by expressing the variation of these quantities over one revolution. Recall that R is the equatorial radius of the Earth, a the semi-major axis of the elliptical orbit, and p the focal parameter of the ellipse, with $p = a(1-e^2)$. These variations carry the subscript 1 to indicate that they refer to the value obtained over one revolution. The variations, be they secular or long-period, are sufficiently slow to be able to avoid integrating over the time t. Indeed, it is enough to multiply the derivative with respect to t of the relevant quantity by $T = 2\pi/n$.

For the angle variables, we write $\Delta_1 \Omega = \dot{\Omega} T$ and $\Delta_1 \omega = \dot{\omega} T$, where the angles are of course given in radians. To first order, we take for $\dot{\Omega}$ and $\dot{\omega}$ the values given by (3.54) and (3.55), respectively, as a function of J_2 alone, when expanding the geopotential:

$$\Delta_1 \Omega = -3\pi J_2 \left(\frac{R}{p}\right)^2 \cos i \; , \tag{3.68}$$

$$\Delta_1 \omega = +3\pi J_2 \left(\frac{R}{p}\right)^2 2\left(1 - \frac{5}{4}\sin^2 i\right) \; . \tag{3.69}$$

For the metric variables, the case for a can be dealt with immediately since we may consider that $\dot{a}$ is zero (invariability of the major axis). Concerning the expressions for $\dot{e}$ and $\dot{i}$, related by (3.64), there is the term in $J_3 \cos\omega$, of degree 3 in R/a, given by (3.65), followed by terms in $J_2^2 \sin 2\omega$ and $J_4 \sin 2\omega$, of degree 4 in R/a, and other terms of higher degree. We consider only the first term in the expansion, referring to J_3. We obtain for the variations over one period:

$$\Delta_1 a \approx 0 \; , \tag{3.70}$$

$$\Delta_1 e = -3\pi J_3 \left(\frac{R}{p}\right)^3 (1 - e^2)\left(1 - \frac{5}{4}\sin^2 i\right) \sin i \cos\omega \; , \tag{3.71}$$

$$\Delta_1 i = +3\pi J_3 \left(\frac{R}{p}\right)^3 e\left(1 - \frac{5}{4}\sin^2 i\right) \cos i \; \cos\omega \; . \tag{3.72}$$

The critical inclination plays an important role for $\Delta_1\omega$, $\Delta_1 e$ and $\Delta_1 i$.

3.7.2 Full Geopotential

When we consider the geopotential $U(r, \lambda, \phi)$ given by (3.10), it is extremely difficult to calculate the effects of the perturbative potential $\mathcal{R}$. However, let us just mention the existence of orbital resonance phenomena: the influence of certain tesseral coefficients C_{lm} and S_{lm} defined by (3.13) and (3.14) can, for one specific orbit, significantly exceeds the effects of coefficients with higher or lower order and degree. For these values, the periodic perturbations have relatively large amplitude. W. Kaula has developed a formalism which predicts the resonances associated with these specific orbits.

Table 3.5. Sources of perturbation and the effect (secular or periodic variation) induced on the five orbital elements a, e, i, Ω and ω, for a satellite in low Earth orbit, according to the King-Hele theory

Source	Secular variation			Periodic variation	
of perturbation	Large	Small	Indirect	Moderate	Small
Earth's gravity	Ω, ω	–	–	e	i, Ω, ω
Atmosphere	a, e	i	Ω, ω	–	Ω, ω
Lunisolar gravity	–	–	–	–	a, e, i, Ω, ω

This resonance phenomenon is particularly important for recurrent satellites with a one-day recurrence cycle, i.e., effecting a whole number ν of revolutions per day (see Chap. 7). Tesseral harmonics with order a multiple of ν must be taken into account in orbital extrapolations (i.e., precise determination of satellite position at a given time from its osculating elements).

Particularly important are the following:

- For low-orbit satellites: harmonics of order $14, 28, \ldots$, (for $\nu = 14$), $13, 26, \ldots$, (for $\nu = 13$).
- For satellites of medium altitude: harmonics of order $2, 4, 6, \ldots$, in the case where $\nu = 2$ (two revolutions per day), whether the orbit is circular, as for NAVSTAR/GPS, or highly elliptical, as for Molniya.[14]
- For geosynchronous satellites:[15] harmonics of order $1, 2, 3, \ldots$.

More generally, this phenomenon affects all recurrent satellites, where ν is a rational number, i.e., after a certain number of days, the satellite ground track is repeated (see Chap. 7). For example, in the case of the SPOT satellites, where $\nu = 14 + 5/26$, there is a resonance for the tesseral terms of orders 15 and 29.

3.7.3 Terrestrial and Non-Terrestrial Perturbations

Atmospheric drag has a greater effect on a satellite as its orbit is low. It can be shown that e tends to decrease (drag is greater at the perigee than at the apogee, so the orbit tends to become more circular) and a also tends to decrease (it is easy to understand that drag will tend to make a satellite fall back to Earth). The other elements are not significantly affected, at least not directly. However, the variation of a and e, which can be considered

[14] For Molniya, the tesseral harmonic coefficients with the most effect are C_{22}, C_{32}, C_{52}; C_{44}, C_{54}, C_{64}; C_{66}, C_{76}, C_{86}; C_{98} and the corresponding S_{lm}.

[15] For a geostationary satellite, resonance periods of 24 hr and 48 hr are associated with tesseral harmonics of order 1 (C_{31}, C_{41} and the corresponding S_{l1}) and resonance periods of 12 hr, 24 hr, 36 hr, and 48 hr are associated with those of order 2 (C_{22}, C_{32}, C_{42} and the corresponding S_{l2}).

as proportional to the time t (secular variation), induces a variation in the angular parameters which rapidly dominates[16] over the periodic variations, because it is proportional to t^2. The forces induced by radiation pressure (solar or terrestrial) affect variations of the orbital elements.

Non-terrestrial gravitational potential is due almost entirely to the lunisolar potential. The elements a and e are not affected by secular variations. The various orbital elements undergo slight variations with long period. For certain orbits, we observe a very slight secular drift in Ω. These results are summarised in Table 3.5, for a low-orbiting satellite.

3.8 Different Definitions of Period

The three angles associated with Keplerian motion, the anomalies v, E and M, increase by 2π when the time increases by $2\pi/n$, where n is the mean motion. The origin from which these angles are measured is taken at the time

---

[16] For a precessional motion with angular speed $\dot{\Omega}$, the ascending node has longitude Ω given by

$$\Omega = \Omega_0 + \dot{\Omega} t = \Omega_0 + \dot{\Omega}(a_0, e_0, i_0) t \ .$$

It can be shown that atmospheric drag introduces a secular decrease in a and e, whilst the inclination i remains practically unchanged. We write these variable parameters as a function of time t:

$$a = a_0 - \dot{a} t \ , \quad e = e_0 - \dot{e} t \ , \quad i = i_0 \ .$$

The longitude of the ascending node is thus Ω', where

$$\Omega' = \Omega_0 + \dot{\Omega}(a, e, i_0) t \ .$$

Expanding to first order, we write

$$\dot{\Omega}(a_0 - \dot{a}t, e_0 - \dot{e}t, i_0) = \dot{\Omega}(a_0, e_0, i_0) - \left(\frac{\partial \dot{\Omega}}{\partial a} \dot{a} t + \frac{\partial \dot{\Omega}}{\partial e} \dot{e} t \right) \ ,$$

which gives

$$\Omega' = \Omega - \left(\frac{\partial \dot{\Omega}}{\partial a} \dot{a} + \frac{\partial \dot{\Omega}}{\partial e} \dot{e} \right) t^2 \ ,$$

or, putting $\delta \ddot{\Omega}$ for the quantity in brackets in the above formula,

$$\Omega' - \Omega = -(\delta \ddot{\Omega}) t^2 \ .$$

For the argument of the perigee, similar reasoning leads to

$$\omega' - \omega = -(\delta \ddot{\omega}) t^2 \ .$$

$t = t_p$ when the satellite passes the perigee, as can be seen from (1.58) and (1.66). The time interval between two successive passages at perigee is called the anomalistic period.[17] It is therefore the anomalistic period T_a which is obtained with the mean motion n, and this differs from the period T_0 obtained with the Keplerian mean motion n_0. We have the definition

$$nT_a = n_0 T_0 . \tag{3.73}$$

With $n = n_0 + \Delta n$, calculated using (3.56), this implies

$$T_a = \frac{1}{1 + \dfrac{\Delta n}{n_0}} T_0 . \tag{3.74}$$

We also wish to know the time interval between two passages at the ascending node (or the descending node). This is the nodal period or draconitic[18] period T_d. This differs from T_a because the perigee moves through apsidal precession. This happens even in the case of a circular orbit, since $\dot{\omega}$ does not vanish when e is zero. We have the relation

$$n' T_d = n T_a , \tag{3.75}$$

where n' is the mean motion when the ascending node is taken as origin. Composing the two motions, we have

$$n' = n + \dot{\omega} , \tag{3.76}$$

yielding one period in terms of the other:

$$T_d = \frac{1}{1 + \dfrac{\dot{\omega}}{n}} T_a , \tag{3.77}$$

$$T_a = \left(1 + \frac{\dot{\omega}}{n}\right) T_d . \tag{3.78}$$

The draconitic period T_d is given in terms of the Keplerian period T_0 by

$$T_d = \frac{1}{\left(1 + \dfrac{\dot{\omega}}{n}\right)\left(1 + \dfrac{\Delta n}{n_0}\right)} T_0 , \tag{3.79}$$

---

[17] The adjective 'anomalistic' derives from the word 'anomaly', since the three anomalies are all zero (modulo 2π) at the perigee.

[18] The adjective 'draconitic' was originally used for the draconitic period or month, which refers to the passage of the Moon at its ascending node. The word comes from the Greek (ὁ δράκων, οντος), meaning 'dragon' (literally, 'which stares'). Eclipses only occur when the Moon passes through a node of its orbit. In ancient times, the Greeks thought that, during an eclipse, the Moon was swallowed up by a dragon, hiding near the nodes of the lunar orbit.

or, to a first approximation,

$$T_{\mathrm{d}} \approx \left(1 - \frac{\dot{\omega} + \Delta n}{n_0}\right) T_0 \,. \tag{3.80}$$

With this approximation, and for a circular orbit [with $e = 0$ in (3.55) and (3.56)], we obtain

$$T_{\mathrm{d}} \approx \left[1 - \frac{3}{2} J_2 \left(\frac{R}{a}\right)^2 \left(4\cos^2 i - 1\right)\right] T_0 \,. \tag{3.81}$$

The daily orbital frequency ν, defined using the mean motion, is established with the draconitic period. It represents the number of round trips, from one north–south equatorial crossing to the next, per day:

$$\nu = \nu_{\mathrm{d}} = \frac{86\,400}{T_{\mathrm{d}}(\mathrm{sec})} \,. \tag{3.82}$$

We also define the frequency ν_{a}, which represents the number of revolutions (passages at perigee) per day:

$$\nu_{\mathrm{a}} = \frac{86\,400}{T_{\mathrm{a}}(\mathrm{sec})} \,. \tag{3.83}$$

This quantity is provided by space organisations such as NORAD to calculate the orbital elements. In fact, the period T_{a} is mainly used to calculate the semi-major axis of the orbit. Concerning the motion of the satellite relative to a frame fixed with respect to the Earth, it is the draconitic (or nodal) period T_{d} which comes into play. For Keplerian motion, all these periods coincide: $T = T_0 = T_{\mathrm{a}} = T_{\mathrm{d}}$.

3.9 Precessional Motion

3.9.1 Precession of the Equinoxes

A simplified model of the terrestrial ellipsoid is given by a sphere with an equatorial bulge. The mass of this bulge, uniformly distributed around the equator, is such that the moments of inertia I_x and I_z with respect to an equatorial diameter and the polar axis, respectively, are the same as for the ellipsoid. This is a standard method for calculating the precession of the equinoxes.

If the Earth were a sphere composed of homogeneous concentric layers, the gravitational effect exerted by the bodies of the Solar System (in particular, the Sun and Moon) would reduce to a force through its centre. The Earth's motion would then be a uniform motion around a fixed axis. The equatorial plane would be fixed relative to the plane of the ecliptic. The straight line at the intersection of these two planes, the line of nodes (called here the line of the equinoxes), would thus be fixed with respect to a Copernican frame.

Let us consider the bulging Earth model. As the precessional motion here is very slow compared with the Moon's revolutions, or the apparent revolutions of the Sun around the Earth, we may replace the Moon and the Sun by an equivalent mass distribution along their orbits, considered as circular, in the plane of the ecliptic (which is an approximation for the Moon). This is Gauss' method.[19] It can be shown (using Lagrange and Poisson's theory of motion with the gyroscopic approximation) that the the moment of the gravitational forces exerted by the mass of the Sun and Moon on this equatorial bulge causes a motion of the Earth's axis of rotation, whilst the angle between the equatorial and ecliptic planes remains fixed. This angle ε is the obliquity. The line of nodes moves in the retrograde direction (opposite to the direction of the Earth's rotation). This motion, known as the precession of the equinoxes, is very slow on the human time scale: one round trip every 25 800 years, or $50''.29$ per year, with the Moon contributing $34''$ and the Sun $16''$. The angular momentum, and hence the rate of precession, is proportional to the difference $(I_x - I_z)$ between the moments of inertia, related to the J_2 term. The rate of precession is also proportional to $\cos\varepsilon$.

The precession of the equinoxes has been known since ancient times, thanks to Hipparchos.[20]

3.9.2 Precession of the Line of Nodes of a Satellite

With the same type of argument, one can calculate approximately the precessional motion of the circular orbit of a satellite. Consider a satellite in

[19] Carl Friedrich Gauß (1777–1855) was a German astronomer, mathematician and physicist. Gauss was interested in astronomy from very early on. In 1805, he published a treatise on celestial mechanics, *Theoria motus corporum coelestium.* In 1818, he published a monograph on the secular perturbations of a planet's motion when it is affected by another planet. The method consists in replacing the perturbing body by a torus, distributing the matter of the body along its trajectory. This immense work covers mathematics (geometry, cartography, conformal representations, probability), physics (magnetism, optics, field theory), and geodesy. See also the note on Piazzi.

[20] Hipparchos of Nicaea (second century BC), (ὁ Ἵππαρχος, ου), was a Greek astronomer. By his observations made in Rhodes, he was the first astronomer to make truly accurate measurements of the positions of the stars, specifying their positions on the celestial sphere using meridians and parallels. He introduced into Greece the Babylonian idea of dividing the circle into degrees, minutes and seconds. He can be considered as the inventor of trigonometry. He invented the stereographic projection. He discovered the precession of the equinoxes by comparing his measurements of stellar positions with those made by Timocharis a century and a half earlier, and also with those made much earlier by the Babylonians. The works of Hipparchos did not come directly to us, but were mentioned by the geographer Strabo and the astronomer Ptolemy.

circular orbit around the Earth with its equatorial bulge. We distribute the mass of the satellite uniformly around its trajectory (the rate of precession is roughly one ten thousandth of the orbital angular speed of the satellite), and calculate the moment of the gravitational forces between these two rings. We then show that the orbit undergoes a precessional motion in which the angle between the orbital plane of the satellite and the equatorial plane of the Earth (the inclination i of the satellite) remains constant. We calculate the moment of the gravitational forces, then the rate of precession $\dot{\Omega}$, which is proportional to $(I_x - I_z)$, hence to J_2, and to $\cos i$.

3.9.3 The Earth as a Satellite

One can use the analogy between an artificial satellite in orbit around the Earth and the Earth viewed as a satellite of the Sun. Let us first note a basic difference between the two problems. The artificial satellite is considered as a point with respect to the Earth and its Keplerian motion is mainly perturbed by the ellipticity of the attracting body (via the J_2 term of the Earth). As a satellite of the Sun, the Earth cannot be treated as a point, but must be considered as a rotating solid, with its polar axis of rotation defining the equatorial plane, and non-spherical nature induced by perturbations.

For the Keplerian elements a, e and ω, the analogy between the two approaches is clear. The metric elements a and e undergo no secular variations. Over a long period of time, there are none for a (stability of the major axis, studied by Laplace, Poisson, and Poincaré), but there is one for e : the eccentricity varies between $e = 0.005$ and $e = 0.050$, with a period of 100 000 yr. The angle element ω undergoes a secular variation, $\dot{\omega} = 11''.6$ per year, which corresponds to a cycle[21] of 110 000 yr.

The variations of the other Keplerian elements can be understood by regarding the Earth as a spinning top. (They are calculated by considering the gyroscopic effect.) The third Keplerian metric element (equivalent to the inclination) is the obliquity ε. It undergoes no secular variations, but it does have a long-period variation in which ε varies between 22° and 25°, with a period of 40 000 yr, upon which is superposed a very small short-period variation known as nutation, with amplitude $9''$ and period 18.6 yr due to the Moon. The angle element Ω undergoes a secular variation, $\dot{\Omega} = -50''.29$ per year, mentioned above, representing the precession of the equinoxes.

The theory of Milankovitch (established between 1920 and 1940) is based on a combination of these periods (100 and 40 thousand years) and cycles (26 and 110 thousand years). This explains the succession of warm and cold periods (ice ages) affecting the Earth's climate. The theory was taken up

[21] We use the word 'cycle' for a period of time at the end of which the relevant point (here the perigee) is located in the same position with respect to a Copernican frame. A cycle is related to a secular variation. For short or long periodic motions, we shall use the word 'period'.

again by A. Berger in 1980 and obtained a striking experimental confirmation with the analysis of air bubbles trapped in the Antarctic ice over a period of 400 000 yr.

3.10 Historical Note on Geodesy

3.10.1 Calculating the J_2 Term Using Geodesy

The J_2 term in the expansion of the geopotential, which can be related to a difference between the moments of inertia of the Earth with respect to the polar axis and an equatorial axis, as explained by (3.17) and (3.18) discussed earlier, is not directly measurable. Without waiting for the advent of the satellite, it was possible to determine it from geodetic considerations.

The geoid is an equipotential surface of the gravitational field which coincides with the mean sea level (disregarding the tides, currents and waves). The direction given by a plumbline at any location on this surface is orthogonal to it. The surface of a lake at rest represents this equipotential surface at a given altitude. To study the field of gravity $\boldsymbol{g}$ at the surface of the Earth, one must consider the gravitational force field in a frame $\Re_{\mathrm{T}}$ moving with the Earth, rather than in the Galilean frame $\Re$. To obtain the relevant relations in $\Re_{\mathrm{T}}$, one must take into account the acceleration of transport a_{t} due to the Earth's rotation as well as the acceleration calculated in $\Re$:

$$\boldsymbol{a}_{\mathrm{t}} = -\varpi^2 \boldsymbol{HS}_0 \ ,$$

where ϖ is the angular speed of rotation of the Earth[22] and H is the projection of S_0, the point of latitude ϕ at the surface of the Earth, on the polar axis. The geocentre is O and we set $r = OS_0$. We thus have $HS_0 = r\cos\phi$. The unit vector in the direction $\boldsymbol{OS}_0$ is $\boldsymbol{e}_r$.

The acceleration $\boldsymbol{g}$ in $\Re_{\mathrm{T}}$ is obtained using the rule for combining accelerations:

$$\boldsymbol{g}_{\Re_{\mathrm{T}}} = \boldsymbol{g}_{\Re} + \varpi^2 \boldsymbol{HS}_0 \ . \tag{3.84}$$

The vector $\boldsymbol{g}_{\Re_{\mathrm{T}}}$ corresponds to the weight: this is what defines the weight of a body at a given location. The weight is the vector sum of the gravity $\boldsymbol{g}_{\Re}$ or $\boldsymbol{g}$ and the so-called centrifugal acceleration.

The vector $\boldsymbol{g}_{\Re}$ lies along $\boldsymbol{OS}_0$ and the vector $\boldsymbol{g}_{\Re_{\mathrm{T}}}$ makes a very small angle with $\boldsymbol{g}_{\Re}$ (at most 0.1°). We may thus write

$$\boldsymbol{g}_{\Re} = -g_{\Re}\boldsymbol{e}_r \ , \quad \boldsymbol{g}_{\Re_{\mathrm{T}}} = -g_{\Re_{\mathrm{T}}}\boldsymbol{e}_r \ .$$

[22] The notation ϖ will only be used in this paragraph. In the following chapters, we use another notation for this quantity, namely $\dot{\Omega}_{\mathrm{T}}$. This notation will be explained later.

The angle $(\boldsymbol{HS}, \boldsymbol{OS}_0)$ represents the latitude ϕ. Projecting (3.84) onto $\boldsymbol{OS}_0$, we thus have

$$-g_{\Re_{\mathrm{T}}}\boldsymbol{e}_r = (-g_{\Re} + \varpi^2 r \cos^2\phi)\boldsymbol{e}_r \,. \tag{3.85}$$

Expressing the fields $g_{\Re}$ and $g_{\Re_{\mathrm{T}}}$ in terms of the respective potentials U and U_{T} and integrating (3.85) with respect to r, we obtain

$$U_{\mathrm{T}} = U + \frac{1}{2}\varpi^2 r^2 \cos^2\phi \,.$$

Cutting off the expansion of U at the second order (geoid = ellipsoid), given by (3.19), we obtain

$$U_{\mathrm{T}}(r,\phi) = \frac{\mu}{r}\left[1 - \left(\frac{R}{r}\right)^2 J_2 \frac{3\sin^2\phi - 1}{2}\right] + \frac{\varpi^2}{2} r^2 \cos^2\phi \,. \tag{3.86}$$

We consider the terrestrial ellipsoid, with semi-major axis $a = R = R_{\mathrm{e}}$, and semi-minor axis b. Now the geometric flattening, or flattening, is given by (1.92)

$$f = \frac{a-b}{a} \,.$$

For an arbitrary point on the Earth's surface (on the ellipsoid), the potential is the same. Choose a point at the pole and a point on the equator:

$$U_{\mathrm{T}}(r = a, \phi = 0) = U_{\mathrm{T}}(r = b, \phi = \pi/2) \,.$$

The relation (3.86) gives

$$\frac{\mu}{a}\left(1 + \frac{1}{2}J_2\right) + \frac{\varpi^2}{2}a^2 = \frac{\mu}{b}\left(1 - \frac{a^2}{b^2}J_2\right) \,.$$

The quantities f and J_2 are small compared to unity. Neglecting small quantities of second order, we obtain on the right-hand side

$$\frac{\mu}{a}(1+f)\big[1 - J_2(1+2f)\big] \approx \frac{\mu}{a}(1 + f - J_2) \,,$$

which yields

$$J_2 = \frac{2}{3}f - \frac{1}{3}m_a \,, \tag{3.87}$$

where

$$m_a = \frac{\varpi^2 a^3}{\mu} \,. \tag{3.88}$$

The dimensionless quantity m_a is easily calculated since all the elements are known and f is known ($f = 1/298.3$). We obtain

$$f = 3.353 \times 10^{-3} \;, \quad m_a = 3.461 \times 10^{-3} \;,$$

$$J_2 = 1.0814 \times 10^{-3} \;,$$

which is very close to the value of J_2 given in Table 3.2.

Carrying out the calculations with the small quantities of second order, we find that

$$J_2 = \frac{2}{3}f - \frac{1}{3}m_b - \frac{1}{3}f^2 + \frac{2}{21}fm_b \;,$$

where we have set

$$m_b = \frac{\varpi^2 a^2 b}{\mu} = m_a(1-f) \;.$$

The numerical result gives $J_2 = 1.082\,634 \times 10^{-3}$, which leads to a relative error of 8×10^{-6} with the value retained for J_2.

3.10.2 Clairaut's Formula

Historically, it was the quantity f that scientists sought to calculate. That is, they wished to know the flattening without measuring the Earth's meridian. They had to find some way of expressing J_2, and this can be done by measuring $g_{\Re_\mathrm{T}}$, the acceleration due to weight, at various points around the Earth. Equation (3.85) gives

$$g_{\Re_\mathrm{T}} = \frac{\mu}{r^2}\left[1 - 3\left(\frac{a}{r}\right)^2 J_2 \frac{3\sin^2\phi - 1}{2}\right] - \varpi^2 r\cos^2\phi \;. \tag{3.89}$$

We calculate $g_{\Re_\mathrm{T}}$ at the equator (g_e) and the pole (g_p):

$$g_\mathrm{e} = \frac{\mu}{a^2}\left(1 + \frac{3}{2}J_2\right) - \varpi^2 a \;,$$

$$g_\mathrm{p} = \frac{\mu}{b^2}\left(1 - 3\frac{a^2}{b^2}J_2\right) \approx \frac{\mu}{a^2}(1 + 2f - 3J_2) \;.$$

Neglecting small quantities of second order, the difference yields

$$g_\mathrm{p} - g_\mathrm{e} = \frac{\mu}{a^2}(2f - \frac{9}{2}J_2) + \varpi^2 a \;.$$

Replacing g_e by (μ/a^2) in the small terms, since $g = \mu/R^2$ to a first approximation, we obtain

$$\frac{g_{\mathrm{p}} - g_{\mathrm{e}}}{g_{\mathrm{e}}} = 2f - \frac{9}{2}J_2 + m_{\mathrm{g}} \ ,$$

where

$$m_{\mathrm{g}} = \frac{\varpi^2 a}{g_{\mathrm{e}}} \ . \tag{3.90}$$

The value of m_{g} is equivalent under the given approximations to that of m_a in (3.88). We set

$$\beta = \frac{g_{\mathrm{p}} - g_{\mathrm{e}}}{g_{\mathrm{e}}} \ ,$$

a dimensionless quantity which might be called the flattening of gravity. This can be accurately obtained by measuring g at the equator and the pole. We thus have another relation for J_2, viz.,

$$J_2 = \frac{2}{9}(2f + m_a - \beta) \ . \tag{3.91}$$

Comparing the two equations (3.87) and (3.91) giving J_2, we obtain Clairaut's formula[23]

$$f = \frac{5}{2}m_a - \beta \ . \tag{3.92}$$

This formula gave f in terms of two measurements of g. It was then possible to check the value for the flattening obtained using measurements of the Earth's meridians (see the next section). Numerical calculations give (with $g_{\mathrm{e}} = 9.7804\ \mathrm{m\,s^{-2}}$ and $g_{\mathrm{p}} = 9.8322\ \mathrm{m\,s^{-2}}$) :

$$\beta = 5.296 \times 10^{-3} \approx 1/189 \ , \quad m_a = 3.467 \times 10^{-3} \approx 1/288 \ ,$$

$$f = 3.373 \times 10^{-3} \approx 1/297 \ .$$

Given the approximations made here, this result can be considered to be highly satisfactory.

To sum up, the gravity $g = g_{\Re}$ varies between 9.814 at the equator to 9.832 at the pole, in SI units, due to the flattening of the Earth. Measured experimentally, the weight $g_{\Re_{\mathrm{T}}}$ varies from 9.780 at the equator to the same value 9.832 at the pole. This is because one must add the variation caused by the Earth's rotation (zero at the pole) to the variation of $g_{\Re}$. The amplitudes of these two variations are of the same order of magnitude.

---

[23] Alexis Claude Clairaut (1713–1765) was a French astronomer and mathematician. Accepted into the French Academy of Sciences at the age of eighteen, he published his *Théorie de la figure de la Terre* in 1743, using the different accelerations due to the weight at the poles and the equator. He then studied the three-body problem and published his *Théorie de la Lune* in 1752.

3.10.3 Terrestrial Ellipsoid

The first precise measurement of the Earth's radius was made by Picard.[24] Newton put forward the idea that the Earth was an ellipsoid flattened at the poles and calculated the flattening to be $f = 1/578$, then $f = 1/231$. Huygens obtained $f = 1/230$. A lively debate then ensued with Cassini,[25] who maintained that the ellipsoid was stretched out along the polar axis, with $f = -1/95$. Newton's calculations were confirmed[26] by measurements

---

[24] Father Jean Picard (1620–1683) was a French astronomer and geodesist. He carried out the first scientific (and highly accurate) determination of the Earth's radius. In 1670, he measured the length of an arc of $1°22'$ on the Earth's meridian (from Amiens to Paris, or more precisely, from Sourdon to Malvoisine) using triangulation. He found that an arc of $1°$ (geodetic unit of measurement) corresponded at this latitude to 57 060 toises (1 toise = 6 feet). He immediately communicated his result to Newton, who was thus able to check the relation between the accelerations and the squares of the distances, thereby obtained a clear confirmation of his universal theory of gravitation.
In another area, Picard was the first to carry out systematic measurements of the diameter of the solar disk. He observed its variations and sought the connection with climate change on Earth. His series of measurements between 1666 and 1682 was continued by La Hire from 1683 to 1718.

[25] Jacques Cassini (1677–1756), known as Cassini II, was a French astronomer, son of Jean-Dominique Cassini. He contributed to the measurement of the meridian at the Paris Observatory (1700–1718) and concluded that the polar radius was longer than the equatorial radius.

[26] Under the recommendations of the French Academy of Sciences, the Count of Maurepas, Naval Minister, organised an expedition to the equinoctial and polar regions in 1734 to determine the flattening of the Earth. A group of academics went to Peru (to a region which now belongs to Ecuador), and another to Lapland, in the outer reaches of Sweden and Finland. The unit of length was the *toise* of Châtelet (Paris). The equatorial group (L. Godin, C. M. de la Condamine and P. Bouguer) obtained 56 746 toises for $1°$ at a mean latitude of $-01°30'$. The arctic group (P.L.M. de Maupertuis, A.C. Clairaut and A. Celsius) obtained 57 438 toises for $1°$ at a mean latitude of $66°20'$. In France, the official value at this period was 57 074 toises for $1°$ at a mean latitude of $49°29'$.
Naturally, the angles of $1°$ are not geocentric, but are measured from the normal to the surface (taken as an ellipsoid). The radius of curvature ρ_c of an ellipse (determined by a and e) as a function of the latitude ϕ is

$$\rho_c = \frac{a(1-e^2)}{(1-e^2\sin^2\phi)^{3/2}} .$$

Taking two measurements of arc and hence two radii of curvature ρ_{c1} and ρ_{c2} at two different latitudes ϕ_1 and ϕ_2, we obtain e^2 without needing to know a. For a small flattening f, expanding f to first order (in particular, taking $e^2 \approx 2f$), we obtain

$$f = \frac{1-(\rho_{c1}/\rho_{c2})^{2/3}}{\cos 2\phi_1 - \cos 2\phi_2} .$$

carried out in person by the great scholars of the Age of Enlightenment, such as Maupertuis,[27] in Peru and Lapland.

From this time on, with Clairaut, the founder of modern geodesy, and Cassini de Thury,[28] the flattening of the Earth was taken to be around $f = 1/300$, which is very close to the currently accepted value.

Alongside these purely geodetic measurements, using measurements in the field, knowledge of the Earth's shape moved forward hand in hand with mathematical progress in celestial mechanics. D'Alembert[29] calculated the flattening of the Earth in 1749 by studying the precession of the equinoxes.

Taking the arcs in France and Peru, this gives $f = 1/300$, which agrees with the theories of the day. However, using the arc in Lapland, we obtain $f = 1/207$ when we combine it with the arc in Peru, and $f = 1/123$ when we combine it with the arc in France. In the face of these results, the Swedish Academy organised a further expedition to Lapland in 1803.

The metre was defined as the ten-millionth part of one quarter of the Earth meridian. Measurements were carried out by Delambre and Méchin between 1792 and 1798, using triangulation along the meridian from Dunkerque to Perpignan. The characteristics of the Earth's ellipsoid thereby defined were $a = 6\,375.653$ km, $f = 1/334$.

When the metric system was officially established in France (10 December 1799 or *19 frimaire An VIII*), the official conversion from the *toise* to the metre was: 1 toise of Châtelet $= 1.949\,036\,3$ m .

27 Pierre Louis Moreau de Maupertuis (1698–1759) was a French physicist. He led the expedition to Lapland, thereby earning himself the following verse from Voltaire:

Vous allâtes vérifier en ces lieux pleins d'ennui
Ce que Newton trouva sans sortir de chez lui.

In these dull and dismal quarters you will only ascertain
What Newton always knew without leaving his domain.

There may be no connection with this ironic tribute, but Maupertuis subsequently discovered his famous principle of least action in 1744.

28 César-François Cassini (1714–1784), known as Cassini III or Cassini de Thury, was a French astronomer. With his father Cassini II, he contributed to the measurement of the perpendicular to the meridian at the Paris Observatory in 1733. The new measurement of the meridian (1739–1740) convinced him of his father's mistake with regard to the flattening of the Earth. He published a map of France known as the Cassini map, the first basic map of the country, a topographical map with scale 1:86 400 (1 line for 100 toises), using the Cassini projection. It was begun in 1744 and completed in 1790 by his son Jean-Dominique Cassini (1748–1845), known as Cassini IV.

The four Cassinis were successive directors of the Paris Observatory from its foundation by Cassini I in 1671 up to the French Revolution, with Cassini IV, in 1793, the last three Cassinis taking up the succession at the death of their fathers.

29 Jean le Rond D'Alembert (1717–1783) was a French mathematician, physicist and philosopher. He published *Recherche sur la précession des équinoxes et sur la nutation de l'axe de la Terre dans le système newtonien* in 1749. In 1743, he

Then, in 1802, Laplace found practically the same value by studying the libration and the motion of the Moon: $f = 1/305$ from the first, $f = 1/303$ from the second.

However, it was with the advent of the satellite that the most accurate studies of the geoid could be accomplished.

3.11 Geoid

3.11.1 Satellites and Geodesy

The first artificial satellite artificiel, Sputnik-1, was launched by the USSR on 4 October 1957 and only emitted its signal for three weeks. [30]

However, by studying the trajectory of the following satellites,[31] launched shortly afterwards, the zonal coefficient J_2 was determined by the British geodesist D. King-Hele in 1958. The value[32] was quite close to predictions calculated from Earth-based measurements.

The US satellite Vanguard-1, launched on 17 March 1958, made it possible for the first time to evaluate the discrepancy between the ellipsoid and the geoid. The value of J_4 was obtained in the same year, and that of J_3, the first odd zonal term, in 1959 by Y. Kozai. This important harmonic of degree 3 corresponds to a raising of the North Pole by 15 m above the ellipsoid and a lowering of the South Pole by the same amount.

In 1961, W. Kaula produced a complete model of degree 4, i.e., involving all the coefficients C_{lm} and S_{lm}: a sectorial harmonic, coefficient of the associated Legendre function P_{22}, accounts for an elevation of the geoid around 165°E and 15°W and a depression around 75°E and 105°W. These points mark out four equal sectors on the equator[33] since we are dealing with P_{lm} in the case $l = 2$ and $m = 2$.

---

put forward the principle now known as D'Alembert's principle, in his *Traité de dynamique*. With Diderot and others, he wrote the celebrated *Encyclopédie* [or *Dictionnaire raisonn des sciences, des arts et des métiers (1751-1772)*].

[30] Sputnik-1 was spherical with mass 84 kg and diameter 58 cm. Altitude at perigee $h_{\mathrm{p}} = 228$ km, altitude at apogee $h_{\mathrm{a}} = 947$ km, inclination $i = 65.128°$, period $T = 96.17$ min ($\Delta T = 1.80$ s/day), argument of the perigee $\omega = 41°$N, secular variations (degrees/day): $\dot{\Omega} = 3.157$, $\dot{\omega} = 0.432$.

[31] Sputnik-2 (mass: 508 kg) was launched on 4 November 1957: $h_{\mathrm{p}} = 225$ km, $h_{\mathrm{a}} = 1\,671$ km, $i = 65.310°$, $T = 103.75$ min ($\Delta T = 3.08$ s/day), $\omega = 40.45°$N, $\dot{\Omega} = 2.663$, $\dot{\omega} = 0.407$.
Sputnik-3 (mass: 1 327 kg) was launched on 15 May 1958: $h_{\mathrm{p}} = 226$ km, $h_{\mathrm{a}} = 1\,881$ km, $i = 65.188°$, $T = 105.95$ min ($\Delta T = 0.75$ s/day), $\omega = 44.57°$N, $\dot{\Omega} = 2.528$, $\dot{\omega} = 0.326$.

[32] The results from Sputnik-2 gave $J_2 = 1.084 \times 10^{-3}$. At this point, the notation J_n was adopted in homage to the British geophysicist Sir Harold Jeffreys (1891–1989).

[33] D. King-Hele summed this up in a little verse of his own invention:

Our knowledge of the Earth's gravitational potential moved ahead very quickly from this time. Geodesists not only took advantage of all available satellites, but they also sent up their own dedicated satellites.[34] Space-based geodesy has now become a 'dialectic' science in the sense that geopotential models are better known by localising satellites and studying their trajectories, and the position of the satellites is better determined by improved potential models.

The US satellites NAVSTAR, launched as part of the GPS programme (Global Positioning System), or the Soviet then Russian satellites, for the equivalent GLONASS programme, allow us to determine positions on Earth with accuracies between a few metres and a few millimetres depending on the receiver used and the chosen operational mode.

The French–US satellite TOPEX/Poseidon, launched in 1992 for oceanographic and geodetic purposes, measures the sea level (whose mean represents the geoid) relative to the reference ellipsoid with an accuracy between 2 and 3 cm. This means that the altitude of the satellite is known to an even better accuracy. The quality of the altimeters used is important, but the accuracy with which the satellite orbit is established by the DORIS system is just as crucial. And this in turn is possible thanks to potential models involving spherical harmonics of a very high degree.

To sum up, the geoid has been defined with great accuracy with respect to the Earth ellipsoid. The North Pole is 10 m above and the South Pole 30 m below. Apart from this deformation along the Earth's axis of rotation, there are two diametrically opposed bulges and two diametrically opposed hollows. The largest bulge is 76 m above the ellipsoid, centered on the island of New Guinea and extending across the whole of the western Pacific. The largest hollow, measuring 93 m, is situated just south of India. Another is located in

---

When you cut a slice
Through the polar ice
    The Earth is like a pear.
But sliced along the equator
She looks like a potato –
    A giant *pomme de terre.*

[34] Among these, the US series GEOS (Geodetic Earth Orbiting Satellite), GEOS-1 (Explorer-29), GEOS-2 (Explorer-36), PAGEOS, LAGEOS, with passive ranging (PA) or laser ranging (LA) of the satellite, which followed on from the satellites Echo-1 and Echo-2 (balloon satellites), Anna-1B (Army, Navy, Nasa, Air Force, first satellite to emit flashes), ADE-A (Atmospheric Density Explorer, Explorer-19), Beacon-Explorer-1 (Explorer-22 or BE-B, first satellite equipped with laser reflectors), and Beacon-Explorer-2 (Explorer-27 or BE-C). After 1970 came the French satellites Starlette and Stella, launched in 1975 and 1993, the Japanese satellite EGS-1 (Earth Geodetic Satellite, or EGP (Experimental Geodetic Payload, also called Ajisai, 'hydrangea' in Japanese), launched in 1986, and the Russian satellite Fizeau (Meteor-2-21), launched in 1993.

the Hudson Bay in Canada. These anomalies bear no relation to the relief of the Earth's surface. They are explained as manifestations of mass anomalies in the Earth's mantle. A map of these anomalies is given in Colour Plate XVIa.

3.11.2 Development of Geopotential Models

The Main Models

The first satellite data were integrated into existing models and, from 1970, certain models were established exclusively on the basis of space data (the so-called satellite-only models). The SAO-SE model (Smithsonian Astrophysical Observatory – Standard Earth), established[35] in 1966, used in 1972 the first laser-ranging measurements to establish satellite distances. The NSWC model (Naval Surface Warfare Center, ex-NWL or Naval Weapon Laboratory) was mainly based on satellites in the Transit series.

Models integrate space-based measurements (satellite trajectories, altimetric measurements made by satellite) and Earth-based measurements (surface gravimetric measurements, currently accurate to $10^{-9} g_0$). The main models of this type currently in use are GEM, JGM, EGM and GRIM.

The GEM model (Goddard Earth Model) was established by NASA's GSFC (Goddard Space Flight Center) in the United States as a reaction to the classified US military models. The first model, GEM-1, was published in 1972, expanding the potential to degree 12. Then came GEM-2, and a whole series up to GEM-10 (expansion to degree 20) in 1977, followed by GEM-L in 1983 (using LAGEOS[36] data) and GEM-T in 1984 (preparation for TOPEX/Poseidon). The GEM-T2 model, published in 1990, was obtained using data from 31 satellites and 2.4 million observations (on 1130 orbital arcs). It gave a model with all coefficients up to degree 36 and some up to degree 50. It also gave a very high order expansion for the tides. This model uses satellites with a wide range of orbits, some dedicated to geodesy and others not.[37] It will be further refined in GEM-T3.

[35] The satellites used here, launched between 1959 and 1961, were, in order of inclination, Explorer-2, Vanguard-2 and -3, Telstar, Echo-1, Relay-1, Anna-1B, Transit-4A, Injun-1, Solrad-3 (or GREB-3), with $i = 29°$ to $67°$, and MIDAS-4, with $i = 96°$. Note that Injun-1 and Solrad-3, launched as auxiliary passengers on Transit-4A, were unable to separate from one another.

[36] The name LAGEOS-1 was only given to this satellite after LAGEOS-2 was launched.

[37] One finds the complete range of inclinations, from the lowest (less than 40°), represented by PEOLE, and Diadème-1 and 2 (D1-C and D1-D), up to the highest (over 140°), represented by OV1-2, through satellites with strictly polar orbits ($i = 90.0°$), represented by Transit-5B-2 and Nova-1. There are satellites that have been launched recently, just as there are satellites launched at the very beginning of space exploration, between 1959 and 1961, such as Courier-1B and all

The JGM model (Joint Gravity Model) was produced jointly by NASA and the University of Texas. In 1994, JGM-2 amended GEM-T3 with the first results from TOPEX/Poseidon. In 1996, JGM-3 integrated data from other satellites, such as LAGEOS-2.

The EGM model (Earth Gravity Model) is the result of collaboration between GSFC-NASA, NIMA (National Imagery and Mapping Agency), and OSU (Ohio State University). In 1996 came EGM96S, of degree and order 70, with data provided solely by satellites, and EGM96, of degree and order 360, adjoining geophysical data (see Colour Plate XVIa). They used data from 40 satellites,[38] including satellite-to-satellite measurements, with the GPS constellations[39] and TDRSS.

The GRIM model was the result of a joint project between GRGS (*Groupe de Recherche en Géodésie Spatiale*) in France and DGFI (*Deutsches Geodätisches Forschungsinstitut*) in Germany. The first model, GRIM1, in 1976 (up to degree 10), was followed by GRIM2, GRIM3 and GRIM4. The GRIM4-S2 model, obtained using 27 satellites, gave the harmonics up to degree 50 and high order expansions for the tides, going even further with the GRIM4-S4 model. In 2000, there followed the GRIM5-S1 and GRIM5-C1 models, the first based solely on satellite data,[40] and the second using all data. The -C1 model provides a complete matrix of degree and order 120 and an incomplete matrix for degrees 121 to 360. This corresponds to an accuracy of 50 cm for the height of the geoid and 5.5 mgal for the surface gravity field.[41] GRIM5 was established to prepare for two oceanographic missions, the French–US satellite Jason-1 and the European Envisat, and other missions dedicated to studies of the geopotential, namely, the satellites CHAMP, GRACE-A and -B, and GOCE. Geodesy is moving ahead very quickly in this interplay with

those used by the SAO-SE model, remote-sensing and communications satellites, such as Landsat-1, Oscar-7 and -14 (Transit-O-7 and -O-14), and of course satellites with geodetic missions, such as OGO-2, the two Beacon-Explorers (called BE-B and BE-C by geodesists), SECOR-5 (EGRS-5), LAGEOS-1, GEOS-1, -2, -3, Ajisai, Starlette, Seasat, and Geosat. Note also the use of laser ranging between the two satellites ATS-6 (geostationary) and GEOS-3.

[38] In addition to those providing data for GEM-T2, we find the following more recent satellites: ERS-1, Etalon-1, EUVE (BerkSat), GFZ-1, HILAT (NNS-O-16), LAGEOS-2, Radcal, Stella et TOPEX/Poseidon.

[39] The satellites NAVSTAR-35 and -36 (or USA-96, -100), launched in 1993 and 1994, are equipped with laser reflectors.

[40] The satellites used were Starlette, Ajisai, LAGEOS-1 and -2, Geosat, SPOT-2 and -3, ERS-1 and -2, Stella, Westpac-1 (WPLTN-1, West Pacific Laser Tracking Network), TOPEX/Poseidon, GFZ-1, D1-C, D1-D, GEOS-3, Meteor-3-07, Nova-3, Etalon-1 and -2 (Kosmos-1989 and -2024), and PEOLE.

[41] The gal is the unit of acceleration in the old CGS system, named after Galileo: 1 gal $= 1\ \mathrm{cm\,s^{-2}}$. The gal is still used in geophysics. One milligal (mgal) corresponds to about one millionth of the Earth's gravity field g_0.

Table 3.6. Comparison between different models. *Upper*: normalised zonal coefficients C^*_{l0}. All values to be multiplied by 10^{-6}. *Lower*: values of the geocentric gravitational constant μ, equatorial radius R, and flattening $(1/f)$. The whole number part is the same for all models, i.e., only the decimal changes

Coefficient	GEM-T2	JGM-3	GRIM5-C1	GRIM5-S1
$C^*_{2\,0}$	−484.1652998	−484.165368	−484.16511551	−484.16511551
$C^*_{3\,0}$	0.9570331	0.957171	0.95857491	0.95857492
$C^*_{4\,0}$	0.5399078	0.539777	0.53978784	0.53978784
$C^*_{5\,0}$	0.0686883	0.068659	0.06726760	0.06720440
$C^*_{6\,0}$	−0.1496092	−0.149672	−0.14984936	−0.14985240
$C^*_{7\,0}$	0.0900847	0.090723	0.09301877	0.09311367
$C^*_{8\,0}$	0.0483835	0.049118	0.05039091	0.05046451
$C^*_{9\,0}$	0.0284403	0.027385	0.02628356	0.02620763
$C^*_{10\,0}$	0.0549673	0.054130	0.05101952	0.05076191
$C^*_{20\,0}$	0.0199685	0.018790	0.02340848	0.02342817
$C^*_{99\,0}$			−0.00128836	−0.00001554

Quantity	Unit	Integer part	GEM-T2	JGM-3	GRIM5
$\mu = \mathcal{G}M$	$\mathrm{km^3 s^{-2}}$	398 600	.436	.441 5	.441 5
R	km	6 378	.137	.136 30	.136 46
$1/f$	(dimensionless)	298	.257	.257 65	.257 65

satellite development, and should witness still more spectacular results in the near future as data from geodetic satellites is put to use.

Comparison of Geopotential Models

Models of the geopotential cannot be compared term by term (beyond degree 5). Two models can have rather different terms and yet still have a very close final result: different weightings of the spherical harmonics can lead to very close results. Beyond degree 16, even the signs of the coefficients can change from one model to another, without there being any harmful effect on the restitution of the geoid and satellite tracking. This highlights a problem when truncating series: the coefficients of a model of degree 10 do not correspond to the coefficients of the first 10 degrees in a model of degree 20.

As an example, Table 3.6 (upper) gives the coefficients $C^*_{l\,0}$ for four models mentioned above. (Coefficients of order 0 and degree l between 2 and 10. We recall that, for $l = 0$, the coefficient is equal to unity and, for $l = 1$, it is zero. All coefficients in the table are in units of 10^{-6}.) We have the following relation between the harmonic coefficients $C_{l\,0}$ defined by (3.11) and the so-called normalised zonal coefficients $C^*_{l\,0}$:

$$C_{l\,0} = \sqrt{2l+1}C^*_{l\,0} \,.$$

These coefficients are associated with values of μ, R and f corresponding to each model [see Table 3.6 (lower)].

3.11.3 Evaluation of the Geocentric Gravitational Constant

It is easy to see that the geocentric gravitational constant $\mu = \mathcal{G}M$ plays a key role in space mechanics. It can be obtained to very high accuracy, well above what can be achieved for the universal constant of gravitation $\mathcal{G}$.[42]

The first values for μ were given by Kepler's third law applied to the lunar orbit. More and more precise values were obtained using the space probes Ranger, Mariner, and Venera, then satellites, preferably with high altitude, since lower satellites are subject to non-gravitational effects. Today, the most precise measurements are obtained by laser ranging on the satellite LAGEOS. Table 3.7 shows the values obtained for μ, together with the estimated error, by various methods (mentioning the year).

3.12 Appendix: Astronomical Constants

The astronomical constants can be put into three groups: defining constants, primary constants, and derived constants. The numerical values of these constants are given in Table 3.8. These are the so-called IERS 1992 values, named

[42] Henry Cavendish (1731–1810) was a British physicist and chemist. He was the first to obtain a precise value for $\mathcal{G}$, published in 1798 in his famous paper entitled *Experiments to determine the density of the Earth.* He used a subtle method: instead of taking advantage of very large masses (like those who, at the time, sought to measure the deviation of a plumbline by a mountain), he used a torsion balance with a very fine thread, suspending two small metal weights (50 g). Bringing two large lead balls (30 kg) to a distance of 15 cm, he measured the torsion of the thread using a mirror to create a 'light lever' and deduced $\mathcal{G}$ from the period of the motion ($\sim$ 2 hr). He thereby calculated the density of the Earth and found $d = 5.48$ (current value: 5.52). This density is greater than that of the rocks in the Earth's crust ($\sim$ 2.7), and Cavendish thus demonstrated that the Earth contained a very dense central part.
The method was later refined by Charles Boys (1895) using a very fine quartz thread (2 μm) and still smaller masses (2.7 g, 7.5 kg at 15 cm), on a short period (3 min). This type of experiment is still used to measure $\mathcal{G}$, but the relative accuracy does not exceed $\delta\mathcal{G}/\mathcal{G} = 10^{-4}$. Other ways are now sought to improve accuracy. Current recommendations (CODATA 1998) give the value

$$\mathcal{G} = (6.673 \pm 0.010) \times 10^{-11}\ \mathrm{m^3 s^{-2} kg^{-1}} \,.$$

Table 3.7. Measured geocentric gravitational constant $\mu = \mathcal{G}M$ and estimated error. Historical evolution indicating method used and year

Method	Year	μ [km^{3}s^{-2}]	Error
Lunar orbit	1959	398 620.	±6.
Explorer-27	1965	398 602.	±4.
Ranger-6, -7, -8, -9	1966	398 601.0	±0.7
Mariner-9	1971	398 601.2	±2.5
Venera-8	1972	398 600.4	±1.0
ATS-6/GEOS-3	1979	398 600.40	±0.1
Laser/Moon	1985	398 600.444	±0.010
Laser/LAGEOS	1992	398 600.441 5	±0.000 8
Laser/LAGEOS	2000	398 600.441 9	±0.000 2

after the system defined in 1992 by the International Earth Rotation Service (IERS). We have only included those constants used in this book.

Defining Constants

The two defining constants are the Gaussian gravitational constant and the speed of light.

- We consider a body of mass m in gravitational interaction with the Sun, of mass M_S, in the framework of the two-body problem. We also assume that m is negligible compared with M_S. The mean motion of the Keplerian motion, with semi-major axis a_S, is given by $n = \sqrt{\mu_S/a_S^3}$. We set $k^2 = \mu_S = \mathcal{G}M_S$. Putting $a_S = 1$ (this length is defined to be unity), we thus obtain $k = n$. With this choice of units, the Gauss gravitational constant k, is equal to the mean motion. Historically, it was calculated from values measured for the Earth in orbit around the Sun. Using (2.8), we obtain $n^2a_S^3 = k^2(1 + M/M_S)$, and hence the value of k, with the mass of the Earth M (which is not neglected in comparison with M_S) and $n = 2\pi/N_{\rm sid}$, where $N_{\rm sid}$ is the value for the sidereal year in days. As the period of revolution of the Earth around the Sun is very stable, the value of k is then defined to be $k = 0.017\,202\,098\,95$, with a_S as unit of length and the mean day as unit of time. The dimensions of k are the same as those of $(\mathcal{G}M_S)^{1/2}$. Expressing angles in degrees, we have $k = 0.985\,607\,668\,425$.
- The speed of light has been a defining constant since 1983. At this date, the definition of the unit of length became the following: the metre is the length travelled by light in a time of 1/299 792 458 seconds. The speed of light is thus given as a defined value, with no error margin. This development was made because times are now measured to higher accuracy than lengths.

Table 3.8. Astronomical constants: defining constants, primary constants and derived constants (in that order)

Constant	Value	Units
Gaussian gravitational constant	$k = 0.017\,202\,098\,95$	SI unit
Speed of light	$c = 299\,792\,458$	$\mathrm{m\,s^{-1}}$
Light time for unit distance	$t_{\mathrm{A}} = 499.004\,783\,53$	s
Equatorial radius of Earth	$R = 6\,378\,136.3$	m
Geopotential ellipticity factor	$J_2 = 0.001\,082\,6362$	(dimensionless)
Geocentric gravitational constant	$\mu = 3.986\,004\,415 \times 10^{14}$	$\mathrm{m^3s^{-2}}$
Mass ratio $M_{\mathrm{Moon}}/M_{\mathrm{Earth}}$	$\mu_{\mathrm{L}}/\mu = 0.012\,300\,34$	(dimensionless)
Obliquity of the ecliptic for J2000 (2000 01 01 12:00)	$\varepsilon_0 = 23°26'\,21''.4119$	degrees
Mean angular speed of rotation of the Earth	$\varpi = 7.292\,115 \times 10^{-5}$	$\mathrm{rad\,s^{-1}}$
Unit of distance: astronomical unit 1 a.u. $= ct_{\mathrm{A}}$	$ct_{\mathrm{A}} = 1.495\,978\,7061 \times 10^{11}$	m
Heliocentric gravitational constant	$\mu_{\mathrm{S}} = 1.327\,124\,40 \times 10^{20}$	$\mathrm{m^3s^{-2}}$
Mass ratio $M_{\mathrm{Sun}}/M_{\mathrm{Earth}}$	$\mu_{\mathrm{S}}/\mu = 332\,946.045$	(dimensionless)
Flattening of the Earth	$f = 1/298.257$	(dimensionless)

Primary and Derived Constants

The primary constants are determined from observations. Derived constants follow from simple relations using the first two types of constant.

Note. For some of these constants related to geodesy, such as R, J_2 or μ, there is a slight difference between the IERS 1992 values and those used in geopotential models.

3.13 Appendix: Gravitational Sphere of Influence

3.13.1 Attraction of the Sun and Earth

It is easy to understand why a low-orbiting satellite should feel only the Earth's attraction, since the Sun's attraction is extremely weak. But one must ask how far out one may continue to ignore the influence of this third body. In the following, we shall define the radius of the sphere beyond which we may consider that a satellite of the Earth escapes to become a satellite of the Sun. This idea of the sphere of gravitational influence was developed by the astronomer F. Tisserand.[43]

Consider three points A (Sun), B (Earth), and C (satellite). The gravitational constant is μ_S for the Sun and μ for the Earth. Consider the special case when C lies between A and B. We set

$$r = CB\ , \quad \text{satellite–Earth distance}\ ,$$

$$a_S = AB\ , \quad \text{Sun–Earth distance}\ ,$$

$$a_S - r = CA\ , \quad \text{satellite–Sun distance}\ .$$

Consider now the reduced (dimensionless) variables k and x defined by

$$k = \frac{\mu_S}{\mu}\ , \qquad x = \frac{r}{a_S}\ . \tag{3.93}$$

Note that $k \gg 1$ and $x \ll 1$. For the values of the astronomical quantities, see Sect. 3.12 and Table 3.8.

Satellite Close to Earth

The main acceleration here is the central acceleration γ_{CCC}, which we shall write γ_{T_0}, due to the Earth (Keplerian motion):

$$\gamma_{\mathrm{T}_0} = \frac{\mu}{r^2}\ . \tag{3.94}$$

The perturbing acceleration for the satellite is the differental attraction γ_{CS}, here denoted by γ_{T_1}, due to the Sun:

$$\gamma_{\mathrm{T}_1} = \frac{\mu_S}{(a_S - r)^2} - \frac{\mu_S}{a_S^2}\ . \tag{3.95}$$

In the expression for γ_{T_1}, the first term refers to the satellite and the second to the Earth (since the Sun acts on the satellite and on the Earth). Since r is small compared with a_S, we obtain

$$\gamma_{\mathrm{T}_1} \approx \frac{\mu_S}{a_S^2}\left[\left(1 + \frac{r}{a_S}\right)^2 - 1\right] \approx 2\frac{\mu_S}{a_S^3} r\ . \tag{3.96}$$

[43] Félix François Tisserand (1845–1896) was a French astronomer. He continued Delaunay's work on the motion of the Moon and contributed to the *Catalogue photographique de la carte du ciel*. He then published his *Traité de mécanique céleste* in four volumes (1889–1896), in the spirit of Laplace's work. See also the note on Einstein.

The ratio of the accelerations is

$$Q_{\mathrm{T}} = \frac{\gamma_{\mathrm{T}_1}}{\gamma_{\mathrm{T}_0}} = 2\frac{\mu_{\mathrm{S}}}{\mu}\left(\frac{r}{a_{\mathrm{S}}}\right)^3 = 2kx^3 \ . \tag{3.97}$$

Note. By bringing in the expression for the periods, we obtain

$$Q_{\mathrm{T}} = 2\frac{\mu_{\mathrm{S}}/a_{\mathrm{S}}^3}{\mu/r^3} = 2\left(\frac{T_0}{T_{\mathrm{S}}}\right)^2 \ , \tag{3.98}$$

where T_0 is the Keplerian period of the satellite and T_{S} is the period of revolution of the Earth around the Sun, i.e., $T_{\mathrm{S}} = 1$ yr.

Satellite Far from Earth

If a satellite is very far from Earth, so that it is in fact a space probe, the Earth's attraction becomes very small compared with the Sun's. The central acceleration due to the Sun can be written

$$\gamma_{\mathrm{S}_0} = \frac{\mu_{\mathrm{S}}}{(a_{\mathrm{S}} - r)^2} \approx \frac{\mu_{\mathrm{S}}}{a_{\mathrm{S}}^2} \ , \tag{3.99}$$

for even in this case, r is small compared with a_{S}. The perturbing acceleration for the satellite is the differential attraction due to the Earth, viz.,

$$\gamma_{\mathrm{S}_1} = \frac{\mu}{r^2} - \frac{\mu}{a_{\mathrm{S}}^2} \approx \frac{\mu}{r^2} \ . \tag{3.100}$$

The ratio of the accelerations is

$$Q_{\mathrm{S}} = \frac{\gamma_{\mathrm{S}_1}}{\gamma_{\mathrm{S}_0}} = \frac{\mu}{\mu_{\mathrm{S}}}\left(\frac{a_{\mathrm{S}}}{r}\right)^2 = \frac{1}{kx^2} \ . \tag{3.101}$$

3.13.2 Sphere of Influence

The Earth's sphere of gravitational influence Σ is a sphere centred on the Earth with radius ρ_Σ defined by the point on the straight line joining the Sun and Earth such that $Q_{\mathrm{T}} = Q_{\mathrm{S}}$. This gives

$$2k^2x^5 = 1 \ , \tag{3.102}$$

or[44]

$$\rho_\Sigma = 2^{-1/5}\left(\frac{\mu}{\mu_{\mathrm{S}}}\right)^{2/5} a_{\mathrm{S}} \ . \tag{3.103}$$

---

[44] This demonstration is schematic insofar as we are considering the case of three bodies lying along a straight line. The full proof due to Tisserand shows that the surface we seek here is given by

For the Earth, one finds

$$\rho_\Sigma = 0.805 \times 10^6 \text{ km} , \quad \frac{\rho_\Sigma}{R} = 126 , \quad \frac{\rho_\Sigma}{a_S} = \frac{1}{186} ,$$

using the numerical values of the astronomical quantities μ_S, the heliocentric gravitational constant, and a_S, the astronomical unit. It is thus possible to treat r as small compared with a_S.

In Chaps. 10 and 11, we shall apply this idea to other celestial bodies apart from the Earth. For planets in the Solar System, the values of ρ_Σ are given in Table 11.2 (lower).

To end this section, let us just note that we can define three spheres around a celestial body:

- the sphere of influence discussed above, beyond which the planet can no longer hold on to a satellite,
- the sphere of the synchronous orbit,
- the sphere relating to the Roche limit.

The last two spheres refer to the tidal effect. We return briefly to them in Chap. 10, in the context of the natural satellites of Mars. The planetosynchronous orbit for artificial satellites is discussed in detail in Chap. 4.

3.14 Appendix: Lagrange Points

3.14.1 Restricted Three-Body Problem

The Lagrange points arise in the context of the restricted three-body problem, in which one of the bodies (here, the satellite) has negligible mass compared with the two others. The two 'massive' bodies A and B revolve around their centre of mass O (two-body problem) with constant angular speed $\dot{\theta}$. The third, much lighter body C feels the gravitational attraction of A and B.

Lagrange showed that there are five special positions in space at which the body C rotates about O with the same angular speed $\dot{\theta}$. In this situation,

---

$$\rho_\Sigma = \rho_\Sigma(\theta) = \left[\left(\frac{\mu}{\mu_S}\right)^2 \frac{1}{\sqrt{1+3\cos^2\theta}}\right]^{1/5} a_S ,$$

where the polar axis is the straight line Earth–Sun with origin at the centre of the Earth. When $\theta = 0$, we retrieve (3.103). This surface of revolution around the polar axis differs only slightly from a sphere, since the polar radius varies by a factor of 1 to 0.87 ($= 2^{-1/5}$). Tisserand's calculations were made to study the trajectories of comets in the vicinity of Jupiter. The process had already been suggested by D'Alembert, Laplace, and Le Verrier. Laplace used the term sphere of activity.

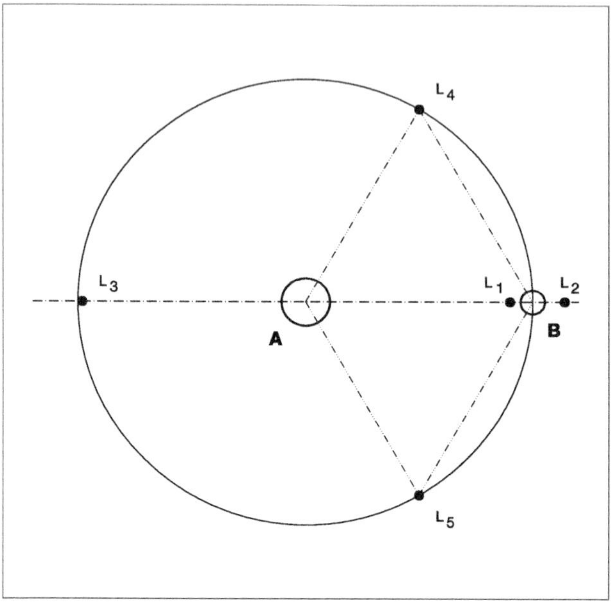

Figure 3.4. Schematic illustration of the five Lagrange points. The five points lie in the orbital plane of B around A. The mass of B is considered to be small compared with the mass of A

the point C is stationary in a Sun–Earth frame. The five points, traditionally denoted by L_1 to L_5, are known as the Lagrange points or libration points.[45]

3.14.2 Simplified Study of Points L_1 and L_2

We shall now find in a schematic manner the position of the first two Lagrange points in the case when B has much smaller mass than A (which is generally the case). These equilibrium points are unstable. We use the notation of Sect. 3.13. The centre of mass is at A and B revolves around A in a circular orbit with constant angular speed $\dot{\theta}$. According to Kepler's third law, we have

$$\mu_S = \dot{\theta}^2 a_S^3 \, . \tag{3.104}$$

Consider a satellite C at L_i, close to B, on the straight line AB. The distance from C to A is thus $(a_S + \varepsilon r)$, where $\varepsilon = -1$ for L_1 and $\varepsilon = +1$ for L_2. For

[45] The word 'libration' comes from the Latin *libratio*, which itself comes from *libra*, meaning 'balance'. The Moon's libration is a complex nodding motion around its central position, composed of a physical and a geometric libration (in longitude and in latitude). It is through this motion that we are able to see 59% of the Moon's surface from Earth, instead of just a half. This term, generally used for lunar libration (also studied by Lagrange), is also used to refer to the five Lagrange points.

each point L_1 and L_2, we express the fact that the resultant of the attractive accelerations is equal to the radial acceleration. Projecting onto the AB axis and using the notation ε, we obtain

$$\frac{\mu_S}{(a_S+\varepsilon r)^2}+\varepsilon\frac{\mu}{r^2}=(a_S+\varepsilon r)\dot{\theta}^2 \ . \tag{3.105}$$

Dividing the left- and right-hand sides of (3.105) and (3.104) and using the reduced variables defined in (3.93), this yields

$$\frac{1}{(1+\varepsilon x)^2}+\varepsilon\frac{1}{kx^2}=(1+\varepsilon x) \ . \tag{3.106}$$

We can expand $(1+\varepsilon x)^2$ to first order in x since $x \ll 1$, whereupon

$$3\varepsilon x \approx \varepsilon\frac{1}{kx^2} \ .$$

Cancelling out the device ε, we find that x has the same value in both cases, namely,

$$x=(3k)^{-1/3} \ . \tag{3.107}$$

For the distances ρ_{L_i}, we now have

$$BL_1=BL_2=\rho_{L_i}=\left(\frac{1}{3}\frac{\mu}{\mu_S}\right)^{1/3} a_S \ . \tag{3.108}$$

Example 3.2. *Calculate the positions of the Lagrange points L_1 and L_2 for various astronomical systems.*

For the Sun–Earth system, where $k = 3.329\times10^5$, or $3k\approx10^6$, we have

$$\text{distance from centre of Earth to } L_{1,2}: \ \rho_{L_i}=10^{-2}a_S\approx1.5\times10^6 \text{ km} \ .$$

For the Earth–Moon system, where $k = 81.3$, replacing a_S by a_L, the mean radius of the lunar orbit, we find

$$\text{distance from centre of Moon to } L_{1,2}: \ \rho_{L_i}=\frac{a_L}{(243.9)^{1/3}}=0.16a_L\approx6\times10^4 \text{ km} \ .$$

For the system of Mars and its moon Phobos, where $k = 5.05\times10^7$ and $a_P = 9.38\times10^3$ km, we obtain

$$\text{distance from centre of Phobos to } L_{1,2}: \ \rho_{L_i}=1.88\times10^{-3}a_P\approx17.6 \text{ km} \ .$$

As the subplanetary equatorial radius of Phobos is 13.4 km, points L_1 and L_2 are only 4.2 km from the surface of Phobos.

3.14.3 The Five Lagrange Points

A complete analysis to find the five points and the equilibrium conditions is much more involved and goes beyond the scope of this book. The classic method consists in writing the equations in a frame rotating about O. This produces two equations: one involves the first three points, and the other the last two. The position of the points is shown schematically in Fig. 3.4 (where O coincides with A).

- Points L_1, L_2 and L_3 lie on the straight line AB. Let y be the reduced distance, i.e., $y = AL_i/AB$ (so that $y = 1 - x$ for L_1), and α the reduced mass of B, $\alpha = \mu/(\mu + \mu_{\mathrm{S}}) = 1/(1+k)$. It can be shown that the three possible values of y are given by the quintic equation

$$\varepsilon_1 \left[(1-\alpha)(1-\alpha+y)^2\right] + \varepsilon_2 \alpha(\alpha+y)^2 + y(1-\alpha+y)^2(\alpha+y)^2 = 0 ,$$

  with suitable values of ε at each point: for L_1 ($\varepsilon_1 = -1$, $\varepsilon_2 = +1$), for L_2 ($\varepsilon_1 = \varepsilon_2 = -1$), or for L_3 ($\varepsilon_1 = \varepsilon_2 = +1$). These three equilibrium points are unstable.
- Points L_4 and L_5 can be shown to lie in the orbital plane of B and A around O in such a way that triangles ABL_4 and ABL_5 are equilateral. The positions of L_4 and L_5 do not depend on the masses of the bodies A and B. It can also be shown that these positions are stable, provided that the mass of A is big enough compared with the mass of B, i.e., 25 times bigger.[46] L_4 and L_5 are distinguished by the convention that L_4 is the point preceding the body B in its revolution, and L_5 is the point following it.

3.14.4 Lagrange Points in Astronomy

Lagrange's theory was brilliantly confirmed in the field of planetary astronomy with the discovery of asteroids at the stable points L_4 and L_5 of the Sun–Jupiter system. The first, 588 Achilles, was discovered in 1906, and this was followed by 617 Patroclus at L_5, and 624 Hektor and 659 Nestor at L_4.

[46] The exact calculation gives

$$k > k_0 , \quad \text{with} , \ k_0 = \frac{25}{2}\left[1 + \sqrt{1 - \left(\frac{2}{25}\right)^2}\right] = 24.96 .$$

The numerical value of k_0 is called the Routh value. For the planes in the Solar System, this condition always holds by a large margin as far as the Sun is concerned. For the Earth–Moon system, it also holds, since $k = 81$ in this case. The only known exception in the Solar System is provided by the Pluto–Charon system.

Table 3.9. Distances of bodies A and B from the Lagrange points L_i ($i = 1, \ldots, 5$). Distances are given in reduced units x and y (dimensionless), in km, and in units of R (Earth radius)

System A–B		L_1	L_2	L_3	L_4, L_5
Sun–Earth					
AL_i	y	0.990	1.010	0.999	1.000
BL_i	x	0.00997	0.010	1.999	1.000
AL_i	10^6 km	148.10	151.10	149.4	149.6
BL_i	10^6 km	1.49	1.50	299.0	149.6
BL_i	R	234	236		
Earth–Moon					
AL_i	y	0.85	1.17	0.99	1.00
BL_i	x	0.15	0.17	1.99	1.00
AL_i	10^3 km	326.7	449.7	380.6	384.4
AL_i	R	51	71	60	60
BL_i	10^3 km	57.7	65.3	765.0	384.4

Several hundred asteroids are now known at the two stable points[47] of this system, as illustrated in Fig. 3.5. The asteroids at L_4 are called Greeks, whilst those at L_5 are the Trojans. The current trend is to use the term Trojans[48] for the asteroids in both groups and to extend the term to other systems. In 1990, a 'Trojan' asteroid was discovered in the Sun–Mars system.[49]

In 1980, several new moons of Saturn were discovered by Earth-based observation at the Lagrange points of the Saturn–Dione and Saturn–Tethys systems. These are known as the Lagrangian moons. These moons librate[50] around the stable positions L_4 and L_5. The various cases are summarised in Table 3.10.

[47] The libration of these asteroids is 14° on average. It cannot exceed 30°.

[48] The duality between Greeks and Trojans is intended to illustrate the unending pursuit, immortalising the *Iliad* in the skies. However, there seems to have been some misunderstanding of Homer's tale, for we find Patroclus with the Trojans and Hektor with the Greeks – enough to make Achilles writhe on his funeral pyre!

[49] This was the asteroid 5261 Eureka, in the L_5 region. Five others are currently known, of which one (1999UJ7) moves in the L_4 region.

[50] This libration can reach 30° for Helene. The Saturn system with its satellites and rings is extremely rich and complex. One might mention the coorbital satellites Janus and Epimetheus on a so-called horseshoe orbit, or the shepherd satellites, which constrain the F ring and prevent it from breaking up.

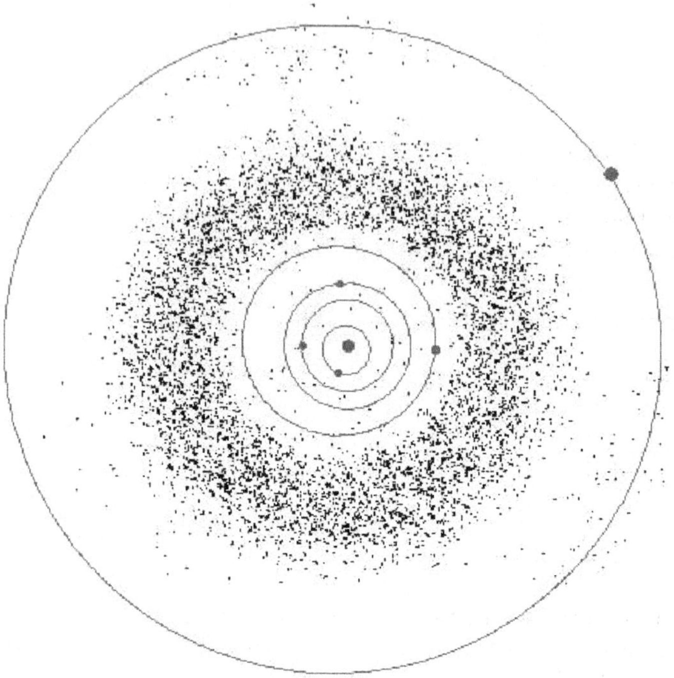

Figure 3.5. Representation of the 7 722 currently known asteroids (*dots*) and planets (*disks*) by their projection onto the plane of the ecliptic on 1 January 2000. Moving out from the Sun at the centre: orbits of Mercury, Venus, Earth, Mars and Jupiter. The Main Belt lies between the orbits of Mars and Jupiter, but it is easy to make out the accumulation of Trojan asteroids on the orbit of Jupiter, close to the points L_4 and L_5 of the Sun–Jupiter system. Credit: SMCS, University of St Andrews

3.14.5 Artificial Satellites at Lagrange Points

Table 3.9 shows the positions of Lagrange points for the Sun–Earth and Earth–Moon systems. We note that, in the latter system, the exact calculation of x gives 0.15 and 0.17 for BL_i, whereas the approximate calculation gives 0.16 for the same two points.

Some artificial satellites have been placed at the point L_1 of the Sun–Earth system. When the satellite arrives in the vicinity of the point L_1, about 1.5 million kilometres from Earth, it is placed in a so-called halo orbit, also denoted by L1LO, the L_1 Lissajous orbit, since the trajectory looks like a Lissajous curve.

The first satellite to be placed in a halo orbit about the L_1 point was ISEE-3, between 1978 and 1982. It was followed by Wind, SOHO, ACE, and Genesis. All these probes had astronomical missions. An Earth-observation satellite DSCOVR is also planned for this orbit.

Table 3.10. Examples of celestial bodies at Lagrange points L_4 and L_5: the so-called Trojan asteroids for a Sun–planet system and natural satellites for a Saturn–natural satellite system

System A–B	Point L_4	Point L_5
Sun–Jupiter	Greeks	Trojans
	588 Achilles	617 Patroclus
	624 Hektor	884 Priam
	659 Nestor	1172 Aenea
	911 Agamemnon	1173 Anchises
	⋮	⋮
Sun–Mars		5261 Eureka
Saturn–Dione	Helene	
Saturn–Tethys	Telesto	Calypso

The L1LO orbit lies roughly in a plane tilted with respect to the ecliptic and has elliptical shape. It measures several hundred thousand kilometres across and the period of motion of the satellite around the Lagrange point is very long: 211 days for Wind, 180 days for SOHO, 179 days for Genesis.[51] Since the Earth–satellite axis does not lie in the plane of the ecliptic, data transmission is not too seriously perturbed by electromagnetic or particle emissions from the Sun.

The point L_2 of the Sun–Earth system was visited for the first time by the probe MAP. Many other missions are destined for a halo orbit around this point (the L2LO orbit). Following MAP, these include the satellites Planck, FIRST renamed Herschel (Planck and Herschel should travel together to the L_2 point), GAIA and JWST, the successors of Hipparcos and Hubble, respectively, and longer-term projects such as Eddington and Darwin.

The stable points should be occupied by solar observation satellites: a Japanese project at L_5 called L5-Mission and a US stereographic observation project at L_4 and L_5 (STEREO-Ahead and STEREO-Behind). Looking at Fig. 3.4 and recalling that the Earth has a prograde orbit around the Sun, it is not difficult to understand the terms 'ahead' and 'behind'.

Concerning the point L_3 of the Sun–Earth system, it is not obvious what kind of project might be located there, except possibly a film of the type Planet X.

[51] As an example, the orbital characteristics of Genesis are as follows. It has an elliptical orbit in a plane making an angle of 46° with the plane of the ecliptic. The latter is denoted by $(L_1; x, y)$, whilst the axis L_1x lies along the Sun–Earth direction. The projection of the semi-major axis along is L_1x is 0.71×10^6 km and the projection of the semi-minor axis along L_1y is 0.56×10^6 km. These quantities should be compared with the distance L_1–Earth, which is 1.49×10^6 km.

3.15 Appendix: Summary of Legendre Functions

Legendre Polynomials

The generating function for the Legendre[52] polynomials is

$$\frac{1}{\sqrt{1-2tx+t^2}} = \sum_{n=0}^{\infty} P_n(x) t^n \, .$$

These polynomials are defined for any $n \geqslant 0$ by

$$P_n(x) = \frac{1}{2^n n!} \frac{\mathrm{d}^n \left[\left(x^2 - 1 \right)^n \right]}{\mathrm{d}x^n} \, .$$

The first few polynomials are:

$$P_0(x) = 1 \, , \quad P_1(x) = x \, ,$$
$$P_2(x) = \frac{3x^2 - 1}{2} \, , \quad P_3(x) = \frac{5x^3 - 3x}{2} \, ,$$
$$P_4(x) = \frac{35x^4 - 30x^2 + 3}{8} \, , \quad P_5(x) = \frac{63x^5 - 70x^3 + 15x}{8} \, .$$

Associated Legendre Functions

The associated Legendre functions are defined in terms of the Legendre polynomials on the interval $[-1, +1]$, for any $l \geqslant 0$ and $0 \leqslant m \leqslant l$, by

$$P_{lm}(x) = \left(1 - x^2\right)^{m/2} \frac{\mathrm{d}^m P_l(x)}{\mathrm{d}x^m} \, .$$

This gives

$$P_{l0}(x) = P_l(x) \, , \qquad P_{ll}(x) = \frac{(2l)!}{2^l l!} \left(1 - x^2\right)^{l/2} \, ,$$

and the first few associated Legendre functions are

$$P_{00}(x) = 1 \, , \quad P_{10}(x) = x \, , \quad P_{11}(x) = \sqrt{1 - x^2} \, ,$$
$$P_{20}(x) = \tfrac{1}{2}(3x^2 - 1) \, , \quad P_{21}(x) = 3x\sqrt{1 - x^2} \, , \quad P_{22}(x) = 3(1 - x^2) \, .$$

[52] Adrien Marie Legendre (1752–1833) was a French mathematician. He introduced the polynomials which are now named after him in his *Recherches sur la figure des planètes* (1784). When put in charge of geodetic measurements (the distance between the Paris and Greenwich meridians) by the revolutionary government known as the *Convention*, he made significant contributions to spherical trigonometry. He obtained new results concerning the elliptic functions, the beta and gamma functions, and the Euler integrals. His work *Eléments de Géométrie* was reprinted thirteen times between 1794 and 1827.

3.16 Appendix: Spherical Trigonometry

3.16.1 Gauss' Relations

A spherical triangle is a triangle on a sphere of unit radius, whose sides are arcs of great circles (or angles at the centre). The angles of the triangle are defined at each vertex in the tangent plane to the sphere (i.e., they are dihedral angles). The angles are usually denoted by A, B, C and the opposite sides by a, b, c, as shown in Fig. 3.6. It can be shown that a spherical triangle is determined by specifying three elements. A fourth element can be calculated from the three known elements. We then have $(6 \times 5)/2 = 15$ relations.

These trigonometric relations are easily obtained by considering the following change of frame. Consider three points A, B, C on a sphere, forming a (non-flat) spherical triangle. We consider two orthonormal frames $\Re_1$ and $\Re'_1$ such that

$$\Re_1(O;\boldsymbol{i},\boldsymbol{j},\boldsymbol{k})\ , \quad \Re'_1(O;\boldsymbol{i}',\boldsymbol{j}',\boldsymbol{k}')\ , \quad \boldsymbol{i}=\boldsymbol{OA}\ , \quad \boldsymbol{i}'=\boldsymbol{OB}\ ,$$

and such that $\boldsymbol{k}$ and $\boldsymbol{k}'$ coincide. We then have $(\boldsymbol{i},\boldsymbol{i}')=c$.

The frame $\Re'_1$ is thus obtained from $\Re_1$ by rotating through an angle c about $\boldsymbol{k}$. In each of the two frames $\Re_1$ and $\Re'_1$, $\boldsymbol{OC}$ can be written in Cartesian coordinates

$$\boldsymbol{OC}=\begin{pmatrix}\cos b\\ \sin b\,\cos A\\ \sin b\,\sin A\end{pmatrix}\ , \qquad \boldsymbol{OC}=\begin{pmatrix}\cos a\\ -\sin a\,\cos B\\ \sin a\,\sin B\end{pmatrix}\ . \tag{3.109}$$

Using the matrix for the rotation through angle c about $\boldsymbol{k}$, we thus obtain

$$\begin{pmatrix}\cos a\\ -\sin a\,\cos B\\ \sin a\,\sin B\end{pmatrix}=\begin{pmatrix}\cos c & \sin c & 0\\ -\sin c & \cos c & 0\\ 0 & 0 & 1\end{pmatrix}\times\begin{pmatrix}\cos b\\ \sin b\,\cos A\\ \sin b\,\sin A\end{pmatrix}\ . \tag{3.110}$$

We thus obtain the three relations known as Gauss' relations:

$$\cos a = \cos b\,\cos c + \sin b\,\sin c\,\cos A\ , \tag{3.111}$$

$$\sin a\,\cos B = \cos b\,\sin c - \sin b\,\cos c\,\cos A\ , \tag{3.112}$$

$$\sin a\,\sin B = \sin b\,\sin A\ . \tag{3.113}$$

3.16.2 Fifteen Relations for the Spherical Triangle

These three equations lead us to the 15 required relations, which are generally grouped as follows. The relations are numbered with Roman numerals from I to XV and the annotation ST, which stands for 'spherical trigonometry'.

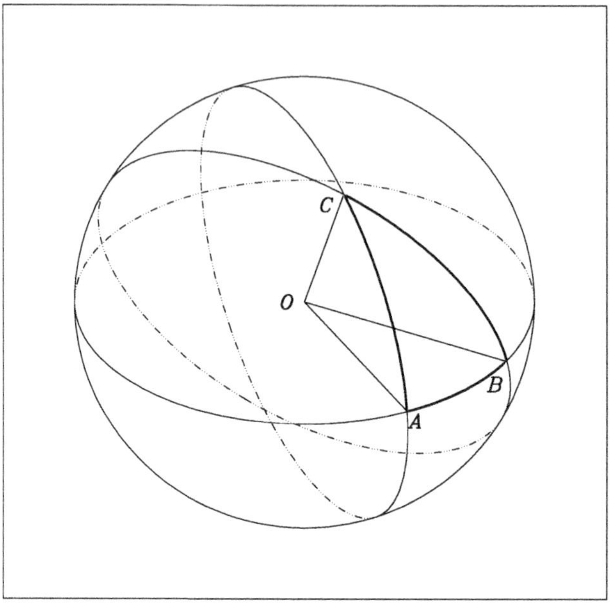

Figure 3.6. Spherical triangle ABC on a sphere with centre O and unit radius. The angles A, B, C are dihedral angles. The corresponding sides a, b, c are arcs of great circles defined by $a = BC$, $b = CA$, and $c = AB$

- Fundamental relations. These are deduced from (3.111) by cyclic permutations. They relate three sides and an angle:

$$\cos a = \cos b \, \cos c + \sin b \, \sin c \, \cos A \,, \tag{ST I}$$
$$\cos b = \cos c \, \cos a + \sin c \, \sin a \, \cos B \,, \tag{ST II}$$
$$\cos c = \cos a \, \cos b + \sin a \, \sin b \, \cos C \,. \tag{ST III}$$

As a corollary, we obtain formulas relating three angles and one side:

$$\cos A = -\cos B \, \cos C + \sin B \, \sin C \, \cos a \,, \tag{ST IV}$$
$$\cos B = -\cos C \, \cos A + \sin C \, \sin A \, \cos b \,, \tag{ST V}$$
$$\cos C = -\cos A \, \cos B + \sin A \, \sin B \, \cos c \,. \tag{ST VI}$$

- Sine formulas. These are deduced from (3.113). These three formulas relate two angles and the opposite sides:

(ST VII)

$$\frac{\sin a}{\sin A} = \frac{\sin b}{\sin B} = \frac{\sin c}{\sin C} \,. \tag{ST VIII}$$

(ST IX)

- Cotangent formulas. These are deduced from (3.112), which involves five elements, and the two other Gauss relations. These formulas relate four consecutive elements of the triangle:

$$\cot a \, \sin b = \cos b \, \cos C + \sin C \, \cot A \,, \qquad \text{(ST X)}$$

$$\cot b \, \sin a = \cos a \, \cos C + \sin C \, \cot B \,. \qquad \text{(ST XI)}$$

The second of these follows from the first by fixing the angle C and permuting a and b, A and B. By cyclic permutation, we now obtain

$$\cot b \, \sin c = \cos c \, \cos A + \sin A \, \cot B \,, \qquad \text{(ST XII)}$$

$$\cot c \, \sin b = \cos b \, \cos A + \sin A \, \cot C \,, \qquad \text{(ST XIII)}$$

$$\cot c \, \sin a = \cos a \, \cos B + \sin B \, \cot C \,, \qquad \text{(ST XIV)}$$

$$\cot a \, \sin c = \cos c \, \cos B + \sin B \, \cot A \,. \qquad \text{(ST XV)}$$

We thus obtain $6+3+6=15$ relations. In the case of a right-angled spherical triangle, where one of the three angles is a right angle, the above formulas reduce immediately to $(5 \times 4)/2 = 10$ relations.

Example 3.3. *Calculate the distance D between two points $M(\lambda, \phi)$ and $M'(\lambda', \phi')$ on the Earth, defined by their longitude and latitude.*

Consider the spherical triangle NMM', where point N represents the North Pole. If N corresponds to A, M to B and M' to C, the angle A and sides b and c can be written as follows in terms of the geographical data:

$$A = \lambda - \lambda' \,, \quad c = \frac{\pi}{2} - \phi \,, \quad b = \frac{\pi}{2} - \phi' \,.$$

The required distance is then a, the length of the arc of great circle MM'. The first Gauss relation (3.111) or (ST I) then gives

$$\cos a = \sin \phi \, \sin \phi' + \cos \phi \, \cos \phi' \, \cos(\lambda - \lambda') \,.$$

The required distance is $D = Ra$, if we consider the Earth as spherical with radius R and express a in radians.

Application: Calculate the distance from Paris to New York.

The geographical coordinates of Paris and New York are (48°50′N; 2°20′E) and (40°42′N; 74°00′W), respectively, which gives, for the points M and M',

$$\phi = +48.87 \,, \quad \lambda = +2.33 \,, \quad \phi' = +40.70 \,, \quad \lambda' = -74.00 \,.$$

The calculation yields

$$a = 0.91597 \text{ rad} = 52.48° \,.$$

We thus obtain the distance directly in nautical miles (1 nautical mile is equivalent to 1 arcsec on the Earth's surface), viz.,

$$D = 52.48 \times 60 = 3149 \text{ nautical miles} ,$$

or in kilometres, if we introduce the Earth's radius, viz.,

$$D = 0.91597R = 5842 \text{ km} .$$

The curve (arc of a great circle) joining two points on the surface of the Earth is said to be orthodromic.

Note. When two points are very close, it is better to use a formula involving the differences between the latitudes and the longitudes, and valid if $a \in [0, \pi)$:

$$\sin^2 \frac{a}{2} = \sin^2 \frac{\phi - \phi'}{2} + \cos\phi \cos\phi' \sin^2 \frac{\lambda - \lambda'}{2} .$$

4 Motion of Orbit, Earth and Sun

In the last chapter, we calculated the motion of the orbital plane of a satellite with respect to a Galilean frame, via the rate of nodal precession, and the motion of the orbit in this plane. At the beginning of this chapter, we shall review the way the Earth moves with respect to a Galilean frame. By composing the two motions, we will then be able to follow the motion of the satellite relative to the Earth, which was indeed our original aim.

We shall then study the apparent motion of the Sun relative to the Earth, so that we may subsequently study the cycles of the satellite in relation to the Sun. The aim here will ultimately be to specify the geometry of the satellite, its target, and the Sun: we consider a point on the Earth's surface and determine how this point is seen by the satellite and under what conditions of solar illumination.

In the last two sections, we examine two types of satellite orbit for which two of the quantities studied here play a key role. These quantities are the mean motion n and the nodal precession rate $\dot{\Omega}$. We shall find that they can take certain values of particular importance for the satellite. The first quantity n determines the geosynchronous orbits, and the second $\dot{\Omega}$ the Sun-synchronous orbits.

4.1 Motion of the Orbit

4.1.1 Secular Variations. Simplified Case

We reconsider here the equations giving the secular variations of the orbital elements for a circular orbit, in the context of a simplified geopotential (expansion up to degree 2). We then treat the general case.

It is reasonable to ask why we should be concerned with the special case of circular orbits when we could just set $e = 0$ in the general equations. We have chosen to do so because it will allow us (at last!) to present a selection of simple formulas, and the reader will agree that they are easier to use than the general ones. Moreover, this is a particularly important special case, not to mention the fact that the errors involved in setting the eccentricity to zero are very small, as long as e is itself small. Indeed, for precession rates, e arises via a term $(1 - e^2)^{-2}$, which is roughly $(1 + 2e^2)$, if $e < 0.1$. As an example, the error in $\dot{\Omega}$ will be of the order of 10^{-5} if $e = 2 \times 10^{-3}$.

Secular Variation of Orbital Elements

Keeping only the J_2 term in the relative perturbation of the geopotential, we showed in Chap. 3 that the metric elements remained constant, whilst the angle elements underwent secular variations. Using (3.54) through (3.56), we obtain the values of $\dot{\Omega}$, $\dot{\omega}$ and $\dot{M}$ as a function of the metric elements and the mean motion $n = \sqrt{\mu/a^3}$:

$$\frac{\dot{\Omega}}{n} = -\frac{3}{2} J_2 \left(\frac{R}{a}\right)^2 \cos i \,, \tag{4.1}$$

$$\frac{\dot{\omega}}{n} = \frac{3}{4} J_2 \left(\frac{R}{a}\right)^2 \left(5 \cos^2 i - 1\right) \,, \tag{4.2}$$

$$\frac{\dot{M} - n}{n} = \frac{\Delta n}{n} = \frac{3}{4} J_2 \left(\frac{R}{a}\right)^2 \left(3 \cos^2 i - 1\right) \,. \tag{4.3}$$

The secular variation of the orbital element Ω will play a key role when we come to study the trajectory of the satellite, as will the secular variations of the elements ω and M when we calculate the period of the motion.

Concerning the parameter ω, the secular variation of $\dot{\omega}$ is perfectly well defined by (3.55) when $e = 0$. However, the position of the perigee, determined by ω, is not defined for a perfectly circular orbit (with $e = 0$), and it is poorly defined in the case of a near-circular orbit. Concerning the parameter M, whose secular variation $\dot{M} - n_0$ is perfectly well defined by (3.56) for $e = 0$, we encounter the same problem in defining an origin, for both circular and near-circular orbits. In these cases, we generally choose the ascending node as origin (see Sect. 2.3.4 on adapted orbital elements).

Nodal Precession Rate

Using (4.1) and expressing the mean motion, the nodal precession rate can be written

$$\dot{\Omega} = -\frac{3}{2} J_2 \sqrt{\frac{\mu}{R^3}} \left(\frac{R}{a}\right)^{7/2} \cos i \,, \tag{4.4}$$

and this can in turn be set in the form

$$\dot{\Omega}(a, i) = -K_0 \left(\frac{R}{a}\right)^{7/2} \cos i \,, \tag{4.5}$$

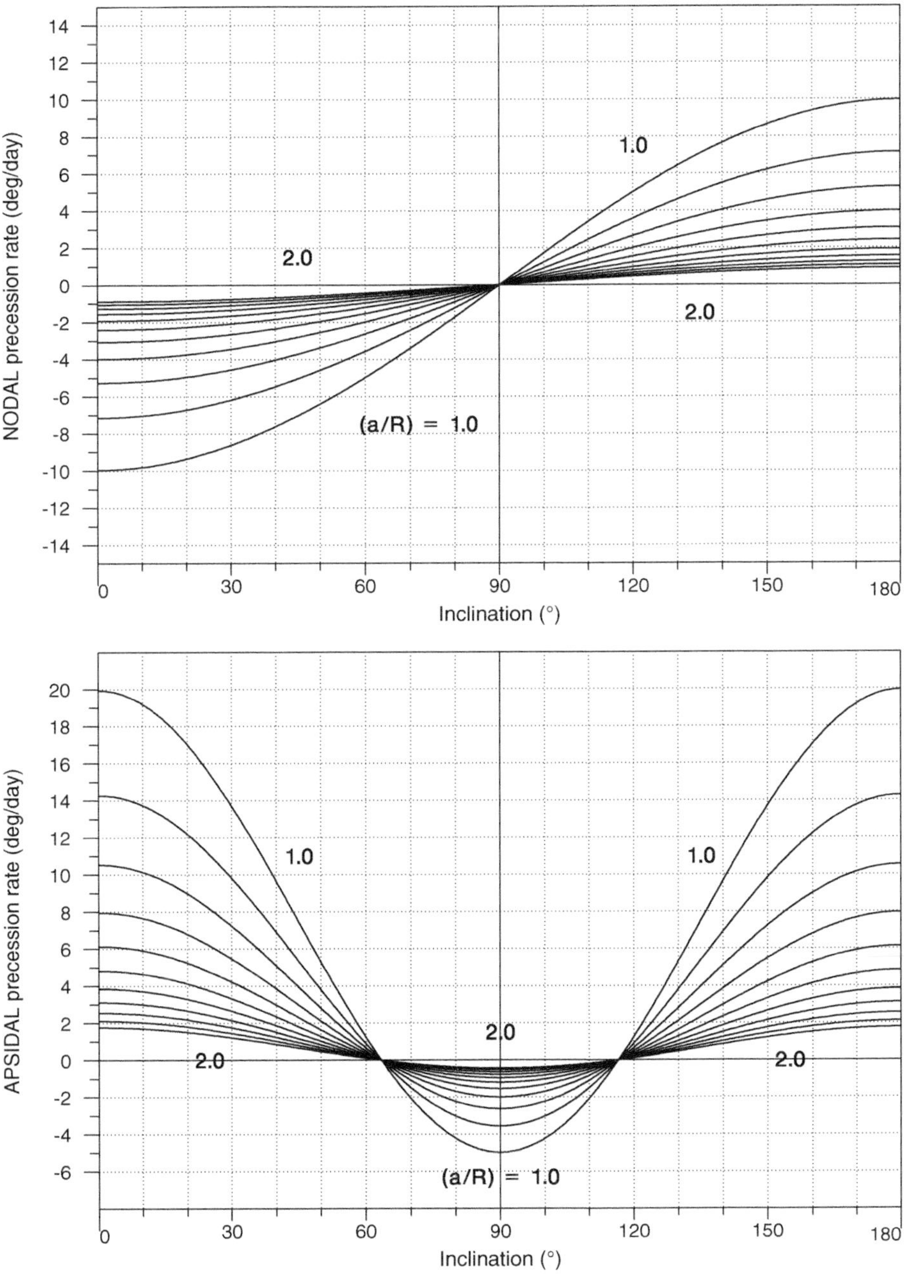

Figure 4.1. Precession rate (in degree/day) as a function of inclination i for various values of the ratio $\eta = a/R$ from $\eta = 1.0$ to $\eta = 2.0$, in steps of 0.1. *Upper*: nodal precession $\dot{\Omega}$. *Lower*: apsidal precession $\dot{\omega}$

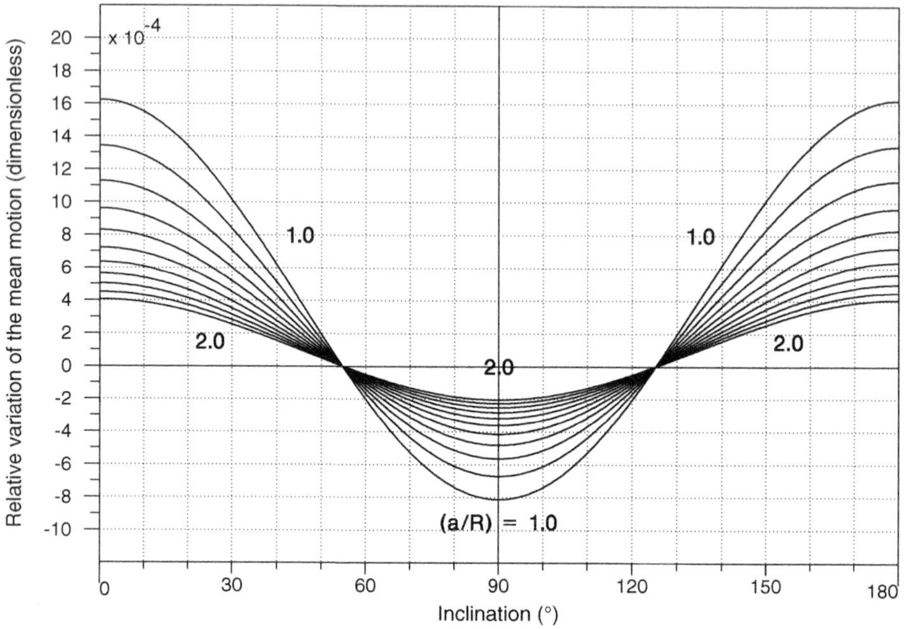

Figure 4.2. Relative variation $\Delta n/n_0$ of the mean motion (dimensionless) as a function of inclination i for various values of the ratio $\eta = a/R$ from $\eta = 1.0$ to $\eta = 2.0$, in steps of 0.1. For example, for $i = 0$ and $a/R = 1.8$, we may read off $\Delta n/n_0 = 5.0 \times 10^{-4}$

or again, using the reduced distance defined by (2.18),

$$\dot{\Omega} = -K_0 \eta^{-7/2} \cos i \ ,$$

where

$$K_0 = \frac{3}{2} J_2 \sqrt{\frac{\mu}{R^3}} \ . \tag{4.6}$$

We can also write K_0 in the following form, using (2.16) and (2.17):

$$K_0 = \frac{3\pi}{T_{0(h=0)}} J_2 \ .$$

Concerning the units of the angular velocity, apart from radians per second (SI units), the units degrees per day and revolutions per year are also found in the literature. With these three units, K_0 is expressed as follows:

$$K_0 = 2.012\,788 \times 10^{-6} \ \mathrm{rad\,s^{-1}} \ , \tag{4.7}$$

$$K_0 = 9.964\,014^\circ \mathrm{day}^{-1} \ , \tag{4.8}$$

$$K_0 = 10.109\,49 \ \mathrm{rev\,yr^{-1}} \ . \tag{4.9}$$

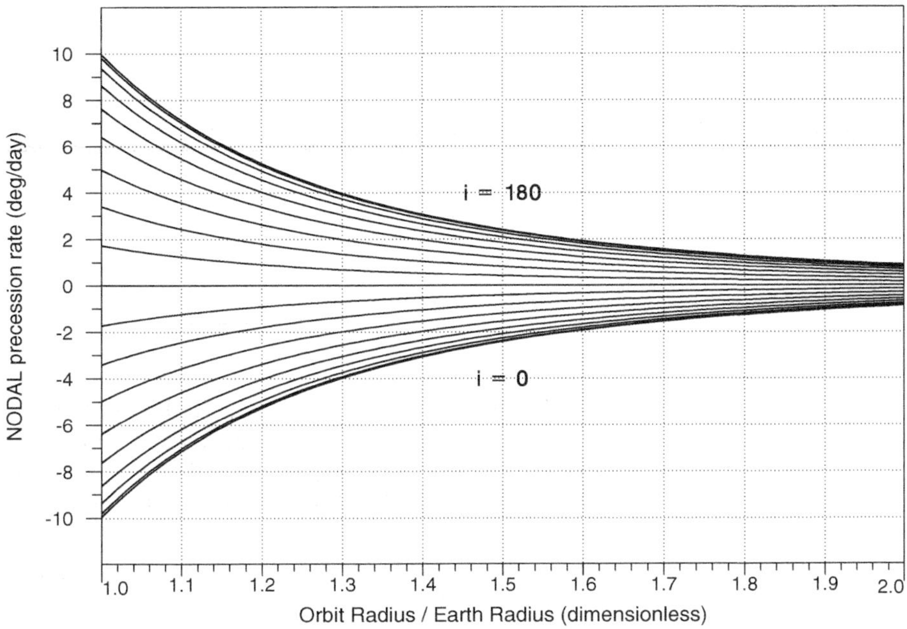

Figure 4.3. Nodal precession rate $\dot{\Omega}$ (in degree/day) as a function of the ratio $\eta = a/R$ for various values of the inclination between $i = 0°$ and $i = 180°$, in steps of 10°. For $i = 90°$, $\dot{\Omega} = 0$. Altitude $h = a - R$

Figure 4.1 (upper) graphs the variation of $\dot{\Omega}(a, i)$ as a function of the inclination i for various values of the ratio a/R. The altitude thus varies from $h = 0$ for $a/R = 1$ to $h = R = 6378$ km for $a/R = 2$ in steps of $0.1R = 637.8$ km. The value of $\dot{\Omega}$ is given in degree/day.

Figure 4.3 graphs $\dot{\Omega}(a, i)$ as a function of the semi-major axis a for various inclinations. These graphs are combined in the upper part of Fig. 4.4. We have plotted curves of constant nodal precession $\dot{\Omega}$ (nodal isoprecession curves) as a function of the inclination i and the altitude h (or a/R).

From these graphs, it is quite clear that, when h increases for a given inclination, $\dot{\Omega}$ decreases. The further the satellite moves from the centre of the Earth, the less it is affected by irregularities in the geopotential. We also see that, in the case of prograde orbits, $\dot{\Omega}$ is negative, i.e., precession is retrograde, whereas in the case of retrograde orbits, $\dot{\Omega}$ is positive. For a strictly polar orbit, $\dot{\Omega}$ is always zero, at all altitudes.

The maximum value of $|\dot{\Omega}|$ is obtained for $i = 0°$ or $i = 180°$, with $h = 0$, and it is equal to $K_0 = 9.96°$/day, or almost 10° per day. The value of $\dot{\Omega}$, close to 1, and denoted by HEL on the two graphs of Fig. 4.4, is relevant for the so-called Sun-synchronous satellites, which we shall discuss at the end of this chapter. (In this case, $\dot{\Omega} = 0.986°$/day.)

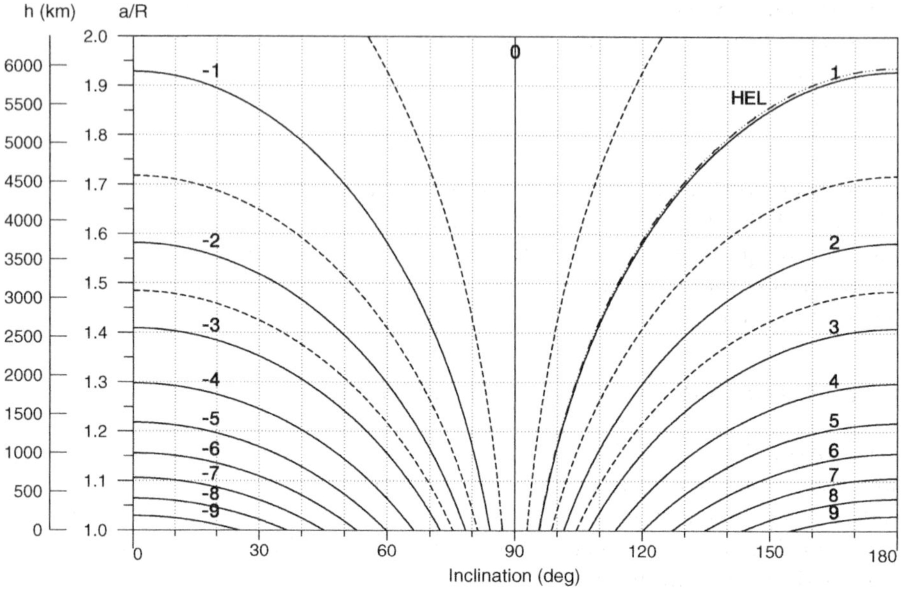

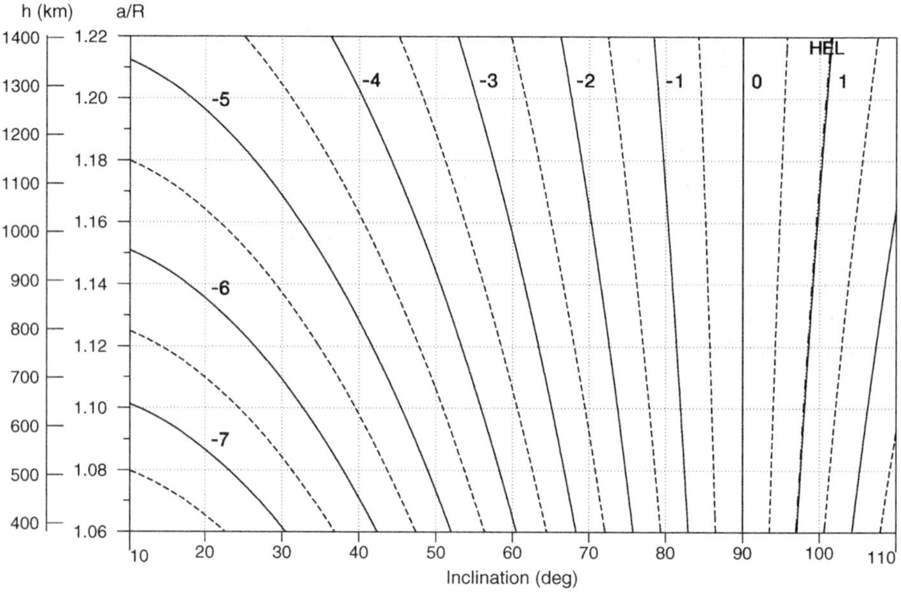

Figure 4.4. Nodal precession rate $\dot{\Omega}$ (in degree/day) as a function of the characteristics of the circular orbit. *Upper*: altitude h between 0 and R, all inclinations. *Lower*: usual range of operation of LEO satellites

Figure 4.4 (lower) enlarges the part of the graph relevant to satellites placed in low orbit, as we shall see shortly. For example, the satellite Meteor-3-07 has $h \approx 1200$ km and $i \approx 83°$, for which we find the value $\dot{\Omega} = -0.71°/\text{day}$ on the graph. We shall carry out this calculation in Example 4.2.

Apsidal Precession Rate

The apsidal precession rate is given in terms of the constant K_0, defined in (4.6):

$$\dot{\omega}(a, i) = \frac{1}{2} K_0 \left(\frac{R}{a}\right)^{7/2} \left(5\cos^2 i - 1\right) , \tag{4.10}$$

or alternatively,

$$\dot{\omega} = \frac{1}{2} K_0 \eta^{-7/2} \left(5\cos^2 i - 1\right) .$$

This is zero at the critical inclination defined in Chap. 3.

Figure 4.1 (lower) is a graph of $\dot{\omega}(a, i)$ as a function of the inclination i for various values of the ratio a/R. The value of $\dot{\omega}$ is given with the same units and the same scale as $\dot{\Omega}$. The two values of the critical inclination appear quite clearly on the graph at 63.4° and 116.6°. Example 4.3 illustrates this question further.

Variation of the Mean Motion

The relative variation of the mean motion is defined by (4.3) and this gives, putting in the numerical factor explicitly,

$$\frac{\Delta n}{n} = 8.119\,701 \times 10^{-4} \eta^{-2} (3\cos^2 i - 1) . \tag{4.11}$$

Figure 4.2 graphs the relative variation $\Delta n / n_0$ of the mean motion as a function of the inclination i for various values of the ratio $\eta = a/R$. We observe that for i between 57.7° and 125.3°, the true motion is slower than the motion relative to a spherical Earth.

4.1.2 Secular Variations up to J_4

If we consider an elliptical orbit and the expansion of the geopotential to a high degree, the expressions for the variation of the orbital elements become extremely complex. We shall not be concerned here with the periodic variations, affecting all the orbital elements. The secular variations concern only the angle elements and it can be shown that only the even zonal harmonics J_{2n} are involved.

Table 4.1. Secular variations of the angle elements for expansion of the geopotential up to order 4. The parameter p and the mean motion n are given by $p = a(1-e^2)$, $n = \sqrt{\mu/a^3}$. Abbreviations: $s = \sin i$, $e' = \sqrt{1-e^2}$. Referring to Table 3.4, we have $s = 1/\sigma$, $e' = 1/\overline{\tau}$

$$\begin{aligned}\frac{\dot{\Omega}}{n} &= J_2\left(\frac{R}{p}\right)^2 \cos i\left(-\frac{3}{2}\right)\\ &+J_2^2\left(\frac{R}{p}\right)^4 \cos i\left[\left(-\frac{45}{8}+\frac{3}{4}e^2+\frac{9}{32}e^4\right)+\left(\frac{57}{8}-\frac{69}{32}e^2-\frac{27}{64}e^4\right)s^2\right]\\ &+J_4\left(\frac{R}{p}\right)^4 \cos i\left(\frac{15}{4}-\frac{105}{16}s^2\right)\left(1+\frac{3}{2}e^2\right)\,. \end{aligned} \tag{4.12}$$

$$\begin{aligned}\frac{\dot{\omega}}{n} &= J_2\left(\frac{R}{p}\right)^2\left[3-\frac{15}{4}s^2\right]\\ &+J_2^2\left(\frac{R}{p}\right)^4\left[\left(\frac{27}{2}-\frac{15}{16}e^2-\frac{9}{16}e^4\right)\right.\\ &\qquad+\left(-\frac{507}{16}+\frac{171}{31}e^2+\frac{99}{64}e^4\right)s^2\\ &\qquad\left.+\left(\frac{1185}{64}-\frac{675}{128}e^2-\frac{135}{128}e^4\right)s^4\right]\\ &+J_4\left(\frac{R}{p}\right)^4\left[\left(-\frac{3}{8}+\frac{15}{8}s^2-\frac{105}{64}s^4\right)\left(10+\frac{15}{2}e^2\right)\right.\\ &\qquad\left.+\left(-\frac{15}{4}+\frac{165}{16}s^2-\frac{105}{16}s^4\right)\left(1+\frac{3}{2}e^2\right)\right]\,. \end{aligned} \tag{4.13}$$

$$\begin{aligned}\frac{\Delta n}{n} = \frac{\dot{M}-n}{n} &= J_2\left(\frac{R}{p}\right)^2 e'\frac{3}{4}\left(2-3s^2\right)\\ &\times\left\{1+J_2\left(\frac{R}{p}\right)^2\frac{1}{8}\left[10+5e^2+8e'-\left(\frac{65}{6}-\frac{25}{12}e^2+12e'\right)s^2\right]\right\}\\ &-J_2^2\left(\frac{R}{p}\right)^4 e'\frac{5}{64}\left(2-e^2\right)s^2\\ &-J_4\left(\frac{R}{p}\right)^4 e'\frac{45}{128}e^2\left(8-40s^2+35s^4\right)\,. \end{aligned} \tag{4.14}$$

Table 4.1 gives expressions for these secular variations in terms of a, e, i, up to degree 4 in R/p. As well as the terms of degree 2 for J_2, there are two terms of degree 4, one for J_4 and one for J_2^2. These quantities are expressed in terms of their quotient by n, the mean motion. One thus obtains in each case a ratio of angular speeds, which is a dimensionless quantity.

The expressions for $\dot{\Omega}$ and $\dot{\omega}$ in (4.12) and (4.13), respectively, were obtained from J. Kovalevsky's analytical theory of satellite motion in [27], whilst the expression for Δn in (4.14) was derived using P.E. Koskela's theory in [12].

4.1.3 Applications: Period and Altitude

We shall now give two examples of calculations involving the relation between the period of a satellite and its altitude. In the first, we calculate the period of a satellite of known altitude, and in the second, the converse (and more difficult) problem, that is, we determine the altitude of a satellite of known period. Other examples of this type will be discussed in Chap. 7 (where the period will be defined by the recurrence condition).

When we speak of the altitude of a satellite in this context, we are in fact referring to the difference between the semi-major axis a of the orbit and the equatorial radius R of the Earth, i.e., $h = a - R$. The quantity h is usually used to describe the satellite, but it is the quantity a that is used in orbital calculations.

The third example in this section is concerned with the apsidal precession rate.

Example 4.1. *Calculate the period of the satellite TRMM in near-circular orbit at an altitude of 350 km, and with inclination 35°. Calculate also the nodal precession rate.*

For this satellite, on 1 June 2001, we have $a = 6\,728.216\,8$ km (or $h = 350.0$ km), $i = 34.9817°$, $e = 9.96 \times 10^{-5}$.
We begin by calculating the Keplerian mean motion for $a = 6\,728\,216.8$ m:

$$n_0 = \left[\frac{3.986\,004\,42 \times 10^{14}}{(6.728\,216\,8 \times 10^6)^3}\right]^{1/2} = 1.143\,981 \times 10^{-3} \text{ rad s}^{-1} ,$$

which gives the Keplerian period

$$T_0 = \frac{2\pi}{n_0} = 5492.38 \text{ s} = 91.540 \text{ min} .$$

The fractional variation $\Delta n/n_0$ is then found from (4.11) to be

$$\frac{\Delta n}{n_0} = 8.1197 \times 10^{-4} \left(\frac{6.378\,137 \times 10^6}{6.728\,217 \times 10^6}\right)^2 \left[3(0.819\,34)^2 - 1\right] = 7.398 \times 10^{-4} .$$

Using the more complex relation (4.14), we obtain 7.407×10^{-4}, which gives the true mean motion as

$$n = n_0 + \Delta n = 1.144\,828 \times 10^{-3} \ \mathrm{rad\,s^{-1}} \ .$$

The true motion is thus faster than the Keplerian motion ($\Delta n > 0$), since $i < 57.7°$. The anomalistic period T_a is obtained from (3.74) as

$$T_a = \frac{91.540}{1 + 7.407 \times 10^{-4}} = 91.472 \ \mathrm{min} \ .$$

We now consider the apsidal precession $\dot{\omega}$. Using (4.2) or (4.10), we have $\dot{\omega}/n = 1.720 \times 10^{-4}$, and using the more complex relation (4.13), $\dot{\omega}/n = 1.725 \times 10^{-4}$. The nodal period T_d is found from (3.79) or (3.80) to be

$$T_d = 91.314 \ \mathrm{min} \ .$$

Note that the apsidal precession $\dot{\omega}$ is positive, since $i < 63.4°$ and that it represents a significant level at $\dot{\omega} = 9.77°$ per day.
Finally,

$$T_d < T_a < T_0 \ ,$$

with

$$T_d - T_a = -9.47 \ \mathrm{s} \ , \quad T_d - T_0 = -13.54 \ \mathrm{s} \ .$$

It is often useful to know the nodal precession $\dot{\Omega}$. It is negative since $i < 90°$. From (4.1) or (4.5), we have $\dot{\Omega}/n = -1.196 \times 10^{-4}$ and from the more complex relation (4.12), $\dot{\Omega}/n = -1.200 \times 10^{-4}$. This gives, in the various different units:

$$\dot{\Omega} = -1.372 \times 10^{-6} \ \mathrm{rad\,s^{-1}} = -6.80° \mathrm{day}^{-1} = -6.89 \ \mathrm{rev\,yr^{-1}} \ .$$

The nodal precession is irrelevant when calculating the period, but concerns the motion of the orbital plane.
The satellite TRMM (Tropical Rainfall Measurement Mission) is a Japanese project in collaboration with NASA. It has a low inclination so as to cover the intertropical region and flies at low altitude to improve the efficiency of its radar instrumentation. From its launch in September 1997 until August 2001, it flew at an altitude of 350 km. Subsequently, in order to extend its lifetime, it was raised to an altitude of 402 km.

Example 4.2. *Calculate the altitude of the satellite Meteor-3-07, in near-circular orbit with draconitic period 109.421 min and inclination 82.56°.*

It is more difficult to calculate the altitude from the period than vice versa, as in Example 4.1. We proceed by iteration. Secular variations are calculated using relations up to degree 4.

- We begin by calculating a_0, the value of a corresponding to the Keplerian motion with the given period. The value of T_d is given by the satellite orbit bulletin as $T_d = 109.421\,425$ min. In a first stage, we thus set $T_0 = T_d = 6\,565.285\,6$ s. The calculation gives

$$a_0^3 = \frac{\mu}{4\pi^2} T_0^2 = \frac{3.986\,004\,36 \times 10^{14}}{4\pi^2} (6.565\,285\,6)^2 \times 10^6 \ ,$$

whence

$$a_0 = 7\,578.129 \text{ km} \ .$$

With this value for a_0, the inclination i and $n_0 = 2\pi/T_0$, we can calculate Δn and $\dot{\omega}$. We find

$$\frac{\Delta n}{n_0} = -5.469 \times 10^{-4} \ , \qquad \frac{\dot{\omega}}{n_0} = -5.267 \times 10^{-4} \ .$$

The true motion here is slower than the Keplerian motion ($\Delta n < 0$), so that $T_a > T_0$, and the perigee revolves in the retrograde direction ($\dot{\omega} < 0$), so that $T_a < T_d$. We thus have

$$T_d > T_a > T_0 \ .$$

- Considering the approximate formula (3.80), we see that this value a_0 for the orbital radius corresponds to the Keplerian period T_0 and a draconitic period T_d given by

$$T_d = \left(1 + \frac{\Delta T}{T_0}\right) T_0 \ ,$$

with

$$\Delta T = T_d - T_0 \ , \qquad \frac{\Delta T}{T_0} \approx -\frac{\dot{\omega} + \Delta n}{n_0} \ .$$

Since the fractional differences are much smaller than unity, this yields

$$T_0 \approx \left(1 - \frac{\Delta T}{T_d}\right) T_d \ .$$

As the value of T_d is known, we obtain T_0 from $T_0 = T_d - \Delta T$. Now for this value of T_0, there corresponds an orbit of radius a_1, obtained from a_0 by $a_1 = a_0 + \Delta a$. The differential relation between a and T is

$$\frac{dT}{T} = \frac{3}{2}\frac{da}{a} \ . \tag{4.15}$$

For these finite increments, da corresponds to $a_1 - a_0 = \Delta a$ and dT corresponds to $T_0 - T_d = -\Delta T$. We obtain

$$\Delta a = \frac{2}{3}\frac{\dot{\omega} + \Delta n}{n_0} a_0 = -\frac{2}{3} 1.073\,5 \times 10^{-3} \times 7\,578.129 = -5.423 \text{ km} \ .$$

We thus find

$$a_1 = 7\,572.706 \text{ km} \ .$$

- The iteration continues in this way, using (3.79), and the results converge very rapidly to give

$$a = 7\,572.704 \text{ km} \ ,$$

which corresponds to an altitude $h = 1194.6$ km.

For the values of the periods, we obtain

$$T_{\mathrm{d}} = 109.421 \text{ min}\,, \quad T_{\mathrm{a}} = 109.364 \text{ min}\,, \quad T_{0} = 109.304 \text{ min}\,.$$

This shows how important it is to distinguish the various periods. The differences found here provide clear evidence for this: $T_{\mathrm{d}} - T_{\mathrm{a}} = 3.46$ s and $T_{\mathrm{d}} - T_{0} = 7.04$ s. Regarding the nodal precession, we have in the various different units:

$$\dot{\Omega} = -1.429 \times 10^{-7} \text{ rad s}^{-1} = -0.71^{\circ}\text{day}^{-1} = -0.72 \text{ rev yr}^{-1}\,.$$

This is small because the orbit is near-polar.

Example 4.3. *Calculate the apsidal precession rate for the following satellites at different inclinations: Megha-Tropiques, TOPEX/Poseidon, ADEOS-1 and Okean-3. Calculate the critical inclination for Molniya.*

We use (4.13) to obtain the apsidal precession rate. For Megha-Tropiques (h = 866 km, $i = 20.0^{\circ}$), we have in the various different units:

$$\dot{\omega} = +22.133 \times 10^{-7} \text{ rad s}^{-1} = +10.96^{\circ}\text{day}^{-1} = +11.12 \text{ rev yr}^{-1}\,.$$

For TOPEX/Poseidon (h = 1336 km, $i = 66.0^{\circ}$):

$$\dot{\omega} = -0.912 \times 10^{-7} \text{ rad s}^{-1} = -0.452^{\circ}\text{day}^{-1} = -0.458 \text{ rev yr}^{-1}\,.$$

For ADEOS-1 (h = 797 km, $i = 98.6^{\circ}$):

$$\dot{\omega} = -5.905 \times 10^{-7} \text{ rad s}^{-1} = -2.92^{\circ}\text{day}^{-1} = -2.97 \text{ rev yr}^{-1}\,.$$

For Okean-3 (h = 636 km, $i = 82.5^{\circ}$):

$$\dot{\omega} = -6.590 \times 10^{-7} \text{ rad s}^{-1} = -3.26^{\circ}\text{day}^{-1} = -3.31 \text{ rev yr}^{-1}\,.$$

For the first of these, Megha-Tropiques, the apsidal precession rate is positive, since the inclination i is less than the critical inclination $i_{\mathrm{C}} = 63.43^{\circ}$. It is a very high rate, since the perigee makes a complete round trip every 33 days, in the prograde direction.
For TOPEX/Poseidon, on the other hand, (4.10) shows that the perigee moves around very slowly because the inclination of its orbit is very close to the critical inclination. The perigee takes more than two years to complete one round trip. However, for inclinations close to i_{C}, the J_2 term of the geopotential is no longer the crucial factor. Other terms in the potential must then be taken into account in order to determine the (small) value of $\dot{\omega}$.
For ADEOS-1, the perigee takes 123 days to complete a round trip in the retrograde direction. In fact, for the two satellites ADEOS-1 and TOPEX/Poseidon, (4.10) and (4.13) are not sufficient to determine $\dot{\omega}$. Periodic variations must also be taken into account, in addition to secular variations. The orbital parameters are chosen in such a way that the perigee is practically stationary. The orbit is then said to be frozen, a situation described further at the end of Chap. 7.
For Okean-3, the values of $\dot{\omega}$ calculated here agree perfectly with the true values, such as those given in Example 7.16 (see Fig. 7.21).

For Molniya, with $a = 26\,552.9$ km, $e = 0.75$, and when the expansion of $\dot{\omega}$ is cut off at J_2, (3.57) gives:

$$i_C = 63.4349° = i_C(J_2) .$$

Using (4.13) and an expansion of $\dot{\omega}$ up to J_4, the elements a and e are involved. This yields

$$i_C = 63.4209° = i_C(J_4) .$$

This value is only slightly altered if the expansion is continued beyond J_4. Indeed,

$$i_C(J_2) - i_C(J_4) = 0.014° .$$

4.1.4 Strictly Polar Satellites

Certain satellites have an inclination of 90° (to within a few tenths of a degree). These are said to be strictly polar or on a strictly polar orbit. Equation (4.12) shows that the nodal precession rate is then zero, since the term $\cos i$ appears as a factor in all the terms J_n. The orbit of the satellite then remains fixed in the frame $\mathfrak{R}$, making a constant angle with a fixed direction in space, e.g., the direction of the vernal equinox Υ. This orbit is sometimes said to be inertial.[1]

Strictly polar orbits are generally used by satellites studying remote regions of the Earth environment (e.g., the ionosphere, the exosphere, the solar wind) or indeed aspects of astrophysics, as well as military satellites (e.g., the first military navigation systems), geodetic satellites, and the satellite carrying the Gravity Probe experiment.

In the first group, we find the US satellites Aurora-1 ($h = 3\,850$ km) and OV3-6 ($h = 420$ km), launched in 1967, Explorer-54 (or AE-D, $h_p = 151$ km and $h_a = 3\,819$ km) launched in 1975, the satellites Dynamics Explorer-1 and -2 (also called DE-A and DE-B), in high and low eccentric orbits, respectively, launched in 1981, the satellites REX-1 and -2 (Radiation Experiment), launched in 1991 and 1996, in low orbits ($h = 800$ km). The astronomical satellite Corot will be placed in inertial orbit.

In the second group, we find all the first polar satellites, which were all US military reconnaissance satellites.[2] This category also includes satellites belonging to the Transit system of the US Navy, the first fully operational navigation system. The first polar satellite in this series was Transit-5A,

[1] Note that the direction of the point Υ is fixed in $\mathfrak{R}$ if we neglect the precession of the equinoxes. Galilean and other frames, e.g., those moving with the Earth, are discussed in Chap. 2.

[2] Examples are the first two military reconnaissance satellites in the KH-1 series, Discoverer-1 and -2, launched in 1959, in a low elliptical orbit, with $i = 89.9°$, or the satellite MIDAS-12, in a higher orbit.

launched in 1962, in a low orbit with $h = 340$ km, $i = 90.6°$. The others[3] all had orbits with $h \approx 1100$ km and inclination $i = 90.0 \pm 0.3°$, from Transit-5B-1, launched in 1963, right up to Transit-O-23 and -O-32, then Transit-O-25 and -O-31, launched[4] in pairs in 1988.

The geophysical satellite Polar BEAR, launched on a strictly polar orbit in 1986 ($h = 1000$ km), belongs to the NNS (Navy Navigation Satellite) series of the Transit family, under the name NNS-O-17. Also in this family are the three satellites[5] in the Nova series, with $h = 1170$ km, $i = 90.0 \pm 0.1°$. The Transit navigation system[6] is the forerunner of the NAVSTAR/GPS system.

The orbit of the satellite Nova-2 made an angle of $+70.2°$ with the point Υ on 1 January 1992. The nodal precession rate was $+0.19°$ per year, $+1.12°$ for Nova-1, and $+0.30°$ for Nova-3. For the Transit–Oscar satellites it was $+11.57°$ for Oscar-29, $-5.00°$ for Oscar-30, and $-1.13°$ for Oscar-31. These values of $\dot{\Omega}$ are not exactly zero (perturbations other than the zonal terms J_n act on the satellite), but they are extremely low. They can be compared with the corresponding values for other satellites: $\dot{\Omega} = -258°$ per year for Meteor-3-07, $\dot{\Omega} = +360°$ per year for Sun-synchronous satellites, as we shall see below, and $\dot{\Omega} = -2\,482°$ per year for the satellite TRMM (see Example 4.1).

The Gravity Probe Relativity Mission aims to carry out very accurate experiments to test Einstein's theory of relativity. The satellite GP-B has orbital characteristics $h = 650$ km, $i = 90.000°$, $e = 0.0134$. It must guarantee a fixed direction to an accuracy of 20 milliseconds of arc per year.

4.2 Motion of the Earth

We consider here the motion of the Earth's axis around the Sun, then the motion of the Earth relative to this axis. We shall not take into account the tiny motion of the Earth relative to its axis of rotation, called the motion of the poles.

---

[3] On 1 January 2000, the satellite Transit-5B-5 (or OPS/6582), launched on 13 December 1964, was the oldest satellite still emitting.

[4] Satellites in the -O series (Operational) are also referred to under the name Oscar, as for the two cited above, viz., Oscar-23, -32, -25 and -31 (Oscar representing O in the aeronautic alphabet).

[5] Launch dates in chronological order: Nova-1 on 25 May 1981, Nova-3 on 12 October 1984, Nova-2 on 16 June 1988.

[6] One can add the US Navy's SECOR satellites (Sequential Collation of Range), also known as EGRS (Experimental Geodetic Research Satellites), most of which were in polar orbit: SECOR-2 and -4, launched into low orbit in 1965, with $i = 90.0 \pm 0.3°$, and SECOR-6, -7, -8 and -9, launched into higher orbits in 1966 and 1967, with $h = 3700$ km, $i = 90.0 \pm 0.3°$, and the three Prototypes of Improved Transit, Triad-1 (or TIP), Triad-2 (or TIP-2), Triad-3 (or TIP-3), launched into low strictly polar orbits in 1972 and 1976. The first drag-free satellite to fly a complety gravitational orbit was Triad-1.

4.2.1 Motion of the Earth about the Sun

The Earth takes one year to cover its elliptical orbit around the Sun. This is the period of revolution. There are several definitions of the year:

- **Sidereal Year.** This is the duration of the Earth's orbit about the Sun referred to a fixed (Copernican) frame. It is equal to 365 d 6 h 9 m 10 s.
- **Tropical Year.** This is the duration of the Earth's orbit about the Sun referred to a moving frame. In fact, it is the time elapsed between two consecutive transits of the Sun through the vernal equinox (the spring equinox). It is equal to 365 d 5 h 48 m 46 s.

The sidereal year is used for all calculations in the Galilean frame $\mathfrak{R}$. The tropical year determines the cycle of the seasons.[7] The tropical year is 20 minutes shorter than the sidereal year. This difference is due to the retrograde motion of the vernal equinox, known as the precession of the equinoxes (see Chap. 3).

- **Anomalistic Year.** This is the time elapsed between two transits at perihelion. It is equal to 365 d 6 h 13 m 53 s. It is used in calculations relating to the Keplerian motion of the Earth around the Sun.[8]

If N_{sid} and N_{tro} are the number of days in the sidereal and tropical years, respectively, we have

$$N_{\text{sid}} = 365.256\,360 \text{ day} , \qquad N_{\text{tro}} = 365.242\,199 \text{ day} .$$

J_{M} denotes the length of the mean day, i.e., by definition, exactly 24 hr,

$$J_{\text{M}} = 24 \text{ hr } 00 \text{ min } 00 \text{ s} = 86\,400 \text{ s} .$$

T_{sid} denotes the number of seconds in a sidereal year, so that

$$T_{\text{sid}} = J_{\text{M}} N_{\text{sid}} = 31\,558\,149.504 \text{ s} .$$

The angular speed of the Earth's axis around the Sun is $\dot{\Omega}_{\text{S}}$ (using the notation $\dot{\Omega}$ by analogy with the precession rate of the longitude of the ascending node, and affixing the subscript S to indicate the Sun), which is thus equal to

$$\dot{\Omega}_{\text{S}} = \frac{2\pi}{T_{\text{sid}}} = \frac{2\pi}{J_{\text{M}}} \frac{1}{N_{\text{sid}}} . \tag{4.16}$$

---

[7] The Julian year of 365.25 days (with one leap year every 4 years), and the Gregorian year of 365.2425 days ($= 365 + 1/4 - 3/400$) are designed to approximate the tropical year.

[8] If we compare these definitions of the year with the definitions of the different periods of a satellite discussed in Chap. 3, the anomalistic year corresponds to the anomalistic period T_{a}, whilst the tropical year corresponds to the nodal or draconitic period T_{d}. The draconitic year is defined in terms of the lunar motion.

With the usual conventions, this value is positive because the Earth revolves around the Sun in the positive trigonometric direction. With the three units already mentioned, we have:

$$\dot{\Omega}_S = 1.990\,992\,99 \times 10^{-7}\ \mathrm{rad\,s^{-1}}\ , \tag{4.17}$$

$$\dot{\Omega}_S = \frac{360}{N_{\mathrm{sid}}} = 0.985\,61^\circ \mathrm{day}^{-1}\ , \tag{4.18}$$

$$\dot{\Omega}_S = 1\ \mathrm{rev\,yr^{-1}}\ . \tag{4.19}$$

The last relation is just the definition of the sidereal year.

4.2.2 Motion of the Earth about the Polar Axis

Relative to a Galilean frame, in one year, the Earth rotates $N_{\mathrm{sid}} = 365.25\ldots$ times about its own axis and completes one round trip about the Sun, which means that it rotates $N_{\mathrm{sid}} + 1$ times in one sidereal year. Let $\dot{\Omega}_T$ be the angular speed of the Earth (using $\dot{\Omega}$ because this is the angular speed of the longitude Ω of the ascending node and affixing the subscript T to indicate 'terrestrial'). Then

$$\dot{\Omega}_T = \frac{2\pi}{T_{\mathrm{sid}}}(N_{\mathrm{sid}} + 1) = \frac{2\pi}{J_M}\frac{N_{\mathrm{sid}} + 1}{N_{\mathrm{sid}}}\ . \tag{4.20}$$

With the usual conventions, this value is positive because the Earth rotates in the positive trigonometric direction about its own axis (the polar axis $\boldsymbol{Oz}$, oriented from south to north).

We thus have, for the three different units:

$$\dot{\Omega}_T = 7.292\,115 \times 10^{-5}\ \mathrm{rad\,s^{-1}}\ , \tag{4.21}$$

$$\dot{\Omega}_T = 360\frac{N_{\mathrm{sid}} + 1}{N_{\mathrm{sid}}} = 360 + \dot{\Omega}_S = 360.985\,61^\circ \mathrm{day}^{-1}\ , \tag{4.22}$$

$$\dot{\Omega}_T = N_{\mathrm{sid}} + 1 = N_{\mathrm{sid}} + \dot{\Omega}_S = 366.242\,199\ \mathrm{rev\,yr^{-1}}\ . \tag{4.23}$$

In the present book, we shall use the idea of the sidereal day[9] with duration 23 h 56 m 04 s in contrast to the mean day. Hence,

$$J_{\mathrm{sid}} = 23.934\,471\ \mathrm{hr}\ ,$$

which represents the Earth's period of rotation.

[9] The sidereal day is the time interval for the Earth to return, in its daily rotation, to the same given direction. However, this direction is not precisely fixed. It follows the motion of the precession of the equinoxes. The length of the sidereal day is thus $J_M N_{\mathrm{tro}}/(N_{\mathrm{tro}} + 1) = 86\,164.0905$ s , or 23 h 56 m 04.0905 s. The notion of the stellar day is sometimes used, defined by $J_M N_{\mathrm{sid}}/(N_{\mathrm{sid}} + 1) = 86\,164.0989$ s , or 23 h 56 m 04.0989 s. The sidereal day is thus related to the tropical year, and the stellar day to the sidereal year.

We note the relation between the angular frequencies of the two motions of the Earth considered here:

$$\dot{\Omega}_{\mathrm{T}} - \dot{\Omega}_{\mathrm{S}} = \frac{2\pi}{J_{\mathrm{M}}} , \tag{4.24}$$

or in degrees per day,

$$\dot{\Omega}_{\mathrm{T}} - \dot{\Omega}_{\mathrm{S}} = 360° \mathrm{day}^{-1} .$$

4.2.3 Motion of the Orbit and Earth

Later, we shall often need to compare the motion of a satellite, mainly characterised by the two angular speeds n (mean motion) and $\dot{\Omega}$ (nodal precession rate), with the two motions of the Earth discussed above.

Daily Orbital Frequency ν

Recall that the daily orbital frequency ν of the satellite, the number of round trips per day, is related to the mean motion n (always expressed here in radians per second) by

$$n = \frac{2\pi}{J_{\mathrm{M}}} \nu . \tag{4.25}$$

We use the symbol P to denote the nodal precession rate in round trips per year.[10] We have

$$\dot{\Omega} = \frac{2\pi}{J_{\mathrm{M}}} \frac{P}{N_{\mathrm{sid}}} . \tag{4.26}$$

We thus obtain the following relations in terms of P:

$$\dot{\Omega}_{\mathrm{T}} - \dot{\Omega} = \frac{2\pi}{J_{\mathrm{M}}} \left(1 + \frac{1-P}{N_{\mathrm{sid}}} \right) , \tag{4.27}$$

$$\dot{\Omega}_{\mathrm{S}} - \dot{\Omega} = \frac{2\pi}{J_{\mathrm{M}}} \frac{1-P}{N_{\mathrm{sid}}} . \tag{4.28}$$

We shall also need to compare $\dot{\Omega}_{\mathrm{T}} - \dot{\Omega}$ with n. We have

$$\frac{\dot{\Omega}_{\mathrm{T}} - \dot{\Omega}}{n} = \frac{1}{\nu} \left(1 + \frac{1-P}{N_{\mathrm{sid}}} \right) . \tag{4.29}$$

[10] This quantity is perhaps more meaningful than $\dot{\Omega}$ expressed in radians per second. To avoid any confusion over units, we have used P for this quantity, expressed in round trips per year, whereas other quantities will be expressed in SI units, unless otherwise stated.

The following two expressions for the quantity P are particularly useful:

$$P = \frac{J_M}{2\pi} N_{sid} \dot{\Omega} , \tag{4.30}$$

$$P = \frac{\dot{\Omega}}{\dot{\Omega}_S} . \tag{4.31}$$

The last relation merely formulates the definition of P.

Daily Recurrence Frequency κ

We use κ to denote the daily recurrence frequency. This quantity, which will be important in the study of recurrent orbits in Chap. 7, is defined by

$$\kappa = \frac{n}{\dot{\Omega}_T - \dot{\Omega}} . \tag{4.32}$$

It is related to the daily orbital frequency ν by

$$\frac{\nu}{\kappa} = 1 + \frac{1 - P}{N_{sid}} . \tag{4.33}$$

4.3 Apparent Motion of the Sun

The aim in studying the apparent motion of the Sun is to represent the direction of the Sun and to understand different notions of solar time, viz., apparent and mean solar time.

4.3.1 Celestial Sphere and Coordinates

On the celestial sphere,[11] illustrated in Fig. 4.5, the equator and the ecliptic (the Sun's trajectory) intersect at two points. The point corresponding to the direction of the Sun when its declination crosses zero from below is the vernal equinox, traditionally denoted by ♈. It corresponds to the spring equinox.[12] The dihedral angle between these two planes, known as the obliquity, is equal to $\varepsilon = 23°26'21'' = 23.44°$.

The direction of the Sun can be defined as follows, in two coordinate systems with origin at the centre of the Earth.

[11] One commonly represents directions in space by means of points on a sphere with arbitrary centre and radius, called the celestial sphere. With any particular direction, one associates the point of intersection of the celestial sphere and the straight line in that direction with origin at the centre of the sphere.

[12] The word 'vernal' comes from the Latin *vernalis*, the adjective derived from *ver, veris*, meaning 'spring'.

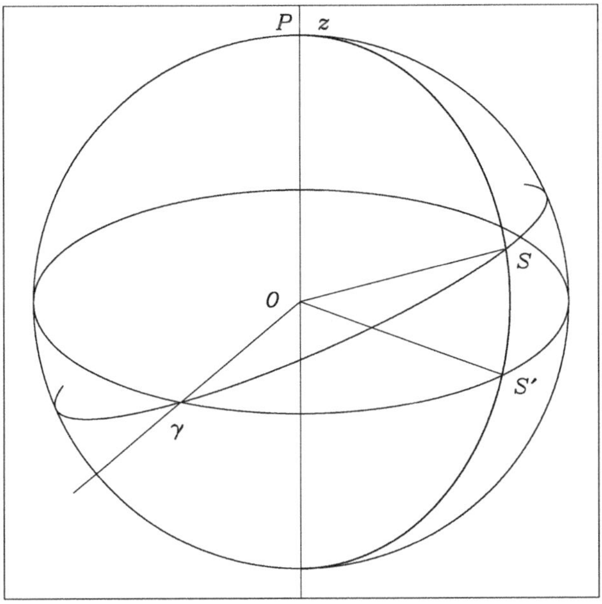

Figure 4.5. Celestial sphere, with vernal equinox Υ, the Sun S, the equatorial plane $O\Upsilon S'$, the ecliptic $O\Upsilon S$, the celestial north pole P, and the meridian PSS'. In the spherical triangle $\Upsilon SS'$, the arc ΥS represents the celestial longitude l, the arc $\Upsilon S'$ the right ascension α, and the arc SS' the declination δ. The angle at Υ is the obliquity ε and the angle at S' is a right-angle

- **Ecliptic Coordinates.** The reference plane is the plane of the Sun's apparent trajectory during its annual revolution, known as the plane of the ecliptic, or just the ecliptic. The origin of these coordinates is taken at the vernal equinox. The angle of azimuth gives the Sun's longitude l, also called the celestial longitude or ecliptic longitude. The height angle gives the Sun's latitude b, called the celestial latitude or ecliptic latitude. By definition, $b = 0$ corresponds to the ecliptic itself.
- **Celestial Equatorial Coordinates.** The plane of reference is the Earth's equatorial plane. The coordinate origin is taken at the vernal equinox. The angle of azimuth is the right ascension α, the dihedral angle between the meridians in the direction of the Sun and the direction of the vernal equinox. The height angle is the declination δ, which is the angle between the direction of the Sun and the reference plane. By definition, the right ascension is related to the hour angle H and the apparent solar time defined below. Concerning the declination, it varies during the year in an almost sinusoidal manner, between the bounds $\delta = \varepsilon$ at the summer solstice and $\delta = -\varepsilon$ at the winter solstice. The values $\delta = 0$ are attained at the equinoxes. We shall return to this point below (see Fig. 4.7).

4.3.2 Hour Angle

Consider an arbitrary point on the Earth (apart from the poles), defined by its coordinates (λ, ϕ) in $\Re_{\mathrm{T}}$. The meridian plane through this location, denoted by $\mathcal{M}$, is the half-plane containing the polar axis and this point. At a given time, we define the meridian plane of the Sun's direction, denoted by $\mathcal{S}$, as the half-plane containing the polar axis and this direction, with longitude λ_{S}. The dihedral angle between these two half-planes is called the hour angle, denoted by H:

$$H = \text{dihedral angle}\,(\mathcal{M}, \mathcal{S}) = \lambda - \lambda_{\mathrm{S}} \ . \tag{4.34}$$

The angle H is measured in the retrograde sense. This convention can be explained as follows. The idea is that the variations in H and the time should occur in the same sense during the day, so that H is negative in the morning, zero at midday, and positive in the evening.

The hour angle (which is in fact an azimuthal angle) can be defined for an arbitrary direction. When it is defined specifically for the direction of the Sun, as here, it is also called the apparent solar time. The hour angle and the apparent solar time are angles, generally given in degrees or hours, where 1 hr corresponds to 15°, since 24 hr corresponds to 360°.

4.3.3 Equation of Time

The apparent solar time defined by the apparent motion of the Sun is a 'natural' idea. It is the time given by a sundial. But this motion does not have the regularity required to form the basis for a time scale. It would be regular, or uniform, if the Earth's orbit were circular and if its plane contained the Earth's equatorial plane.

Equation of Centre

We consider the position of the Sun as defined by ecliptic coordinates. The Sun's trajectory relative to the Earth is not circular but elliptical. The celestial longitude l does not therefore vary in a uniform manner, but corresponds to the true anomaly v, with a different origin. Indeed,

$$l = v - v_{\Upsilon} \ , \tag{4.35}$$

taking the origin of v at the perigee, which corresponds on average to 3 January, and the origin of l at the vernal equinox, which corresponds on average to 21 March. The true anomaly of the vernal equinox is

$$v_{\Upsilon} = 78^{\circ} \ . \tag{4.36}$$

The regular motion is characterised by the mean anomaly M, proportional to the time elapsed since the transit at perigee. The difference, induced by the nonzero eccentricity, between the elliptical and the uniform motions is characterised by $l - M$ or $v - M$, which are equal in value up to a constant. We set

$$E_C = v - M \ , \tag{4.37}$$

a quantity known as the equation of centre, already mentioned in Chap. 1, where it was defined by (1.74) in the discussion of Keplerian motion.

When e is small, as happens for the Earth orbit ($e = 0.0167$), we have seen that this quantity is given by (2.12), viz.,

$$E_C \approx 2e \sin M \ . \tag{4.38}$$

The function E_C is sinusoidal, with period one year and maximum E_{C_m} given by

$$E_{C_m} = 2e = 0.0334 \text{ rad} \ .$$

Its graph is shown by the dashed curve in Fig. 4.6. It is zero twice a year, at the perigee (03 January) and at the apogee (05 July). It reaches its maximum on 03 April and its minimum on 05 October.

Reduction to the Equator

We now consider the Sun's position as defined by celestial equatorial coordinates. The equatorial plane makes an angle ε with the ecliptic, as defined above. It follows that the right ascension does not vary uniformly with the longitude l.

The celestial sphere is represented in Fig. 4.5, where O is the centre of the Earth, P the celestial north pole, Υ the vernal equinox, S the position of the Sun, and S' the intersection of the meridian half-plane of S with the celestial equator. Concerning the spherical triangle $\Upsilon SS'$, the angle at Υ is the obliquity ε, the angle at S' is a right-angle, and the sides (arcs of great circles) have lengths:

$$\overset{\frown}{\gamma S} = l \ , \quad \overset{\frown}{\gamma S'} = \alpha \ , \quad \overset{\frown}{SS'} = \delta \ .$$

Recalling the relations (3.111) and (3.112) from spherical trigonometry (the first two Gauss relations in Sect. 3.16):

$$\begin{cases} \cos\delta \, \cos\alpha = \cos l \ , \\ \cos\delta \, \sin\alpha = \sin l \, \cos\varepsilon \ . \end{cases}$$

Eliminating the solar declination δ, we obtain

$$\sin\alpha \, \cos l = \cos\alpha \, \sin l \, \cos\varepsilon \ .$$

Then, expressing $\cos\varepsilon$ in terms of the tangent of the half-angle,

$$\sin\alpha\,\cos l\left(1+\tan^2\frac{\varepsilon}{2}\right)=\cos\alpha\,\sin l\left(1-\tan^2\frac{\varepsilon}{2}\right).$$

Finally,

$$\sin(\alpha-l)=-\tan^2\frac{\varepsilon}{2}\,\sin(\alpha+l)\,. \tag{4.39}$$

The proximity of α and l is thus expressed in terms of the obliquity ε, and we may say

$$\alpha\approx l-\tan^2\frac{\varepsilon}{2}\,\sin 2l\,. \tag{4.40}$$

The reduction to the equator E_{R} is defined as

$$E_{\mathrm{R}}=\alpha-l\,. \tag{4.41}$$

This characterises the discrepancy, introduced by the obliquity, between the true motion and uniform motion.

In the argument of the sine function, we may put $l\approx M-v_{\Upsilon}$, which yields

$$E_{\mathrm{R}}\approx-\tan^2\frac{\varepsilon}{2}\,\sin 2(M-v_{\Upsilon})\,. \tag{4.42}$$

This function E_{R} is sinusoidal, with biannual period, and has a maximum $E_{\mathrm{R_m}}$ given by

$$E_{\mathrm{R_m}}=\tan^2\frac{\varepsilon}{2}=0.0431\ \mathrm{rad}\,.$$

It goes to zero four times a year, at the two equinoxes (21 March, 23 September) and at the two solstices (21 June, 22 December). Its graph is shown by the dash-dotted curve in Fig. 4.6.

Equation of Time

Consider the apparent motion of the Sun relative to the Earth. Its position is defined by the right ascension α. The position that it would have in a uniform motion of the same period is defined by the mean anomaly M. Taking the same origin, viz., the vernal equinox, we must therefore compare α, which characterises the apparent solar time, with $M-v_{\Upsilon}$, which characterises the mean solar time. The difference between these two angles is called the equation of time E_{T}, given by

$$E_{\mathrm{T}}=\alpha-(M-v_{\Upsilon})=\alpha-l+l-M+v_{\Upsilon}=(\alpha-l)+(v-M)\,.$$

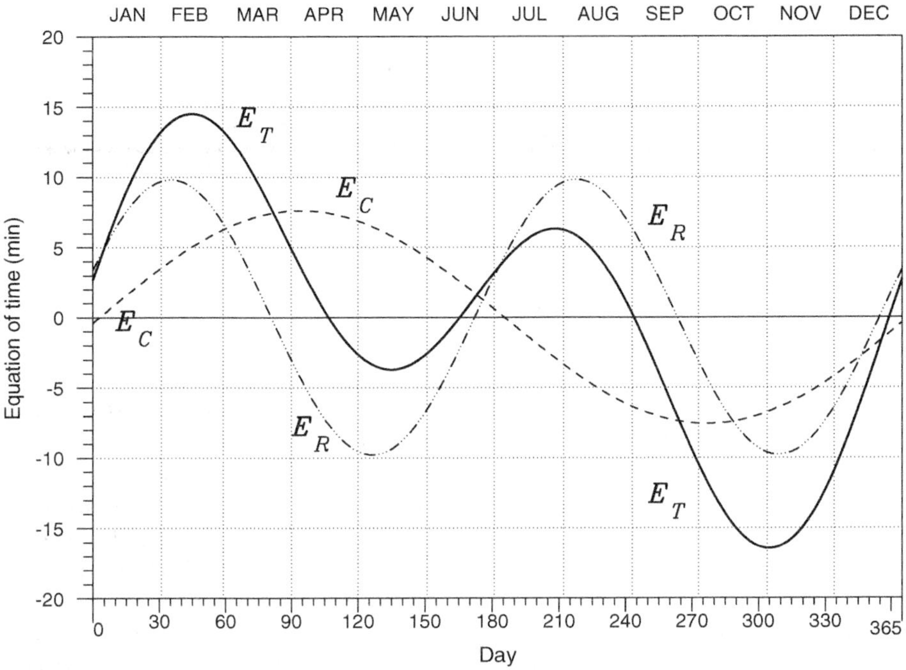

Day J	Day+month	E_T [min]
45	14 Feb	+14.30
107	17 Apr	0
135	15 May	−3.60
166	15 Jun	0
207	26 Jul	+6.37
244	01 Sep	0
305	01 Nov	−16.43
359	25 Dec	0

Figure 4.6. Graph of the equation of time as a function of the day of the year J. The equation of time E_T is the sum of the equation of centre E_C and the reduction to the equator E_R. All these quantities are expressed in minutes. The table shows the maxima, minima, and zeros of E_T with corresponding dates. These dates may vary by one or two days on either side, depending on the year

This shows that the equation of time is the sum of the two quantities E_{C} and E_{R} defined above:

$$E_{\mathrm{T}} = E_{\mathrm{C}} + E_{\mathrm{R}} \,. \tag{4.43}$$

It thus has the value

$$E_{\mathrm{T}} = 2e \sin M - \tan^2 \frac{\varepsilon}{2} \sin 2(M - v_{\Upsilon}) \,. \tag{4.44}$$

Recall that $M = n(t - t_{\mathrm{p}})$, with $n = 2\pi/T$, where T is equal to one year (strictly speaking, one anomalistic year). The mean value of E_{T} over one year is zero by definition.

To the accuracy required here, it is convenient to consider the period equal to one civil year of 365 days and to characterise M by the day J of the year, taking the beginning of the year as zero point (i.e., $J = 1$ is 1 Jan, $J = 2$ is 2 Jan, and so on, up to $J = 365$ for 31 Dec). Then, with the passage at perigee for $J = 3$,

$$M = \frac{360}{365}(J - 3) \tag{4.45}$$

and

$$M - v_{\Upsilon} = \frac{360}{365}(J - 3) - 78 = \frac{360[(J - 3) - 79]}{365} = \frac{360}{365}(J - 82) \,. \tag{4.46}$$

To express E_{T} in time units, we convert radians to minutes of time. In one trip round the orbit, the right ascension changes by 24 hr, so 2π rad is equivalent to 1440 min. Finally, expressing the arguments of the sine function in degrees, we obtain

$$E_{\mathrm{T}}(J)\ [\mathrm{min}] = 7.64 \sin\left[\frac{360}{365}(J - 3)\right] - 9.87 \sin\left[\frac{720}{365}(J - 82)\right] \,. \tag{4.47}$$

With this simplified formula, which is quite adequate for most situations, we observe that the two effects decouple in E_{T}. The eccentricity affects E_{C}, and the obliquity affects E_{R}. This is due to the fact that the quantities e and $\tan^2(\varepsilon/2)$ are much smaller than unity. Extremely detailed and precise expressions for E_{T} can be found in the astronomical literature.

Figure 4.6 shows a graph of E_{T} calculated using (4.47), noting the maxima, minima and zeros. In particular, we see that the equation of time varies with an amplitude of one quarter of an hour and goes to zero four times a year.

Note. The sign convention in the definition of E_{T} sometimes changes, depending on the field of study, as does the convention for geographical longitudes. One must therefore exercise great caution over the sign used for calculations.

4.3.4 Solar Times

Apparent Solar Time, Mean Solar Time

The sign convention adopted is such that the equation of time is the amount by which the mean solar time exceeds the apparent solar time. Hence, if we add E_{T} to the apparent solar time, we obtain the mean solar time:

$$\text{equation of time} = \text{mean solar time} - \text{apparent solar time} .$$

Variations in the duration of the apparent solar day (the time separating two consecutive solar noons) correspond to these irregularities in the apparent solar time. This duration lies between 23 h 59 m 39 s and 24 h 00 m 30 s.

Civil Time, Universal Time UT

With the above definitions, solar noon (hence, $H = 0$) corresponds to 0 h for the apparent solar time. The civil time at a given location is the mean solar time at this location increased by 12 hr.

Universal Time (UT) is civil time on the Greenwich meridian. In the end, it is therefore based on the Earth's rotation, and reflects irregularities in this motion. International Atomic Time (TAI) is the most regular measure of time we have yet been able to achieve. It is obtained in the laboratory using state-of-the-art metrological techniques. This very accurate measurement of time reveals a slowing down of the Earth's rotation. Coordinated Universal Time (UTC) is TAI adjusted by a whole number of seconds to UT.[13]

Solar Times LAT and LMT

In the rest of this book, the solar time scales we shall use are the civil time scales. We reserve the abbreviations LMT for the local mean solar time (or local mean time, for short) increased by 12 hr, and LAT for the local apparent solar time (or local apparent time, for short) increased by 12 hr. We have

$$E_{\mathrm{T}} = \mathrm{LMT} - \mathrm{LAT} , \tag{4.48}$$

noting the value at noon:

$$H = 0 \Longleftrightarrow \mathrm{LAT} = 12 \text{ h} , \quad \mathrm{LMT} = 12 \text{ h} + E_{\mathrm{T}} .$$

The time scales LMT and UT are related by the hour angle:

$$\mathrm{LMT} = \mathrm{UT} + \frac{\lambda}{15} \qquad (\lambda \text{ in degrees, time in hours}) , \tag{4.49}$$

[13] The old name GMT (Greenwich Mean Time) is considered to be inappropriate by astronomers and has been out of use for several decades. One should avoid using it, even though it does still turn up in certain contexts.

with the convention already mentioned for longitudes (− W and + E).

In the following, t will denote the time in UT, and τ the corresponding time in LMT. At a place with longitude λ, we then have

$$\tau(t, \lambda) = \tau = t + \frac{\lambda}{15} \,. \tag{4.50}$$

Time on the LMT scale is also called local time, while time on the LAT scale is called solar time.

Example 4.4. *The ScaRaB (MV2) instrument aboard Resurs-O1-4 was launched on 10 July 1998, at 06:30 UT from the Baikonur base in Kazakhstan. Calculate the time on the LMT and LAT time scales at this location and time.*

The geographical coordinates of the Baikonur space centre are 68° 16′ E and 45° 38′ N. Hence, for the longitude, $\lambda = +68.27°$ and the local mean time is

$$\text{LMT} = \text{UT} + 68.27/15 = \text{UT} + 4.551 = 06\text{ h }30\text{ m} + 4\text{ h }33\text{ m} = 11\text{ h }03\text{ m} \,.$$

The date enters the equation of time: $E_{\text{T}}(J = 191) = 5.2$ min. We thus obtain the apparent solar time as

$$\text{LAT} = \text{LMT} - E_{\text{T}} = 11\text{ h }03\text{ m} - 0\text{ h }05\text{ m} = 10\text{ h }58\text{ m} \,.$$

In conclusion, at Baikonur, the time 06:30 UT corresponds to 11:03 LMT and, on 10 July, to 10:58 LAT.

4.3.5 Historical Note on Time Scales

Prior to 1960, the definition of the second was based on the Earth's rotation. One mean solar day was equal to 86 400 s. The time scale was Universal Time (UT). Between 1960 and 1967, to get around the fact that there were irregularities in the Earth's rotation, the orbital motion of the Earth was chosen to define the second, which thus became a fraction of the tropical year 1900. The time scale was Ephemeris Time (ET).

The year 1967 was historically important in this respect, because it was at this point that the definition of time first left the field of astronomy, to be taken over by the world of physics. The second was defined as the period of a certain type of radiation emitted by the caesium 133 atom. This time scale was called International Atomic Time (TAI). This is currently the legal definition of the second as a unit in the SI system.[14]

[14] The accuracies attained today in observation and time measurement have compelled astronomers to take relativistic effects into account. They have thus defined the so-called dynamical time scales denoted by TDB, TCB, and TCG. For example, one must now distinguish the time given by an Earth-based atomic clock from that supplied by another clock at the barycentre of the Solar System.

In the present book, in which we need to study such things as the crossing time of a satellite and the position of the Sun at that particular moment, we shall only need to use UT, LMT and LAT. It is quite clear that, in order to determine the position of a satellite with sufficient accuracy, we shall not be able to neglect the TAI–UTC adjustment: in one second (added every 12 to 18 months), the satellite will travel some 7 km!

4.3.6 Julian Day, Julian Date

In order to bring the slowing of the Earth's rotation into our equations, to calculate the time difference between two given dates, or to identify a particular date in history without ambiguity, we use the Julian day. The days are counted one after the next, without bringing in reference to month or year, and without discontinuity when a change is made in the calendar. The zero point is taken in a sufficiently remote past to incorporate historical events.[15]

The Julian date corresponds to the Julian day increased by the fraction of the day as counted from 12 h. On 1 January 2000, we then have the following correspondence between the date of the Gregorian calendar and the Julian date:

$$2000\text{-}01\text{-}01 \quad 12:00:00 \Longleftrightarrow \text{JD } 2\,451\,545.0\,.$$

We shall use this notion of Julian date for the NORAD orbital elements in Chap. 5, and for calculations relating to the orbit of Mars in Chap. 10.

[15] Joseph Juste Scaliger (1540–1609), the French scholar, proposed a new chronology in his *De emendatione temporum* (On the correction of time) in 1583. His idea was to produce a continuous count of the years in such as way as to cover all the great civilisations. He called this the Julian system, by analogy with the Julian calendar (introduced by Julius Caesar). The Julian numbering system, quoted by Kepler, was used by astronomers from 1860. They then added the idea of Julian day and Julian date. Scaliger considered the cycles involved in calculating the date of Easter, which was a major concern for astronomers in the Christian world:

- the solar cycle (or dominical cycle) of 28 years (7 times 4, with 7 being the number of days in the week, and there being one leap year every four years),
- the lunar cycle (the golden number, or the Metonic cycle) of 19 years (235 lunations in 19 years),
- the cycle of indiction, of 15 years (a number of historical rather than astronomical significance).

To each year there corresponds a set of three numbers, one for each cycle. Every $28 \times 19 \times 15 = 7980$ yr, the years return to the same values for the three cycles (28, 19 and 15 being coprime). Scaliger chose as origin the year when the numbers of the cycles were all equal to 1. In an imaginary Gregorian calendar, this corresponds to the date Monday 1 January -4712 at 12 h. This year (4713 BC) is a leap year.

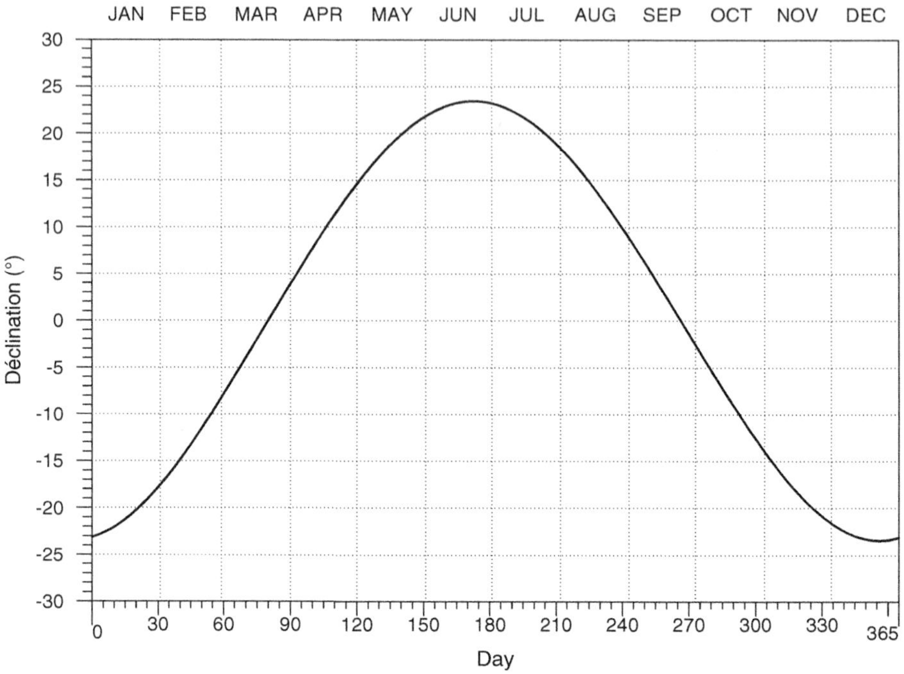

Day J	Day+month	δ [degrees]	δ	Beginning of season	Length of season
21	21 Jan	-20.00			89
54	23 Feb	-10.00			
80	21 Mar	0.00	$\delta = 0$	Spring equinox	
106	16 Apr	$+10.00$			93
141	21 May	$+20.00$			
173	22 Jun	$+23.44$	$\delta = +\varepsilon$	Summer solstice	
205	24 Jul	$+20.00$			93
240	28 Aug	$+10.00$			
266	23 Sep	0.00	$\delta = 0$	Autumn equinox	
293	20 Oct	-10.00			90
326	22 Nov	-20.00			
356	22 Dec	-23.44	$\delta = -\varepsilon$	Winter solstice	

Figure 4.7. Solar declination δ as a function of the day J. The table shows important values of δ with the corresponding value of J. The obliquity of the Earth is $\varepsilon = 23.44°$. The lengths of seasons are given in days

4.3.7 Declination

The solar declination is very simply expressed as a function of the Sun's celestial longitude l and the obliquity ε. Considering the right-angled spherical triangle shown in Fig. 4.5, and using the values of the angles given in the figure caption or the section describing the reduction to the equator, we may use the sine rule (ST VIII) from spherical trigonometry to show that

$$\sin\delta = \sin l \, \sin\varepsilon \, . \tag{4.51}$$

Clearly, we obtain the declination δ in terms of the date (via the mean anomaly M). Using the equation of centre, we have from (4.35), (4.36) and (4.38),

$$l = v - v_{\Upsilon} = M + 2e\sin M - v_{\Upsilon} = (M - v_{\Upsilon}) + 2e\sin M \, .$$

We use the day of the year J ($J = 1$–365). With (4.45) and (4.46), we have seen how M and $M - v_{\Upsilon}$ are related to J. For angles expressed in degrees, as is usual, we must give the eccentricity e in degrees:

$$e = 0.0167 \times 180/\pi = 0.96^\circ \, .$$

This in turn gives

$$\sin\delta = \sin\varepsilon \, \sin\left[\frac{360}{365}(J - 82) + 1.9\sin\frac{360}{365}(J - 3)\right] , \tag{4.52}$$

or putting in the numerical value of the obliquity,

$$\delta(J) = \arcsin\left\{0.39795\sin\left[\frac{360}{365}(J - 82) + 1.9\sin\frac{360}{365}(J - 3)\right]\right\} . \tag{4.53}$$

We recall that the value from (4.38) is approximate (although a very good approximation here). Equation (4.53) gives the declination to within 0.2°, which is quite adequate when studying solar angles, with dates defined in a whole number of days. (The apparent diameter of the Sun is 0.5° and the variation of δ is 0.4° per day near the equinoxes.)

The variation of the declination with date is shown in Fig. 4.7. The significant dates are indicated. Note that the lengths of the seasons, as defined by 90° intervals of solar longitude, are unequal. The passage at perigee (03 Jan) is, in our century, close to the winter solstice (22 Dec), and the seasons close to these dates (autumn and winter) are shorter than the seasons near the passage at apogee. This is just an example of Kepler's second law, the areal law.

4.4 Geosynchronicity

4.4.1 Definition

We consider the Earth's rotation and the rotation of a satellite S in the Galilean frame $\mathfrak{R}$. The satellite is said to be geosynchronous if its motion around the Earth and the rotation of the Earth about its axis have the same angular frequency, i.e., if the mean motion n of the satellite is equal to $\dot{\Omega}_\mathrm{T}$.

$$\text{geosynchronous satellite} \iff n = \dot{\Omega}_\mathrm{T} \,. \tag{4.54}$$

This condition can be met by a satellite whose orbital elements e and i are nonzero. However, in practice, what one usually seeks in a geosynchronous motion is that the subsatellite point S_0 should be immobile on the Earth's surface (in the frame $\mathfrak{R}_\mathrm{T}$). One then says that the satellite is geostationary.

To achieve this, the vectors representing the Earth's rotation and the satellite's rotation should be equal. Concerning their directions, they must therefore be collinear. As the Earth's rotation vector lies along $\boldsymbol{Oz}$, the polar axis, the same must also be true for the rotation vector of S. Since the orbit of S is planar and this plane must contain the centre of attraction O, the centre of the Earth, it must lie in the equatorial plane (whence $i = 0$). The magnitudes of these vectors are equal since the satellite is geosynchronous. As the value of $\dot{\Omega}_\mathrm{T}$ is constant, the motion of S must be uniform, whereupon the subsatellite point will be stationary. The orbit of S must therefore be circular, i.e., with constant altitude:

$$\text{geostationary satellite} \iff \begin{cases} n = \dot{\Omega}_\mathrm{T} \,, \\ i = 0 \,, \\ h = \text{constant} \,. \end{cases} \tag{4.55}$$

A geostationary satellite is thus geosynchronous.[16] The converse is not always true, e.g., Tundra-type orbit. Its position is determined by the longitude of the subsatellite point, called the parking longitude of the geostationary satellite.

4.4.2 Calculating the Orbit

To calculate the radius of the circular orbit of S, we begin by considering the value of the Keplerian mean motion. Setting $n_0 = \dot{\Omega}_\mathrm{T}$, we obtain

$$a_0^3 = \frac{\mu}{\dot{\Omega}_\mathrm{T}^2} = 7.4960128 \times 10^{22} \,,$$

[16] In $\mathfrak{R}$, the satellite is synchronous, whilst in $\mathfrak{R}_\mathrm{T}$, it is stationary. The word 'geosynchronous', meaning 'synchronised with the Earth', takes its origins from two Greek roots and is more satisfying than the word 'geostationary', which is a Greek–Latin hybrid.

$$a_0 = 42\,164.159 \text{ km}\,, \quad h_0 = 35\,786 \text{ km}\,.$$

Using the iterative method to obtain the altitude from the period, as illustrated in Example 4.2, we now obtain

$$a_1 = 42\,164.199 \text{ km}\,, \quad h_1 = 35\,786 \text{ km}\,.$$

At this altitude, as can be seen from Fig. 3.1, the leading perturbation no longer arises from the J_2 term in the geopotential, but from the lunisolar potential. Iterative calculations like those considered previously are no longer suited to the problem. By studying the various perturbations (indeed, rather weak compared with the leading term), one can extract the precise value of the orbital radius, which we shall indicate here with a subscript GS to denote geostationary:

$$a_{\text{GS}} = 42\,165.787 \text{ km}\,, \quad h_{\text{GS}} = 35\,788 \text{ km}\,, \tag{4.56}$$

or as a function of the Earth's equatorial radius R:

$$a_{\text{GS}} = 6.611R\,, \quad h_{\text{GS}} = 5.611R\,. \tag{4.57}$$

In terms of the reduced distance η, we have

$$\eta_{\text{GS}} = \frac{a_{\text{GS}}}{R} = 6.611\,. \tag{4.58}$$

4.4.3 Geostationary Satellites

It is easy to understand the importance of the geostationary satellite. The point is that it always 'views' the same region, and with the same geometry (as we shall see in Chap. 10). For communications satellites and Earth-observation satellites, these are crucial points. For example, such a position allows a weather satellite to make a 'film' in real time, with one image every 15 min, showing the development of cloud formations.

It is just as easy to see the drawbacks of this kind of orbit. A geostationary satellite cannot view the whole of the Earth's surface, either in longitude (which explains why one must arrange several of them at different longitudes), or in latitude (regions above 55° are difficult to attain in this way). Moreover, one is forced to view from a great distance.

Geosynchronous orbits have been a target since the beginnings of the space age, starting with the US satellite series called Syncom[17] (Synchronous Communications Satellite), experimental communications satellites (mass 39 kg). The first, Syncom-1, at $i = 33.3°$, was lost at launch. The next, Syncom-2, at $i = 32.8°$, was the first geosynchronous satellite. It provided

[17] Launch dates: Syncom-1 on 14 February 1961, Syncom-2 on 26 July 1963, Syncom-3 on 19 August 1964.

the first telephone link between the Bight of Benin and the United States, on 31 July 1963 (see the upper part of Fig. 5.10). Syncom-3 can be considered as the first geostationary satellite, since its inclination was $i = 0.1°$. It is thanks to this satellite that the Tokyo Olympic Games of 1964 could be followed live in the United States.

Satellites in the subsequent ATS series (Applications Technology Satellite) were already much bigger (mass 930 kg for ATS-6).[18] The Intelsat series (International Telecommunications Satellite Organisation) was the first family of commercial communications satellites.[19]

The first images from geostationary satellites were those taken by ATS-1 and ATS-3, but the first meteorological satellites on this orbit belonged to the SMS series (Synchronous Meteorological Satellite), launched in 1974 and 1975, SMS-1, $i = 15.5°$ and SMS-2, $i = 12.0°$. These were followed by the GOES series, with GOES-1 at $i = 12.4°$ to GOES-7 at $i = 1.2°$. For the GOES-Next series, starting with GOES-8, orbits were equatorial: $i \approx 0.2°$.

The first Soviet geostationary satellite was placed in orbit much later, such satellites being of little use to a country like Russia. This was Kosmos-637, launched on 26 March 1974 with $i = 14.5°$. Shortly afterwards, the French–German communications satellite Symphonie-1 was launched, on 19 December 1974, with $i = 12.5°$. The first satellite of the European organisation ESA was METEOSAT-1, launched in 1977 with $i = 11.9°$. Since 1988, with METEOSAT-3, these meteorological satellites have been placed on near-equatorial orbits ($i < 1.5°$). In July 2001, there were 850 satellites in near-geostationary geosynchronous orbit (with 52 being launched during the year 2000), of which 320 were operational. The vast majority of these were communications satellites.

In the 1980s, most of these satellites were not equatorial. The equatorial orbit was mainly used for military satellites.[20] These satellites have very low

[18] Apart from two failures, for ATS-2 and -4, all the satellites were placed on slightly inclined orbits. Launch dates: ATS-1, 07 December 1966, $i = 14.5°$ (remained operational for 18 yr, until April 1985), ATS-3, 05 November 1967, $i = 14.5°$, ATS-5, 12 August 1969, $i = 14.5°$, ATS-6, 30 May 1974, $i = 13.1°$.

[19] The first in the series were Intelsat-1 F-1, also known as Early Bird, launched on 6 April 1965, $i = 14.7°$ (stationed over the Atlantic to establish 'fixed' telephone links between Europe and the United States) and the three Intelsat-2 satellites launched in 1967, with $i = 1.6°$ for Intelsat-2 F-2 (Lani Bird), $i = 14.1°$ for Intelsat-2 F-3 (Canary Bird), and $i = 14.5°$ for Intelsat-2 F-4 (Lani Bird-2). Since then, the Intelsat satellites have been launched on a regular basis and placed over the Atlantic, Indian and Pacific oceans. At the end of 2003, with the launch of Intelsat-907 (satellite 07 of generation 9), a total of 65 Intelsat satellites had been launched, 60 successfully and 24 still operating today.

[20] Examples are IMEWS-3 (DSP-F-3), launched in 1972, $i = 0.2°$, and the subsequent satellites in the DSP series (Defense Support Program), such as DSP-F-10, $i = 0.7°$ and DSP-F-11, $i = 0.8°$, used for early warning missions, i.e., rapid detection of enemy missiles.

eccentricity: $e \approx 2 \times 10^{-4}$ for the latest METEOSAT and $e \approx 4 \times 10^{-4}$ for the latest GOES.

Concerning the problem of solar eclipse faced by geostationary satellites, see Sect. 6.3.

4.4.4 Drift of the Geostationary Orbit

The geostationary satellite undergoes the effects of various perturbations, causing it to drift from its course as time goes by. In other words, the subsatellite point S_0 is no longer exactly the assigned reference point. There are two kinds of drift: longitudinal drift and latitudinal drift.

Longitudinal Drift

For meteorological satellites, the longitudinal drift of S_0 can be compensated if it is not too great, by correcting the transmitted image. A very slight variation in a causes this shift and the satellite is then no longer geosynchronous. If a increases, the period also increases and the mean motion decreases. Hence, the satellite rotates less quickly than the Earth in $\Re$, whence the subsatellite point S_0 moves westward in $\Re_\mathrm{T}$. Likewise, if a decreases, S_0 moves eastward. This phenomenon is shown schematically in Fig. 4.8 (see also the lower part of Fig. 2.3).

We shall calculate the displacement $\Delta_1 l$ of the subsatellite point over one day for a variation Δa in the orbital radius (or altitude). In $\Re$, the subsatellite point of a geostationary satellite moves a distance l between times t_0 and t_1, where

$$l = R\dot{\Omega}_\mathrm{T}(t_1 - t_0) .$$

In $\Re$, the subsatellite point of a satellite with mean motion n moves through a distance l' in the same time, where

$$l' = Rn(t_1 - t_0) .$$

The difference measures the deviation in $\Re$, or in $\Re_\mathrm{T}$, which is the drift we seek here:

$$l' - l = R(n - \dot{\Omega}_\mathrm{T})(t_1 - t_0) .$$

If the second satellite is close to the geostationary orbit, with $a = a_\mathrm{GS} + \Delta a$ and $n = \dot{\Omega}_\mathrm{T} + \Delta n$, we have the relation

$$\frac{\mathrm{d}n}{n} = -\frac{3}{2}\frac{\mathrm{d}a}{a} \Longrightarrow \frac{\Delta n}{\dot{\Omega}_\mathrm{T}} \approx -\frac{3}{2}\frac{\Delta a}{a_\mathrm{GS}} .$$

Setting $\Delta l = l' - l$, we have

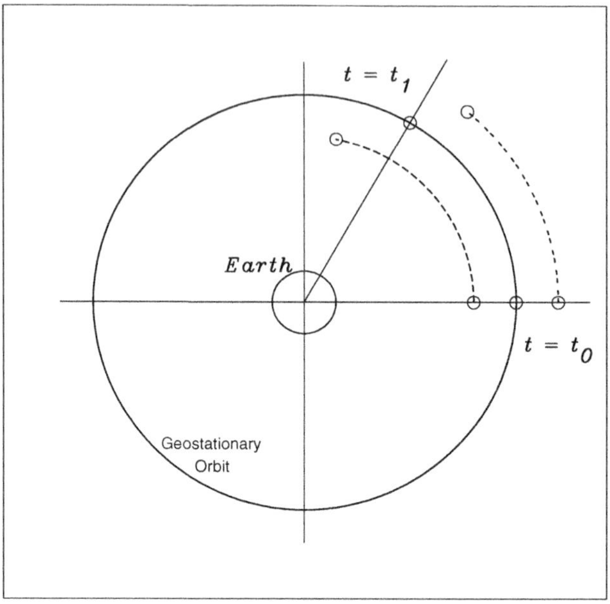

Figure 4.8. Orbit of a geostationary satellite and two other satellites at higher and lower altitudes than those required by geosynchronicity. At time $t = t_0$, the three satellites share the same subsatellite point. At time $t = t_1$, the subsatellite point of the lowest satellite has slipped eastward, whilst that of the highest satellite has slipped westward, relative to the subsatellite point of the geostationary satellite. The diagram shows the Earth viewed from a point located high above the North Pole, in the Galilean frame $\mathfrak{R}$

$$\Delta l = R \Delta n (t_1 - t_0) = -\frac{3}{2} \Delta a \frac{R}{a_{GS}} \dot{\Omega}_T (t_1 - t_0) \, .$$

The value $\eta_{GS} = a_{GS}/R$ is completely determined by (4.58) for a geostationary satellite.

If we consider a time interval of exactly one day, $t_1 - t_0 = J_M$, we may express $\dot{\Omega}_T J_M$ using (4.20). This gives $\Delta_1 l$ (the subscript 1 indicates that we are considering a time interval of 1 day), the displacement of the subsatellite point over one day:

$$\Delta_1 l = -\frac{3}{2} \frac{1}{6.611} 2\pi \frac{366.25}{365.25} \Delta a = -1.4295 \Delta a \quad \text{(per day)} \, . \tag{4.59}$$

The sign here is valid if we apply the usual conventions concerning the longitudes for Δl, i.e., negative for west and positive for east.

We now apply this to two examples. In the first, we calculate the drift of a satellite that is not exactly at the required altitude, and in the second, we show how one may take advantage of this drift to modify the position of the satellite.

Example 4.5. *Calculate the longitudinal drift over one week of the subsatellite point of a geostationary satellite whose altitude has increased by 100 m.*

With $\Delta a = \Delta h = 100$ m, (4.59) gives $\Delta l = -143$ m day^{-1}. The position of the subsatellite point on the ground thus moves 1.0 km westward in one week.

Example 4.6. *As part of the INDOEX project (Indian Ocean Experiment), the European organisation of meteorological satellites EUMETSAT decided to move the satellite METEOSAT-5 from its standby (10° W) to a new position above the Indian ocean (63° E). The method used consisted in shifting the satellite to a lower orbit, whereupon it would drift eastward. Indeed, leaving its initial geostationary position on 14 January 1998, it arrived at its new geostationary position on 19 May 1998. By how much was the radius of the orbit adjusted to make this transfer?*

The departure and arrival dates were $J = 14$ and $J = 139$, respectively, separated by 125 days. Since it takes 12 hr to go from one orbit to the other, we consider that the transfer took 124 days. The distance between the two positions is 73°. The satellite therefore had to move through $73/124 = 0.589°$ per day relative to the Earth, which corresponds to 65.53 km per day on the ground track. Since the shift was eastward, we thus have $\Delta l = 65.53$ km day^{-1}. Applying (4.59), we find

$$\Delta a = -\frac{\Delta l}{1.4295} = -45.84 \text{ km} .$$

The satellite was therefore placed on an orbit 46 km lower. Note that only the two maneuvers at the beginning and the end required energy input, whilst the trip itself cost nothing.[21]

Latitudinal Drift

Latitudinal variations of S_0 (in which S therefore moves slightly outside the equatorial plane under the action of gravitational forces other than the Earth's) show up through a distortion of the ground track. The orbit is slightly tilted with respect to the equatorial plane ($i \neq 0$). During the day, the subsatellite point is not fixed but traces out a figure of 8, between latitudes $\phi = +i$ and $\phi = -i$. If the orbit remains circular, the intersection of the straight line OS (joining the centre of the Earth O to the satellite S) with the horizontal equatorial plane of the node of the orbit traces out a Bernoulli lemniscate, as shown in the upper part of Fig. 5.10.

[21] The speed of the transfer does have an energy cost. The faster one needs to go, the lower the transfer orbit should be. For each transfer maneuver of METEOSAT-5 in the above example, which required two burns, one on the starting orbit and one on the final orbit, EUMETSAT indicate that 300 g of fuel were burnt. The satellite was carrying 6 kg of propellant before the move.

4.4.5 Stationkeeping

The orbit of a geosynchronous satellite evolves as time goes by. The inclination i is affected mainly by the Moon and Sun, the semi-major axis a by the tesseral terms in the geopotential, and the eccentricity e (since the distorted orbit is no longer circular) by the effect of solar radiation pressure.

Concerning variations of the semi-major axis a, it can be shown that there are four special points on the equator, the equilibrium points, with longitudes 75.1°E and 105.3°W for the stable points, and 164°E and 11°W for the unstable points. This effect is a manifestation of the tesseral harmonic P_{22}, of crucial importance amongst the tesseral terms: the terms C_{22} and S_{22} are nonzero, whereas they are taken as zero in any model with cylindrical symmetry [see (3.16)]. Note that these four points divide the equator into almost equal arcs of 90° (see Sect. 3.7.2). Depending on its position, a geostationary satellite will move towards the stable points, or move away from the unstable points. For example, in three months the semi-major axis of the satellite TDF-1, located at 19°W and hence rather close to an unstable point, changed from $a = 42\,164.9$ km to $a = 42\,167.5$ km, values on either side of a_{GS}.

Stationkeeping involves repositioning a satellite within its 'window' after a certain time. (A typical window would constitute about 1° in the east–west direction and $-0.1°$ to $+0.1°$ in the north–south direction.) Stationkeeping operations require maneuvers in which fuel is burnt, and this necessarily puts a limit on the lifetime of the satellite. North–south control represents 95% of the fuel consumption of a satellite like TDF-1.[22]

4.4.6 Geosynchronous Satellites with Highly Eccentric Orbit

Countries like Russia and Canada, situated as they are at high latitudes, have little use for geostationary satellites, which are equatorial. As we shall see shortly, the choice of an orbit that is both inclined and elliptical (to take advantage of the areal law) can be favourable for northerly regions. In order to reduce apsidal precession to a minimum, the critical inclination $i = i_C = 63.4°$ is essential. The period can be fixed at one (sidereal) day so as to obtain a geosynchronous satellite.

The Tundra-type orbit was studied by both Russia and Canada and the idea taken up again by the European Space Agency (ESA) for its Archimedes project. Two different orbits are planned: Tundra (or Tundra 2) and Supertundra (or Tundra 1), with the value $a = 42\,163$ km almost independent of the eccentricity and very close to a_{GS} as given by (4.56). Since the values of the inclination and period are determined, we have:

[22] Launched in 1988, this satellite was maintained in position for its period of use. It was then placed in a graveyard orbit where it was allowed to drift. On 26 April 1999, the altitudes of the perigee and apogee were 36 088 km and 36 093 km, respectively (so that $a = 42\,469$ km), and the inclination was $i = 2.25°$. The drift was $-3.9°$ per day, corresponding to $\Delta a = 304$ km, according to (4.59).

- for Tundra, $e = 0.2668$, $h_p = 24\,536$ km, $h_a = 47\,034$ km,
- for Supertundra, $e = 0.4230$, $h_p = 17\,950$ km, $h_a = 53\,620$ km.

The visibility time, over which the satellite is visible in acceptable geometrical conditions for the relevant regions (see Sect. 5.6), is 8 hr for the first of these orbits and 12 hr for the second, once the position of the apogee has been correctly established. This means that 3 and 2 satellites are needed, respectively, to achieve permanent coverage. In such conditions, we may say that we have obtained the equivalent of one geostationary satellite but at high latitude.

The Tundra orbit has been used successfully since 2000 by the SD-Radio constellation of US communications satellites.[23]

4.5 Sun-Synchronicity

4.5.1 Definition

The orbital plane $\mathcal{P}$ of the satellite rotates in $\Re$, about the polar axis, at a rate $\dot{\Omega}$, which characterises the angular speed of the vector $\boldsymbol{ON}$ in the plane $\mathcal{E}$, where O is the centre of the Earth and N the ascending node, as shown in Fig. 2.1.

We seek a type of orbit in which the transit at the ascending node always occurs at the same solar time. We thus require that $\boldsymbol{ON}$ make a constant angle with the direction of the Sun, since the hour angle (and hence the local mean time) is the dihedral angle between the meridian plane of the relevant point (here N) and the plane containing the polar axis and the Sun. For this to happen, the nodal precession rate $\dot{\Omega}$ must equal the angular speed of the Earth's motion around the Sun. Such a satellite is said to be Sun-synchronous or heliosynchronous:

$$\text{Sun-synchronous satellite} \iff \dot{\Omega} = \dot{\Omega}_S . \tag{4.60}$$

A satellite with an elliptical orbit can be Sun-synchronous, in which case the nodal precession rate is given in the form $\dot{\Omega} = \dot{\Omega}(a, e, i)$. We shall return to this case below, with the example of the satellite Ellipso Borealis. However, in most cases, and in particular for Earth-observation satellites,[24] only circular and near-circular orbits are used, so that $\dot{\Omega} = \dot{\Omega}(a, i)$.

23 The three satellites Sirius-1, -2 and -3 (also called SD-Radio-1, -2 and -3), launched from Kazakhstan on 30 June, 5 September and 30 November 2000, are on a geosynchronous orbit: $e = 0.2700$, $h_p = 24\,400$ km, $h_a = 47\,170$ km. They are operational for North America between longitudes 60°W and 140°W.

24 Satellites devoted to magnetospheric studies are often placed in elliptical Sun-synchronous orbits. Examples are MAGSAT, Ørsted or the two German satellites Aeros-1 and -2. One should also mention those satellites whose orbits, originally intended to be circular, have become elliptical owing to launch errors, e.g., Nimbus-1, considered later.

The ground track of a Sun-synchronous satellite always crosses a given latitude at the same time (local mean time), which is not the crossing time of the ascending node, and which is further away as one moves away from the equator.

If P is the nodal precession rate in round trips per year, the condition for Sun-synchronicity can clearly be written

$$P = 1 . \tag{4.61}$$

4.5.2 Constant of Sun-Synchronicity

The condition (4.60) and the values of K_0 and $\dot{\Omega}_{\mathrm{S}}$ defined by (4.5) and (4.16) lead us to define a quantity k_{h} by

$$k_{\mathrm{h}} = \frac{K_0}{\dot{\Omega}_{\mathrm{S}}} , \tag{4.62}$$

which yields

$$k_{\mathrm{h}} = \frac{3}{4\pi} J_2 \sqrt{\frac{\mu}{R^3}} T_{\mathrm{sid}} . \tag{4.63}$$

This dimensionless constant k_{h}, which we shall call the constant of Sun-synchronicity, plays a very important role in the study of satellites. It depends only on:

- characteristics of the planet playing host to the satellite, such as the mass (via μ), radius (R), flattening (ellipticity factor J_2 of the potential),
- the motion of the planet around the Sun, determining the sidereal year T_{sid}.

It can also be expressed in terms of the Keplerian period of the satellite at altitude 0:

$$k_{\mathrm{h}} = \frac{3}{2} \frac{T_{\mathrm{sid}}}{T_{0(h=0)}} J_2 . \tag{4.64}$$

This constant arises when considering the conditions for Sun-synchronicity, but also more generally, in all aspects of the motion of the satellite orbit relative to its host planet and the Sun.

For the Earth, it has the value

$$k_{\mathrm{h}} = 10.109\,49 . \tag{4.65}$$

This value of the constant, $k_{\mathrm{h}} \approx 10.11$, means that for a satellite of altitude $h = 0$ and inclination $i = 0$, the nodal precession rate is 10.11 times greater than the angular speed of the Earth's axis in its motion around the Sun (absolute value).

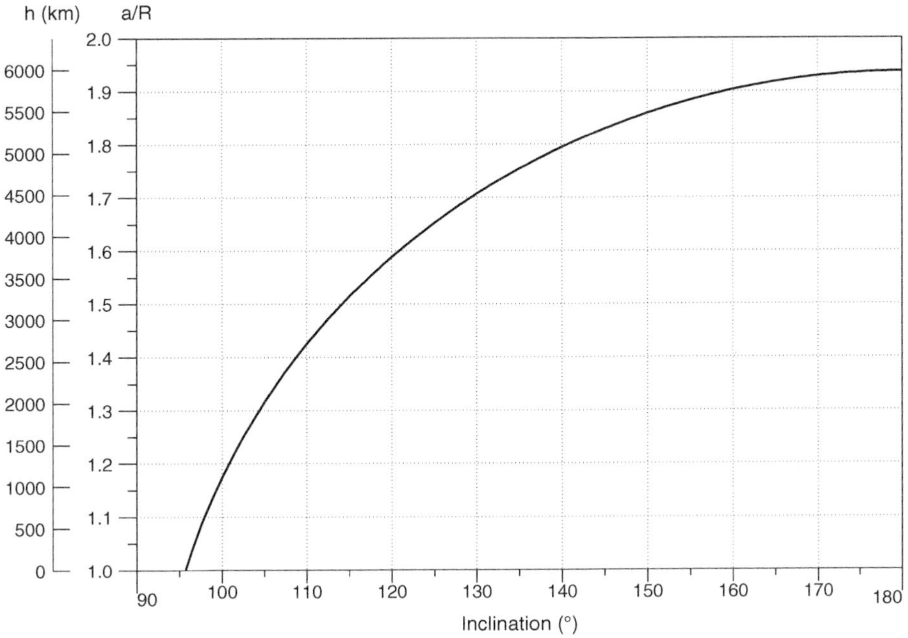

Figure 4.9. Satellite altitude as a function of angle of inclination for Sun-synchronous satellites. The whole range of possible values is shown

4.5.3 Calculating the Orbit

Equations (4.4) to (4.6), (4.60) and (4.62) give

$$\dot{\Omega}(a, i) = \dot{\Omega}_{\mathrm{S}} \iff -\frac{1}{\cos i}\left(\frac{a}{R}\right)^{7/2} = k_{\mathrm{h}} \ . \tag{4.66}$$

For this type of orbit, we have relations giving a in terms of i, and conversely:

$$\left(\frac{a}{R}\right)^{7/2} = -k_{\mathrm{h}} \cos i \ , \tag{4.67}$$

$$i = \arccos\left[-\frac{1}{k_{\mathrm{h}}}\left(\frac{a}{R}\right)^{7/2}\right] \ , \tag{4.68}$$

or, using the reduced distance η,

$$i = \arccos\left(-\frac{\eta^{7/2}}{k_{\mathrm{h}}}\right) \ . \tag{4.69}$$

We thus see that:

- the quantities i and a are related, so that if we choose one, the other is thereby fixed,

- the value of $\cos i$ must be negative for this equality to hold, which shows that a Sun-synchronous orbit is necessarily retrograde.

Figure 4.9 shows the curve of a (or h) as a function of i :

$$\eta = \frac{a}{R} = \frac{R+h}{R} = (-k_\mathrm{h} \cos i)^{2/7} \ . \tag{4.70}$$

In general, the quantity a is chosen and the value of i is then deduced from (4.68), in which case it is called the inclination of the Sun-synchronous orbit, or more elliptically, the Sun-synchronous inclination. In this case, we may attach the subscript HS to i , writing $i_\mathrm{HS} = i_\mathrm{HS}(a) = i(a)$.

Bounds on i and h for a Sun-Synchronous Satellite

The minimum value of i_HS, written $i_\mathrm{HS\,min}$, is obtained for $h = 0$, i.e., for a (fictitious) satellite revolving at ground level. With $\eta = 1$, we have

$$i_\mathrm{HS\,min} = \arccos\left(-\frac{1}{k_\mathrm{h}}\right) = \arccos(-0.0989) = 95.7^\circ \ . \tag{4.71}$$

The maximum value of h, written $i_\mathrm{HS\,max}$, is obtained for $i = 180°$:

$$\eta_\mathrm{HS\,max} = \frac{a}{R} = (k_\mathrm{h})^{2/7} = 1.9367 \ , \tag{4.72}$$

$$a_\mathrm{HS\,max} = 12\,331 \text{ km} \ , \quad h_\mathrm{HS\,max} = 5\,964 \text{ km} \ . \tag{4.73}$$

It is therefore impossible to have a Sun-synchronous satellite (in near-circular orbit) at inclination lower than 96°, or altitude above about 6 000 km.

Calculations for a Standard Sun-Synchronous Satellite

Most Sun-synchronous satellites currently operating have altitude around 800 km (between 700 and 900 km for remote-sensing, and lower for reconnaissance missions). Figure 4.9 shows that, at these altitudes, the relation between i_HS and h is almost linear.

Let us now examine the variation of i near the central value for the range of altitudes mentioned here, i.e., $h_1 = 800$ km. The inclination corresponding to this altitude h_1 is $i_1 = i_{\mathrm{HS}_1} = 98.60°$. Differentiating (4.67), we obtain

$$\frac{7}{2}\frac{\mathrm{d}a}{a} = -\tan i \,\mathrm{d}i \ .$$

Taking i near i_{HS_1} and a near $R + h_1$, we have, for finite increments Δi_HS and Δa, with i_HS in radians and a in metres,

$$\Delta i_\mathrm{HS} = 7.29 \times 10^{-8} \Delta a \ ,$$

or alternatively, with i in degrees and h (or a) in kilometres,

$$\Delta i_\mathrm{HS} = 4.17 \times 10^{-3} \Delta h \ , \tag{4.74}$$

measuring the deviations from the values i_{HS_1} and h_1.

Calculating the Orbit with Expansion up to J_n

Up to now, we have obtained results using (4.66), the basic relation for Sun-synchronicity with the value of $\dot{\Omega}$ limited to the J_2 term. This therefore corresponds to the relation

$$\dot{\Omega} = \dot{\Omega}_S \,, \quad \text{with} \quad \frac{\dot{\Omega}}{n} = J_2 A_2 \cos i \quad \text{and} \quad A_2 = \left(\frac{R}{p}\right)^2 \left(-\frac{3}{2}\right) . \qquad (4.75)$$

When we use an expansion up to degree l for $\dot{\Omega}$, where l is even and we set $l = 2m$, this becomes

$$\dot{\Omega} = \dot{\Omega}_S \,, \quad \text{with} \quad \frac{\dot{\Omega}}{n} = \left[J_2 A_2 + J_2^2 B_2(i) + \sum_{j=2}^{m} J_{2j} A_{2j}(i) \right] \cos i \,, \qquad (4.76)$$

where

$$A_{2j}(i) = \left(\frac{R}{p}\right)^{2j} \sum_{k=0}^{j-1} q_{2k}^{(j)} \sin^{2k} i \,, \quad B_2(i) = \left(\frac{R}{p}\right)^4 (q_0' + q_2' \sin^2 i) \,,$$

and the coefficients $q_{2k}^{(j)}$ and q_{2k}' involve numerical terms and the value of the eccentricity e.

If we solve (4.76) for the inclination i_{HS}, we obtain an equation of degree $(l-1)$ in $\cos i$. This equation is in fact easy to solve because the terms J_2^2 and J_l do not exceed $10^{-3} J_2$. We begin by calculating i in the case when only J_2 is considered, using (4.66) or (4.75). Inserting this value in (4.76), and after several iterations, we obtain the required value. The correction is of the order of 0.03°. For $h = 800$ km, we have $i_{\mathrm{HS}}(J_4) = 98.628°$ and $i_{\mathrm{HS}}(J_2) = 98.603°$, i.e., a difference of 0.025°.

Note. The value of the Sun-synchronous inclination i_{HS} is indicated in all figures in this book showing the ground tracks of Sun-synchronous satellites. Obtained using the expansion to order J_l, they differ by a few hundredths of a degree from the value obtained directly by (4.68).

4.5.4 Sun-Synchronous Satellites

Sun-synchronicity makes judicious use of the nodal precession of the satellite orbit. It is a fundamental advantage in space-based observation to be able to guarantee the passage of a satellite at constant local time for a given latitude, hence in lighting conditions such that the solar zenithal angle varies annually over a well-defined (and rather narrow) range.

The first Sun-synchronous satellite on record is SAMOS-2 (Satellite and Missile Observation System), launched on 31 January 1961, with $h_{\mathrm{p}} =$

474 km, $h_a = 557$ km, $i = 97.4°$, a US military photographic reconnaissance satellite.[25]

The first civilian arena to be interested in Sun-synchronous orbits was meteorology. The satellite Nimbus-1 was launched on 28 August 1964. As the launch was not entirely successful, it ended up on an eccentric, although nevertheless Sun-synchronous orbit, with $h_p = 429$ km, $h_a = 937$ km. It was followed by TIROS-9 and TIROS-10, launched in 1965 into an eccentric orbit for the former and a near-circular orbit with $h \approx 760$ km, $i = 98.8°$ for the latter. All US meteorological satellites were subsequently placed in Sun-synchronous orbit: Nimbus, ESSA, NOAA on the civilian side, and DMSP for the military.

In contrast, the Soviet Union sent up dozens of meteorological satellites, at a rate of three or four per year, as part of the series Meteor-1, -2 and -3, for which they opted for a non-Sun-synchronous direct near-polar orbit. Only the last four of the Meteor-1 series were Sun-synchronous: Meteor-1-28, -1-29, -1-30, -1-31 (also known as Meteor-P-3, -P-4, -P-5, -P-6, respectively), launched between 1977 and 1981, with $h \approx 600$ km, $i = 97.7°$. However, these satellites in the Meteor-P series (Meteor-Priroda, where *priroda* means 'nature' in Russian), adaptations of the meteorological satellites, were more generally devoted to environmental studies and remote-sensing.

Indeed, remote-sensing applications have an even greater interest in Sun-synchronous orbits than meteorology. The first programme in this field was American, with the Landsat programme, which began in 1972 and launched all its satellites into Sun-synchronous orbit. Corresponding programmes, such as the French SPOT, the European ERS, the Indian IRS, and the Russian Resurs-O, were all based on Sun-synchronous satellites, like the environmental observation missions (e.g., EOS, Envisat) and the many programmes implemented since 2000 (e.g., Ikonos) to provide, on a commercial basis, images with resolution of the order of 1 m.

Military reconnaissance satellites intended to operate over long periods (e.g., the French Hélios satellites) are also to be found in the latter category. On the other hand, for reconnaissance satellites on short sporadic missions of a few days or so, the idea of a Sun-synchronous orbit would be meaningless. Indeed, any orbit is Sun-synchronous over a period as short as three days, in the sense that the satellite will overfly a given latitude at almost constant local time. Moreover, near-polar orbits are not necessarily convenient for overflying 'sensitive' regions. Considering the US military reconnaissance programme Key Hole (KH), in which more than three hundred satellites have been launched, there have been very few Sun-synchronous orbits.[26] If most

[25] Among the following satellites, SAMOS-7, -8, -9, -10 and the last of the series, SAMOS-11, launched in 1962, were Sun-synchronous, in very low orbits, with $h \approx 200$ km, $i = 96.2°$.

[26] There have been only a few isolated cases, such as KH-4A-18, launched in 1965, and KH-4B-17, launched in 1972, with $h \approx 300$ km, $i = 96.3°$, and the satellites

of the KH have been Sun-synchronous since the KH-11 series in 1976, this is because the missions have been designed to last longer (months, or even years).

4.5.5 Drift and Stationkeeping

As for any satellite, the orbit of a Sun-synchronous satellite will tend to drift as time goes by, leading to slight modifications in the orbital elements. The two main perturbative effects are the lunisolar attraction and atmospheric drag. The first of these causes a slight variation in the satellite inclination, whilst the second causes braking and hence a reduction in altitude.[27]

In the case of Sun-synchronous satellites, it is of particular importance to achieve a constant crossing time at the ascending node (in local time or LMT). If, like the SPOT, Landsat, ERS and ADEOS satellites, the Sun-synchronous satellite is recurrent, in the sense that its ground track must repeat exactly the same locus after a certain number of revolutions (see Chap. 7), it is even more important to maintain the orbit.[28]

As an example, for the Japanese satellite ADEOS-1 (Advanced Earth Observing Satellite, also called Midori, meaning 'green' in Japanese), the local crossing time at the descending node[29] $\tau_{\rm DN}$ was fixed at $\tau_{\rm DN} = 10{:}30 \pm 0{:}15$.

in the KH-7 and KH-9 series, such as KH-7-01 (Gambit-1), launched in 1963, $h = 190$ km, $i = 95.5°$, and KH-9-01 (Big Bird-1), launched in 1971, $h_{\rm p} = 180$ km, $h_{\rm a} = 300$ km, $i = 96.4°$.

27 The density of molecules in the upper atmosphere is related to the level of solar activity (with a cycle of about 11 yr). During the SPOT-4 mission it was observed that the fall in altitude of this satellite was around 1.20 m per day in periods of low solar activity. However, in periods of high activity, it was between 5 and 10 m per day, and could exceptionally reach 30 m per day.
ESA's satellite ERM (Earth Radiation Mission) was designed for a very low altitude ($h = 362$ km), with low cross-section to reduce atmospheric drag. Moreover, the mission was programmed for 2005 to take advantage of a period of minimal solar activity. In these conditions, orbital readjustments could have been carried out every 10 days. However, the ERM project was abandoned in 1999 and replaced by EarthCARE (Clouds Aerosols Radiation Explorer), in a collaboration with Japan. The launch has been pushed back to around 2008 and the altitude will be increased in consequence to $h = 453$ km (see Table 7.2).

28 This is the case for the Japanese satellite JERS-1 (Japan Earth Resource Satellite, also called Fuyo-1, meaning 'purple rose' in Japanese), launched on 11 February 1992. Its relatively low altitude ($h = 568$ km), Sun-synchronous orbit with tightly tuned recurrence requires orbital readjustment every week.

29 Mission control provided the following value for $\tau_{\rm DN}$ and the expected drift in time: $\tau_{\rm DN} = \tau_0 + a\delta + b\delta^2 = 10.6872 + 4.4329 \times 10^{-6}\delta - 5.8434 \times 10^{-10}\delta^2$, where the time $\tau_{\rm DN}$ is expressed in decimal hours and δ represents the time elapsed in hours since 00:00 on the launch day. The maximum of this parabolic function obtains for $\delta = a/2b = 3.793 \times 10^3$ hr ≈ 158 day.
The satellite was launched on 27 August 1996, with $\tau_{\rm DN} = 10{:}41$. The value went

Another example is the satellite SPOT-4, for which the crossing time at the descending node was fixed at $\tau_{\rm DN} = 10{:}30 \pm 0{:}10$. In order to keep to the recurrence constraint, which required the ground track to pass within 3 km of the equator, altitude corrections were necessary. These maneuvers took place every two to eight weeks depending on the level of solar activity. Every eighteen months or so, the inclination was also reset. Such frequent maneuvers meant that the crossing time at the descending node was in fact $\tau_{\rm DN} = 10{:}30 \pm 0{:}02$ (the maximal discrepancy of 2 min being well below the variation in the equation of time).

Certain Sun-synchronous satellites are no longer maintained in orbit (by necessity or by choice), and the local crossing time at the ascending node drifts in consequence (see Fig. 6.6).

Calculating the Drift in Local Crossing Time

The drift in the local crossing time is due to the secular drift in the inclination i of the orbital plane. The Lagrange equations show that $\mathrm{d}i/\mathrm{d}t = 0$ under the effects of the geopotential. In fact it is the gravitational perturbation due to the Sun which causes this secular variation and, although it is extremely small (a few hundredths of a degree per year), it mounts up like any other such variation.

The local time at the equator, characterised by $\tau_{\rm DN}$, is directly related to the (ascending, or here, descending) node Ω. Differentiating $\dot{\Omega}$ given by (4.1) with respect to time, we thus obtain the second derivative of $\tau_{\rm DN}(t)$ as

$$\frac{\ddot{\tau}_{\rm DN}}{n} = \frac{\mathrm{d}}{\mathrm{d}t}\left(\frac{\dot{\Omega}}{n}\right) = \frac{3}{2} J_2 \left(\frac{R}{a}\right)^2 \sin i \left(\frac{\mathrm{d}i}{\mathrm{d}t}\right) , \tag{4.77}$$

where $\mathrm{d}i/\mathrm{d}t$ is the secular variation of i. This value depends on the angle between the orbit and the direction of the Sun, and hence on $\tau_{\rm DN}$. We have

$$\frac{\mathrm{d}i}{\mathrm{d}t} = -0.032 \text{ degrees/year} \quad \text{for } \tau_{\rm DN} \approx 10{:}30 \, .$$

For a 'typical' Sun-synchronous satellite ($h \approx 800$ km, $\tau_{\rm DN} \approx 10{:}30$), this gives

$$\ddot{\tau}_{\rm DN} = -5.44 \text{ min/yr}^2 \, .$$

This constant value of $\ddot{\tau}_{\rm DN}$ causes a variation in $\tau_{\rm DN}$ going as t^2.

---

through a maximum $\tau_{\rm DN} = 10{:}42$ after 5 months and then returned to $\tau_{\rm DN} = 10{:}41$ after 10 months. The following values were then expected: $\tau_{\rm DN} = 10{:}40$ for 12 months, $\tau_{\rm DN} = 10{:}35$ for 24 months, and $\tau_{\rm DN} = 10{:}24$ for 36 months. The value was $\tau_{\rm DN} = 10{:}15$ after 42 months. At this point, the orbit would have had to be modified because the variation in $\tau_{\rm DN}$ would have been very fast, but the satellite only operated for 10 months. This mission was followed by ADEOS-2 (Midori-2, often written Midori-II), with the same choice for the local crossing time.

4.5.6 Sun-Synchronous Satellites with Highly Eccentric Orbit

For a highly eccentric orbit, the condition for Sun-synchronicity also depends on another constraint. Indeed, to ensure that the perigee does not drift along the orbit, the inclination must be chosen to have its critical value, given by (3.57), which is here greater than 90°, in fact, $i = 116.6°$.

As the inclination is fixed, one next chooses a and e for a Sun-synchronous orbit. This configuration (critical Sun-synchronous inclination) has been envisaged for the constellation of communications satellites Ellipso Borealis, whose orbital elements are given in Chap. 5.

5 Orbit and Ground Track of a Satellite

5.1 Position of the Satellite on its Orbit

Let $\Re(O; x, y, z)$ be the Galilean reference frame already defined. The satellite S is in an elliptical orbit around the centre of attraction O. The orbital plane $\mathcal{P}$ makes a constant angle i with the equatorial plane $\mathcal{E}$. However, although this plane $\mathcal{P}$ is considered as fixed relative to $\Re$ in the Keplerian motion, in a real (perturbed) motion, it will in fact rotate about the polar axis. This is precessional motion,[1] occurring with angular speed $\dot{\Omega}$, as calculated in the last two chapters. A schematic representation of this motion is given in Fig. 5.1. We shall describe the position of S in $\Re$ using the Euler angles.

5.1.1 Position of the Satellite

The three Euler angles ψ, θ and χ were introduced in Sect. 2.3.2 to specify the orbit and its perigee in space. In the present case, we wish to specify S. We obtain the correspondence between the Euler angles and the orbital elements using Fig. 2.1:

$$\psi = \Omega \,, \tag{5.1}$$

$$\theta = i \,, \tag{5.2}$$

$$\chi = \omega + v \,. \tag{5.3}$$

Although they are fixed for the Keplerian orbit, the angles Ω, ω and $M - nt$ vary in time for a real orbit. The inclination i remains constant, however.

The distance from S to the centre of attraction O is given by (1.41), expressed in terms of the true anomaly v:

$$r = \frac{a(1-e^2)}{1 + e\cos v} \,. \tag{5.4}$$

[1] The word ‘precession’, meaning ‘the action of preceding’, was coined by Copernicus around 1530 (*præcessio* in Latin) to speak about the precession of the equinoxes, i.e., the retrograde motion of the equinoctial points. This term was then taken up in mechanics to describe the corresponding Euler angle. In the motion of the satellite orbital plane, the word ‘precession’ clearly refers to a motion that may actually be prograde, as well as retrograde.

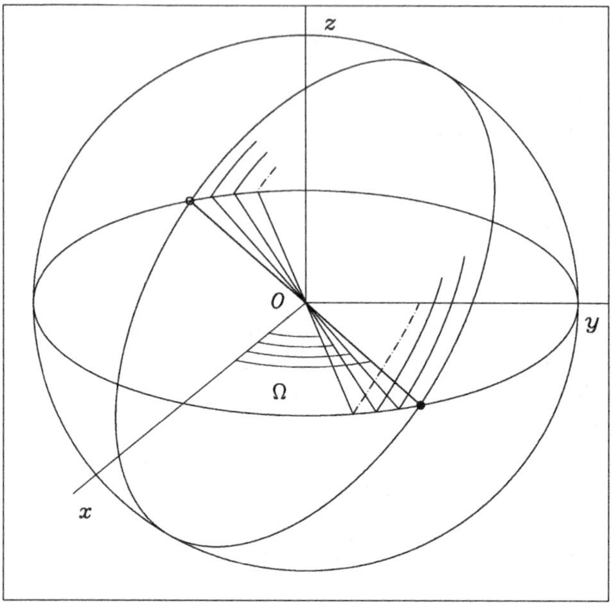

Figure 5.1. Precessional motion of the orbit in the frame $\Re$. The orbital plane rotates about the polar axis, maintaining a fixed inclination relative to the equatorial plane (xOy). Its projection onto the equatorial plane can be used to measure Ω, the longitude of the ascending node, whose variation is given by $\dot{\Omega}$. If the satellite has a prograde orbit (as here, where the ascending node has been indicated by a *small black circle*, the descending node by a *small white circle* and the latest ground track by a *dash-dotted curve*), the precessional motion is retrograde, i.e., $\dot{\Omega} < 0$

Since this distance is specified, the position of S is determined by composing the following three rotations, shown schematically in Fig. 5.2 and described below:

- Precessional motion in $\mathcal{E}$, taking the straight line Ox onto the straight line ON ($= Ox_1$):

$$\Longrightarrow \quad [P_1] : \text{Rotation through angle } (\boldsymbol{Ox}, \boldsymbol{Ox}_1) = \psi \text{ about } \boldsymbol{Oz} \, .$$

- Rotation of $\mathcal{E}$ onto $\mathcal{P}$ about the line of nodes:

$$\Longrightarrow \quad [P_2] : \text{Rotation through angle } (\boldsymbol{Oz}_1, \boldsymbol{Oz}_2) = \theta \text{ about } \boldsymbol{Ox}_1 \, .$$

- Motion in $\mathcal{P}$ which takes the straight line ON ($= Ox_1 = Ox_2$) onto the straight line OS (or OX):

$$\Longrightarrow \quad [P_3] : \text{Rotation through angle } (\boldsymbol{Ox}_2, \boldsymbol{OX}) = \chi \text{ about } \boldsymbol{Oz}_2 = \boldsymbol{OZ} \, .$$

It can be shown that any rotation of a solid can be decomposed into three elementary rotations about suitably chosen axes. In the case of the Euler angles, this decomposition is one-to-one with the following domains:

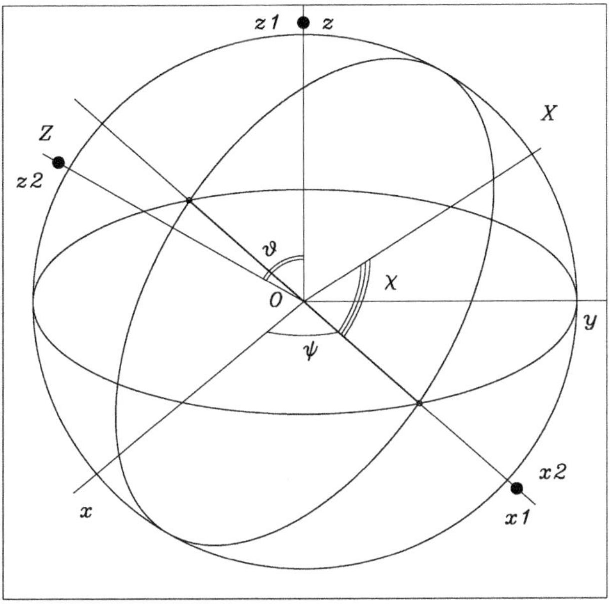

Figure 5.2. The three rotations taking a given point on a sphere to another arbitrary point, using the three Euler angles. *Black circles* indicate the three axes of rotation: $Oz = Oz_1$ for $[P_1]$, $Ox_1 = Ox_2$ for $[P_2]$, and $Oz_2 = OZ$ for $[P_3]$

$$\psi \in [0, 2\pi) \,, \quad \theta \in [0, \pi) \quad \chi \in [0, 2\pi) \,.$$

The axes and angles of rotation are summarised here:

$$\begin{aligned}
(\boldsymbol{Ox}, \boldsymbol{Oy}, \boldsymbol{Oz}) &\overset{P_1}{\longmapsto} (\boldsymbol{Ox}_1, \boldsymbol{Oy}_1, \boldsymbol{Oz}_1 = \boldsymbol{Oz}) \,, \\
(\boldsymbol{Ox}_1, \boldsymbol{Oy}_1, \boldsymbol{Oz}_1) &\overset{P_2}{\longmapsto} (\boldsymbol{Ox}_2 = \boldsymbol{Ox}_1, \boldsymbol{Oy}_2, \boldsymbol{Oz}_2) \,, \\
(\boldsymbol{Ox}_2, \boldsymbol{Oy}_2, \boldsymbol{Oz}_2) &\overset{P_3}{\longmapsto} (\boldsymbol{OX}, \boldsymbol{OY}, \boldsymbol{OZ} = \boldsymbol{Oz}_2) \,.
\end{aligned}$$

We then have the three rotation matrices:

$$P_1 = \begin{pmatrix} \cos\psi & -\sin\psi & 0 \\ \sin\psi & \cos\psi & 0 \\ 0 & 0 & 1 \end{pmatrix} , \tag{5.5}$$

$$P_2 = \begin{pmatrix} 1 & 0 & 0 \\ 0 & \cos\theta & -\sin\theta \\ 0 & \sin\theta & \cos\theta \end{pmatrix} , \tag{5.6}$$

$$P_3 = \begin{pmatrix} \cos\chi & -\sin\chi & 0 \\ \sin\chi & \cos\chi & 0 \\ 0 & 0 & 1 \end{pmatrix} . \tag{5.7}$$

The matrix product of these three matrices gives the matrix P calculated below.

We consider without loss of generality that N is on the axis Ox at the time origin. Its coordinates are thus $(r, 0, 0)$. The coordinates of $S(X, Y, Z)$ are obtained from those of $N(x, y, z)$ via application of P:

$$\begin{pmatrix} X \\ Y \\ Z \end{pmatrix} = P \begin{pmatrix} x \\ y \\ z \end{pmatrix} = P \begin{pmatrix} r \\ 0 \\ 0 \end{pmatrix} .$$

We see that only the first column of the matrix P will be required for this calculation. We shall therefore calculate the matrix product $P = P_1 P_2 P_3$ and write it in the form

$$P = \begin{pmatrix} \cos\psi\,\cos\chi - \sin\psi\,\sin\chi\,\cos\theta & P_{12} & P_{13} \\ \sin\psi\,\cos\chi + \cos\psi\,\sin\chi\,\cos\theta & P_{22} & P_{23} \\ \sin\chi\,\sin\theta & P_{32} & P_{33} \end{pmatrix} , \tag{5.8}$$

which gives

$$\begin{pmatrix} X \\ Y \\ Z \end{pmatrix} = r \begin{pmatrix} \cos\psi\,\cos\chi - \sin\psi\,\sin\chi\,\cos\theta \\ \sin\psi\,\cos\chi + \cos\psi\,\sin\chi\,\cos\theta \\ \sin\chi\,\sin\theta \end{pmatrix} . \tag{5.9}$$

Using the orbital parameters given by (5.1)–(5.4), we obtain

$$\begin{pmatrix} X \\ Y \\ Z \end{pmatrix} = \frac{a(1-e^2)}{1+e\cos v} \begin{pmatrix} \cos\Omega\,\cos(\omega+v) - \sin\Omega\,\sin(\omega+v)\,\cos i \\ \sin\Omega\,\cos(\omega+v) + \cos\Omega\,\sin(\omega+v)\,\cos i \\ \sin(\omega+v)\,\sin i \end{pmatrix} . \tag{5.10}$$

Consider a spherical coordinate system in the Galilean frame $\mathfrak{R}$. The plane of reference is the equatorial plane xOy of the Earth, Oz is the polar axis and the position of Ox is fixed in space.

The point S can be specified in $\mathfrak{R}$ by its spherical coordinates, the longitude λ and the latitude ϕ, measured with the usual convention following from the right-handed trigonometric system. The longitude of Ox (position of N at the time origin) is denoted by λ_0. Hence,

$$\begin{pmatrix} X \\ Y \\ Z \end{pmatrix} = r \begin{pmatrix} \cos\phi\,\cos(\lambda-\lambda_0) \\ \cos\phi\,\sin(\lambda-\lambda_0) \\ \sin\phi \end{pmatrix} . \tag{5.11}$$

We thus obtain the position of $S(\lambda, \phi)$ as a function of time and the other orbital parameters via X, Y, Z:

$$\phi = \arcsin\frac{Z}{r} , \qquad \lambda = \lambda_0 + \arccos\frac{X}{r\cos\phi} , \tag{5.12}$$

$$\lambda - \lambda_0 \text{ from the sign of } Y , \qquad \lambda - \lambda_0 \in (-\pi, +\pi] . \tag{5.13}$$

If $|\phi| = \pi/2$, λ is not determined (and its determination would be pointless).

5.1.2 Equation for the Ground Track

In many circumstances, one needs to know the position of the satellite relative to the Earth. One must therefore represent S in the frame $\Re_T$, whose axes in the equatorial frame rotate with the Earth. The transformation from this frame to the Galilean frame $\Re$ is obtained by a simple rotation about the polar axis $\boldsymbol{Oz}$ with angular speed $(-\dot{\Omega}_T)$, since $\Re_T$ rotates in $\Re$ with angular speed $\dot{\Omega}_T$. Bear in mind that these calculations are carried out in the Galilean frame $\Re$, whilst the results may be expressed in the frame of our choice.

Recalling the above definition of λ_0, the equations of motion of S are the same in $\Re_T$ as in $\Re$, provided that we replace the value of ψ in (5.1) by

$$\psi = \lambda_0 + (\dot{\Omega} - \dot{\Omega}_T)(t - t_{AN}) , \tag{5.14}$$

where the time origin, the crossing time at the ascending node N, is written $t = t_{AN}$.

The satellite ground track is defined as the intersection of the straight line segment OS with the Earth's surface. Its equation is thus obtained by replacing r by R in the above equations. (For this application, we may treat the Earth as a sphere of radius R.)

5.2 Ground Track of Satellite in Circular Orbit

Near-circular orbits, which may be considered as circular in a first approximation, constitute a very important and frequently encountered case. Let us now study some notions developed specifically for these orbits, such as the equatorial shift or the apparent inclination.

The velocity of the satellite will be calculated in Sect. 5.5.

5.2.1 Equation for Satellite Ground Track

When the orbit is circular, the motion is uniform with angular frequency n, the mean motion. Using the notation introduced above, the value of χ in (5.3) can be replaced by

$$\chi = n(t - t_{AN}) . \tag{5.15}$$

We thus obtain the equation for the ground track, with (5.14), (5.2) and (5.15) substituted into (5.9) and (5.11), where r has been changed to R.

In this case, the ground track of the satellite is determined by two quantities relating to the ascending node taken as origin, namely, its longitude λ_0 and the crossing time t_{AN}, which constitute the initial conditions of the uniform motion.

Sun-Synchronous Satellites

For Sun-synchronous satellites, the angle ψ takes a specific value since $\dot{\Omega} = \dot{\Omega}_\mathrm{S}$. We have seen that the two angular frequencies characterising the Earth's (annual and daily) motion are related by (4.24). Hence, according to (5.14),

$$\dot{\psi} = \dot{\Omega}_\mathrm{S} - \dot{\Omega}_\mathrm{T} = -\frac{2\pi}{J_\mathrm{M}} \,. \tag{5.16}$$

Using the daily orbital frequency as given by (4.25), we obtain, for Sun-synchronous satellites, the very simple relation

$$\dot{\psi} = -\frac{n}{\nu} \,. \tag{5.17}$$

We shall see the very important consequences of this relation in the following chapters, in particular, when studying the crossing time of the satellite and the question of recurrent orbits.

5.2.2 Maximum Latitude Attained

The ground track moves between two bounding latitudes, $\phi = +\phi_\mathrm{m}$ in the northern hemisphere and $\phi = -\phi_\mathrm{m}$ in the southern hemisphere. Considering the maximum positive value of Z, obtained for $\sin n(t - t_\mathrm{AN}) = 1$, we obtain

$$\sin \phi_\mathrm{m} = \sin i \,, \tag{5.18}$$

which implies that:

- for prograde satellites ($0 \leqslant i \leqslant \pi/2$), $\phi_\mathrm{m} = i$,
- for retrograde satellites ($\pi/2 \leqslant i \leqslant \pi$), $\phi_\mathrm{m} = \pi - i$.

This value ϕ_m ($\phi_\mathrm{m} \geqslant 0$) is called the maximum attained latitude.

Example 5.1. *Calculate the maximal latitude attained by the Chinese satellite FY-1A, in Sun-synchronous orbit at an altitude of $h = 901$ km.*

Given the altitude, we determine the inclination using (4.67). In the present example, we can use the simplified formula (4.74):

$$\Delta i = (901 - 800) \times 4.17 \times 10^{-3} \approx 0.4^\circ \,, \quad i = i_0 + \Delta i = 98.6 + 0.4 = 99.0^\circ \,,$$

which yields the maximal attained latitude $\phi_\mathrm{m} = 180 - i = 81.0^\circ$. Note that this is the latitude reached by the ground track, not by oblique sightings by instruments carried aboard. The ground track of the satellite thus remains within the bounding latitudes 81°N and 81°S.

5.2.3 Equatorial Shift

The difference in longitude between two consecutive ascending nodes λ_1 and λ_2 is called the equatorial shift and denoted by $\Delta\lambda_{\mathrm{E}}$, i.e.,

$$\Delta\lambda_{\mathrm{E}} = \lambda_2 - \lambda_1 \; .$$

Rough Calculation

It is often sufficient to carry out a quick calculation of the equatorial shift, which is then denoted by $\Delta_0\lambda_{\mathrm{E}}$. Indeed we may say to a first approximation that, during one revolution of period T (and we may take the Keplerian period here), the orbit of the satellite will not have moved relative to $\Re$, whilst the Earth makes one complete turn every day, i.e., it rotates through 15° per hour, or 1° every 4 min relative to this same frame. In this context, we do not bother with the precession of the orbit, or the Earth's motion relative to the Sun over the time taken for the satellite to complete one revolution. This amounts to using the relations $\dot{\psi} \approx -\dot{\Omega}_{\mathrm{T}}$ and $\dot{\Omega}_{\mathrm{T}} \approx 2\pi/J_{\mathrm{M}}$.

We have the simplified relation

$$\Delta_0\lambda_{\mathrm{E}} \text{ [degrees]} = -\frac{T \text{ [min]}}{4} \; , \tag{5.19}$$

where the minus sign indicates a shift westwards. Observing that the value of one degree of longitude is 111.3 km on the equator, we can also write

$$\Delta_0\lambda_{\mathrm{E}} \text{ [km]} = -27.82T \text{ [min]} \; . \tag{5.20}$$

Using the daily orbital frequency ν (number of round trips per day), we obtain $\Delta_0\lambda_{\mathrm{E}}$ [degrees] $= -360/\nu$ and $\Delta_0\lambda_{\mathrm{E}}$ [km] $\approx -4 \times 10^4/\nu$.

Exact Calculation

During one revolution lasting $T = T_{\mathrm{d}}$, the orbital plane will have rotated through an angle ψ with respect to $\Re_{\mathrm{T}}$. The exact value of the equatorial shift as given by (5.14) with $t - t_{\mathrm{AN}} = T$ is therefore

$$\Delta\lambda_{\mathrm{E}} = \dot{\psi}T = -(\dot{\Omega}_{\mathrm{T}} - \dot{\Omega})T \; . \tag{5.21}$$

We note the following points, which follow from (5.21):

- The equatorial shift is always negative, since $\dot{\Omega}_{\mathrm{T}}$ is greater than $\dot{\Omega}$. The shift is westward for a satellite below the geosynchronous orbit.
- For satellites in Sun-synchronous orbit, we saw the specific value of $\dot{\psi}$, according to (5.16). Over one nodal period T, we have

$$\Delta\lambda_{\mathrm{E}} = \dot{\psi}T = -\frac{2\pi}{J_{\mathrm{M}}}T \; . \tag{5.22}$$

Writing the angles in degrees and the time in minutes, we obtain (5.19). For a Sun-synchronous satellite, the approximate formula is identical to the exact one. This is because the two approximations we made in the rough calculation (neglecting precession and the annual motion of the Earth) exactly balance for this type of satellite.

- For satellites in geosynchronous orbit, $T = 2\pi/\dot{\Omega}_{\mathrm{T}}$ and $\dot{\Omega}$ is negligible. (In any case, it is not the leading term in the perturbation treatment for this type of satellite.) In this case, we thus have

$$\Delta\lambda_{\mathrm{E}} = -\dot{\Omega}_{\mathrm{T}} T = -2\pi = 0 \mod (2\pi) . \tag{5.23}$$

There is no equatorial shift for such a satellite. The projection of two consecutive ascending nodes does not move on the Earth. (If the satellite is geostationary, we cannot even speak of an ascending node.)

Example 5.2. *Calculate the equatorial shift for the satellite Meteor-3-07.*

The characteristics of the orbit of this satellite are given in Example 4.2. For the quick calculation, we use (5.20) with $T = 109.4$ min. Then

$$\Delta_0\lambda_{\mathrm{E}} = -27.82 \times 109.4 = -3044 \text{ km} .$$

For the exact calculation, with the values already given, viz.,

$$\begin{aligned} \dot{\Omega}_{\mathrm{T}} &= 729.212 \times 10^{-7} \text{ rad s}^{-1} , \\ \dot{\Omega} &= -1.429 \times 10^{-7} \text{ rad s}^{-1} , \\ \dot{\Omega}_{\mathrm{T}} - \dot{\Omega} &= 730.641 \times 10^{-7} \text{ rad s}^{-1} , \end{aligned}$$

and $T = 6\,565.28$ s, we obtain

$$\Delta\lambda_{\mathrm{E}} = -0.479\,7 \text{ rad} = -27.48^\circ = -3\,059.51 \text{ km} .$$

The equatorial shift of the satellite Meteor-3-07 is thus 3 059.5 km westward (see also Fig. 5.3 and Table 5.1).

5.2.4 Apparent Inclination

Definition and Calculation of Apparent Inclination

The apparent inclination is the angle between the ground track and the equator. This angle i' differs from the angle i representing the inclination of the satellite, which is the inclination of the orbital plane of the satellite with respect to the equatorial plane. This happens because i is measured in $\Re$, whereas i' is measured in $\Re_{\mathrm{T}}$.

To calculate i', we consider in $\Re_{\mathrm{T}}$ the tangent plane to the sphere of radius R at the point on the Earth's surface corresponding to the ascending node, using the orthogonal unit vectors $\boldsymbol{e}_\lambda$ and $\boldsymbol{e}_\phi$ already defined. This ascending node is denoted by N in $\Re$ and N_0 in $\Re_{\mathrm{T}}$. At time $t = t_{\mathrm{AN}}$, the three points

Meteor-3-07 ⟨ Метеор ⟩
Orbit

Terrestrial frame

>>>> Time span shown: 109.42 min = 0.08 day

Altitude = 1194.6 km a = 7572.703 km

Inclination = 82.56 °

Period = 109.42 min * rev/day =13.16

Equat. orbital shift = 3059.5 km (27.5 °)

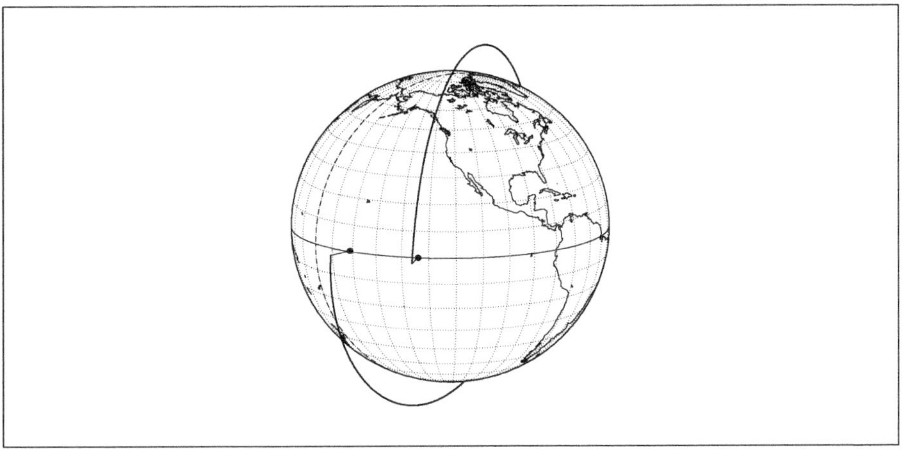

Projection: Orthographic Map centre: 12.0 ° N; 122.0 ° W Asc. node: -133.95 ° [07:43 UTC] Ιξιων

Property: none Aspect: Oblique MC ★ LMD

T.:Azimuthal ⊕ Graticule: 10° [-90.0 / +78.0 / -148.0] Gr.Mod.: GEM-T2 Ατλας

Meteor-3-07 ⟨ Метеор ⟩
Ground track

Terrestrial frame

>>>> Time span shown: 109.42 min = 0.08 day

Altitude = 1194.6 km a = 7572.703 km

Inclination = 82.56 °

Period = 109.42 min * rev/day =13.16

Equat. orbital shift = 3059.5 km (27.5 °)

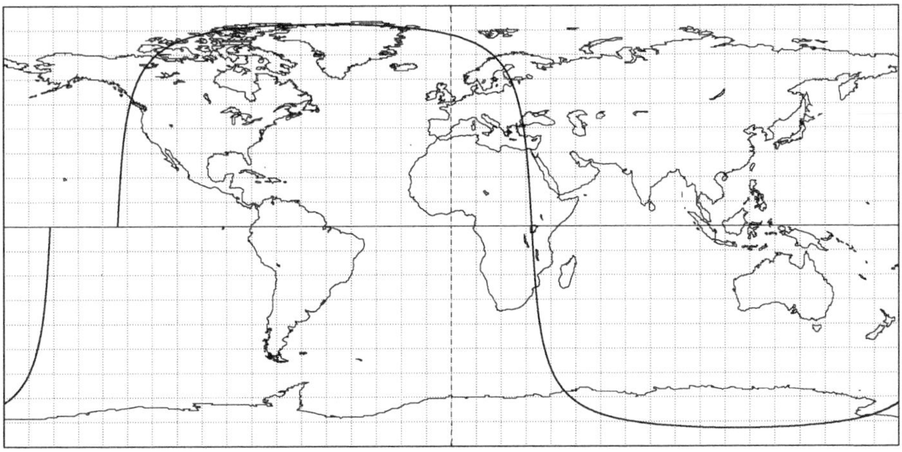

Projection: Plate-carrée Map centre: 0.0 ° ; 0.0 ° Asc. node: -133.95 ° [07:43 UTC] Ιξιων

Property: none Aspect: Direct App. inclin. = 86.93 ° MC ★ LMD

T.:Cylindrical ⊕ Graticule: 10° [+0.0 / +0.0 / +0.0] Gr.Mod.: GEM-T2 Ατλας

Figure 5.3. Orbit and ground track of the satellite Meteor-3-07 over one revolution. The distance between the two successive ascending nodes is the equatorial shift

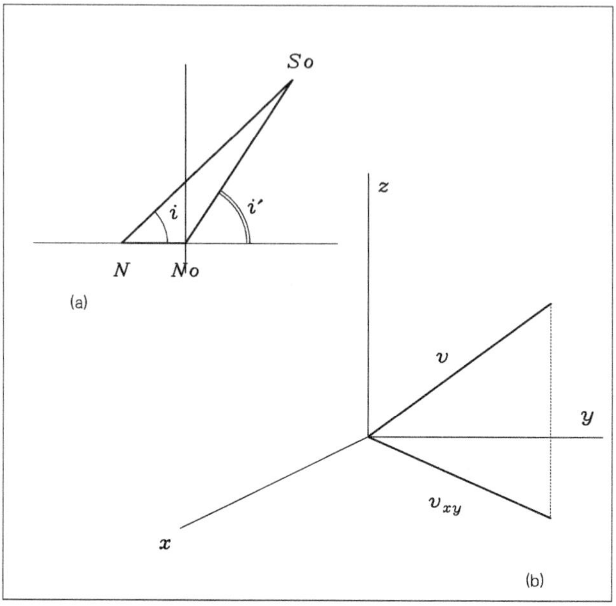

Figure 5.4. Diagrams for the description of apparent inclination. (**a**) Inclination i and apparent inclination i'. The subsatellite point S_0 and the point on the ground track corresponding to the ascending node have been represented in $\Re_T$. (**b**) Components of the satellite velocity in the frame $\Re_T$. The velocity vector v and its projection v_{xy} onto the equatorial plane have been represented in the frame moving with the Earth

S_0 (the subsatellite point), N and N_0 coincide. An infinitesimal time dt later, N and S_0 have moved away from N_0, as shown in Fig. 5.4a. In the frame $\Re_T(N_0, \boldsymbol{e}_\lambda, \boldsymbol{e}_\phi)$, the components of the vectors $\boldsymbol{N}_0\boldsymbol{N}$ and $\boldsymbol{N}\boldsymbol{S}_0$ are, setting $dt = 1$,

$$\boldsymbol{N}_0\boldsymbol{N} = \begin{pmatrix} -(\dot{\Omega}_T - \dot{\Omega}) \\ 0 \end{pmatrix}, \qquad \boldsymbol{N}\boldsymbol{S}_0 = \begin{pmatrix} n\cos i \\ n\sin i \end{pmatrix}.$$

We thus deduce the components of $\boldsymbol{N}_0\boldsymbol{S}$:

$$\boldsymbol{N}_0\boldsymbol{S} = \begin{pmatrix} n\cos i - (\dot{\Omega}_T - \dot{\Omega}) \\ n\sin i \end{pmatrix},$$

and the apparent inclination is given by

$$\tan i' = \frac{n\sin i}{n\cos i - (\dot{\Omega}_T - \dot{\Omega})}. \tag{5.24}$$

In terms of the daily recurrence frequency κ defined by (4.32), we may write

$$\tan i' = \frac{\sin i}{\cos i - (1/\kappa)} \; . \tag{5.25}$$

Note that we always have $i' \geqslant i$.

For a Sun-synchronous satellite, we may replace κ by ν, since according to (4.33), these two daily frequencies are equal.

Example 5.3. *Calculate the angle between the ground track of the TRMM satellite and the equator when the satellite crosses the ascending node.*

For TRMM ($i = 34.98°$), we calculated the nodal precession rate in Example 4.1. We obtain

$$\tan i' = \frac{\sin 34.98}{\cos 34.98 - 0.0649} = 0.7604 \; ,$$

$$i' = 37.25° \; , \quad i' - i = 2.27° \; .$$

Example 5.4. *Calculate the apparent inclination for the ground track of a geosynchronous satellite.*

For a geosynchronous satellite, we have $\dot{\Omega}_T/n = 1$ and the term $\dot{\Omega}$ is negligible. Equation (5.24) becomes

$$\tan i' = \frac{\sin i}{\cos i - 1} = -\frac{\cos(i/2)}{\sin(i/2)} = \tan\left(\frac{\pi}{2} + \frac{i}{2}\right) \; ,$$

$$i' = 90° + \frac{i}{2} \; , \quad i' - i = 90° - \frac{i}{2} \; .$$

When i is very small, e.g., $i = 1°$, we have $i' = 90.5°$: the ground track is not a point but a small line segment almost perpendicular to the equator, between latitudes 1°N and 1°S, which transforms into a figure of 8 (lemniscate) when i increases, growing larger with i. The first operational geosynchronous satellite, Syncom-2, had inclination 32.8°. Its ground track made an angle of 106.4° with the equator, or an angle of 16.4° with the nodal meridian, as can be seen from the upper part of Fig. 5.10.

Calculating the Inclination from the Apparent Inclination

We have obtained i' as a function of i (and a). It is sometimes useful to do the opposite calculation, obtaining i as a function of i' (and a). Using (4.1) and expressing $\dot{\Omega}/n$ by means of J_2 and i, (5.24) can be replaced by the following equation:

$$A \cos i - B = C \sin i \; , \tag{5.26}$$

where

$$A = 1 - \frac{3}{2} J_2 \left(\frac{R}{a} \right)^2 , \quad B = \frac{\dot{\Omega}_{\mathrm{T}}}{n} , \quad C = \frac{1}{\tan i'} . \tag{5.27}$$

The aim now is to solve this for i. To begin with, we take $n = n(a, i) = n_0$ in B so that this term does not depend on i. In a second step, we substitute into B the value of n obtained using the previous value of i. The method converges very quickly.

Equation (5.26) transforms to a second order equation in $\tan(i/2)$. The solution, which is unique since the angle i lies in the interval $[0, \pi]$, is given by

$$\tan \frac{i}{2} = \frac{-C + \sqrt{C^2 + A^2 - B^2}}{A + B} . \tag{5.28}$$

We note that we have $A \approx 1$, $B \approx 1/\nu$ and hence $B < A$, except for satellites in geosynchronous orbit or higher. This method allows one to find the inclination of a satellite of known altitude by measuring the apparent inclination, on a map of the ground track, for example. As we shall see below, it also allows one to calculate the inclination from the components of the velocity vector of the satellite.

Calculating the Inclination from the Satellite Velocity

The position and velocity of a satellite can be found either from the orbit bulletin or from remote-sensing data from the instruments aboard. The position and velocity are given relative to the Earth. Let v_X, v_Y, v_Z be the velocity components in $\Re_{\mathrm{T}}$. At the ascending node, the angle between the velocity vector, and hence the satellite trajectory, and the equatorial plane is the apparent inclination.

Using Fig. 5.4b, we obtain the relation between i and the velocity at the node (using the absolute value for the velocity components if the nodes cannot be distinguished):

$$\tan i' = \frac{v_Z}{\sqrt{v_X^2 + v_Y^2}} . \tag{5.29}$$

From the value for a and the value for i' obtained in this way, we find i using (5.28).

Example 5.5. *Calculate the inclination of the satellite Meteor-3-07 using the components of the velocity vector.*

The values in Table 5.1 were obtained by interpolation of the raw remote-sensing data collected during the first revolution in which the instrument ScaRaB was

Table 5.1. Raw remote-sensing data from the ScaRaB instrument relating to the first operational revolution of Meteor-3-07. The table shows the altitude h (in km), the velocity components v_X, v_Y, v_Z (in $\mathrm{km\,s^{-1}}$), the latitude ϕ and the longitude λ (in degrees), and the crossing time (in UT) at the ascending (A) and descending (D) nodes

Time 1994 02 24	Node	Latitude ϕ	Longitude λ	Altitude h	Velocity components v_X	v_Y	v_Z
07:43:29	A	0.00	−133.95	1211.658	0.280	−0.261	7.182
08:38:10	D	0.00	32.33	1188.318	−0.202	0.331	−7.204
09:32:54	A	0.00	−161.44	1211.657	−0.128	−0.361	7.182

recording. Nodes (1) and (3) were ascending ($v_Z > 0$), whilst node (2) was descending ($v_Z < 0$). We calculate i' from (5.29):

$$\tan i' = \frac{7.182}{0.383} = 18.763\,, \quad \tan i' = \frac{7.204}{0.388} = 18.578\,, \quad \tan i' = \frac{7.182}{0.383} = 18.751\,.$$

We then deduce i in each case and take the average between the ascending and descending nodes:

$$i' = 86.933^\circ\,.$$

We calculate the semi-major axis a from the nodal period, since we know the time elapsed between two consecutive transits at the ascending node (see Examples 4.1 and 4.2). With $T = 109$ min 25 s $= 109.42$ min, we obtain $a = 7\,572.7$ km. We observe that the altitudes given in Table 5.1 vary by 12 km on either side of the average $h = 1200$ km depending on the type of node. This is explained by the fact that the orbit is slightly eccentric and the argument of the perigee is not $\pm 90^\circ$. With (5.28), the values of i' and a give

$$A = 0.998\,85\,, \quad B = 7.619 \times 10^{-2}\,, \quad C = 5.358 \times 10^{-2}\,, \quad \cos i = 0.129\,47\,.$$

Finally we obtain the value of i, viz.,

$$i = 82.561^\circ\,.$$

We thus find the inclination $i = 82.56^\circ$ of the satellite Meteor-3-07, which is precisely the value communicated by the Russian Space Agency.

5.2.5 Angle Between the Ground Track and a Meridian

We calculate the angle between the satellite ground track and a meridian for an arbitrary point on the ground track. The calculation of the angle between the ground track and a line of latitude gives a generalisation of the apparent inclination. However, in practice, it is more useful to known the angle between the ground track and the north–south direction.

In the frame $\Re$, the satellite orbit cuts the meridian at an angle j. Referring to Fig. 6.4, P is the subsatellite point (with latitude ϕ), N is the point on the ground track corresponding to the ascending node (the dihedral angle at N gives the inclination i), and PQ is the meridian through P, where Q is on the equator. The dihedral angle at P is the angle j that we wish to determine. Using the relation (ST V) with PQN for CAB, we obtain

$$\sin j = \frac{\cos i}{\cos \phi} \,. \tag{5.30}$$

To calculate j', we consider in $\Re_{\mathrm{T}}$ the plane tangent to the sphere of radius R at the relevant point, with latitude ϕ, and orthogonal unit vectors $\boldsymbol{e}_\lambda$ and $\boldsymbol{e}_\phi$ as already defined. As in the calculation of the apparent inclination, the relevant point is denoted by P in $\Re$ and P_0 in $\Re_{\mathrm{T}}$. At time $t = t_{\mathrm{AN}}$, the three points S_0 (the subsatellite point), P and P_0 coincide. After an infinitesimal time $\mathrm{d}t$ ($\mathrm{d}t = 1$), we obtain

$$\boldsymbol{P}_0\boldsymbol{P} = \begin{pmatrix} -(\dot{\Omega}_{\mathrm{T}} - \dot{\Omega})\cos\phi \\ 0 \end{pmatrix} , \qquad \boldsymbol{P}\boldsymbol{S}_0 = \begin{pmatrix} n \sin j \\ n \cos j \end{pmatrix} .$$

We deduce the components of $\boldsymbol{P}_0\boldsymbol{S}$:

$$\boldsymbol{P}_0\boldsymbol{S} = \begin{pmatrix} n\sin j - (\dot{\Omega}_{\mathrm{T}} - \dot{\Omega})\cos\phi \\ n\cos j \end{pmatrix} .$$

This should be compared with (5.24).

Using the daily frequency κ, we thus obtain

$$\tan j' = \frac{\sin j - (1/\kappa)\cos\phi}{\cos j} \,, \tag{5.31}$$

and rewriting j,

$$\tan j' = \frac{\cos i - (1/\kappa)\cos^2\phi}{\sqrt{\cos^2\phi - \cos^2 i}} \,. \tag{5.32}$$

When the latitude of the point P equals the maximal attained latitude, one can check that the ground track is in fact normal to the meridian.

5.3 Classifying Orbit Types

Satellite orbits can be classified according to various criteria: the inclination, the altitude, the eccentricity, or various properties.

Classification by Inclination

We have seen that the angle of inclination i of the orbit (angle of nutation θ for the Euler angles) is defined to lie between 0° and 180°. If i is less than

90°, the orbit is prograde, whereas if i is greater than 90°, it is retrograde. When $i = 90°$, the orbit is polar. One may say strictly polar, because when i lies between 80° and 100°, one often describes it as a polar orbit, whereas near-polar would be more appropriate. Figures 5.5 and 5.6 show the ground tracks of these orbits.

If $i = 0°$ (or $i = 180°$, although this has never happened), the orbit is equatorial, and for i less than 10°, it is near-equatorial.

Classification by Altitude

Satellites in near-circular orbit are classified according to their mean altitude. We speak of a Low Earth Orbit (LEO) when the satellite flies at an altitude below 1 500 km, a Medium Earth Orbit (MEO) for GPS satellites at an altitude of around 20 000 km, and a Geostationary Earth Orbit (GEO) (also sometimes called the Clarke orbit) for geostationary satellites at an altitude of 36 000 km. We shall often use these abbreviations, which are concise and consistent.[2] Almost all satellites in orbits with low eccentricity fall into one of these three categories. (For example, it is very rare, to find a satellite at an altitude of 8 000 km.)

For highly elliptical orbits,[3] such as the Molniya or Tundra orbits, we use the abbreviation HEO (Highly Eccentric Orbit). The name GTO (Geostationary Transfer Orbit) is usually a temporary one, because the satellite has been placed on this highly eccentric orbit for transfer towards a GEO orbit. Some satellites can be found in such orbits, some deliberately placed there, others because the apogee thrust used to make the orbit circular has been unsuccessful. Finally, if a satellite has not been correctly placed in orbit, it is sometimes given the title FTO (Failed Transfer Orbit)!

The orbits L1LO and L2LO refer to halo orbits around the Lagrange points, which were discussed in Sect. 3.14.

Classification by Properties

When one needs to specify that a satellite is not geostationary, the term non-geostationary satellite is used. Likewise a Sun-synchronous satellite is contrasted with a non-Sun-synchronous satellite. We shall see other properties later, such as recurrent and frozen orbits.

---

[2] When we are referring to the satellite as LEO rather than the orbit, we understand of course 'low Earth orbiting' satellite. One does occasionally find the term GEO meaning Geosynchronous Earth Orbit, as opposed to GSO for GeoStationary Orbit. In addition, and somewhat unnecessarily, one finds the term IGSO meaning Inclined GeoSynchronous Orbit for geosynchronous orbits that are tilted and therefore not geostationary.

[3] For Molniya-type orbits, the term THEO (Twelve Hour Eccentric Orbit) is sometimes used. For very high orbits, like the orbit of Geotail, we use the term VHO (Very High Orbit).

Orbit and Revolution

Since all scientific enterprise is based on a precise use of language, one must mention a very common error which consists in saying 'orbit' instead of 'revolution' or 'round trip', an error which occurs in English, French and very likely in other languages too. For example, we may read: the satellite Terra, during orbit 7778 .... This confusion is unjustified, and indeed, it is never encountered in astronomy: one never says that the Moon makes one orbit around the Earth every month.

5.4 Classifying Satellites by Mission

Our classification of satellites according to mission, which is of course rather arbitrary, aims to illustrate the various types of orbit. We begin with satellites designed for geophysics and Earth observation, then for navigation and communications, astronomy, technological development, and others that elude straightforward classification. We shall touch briefly upon military satellites and their specific missions, and satellites carrying humans.

The mission of a satellite often covers a range of different areas, e.g., an oceanographic satellite may also take part in a geodesy mission or a mission to develop altimetric techniques, and there has often been a large measure of ideology in satellite missions, especially at the beginning of the space age. However, we shall not be making a special entry for ideology!

With regard to military (or partly military) satellites the nomenclature is often somewhat vague (even confused). From 1984, the United States called some of its satellites USA followed by a number specifying order of launch. Previously, these satellites had been called OPS followed by a four-figure number, without chronological ordering. (Between 1963 and 1984, close on 500 OPS satellites were launched.) The USSR, then Russia, also created confusion with the Kosmos satellites: this name (from the Russian word КОСМОС, originating itself from the Greek word ὁ κόσμος, ου, meaning 'order' or 'well ordered', hence 'universe') groups a whole multitude[4] of satellites (not always military), on every type of orbit and for every available type of mission. The People's Republic of China did likewise with the appellation DFH (Dong Fang Hong, where 'dong fang' means 'Orient' and 'hong' means 'red'), which covers the great majority of Chinese satellites. Without doing anything to

[4] The launch dates were as follows: Kosmos-1 on 16 March 1962, Kosmos-1001 on 4 April 1978, Kosmos-2001 on 14 February 1989. The launch rate then subsided somewhat. We give here the last Kosmos launched in the year: Kosmos-2054 (1989), Kosmos-2120 (1990), Kosmos-2174 (1991), Kosmos-2229 (1992), Kosmos-2267 (1993), Kosmos-2305 (1994), Kosmos-2325 (1995), Kosmos-2336 (1996), Kosmos-2348 (1997), Kosmos-2364 (1998), Kosmos-2368 (1999), Kosmos-2376 (2000), Kosmos-2386 (2001), Kosmos-2396 (2002), Kosmos-2404 (2003).

simplify the situation, these satellites are also recorded by Western organisations under the appellation PRC (People's Republic of China), with another numbering system.

Satellites placed in orbit by the US Space Shuttle are noted: launched by STS-(number). A satellite that is not specified in a series is denoted by -n (e.g., Molniya-n).

Launch dates are given up to 1 November 2004.

5.4.1 Geophysical Satellites

Geodesy

We have already mentioned these satellites in Chap. 3, where we gave a complete list of the satellites used for the geopotential models EGM96S and GRIM5-S1.

The satellite Sputnik-1, the very first of all satellites, can be considered as a geodesy satellite. At the beginning of space geodesy, many satellites were placed above the LEO altitude so as to reduce atmospheric drag. Examples are PAGEOS, launched in 1966, between 3 000 and 5 200 km, with $i = 84.4°$, the two LAGEOS (Laser Geodynamics Satellite), with $h = 5\,900$ km and inclinations $i = 109.8°$ for LAGEOS-1, launched in 1976 and $i = 52.6°$ for LAGEOS-2, launched in 1992 by STS-52. The ground track of the orbit of LAGEOS-1 is shown in Fig. 5.5 (upper).

The satellites SECOR-7, -8, -9 orbit at 3 700 km altitude and the Soviet satellites Etalon-1 and -2 (Kosmos-1989 and -2024), launched in 1989 with GLONASS, are in a circular MEO orbit, $h = 19\,130$ km, $i = 64.8°$. Others are at altitudes between 1500 and 1000 km: the fifteen Soviet satellites Geo-1K, such as Kosmos-2226, the French satellite Starlette and the US pioneer Anna-1B, launched in 1962, $h = 1\,120$ km, $i = 50.1°$.

The Japanese satellite LRE (Laser Ranging Experiment), launched into eccentric orbit in 2001, $h_p = 271$ km, $h_a = 36\,214$ km, $i = 28°$, is equipped with 126 laser reflectors.

There are some Sun-synchronous satellites between 800 and 1 000 km, such as TOPO-1 and those launched after 1993, Stella and Westpac-1 (Sun-synchronous because they are microsatellites that were themselves launched by Sun-synchronous satellites). Since then, geodesy satellites have been placed in lower orbits. An example is GFZ-1 (Geo Forschungs Zentrum), launched in 1995, $h = 380$ km, $i = 51.6°$.

Our knowledge of the geopotential has become so precise that a whole new generation of geodesy satellites[5] was put in operation in 2000. They carry ultra-sensitive accelerometers. Their altitudes must be as low as possible for better detection of gravitational anomalies, whilst a continuous thrust compensates for the higher level of atmospheric drag. The German

[5] Launch dates: CHAMP on 15 June 2000, GRACE-A and -B on 17 March 2002.

satellite CHAMP (Challenging Minisatellite Payload for geophysical research and applications) is on a near-polar orbit $i = 87.3°$ with an initial altitude $h = 454$ km ($h = 300$ km after 5 years, mission lifetime). The US–German system GRACE (Gravity Recovery And Climate Experiment) comprises two satellites, GRACE-A and GRACE-B, 220 km apart on the same orbit at $h = 485$ km, $i = 89.5°$. The altitude should drop down to 250 km over the five-year lifetime of the two satellites. The experiment involves measuring the relative speed of the two satellites to an accuracy of 1 $\mu m\,s^{-1}$. This allows one to detect very weak fluctuations in the Earth's gravitational field and hence to follow the movement of water in the Earth's hydrological cycle.

The European project GOCE (Gravity field and steady state Ocean Circulation Experiment) uses a Sun-synchronous satellite at very low altitude,[6] i.e., $h \approx 250$ km.

Earth Environment: Ionosphere and Magnetosphere

To study the Earth's magnetic field, two satellites are in Sun-synchronous LEO, but elliptical orbit, namely MAGSAT (Magnetic field Satellite, AEM-3), launched in 1979, $h_p = 352$ km, $h_a = 561$ km, $i = 96.8°$, and the Danish Ørsted, launched in 1999, $h_p = 450$ km, $h_a = 850$ km, $i = 96.5°$. To study the radiation belts, the Chinese satellite SJ-5 (Shi Jian-5, DFH-47, where 'shi jian' means 'achievement'), launched in 1999, at the same time as FY-1C, is in a circular Sun-synchronous orbit with $h = 855$ km. SJ-6A and SJ-6B, launched in 2004, are in a lower orbit, $h = 602$ km.

In near-polar LEO orbit, between 800 and 1000 km, are the even-numbered OGO satellites (Orbiting Geophysical Observatory), OGO-2, -4, -6, called POGO (Polar OGO), launched between 1965 and 1969, the Swedish satellites Astrid-1 and -2, launched in 1995 and 1998, and the strictly polar satellite Polar BEAR (Beacon Experiments and Auroral Research).

To study the magnetosphere, that is, the zone of interaction between particles excited by the solar wind and the Earth's magnetic field, satellite orbits have to be very high and highly elliptical. The first US satellite[7] to be placed in orbit, Explorer-1, launched on 11 February 1958, with $h_p = 347$ km, $h_a = 1\,859$ km, $i = 33.2°$, already had some of these features. Its mass was only 5 kg, but it discovered two radiation belts around the Earth, since

[6] At this altitude, and for this 800 kg satellite, the acceleration due to atmospheric drag is 1.5×10^{-5} m s^{-2}, whilst the acceleration due to radiation pressure is 6.1×10^{-8} m s^{-2}. As a comparison, these values are respectively 6.0×10^{-8} and 3.7×10^{-8} for μSCOPE, a 120kg satellite planned for circular orbit at an altitude of 700 km.

[7] Following the Soviet launch of the two Sputniks, the United States wished to react very quickly. The first US satellite was to be a Vanguard, prepared by the US Navy, but in the end it was an Explorer of the US Army that was first placed in orbit. This competition between the two branches of the armed forces came to an end when NASA was created on 1 October 1958.

referred to as the Van Allen belts. This radiation was studied by the Soviet Elektron programme for which four satellites were launched in two pairs in 1964: Elektron-1 and -2, Elektron-3 and -4. They all followed eccentric orbits, with inclination $i \approx 61°$, and with $h_a \approx 6\,500$ km for the odd numbers, $h_a \approx 65\,000$ km for the even numbers.

Magnetospheric studies continued with a great many satellites launched between 1964 and 1968, such as the odd-numbered OGO satellites, OGO-1, -3, -5, known as EOGO (Eccentric OGO), Explorer-34 (IMP-F or IMP-5, Interplanetary Monitoring Platform), launched in 1967 with $h_p = 242$ km, $h_a = 214\,400$ km, $i = 67.1°$, or Explorer-50 (IMP-J or IMP-8), launched into a very high orbit in 1973 with variable inclination between 32° and 55° (and after thirty years, this satellite is still operational).

For the ISEE experiment (International Sun–Earth Explorer), the two satellites ISEE-1 and -2 were launched in 1977, on highly eccentric orbits: $h_p \approx 400$ km, $h_a \approx 138\,000$ km, $i = 12.7°$ and 13.5°. Then in 1978, ISEE-3 was the first satellite placed in a halo orbit[8] around the Lagrange point L_1, i.e., the halo orbit known as L1LO (see Sect. 3.14).

The satellite Wind, launched in 1994, was also placed in an L1LO orbit around the point L_1, where it remained from May 1997 until April 1998. From this location, it was able to observe the solar wind before it became perturbed by the Earth's magnetosphere. It was subsequently placed on a highly complex orbit known as a petal orbit from November 1998 to April 1999.[9] The satellite ACE (Advanced Composition Explorer), launched in 1997, is also in an L1LO orbit.

We should also mention the highly eccentric orbits of the following satellites: Geotail, a Japanese satellite launched in 1992, $h_p = 41\,360$ km, $h_a = 508\,500$ km, $i = 22.4°$; Polar, launched in 1996, on an orbit with variable parameters,[10] $a \approx 60\,000$ km, $e \approx 0.7$, $i \approx 80°$ (several revolutions are

[8] When it had accomplished its mission, the satellite was withdrawn from the point L_1 in June 1982. Using a lunar flyby as a gravity-assist maneuver, it was removed from the Earth's gravitational attraction and sent into heliocentric orbit for the ICE mission (International Cometary Explorer), in an encounter with a comet (perihelion 0.93 a.u., aphelion 1.03 a.u., inclination 0.1°, period 355 day).

[9] The satellite left the point L_1 in the Earthward direction, roughly in the plane of the lunar orbit, before moving into the petal orbit. In this configuration, the satellite moves alternately behind the Earth and the Moon. In this plane and in a frame moving with the Earth, the trajectory sketches out a daisy with the Earth at the centre. The tips of the petals represent the different positions of the Moon in its rotation about the Earth. It has period 17.5 day, radius of ellipse $r_p \approx 6$ to $10R$, $r_a \approx 80R$ (where the Earth–Moon distance is $60R$).

[10] The orbit shown in the figure is that of 13 February 2002. This satellite, the Polar Plasma Laboratory, is part of the GGS mission (Global Geospace Science) with Wind and Geotail, and this is itself just one component of the ISTP programme (International Solar Terrestrial Physics), which includes the European missions SOHO and Cluster and the Russian mission Interball.

shown in the lower part of Fig. 5.22); FAST (Fast Auroral Snapshot Explorer, SMEX-2), launched in 1996, $h_p = 353$ km, $h_a = 4\,163$ km, $i = 83.0°$ (see lower part of Fig. 5.21 for orbit in November 2004); Equator-S, launched in 1997, $h_p = 496$ km, $h_a = 67\,230$ km, $i = 7.0°$ (orbit obtained by transfer via a GTO orbit); IMAGE (Imager for Magnetopause-to-Aurora Global Exploration, MIDEX-1), launched in 2000, on a polar orbit with $h_p \approx 1000$ km and h_a of the order of 7 Earth radii.

The Russian Interball experiment is based on Interball Tail (or Interbol-1, Prognoz-11), launched in 1995 on a highly elliptical orbit with period $T =$ 91 hr, and Interball Aurora (or Interbol-2, Prognoz-12), launched in 1996 on a Molniya orbit. In each case, the Czech satellites Magion (Magnetosphere–Ionosphere), Magion-4, then Magion-5, were launched jointly with an Interball satellite. The orbit of the forthcoming Interbol-3 is planned for $h_a = 400\,000$ km. We also mention the Chinese satellite KF1-SJ-4 (Shi Jian-4, DFH-38), launched in 1994, on a GTO orbit, with $i = 28.6°$.

To study the magnetosphere and phenomena related to the aurora borealis, Japan sent four satellites EXOS (Exospheric Observations) between 1978 and 1989, in alternately low and high eccentric orbits, with $i = 69°$ for EXOS-A (Kyokko), $i = 31°$ for EXOS-B (Jikiken), and $i = 75°$ for EXOS-C (Ohzora) and EXOS-D (Akebono).

The European experiment Cluster-2 comprises four satellites in formation.[11] They have a very high orbit, with $h_p = 17\,200$ km, $h_a = 120\,600$ km, $i = 65°$, $T = 57$ hr.

The Double Star programme comprises two Chinese satellites carrying European satellites similar to those designed for Cluster, in eccentric orbits,[12] with perigee at 600 km altitude. The first, DSP-1, $h_a = 79\,000$ km, is in an equatorial orbit, and the second, DSP-2, $h_a = 39\,000$ km, is in polar orbit.

To study the ionosphere,[13] there are the US satellites UARS (Upper Atmosphere Research Satellite), $h = 570$ km, $i = 56.9°$, and TIMED (Thermo-Iono-Mesosphere Energetics and Dynamics), $h = 625$ km, $i = 74.0°$. In addition, there is the Taiwanese satellite Rocsat-1 (Republic of China Satellite), with a slightly inclined orbit, $h = 630$ km, $i = 35°$ (for oceanographic purposes) and many Interkosmos, such as Interkosmos-12, several Kosmos, such as Kosmos-196, and the Chinese satellites Atmosphere-1 and -2 (DFH-31

[11] These satellites, Rumba, Salsa, Samba and Tango, fly a few hundred kilometres apart. They were launched in two stages, on 16 June and 19 August 2000, to avoid repetition of the disaster when Cluster was launched together on 4 June 1996.

[12] DSP-1 (also called Tan Ce-1 – Explorer-1 in Chinese – or TC-1), $a = 46\,148.1$ km, $e = 0.8494$, $i = 28.5°$, launched on 29 December 2003. DSP-2 (Tan Ce-2 or TC-2), $a = 26\,228.1$ km, $e = 0.7301$, $i = 90°$, launched on 25 July 2004.

[13] Launch dates: UARS on 12 September 1991 (STS-48), TIMED 7 on December 2001 (with Jason-1 but in a different orbit), Rocsat-1 on 27 January 1999, Interkosmos-12 on 30 October 1974, Atmosphere-1 and -2 (DFH-31 and -32) on 3 September 1990, SAMPEX on 3 July 1992, TERRIERS on 18 May 1999.

and -32), which are Sun-synchronous with $h = 800$ km and $h = 610$ km. The satellite SAMPEX (Solar Anomalous and Magnetospheric Particle Explorer, SMEX-1) is near-polar, with $h_p = 506$ km, $h_a = 670$ km, $i = 81.7°$. The US satellite TERRIERS, is Sun-synchronous with $h = 550$ km, but did not function as planned.

5.4.2 Earth-Observation Satellites

Atmosphere and Meteorology

The possibility of observation from space aroused the interest of meteorologists from an early stage. It was their dream to know the global state of the atmosphere at a glance. In order to do so, the orbits used have always been Sun-synchronous LEO orbits (see Fig. 5.7) or GEO orbits (see Fig. 5.8), apart from the first satellites and the Meteor satellites.

LEO Meteorological Satellites. For NASA's Nimbus programme[14] all seven satellites were Sun-synchronous: from Nimbus-1 to Nimbus-6, on a rather high LEO orbit, $h = 1\,100$ km, $i = 99.9°$, whilst Nimbus-7 followed a slightly lower orbit, $h = 950$ km, $i = 99.1°$.

The programme which is known today as the NOAA programme (National Oceanic and Atmospheric Administration), the US meteorological organisation, can be divided into five series: TIROS (Television and InfraRed Observation Satellite), TOS (TIROS Operational System), ITOS (Improved TOS), TIROS-N and ATN (Advanced TIROS-N). The first comprises twelve satellites and began on 1 April 1960 with the launch of the first meteorological satellite,[15] TIROS-1. Up to TIROS-8, launched in 1963, the orbits were similar, $h \approx 680$ km, i between 48° and 58°. Subsequently, all further satellites were Sun-synchronous: TIROS-9 and -10, launched in 1964, and ESSA-1 and -2 (Environmental Science Service Administration), launched in 1966. The TOS series comprised seven satellites, from ESSA-3 to ESSA-9, launched between 1966 and 1969, on the orbit $h = 1\,450$ km, $i = 102°$. The ITOS series used exactly the same orbit for six satellites, ITOS-1, NOAA-1, -2, -3, -4, -5, launched between 1970 and 1976. The last two series[16] adopted a lower orbit: $h = 800$ km, $i = 98.8°$ for TIROS-N, with the satellites TIROS-N and

[14] Launch dates: Nimbus-1 on 28 August 1964, Nimbus-2 on 15 May 1966, Nimbus-3 on 14 April 1969, Nimbus-4 on 8 April 1970, Nimbus-5 on 11 December 1972, Nimbus-6 on 12 June 1975, Nimbus-7 on 24 October 1978.

[15] The three US satellites launched in 1959 had provided useful meteorological data. These were Vanguard-2, Explorer-6 (first photograph of the Earth) and Explorer-7 (first data concerning the Earth radiation budget). However, the first devoted entirely to meteorology was TIROS-1.

[16] Launch dates: TIROS-N on 13 October 1978, NOAA-6 on 27 June 1979, NOAA-7 on 23 June 1981, NOAA-8 on 28 March 1983, NOAA-9 on 12 December 1984, NOAA-10 on 17 December 1986, NOAA-11 on 24 September 1988, NOAA-12 on

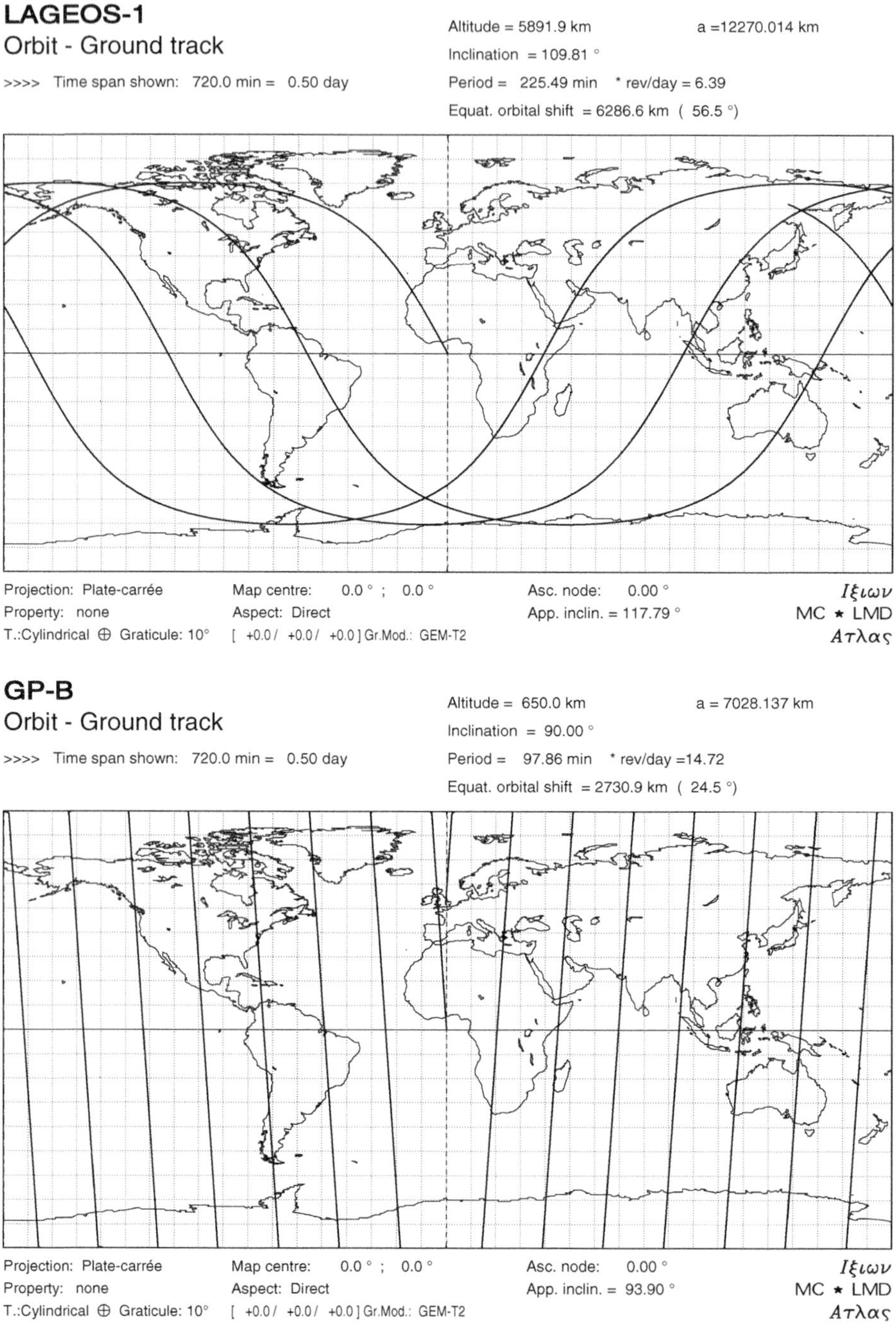

Figure 5.5. Ground tracks of the orbits of a retrograde satellite, LAGEOS-1, and a polar satellite, GP-B, over a time span of half a day

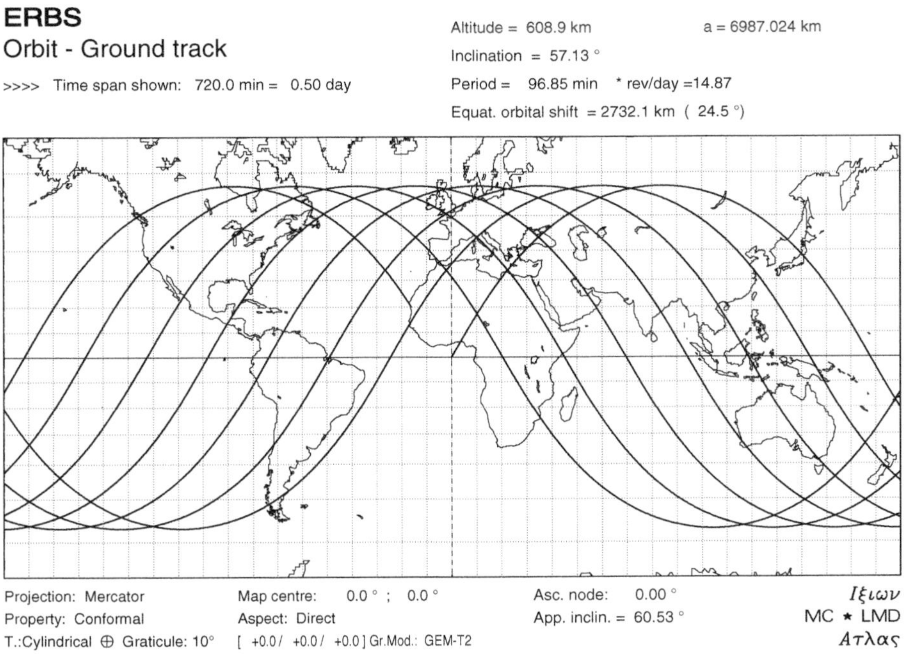

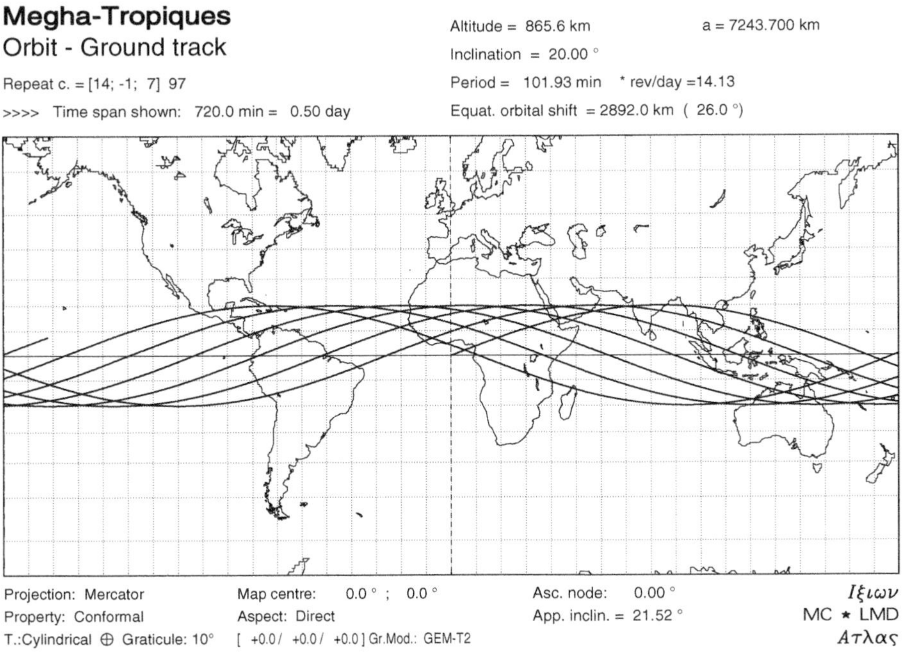

Figure 5.6. Ground tracks of the orbits of two prograde LEO satellites, ERBS and Megha-Tropiques, over a time span of half a day

NOAA-6 and -7; $h = 850$ km, $i = 98.9°$ for ATN, with the satellites NOAA-8 to -16, $h = 812$ km, $i = 98.8°$ for NOAA-17. The programme known as POES (Polar-orbiting Operational Environmental Satellites), comprising the last two series, should be replaced around 2010 by the programme NPOESS (National POES System), a joint project of NOAA and NASA. The programmes will be linked by the satellite NPP (NPOESS Preparatory Project), $h = 824$ km.

The military satellites DMSP (Defense Meteorological Satellite Program) supply some data to the civilian sector. They are all Sun-synchronous, following slightly elliptical orbits, with h between 750 and 850 km, and i between 98.6° and 99.2°. Thirteen satellites were launched between 1965 and 1969 to make up the first block (known as Block 4), from DMSP-4A F-1 (OPS/6026) to DMSP-4A F-13 (also called DMSP-4B F-3, or OPS/1127). The second block (known as Block 5) began in 1970 with DMSP-5A F-1 (OPS/0054) and is still running,[17] with the extension Block 5D3.

Soviet then Russian meteorological satellites[18] were not Sun-synchronous until 2001. They are divided into three Meteor series with near-polar LEO orbits. The first two series involved 48 satellites: Meteor-1, from Meteor-1-01 in 1969 to Meteor-1-27 in 1977, with $h = 870$ km, $i = 81.2°$; Meteor-2, from Meteor-2-01 in 1975 to Meteor-2-21 in 1993, with $h = 940$ km, $i = 82.5°$. The third series involved 6 satellites in slightly higher orbits, with $h = 1\,200$ km, $i = 82.6°$. The new generation, known as Meteor-3M, is Sun-synchronous. The first of the series is Meteor-3M-1, $h = 1\,005$ km, $i = 99.7°$.

The Chinese satellites[19] in the FY-1 series (or Feng Yun-1, where 'feng yun' means 'wind and cloud') are Sun-synchronous, with $h = 858$ km, $i = 98.9°$.

The European Space Agency has a programme of (Sun-synchronous) polar platforms. The initial programme POEM (Polar Orbiting Earth Mission)

---

14 May 1991, NOAA-13 on 9 August 1993, but only operated for a few days, NOAA-14 on 30 December 1994, NOAA-15 on 13 May 1998, NOAA-16 on 21 September 2000, NOAA-17 on 24 June 2002.

[17] Launch dates: DMSP-5D2 F-8 (also called USA-26) on 20 June 1987, DMSP-5D2 F-9 (USA-29) on 3 February 1988, DMSP-5D2 F-10 (USA-68) on 1 December 1990, DMSP-5D2 F-11 (USA-73) on 28 November 1991, DMSP-5D2 F-12 (USA-106) on 29 August 1994, DMSP-5D2 F-13 (USA-109) on 24 March 1995, DMSP-5D2 F-14 (USA-131) on 4 April 1997, DMSP-5D3 F-15 (USA-147) on 12 December 1999, and DMSP-5D3 F-16 (USA-172) on 18 October 2003.

[18] Launch dates: Meteor-3-01 on 24 October 1985, Meteor-3-03 on 26 July 1988, Meteor-3-04 on 25 October 1989, Meteor-3-05 on 24 April 1991, Meteor-3-06 on 15 August 1991, Meteor-3-07 on 25 January 1994, Meteor-3M-1 on 10 December 2001.

[19] Launch dates: FY-1A (DFH-24) on 6 September 1988, FY-1B (DFH-30) on 3 September 1990, FY-1C (DFH-46) on 10 May 1999, FY-1D (DFH-53) on 15 May 2002. The next series of Sun-synchronous LEO satellites is FY-3, from FY-3A to FY-3D.

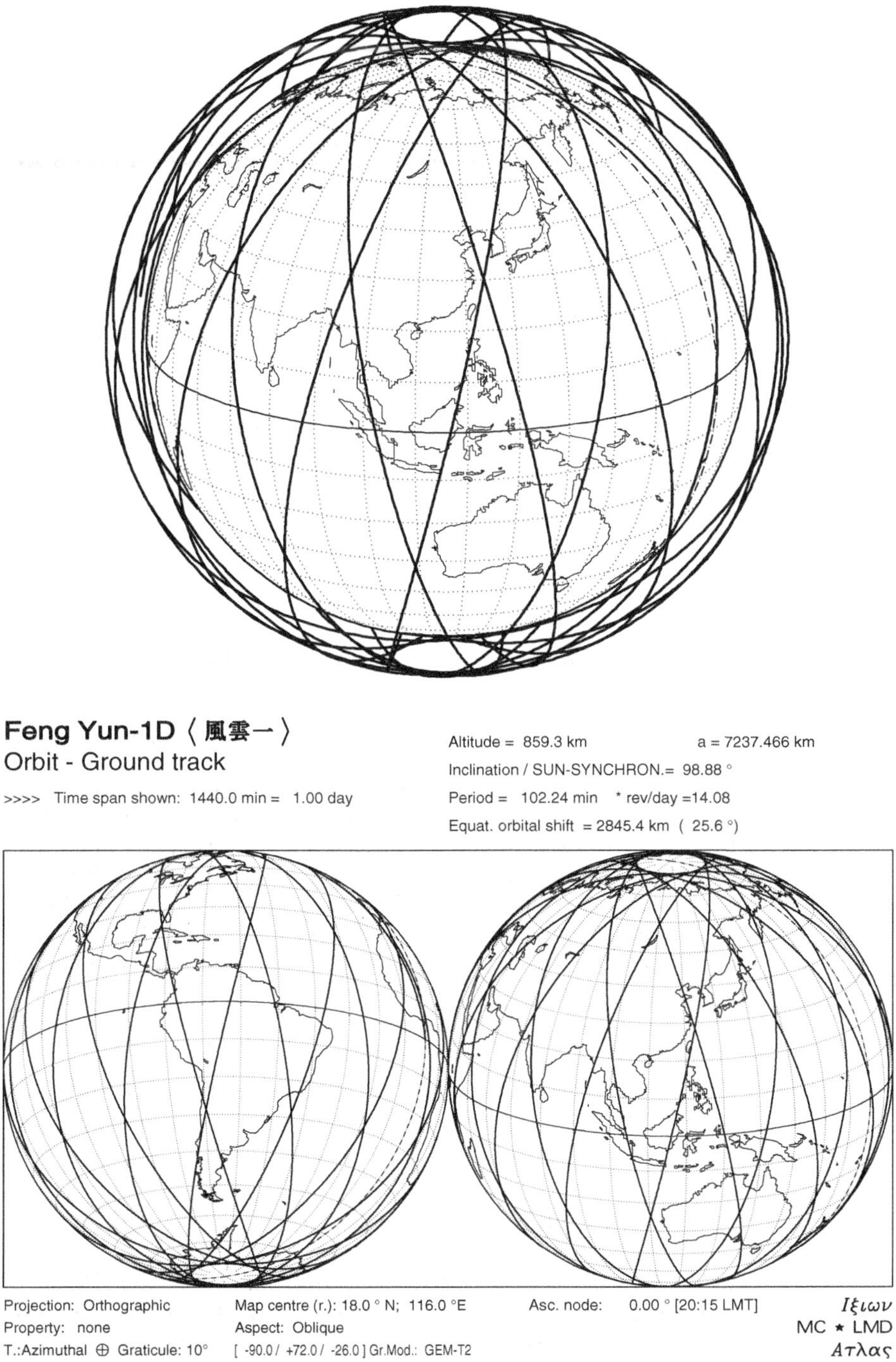

Figure 5.7. Orbit and ground track of a Sun-synchronous satellite over a time span of one day

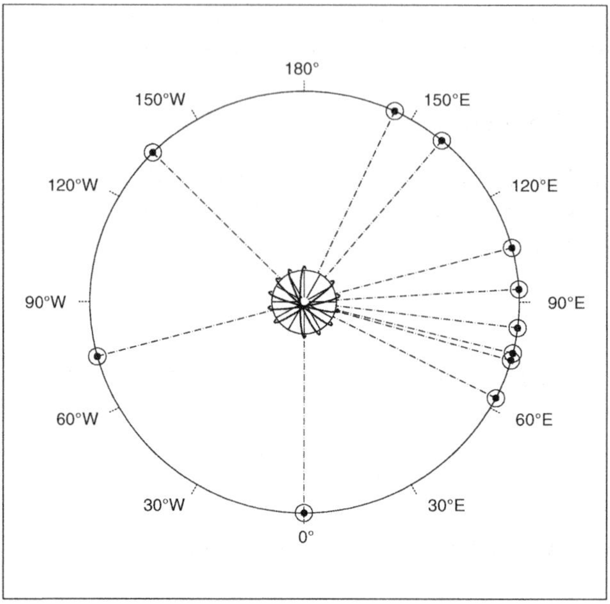

Location λ_S	Operator	Satellite (series)	Satellite in orbit	Type
0.0°–	EUMETSAT	METEOSAT	METEOSAT-7	O
0.0°–	EUMETSAT	METEOSAT	METEOSAT-8	P
63.0°E	EUMETSAT	METEOSAT	METEOSAT-5	O
74.0°E	India	METSAT	Kalpana-1	O
76.0°E	Russia	GOMS	Elektro-1	B
83.0°E	India	INSAT	INSAT-2E	O
93.5°E	India	INSAT	INSAT-3A	O
105.0°E	China	Feng Yun-2	FY-2B	O
140.0°E	Japan	GMS	GMS-5	B
155.0°E	US/NOAA	GOES	GOES-9	O
135.0°W	US/NOAA	GOES-W	GOES-10	O
75.0°W	US/NOAA	GOES-E	GOES-12	O

Figure 5.8. WMO Space Programme: Current geostationary satellites coordinated within CGMS. (WMO: World Meteorological Organization; CGMS: Coordination Group for Meteorological Satellites). List with parking positions of geostationary meteorological satellites as of 1 February 2004. Satellite status: O (operational), P (pre-operational), B (back-up). The geostationary orbit and the Earth are drawn on the same scale. Orbits of Sun-synchronous satellites at altitude 800 km are also plotted on the same scale. The viewpoint is located very high up on the polar axis above the North Pole

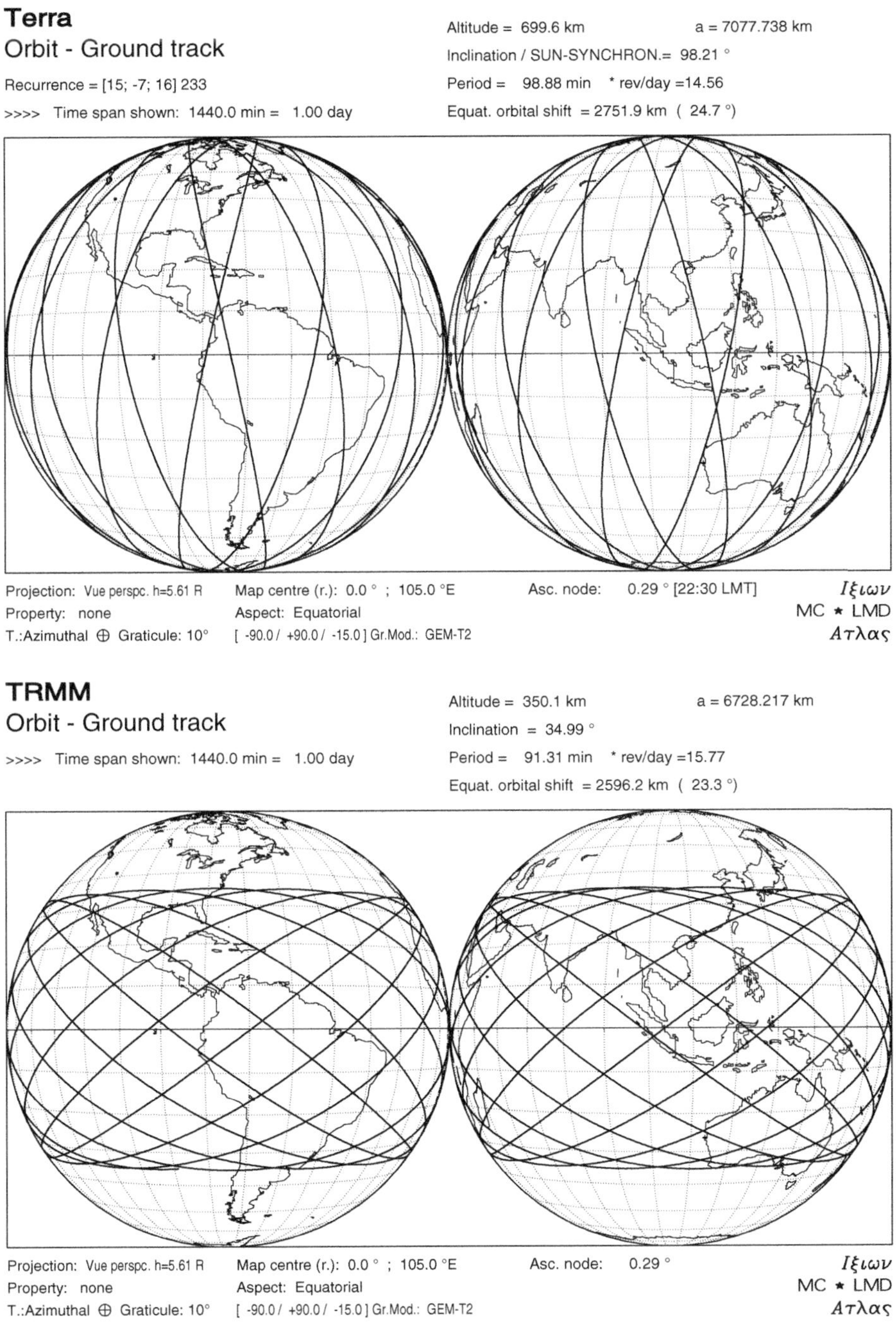

Figure 5.9. Ground tracks of the orbits of Terra and TRMM over a time span of one day, as viewed from a geostationary satellite with parking longitude 75°W (*left*), 105°E (*right*)

was divided into two parts, one for the environment with Envisat, the other for operational meteorology (EUMETSAT) with MetOp-1, -2 and -3. The MetOp satellites (Meteorological Operational satellites) are planned for a mean altitude of $h = 830$ km.

The project known as Rocsat-3/COSMIC (Constellation Observing System for Meteorology, Ionosphere and Climate), a collaboration between Taiwan and the United States, comprises a constellation of microsatellites, $h = 700$ km, $i = 72°$, with 3 planes containing 2 satellites each.

GEO Meteorological Satellites. The geostationary programme has been very widely developed for the purposes of operational meteorology. In order to avoid large distortions due to the viewing angle, the various meteorological institutions have sought to distribute their satellites as harmoniously as possible around the geostationary orbit, under the coordination of the World Meteorological Organization (WMO).

In the United States, these satellites are placed alternately on the longitudes of the east and west coasts. This approach was already in use for the SMS satellites (SMS-1 with $\lambda_S = 75°$W, SMS-2 with $\lambda_S = 115°$W) and was continued with the GOES series[20] (Geostationary Operational Environmental Satellite) and GOES-Next, the satellites being designated GOES-East or GOES-West depending on the case. The satellite GIFTS (Geosynchronous Imaging Fourier Transform Spectrometer, or EO-3 NMP/NASA) will be placed over the Indian Ocean.

For Europe, the geostationary programme is run by EUMETSAT with the METEOSAT satellites. The various operational METEOSAT spacecraft[21] have all been placed at longitude $\lambda_S = 0°$. Some of them can be reserved, or loaned (like METEOSAT-3 to replace GOES-E from February 1993 to May 1995), or sent on missions (such as METEOSAT-5 for the INDOEX experiment, see Example 4.6).

Although Russia generally prefers Molniya orbits to equatorial orbits, it nevertheless launched the GOMS programme (Geostationary Operational

[20] Launch dates: SMS-1 on 17 May 1974, SMS-2 on 6 February 1975, GOES-1 (SMS-3) on 16 October 1975, GOES-2 on 16 June 1977, GOES-3 on 16 June 1978, GOES-4 on 9 September 1980, GOES-5 on 22 May 1981, GOES-6 on 28 April 1983, GOES-7 on 26 February 1987, GOES-8 on 13 April 1994, GOES-9 on 23 May 1995, GOES-10 on 25 April 1997, GOES-11 on 3 May 2000, GOES-12 on 23 July 2001.

[21] Launch dates: METEOSAT-1 on 23 November 1977, METEOSAT-2 on 19 June 1981, METEOSAT-3 on 15 June 1988, METEOSAT-4 on 6 March 1989, METEOSAT-5 on 2 March 1991, METEOSAT-6 on 20 November 1993, METEOSAT-7 on 2 September 1997, MGS-1 (METEOSAT-8) on 28 August 2002. The satellite MSG-1, the first in the MSG series (METEOSAT Second Generation), was renamed METEOSAT-8 when it became operational. The satellites MSG-2, -3 and -4 are programmed, and MTG-1 (METEOSAT Third Generation) is planned for 2015.

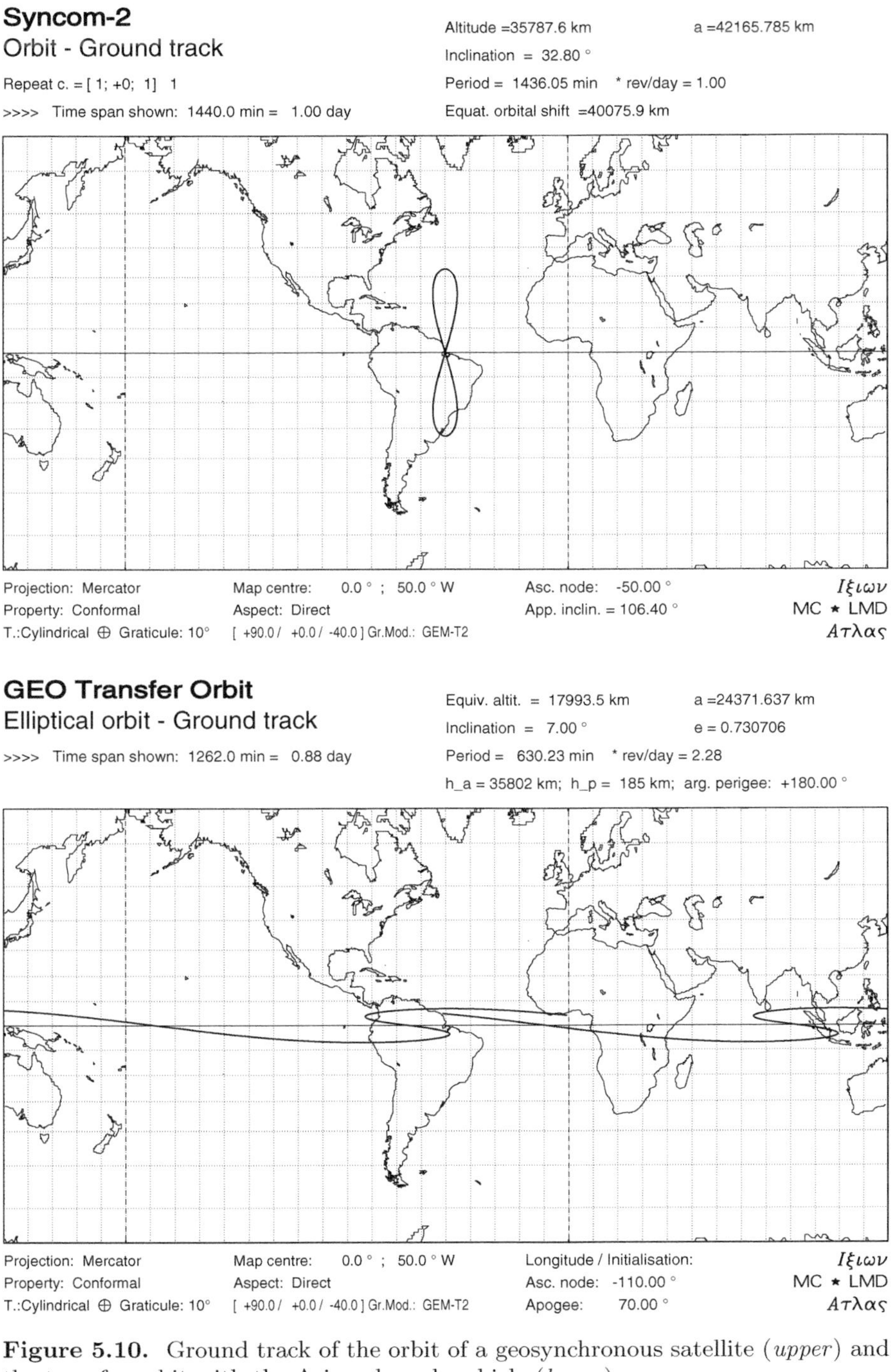

Figure 5.10. Ground track of the orbit of a geosynchronous satellite (*upper*) and the transfer orbit with the Ariane launch vehicle (*lower*)

Meteorological Satellite) of geostationary satellites.[22] For India, the INSAT series (Indian Satellite) contains satellites for the purposes of meteorology[23] and communications. China has launched satellites[24] in the FY-2 series (Feng Yun-2, not to be confused with the LEO satellite series FY-1 and FY-3 already mentioned) since 1997. Since 1977, Japan has been launching its geostationary GMS satellites[25] (Geostationary Meteorological Satellite), also known as Himawari ('himawari' means 'sunflower').

Figure 5.8 shows the (official) positions of the operational satellites as of 1 February 2004. In this distribution, one observes a large 'hole' above the Pacific, and very closely spaced satellites at Asian longitudes. China and India prefer to control their own data.

Satellites for Atmospheric Studies. Satellites devoted to atmospheric research fly in low orbits, that may be Sun-synchronous or otherwise.[26] The following are Sun-synchronous: HCMM (Heat Capacity Mapping Mission, also called AEM-1, Application Explorer Mission), $h = 600$ km, ADEOS-1 and ADEOS-2, and the Swedish satellite Odin (atmosphere and astrophysics), $h = 622$ km. Satellites for ozone studies are Sun-synchronous: TOMS-EP (Total Ozone Mapping Spectrometer and Earth Probe), $h = 750$ km, was Sun-synchronous, and its successor[27] QuikTOMS should have been.

Non-Sun-synchronous satellites include the US satellites SAGE (Stratospheric Aerosols and Gas Experiment or AEM-2), $h \approx 600$ km, $i = 55°$, ERBS (Earth Radiation Budget Satellite), launched by STS-17 (STS-41-G), $h = 600$ km, $i = 57°$, TRMM, $h = 350$ km, $i = 35°$, and the Cana-

[22] Launch date: GOMS-1 on 31 October 1994. The series is also called Elektro, and this satellite thus carries the names Elektro-1 and GOMS-Elektro-1 as well as GOMS-1.

[23] Launch dates: INSAT-1A on 10 April 1982, INSAT-1B on 30 August 1983 (STS-8), INSAT-1C on 21 July 1988, INSAT-1D on 12 June 1990, INSAT-2A on 10 July 1992, INSAT-2B on 23 July 1993, INSAT-2E on 3 April 1999, METSAT-1 (subsequently called Kalpana-1) on 12 September 2002, INSAT-3A on 9 April 2003, INSAT-3E on 27 September 2003.

[24] Launch dates: FY-2A (DFH-45) on 10 June 1997, FY-2B (DFH-49) on 25 June 2000, FY-2C on 19 October 2004.

[25] Launch dates: GMS-1 on 14 July 1977, GMS-2 on 10 August 1981, GMS-3 on 2 August 1984, GMS-4 on 5 September 1989, GMS-5 on 18 March 1995. The next satellites will belong to the MTSAT generation (Multi-functional Transport Satellite). The first will be MTSAT-1R (to replace MTSAT-1, destroyed during launch on 15 November 1999), followed by MTSAT-2.

[26] Launch dates: HCMM on 26 April 1978, ADEOS-1 on 27 August 1996, TOMS-EP on 2 July 1996, Odin on 20 February 2001, QuikTOMS on 21 September 2001 (failed), SAGE on 18 February 1979, ERBS on 5 October 1984, TRMM on 28 November 1997, ADEOS-2 on 14 December 2002, SciSat-1 on 13 August 2003.

[27] Depending on how fast one writes it! The NASA satellites QuikTOMS and QuikScat are spelt like this, whereas those of DigitalGlobe are written QuickBird.

dian satellite SciSat-1, $h = 650$ km, $i = 73.9°$. The Italian satellites San Marco-2, -3, -4 and -5, launched between 1967 and 1988, are in equatorial LEO orbit ($i = 3°$), as will be FBM (French–Brazilian Microsatellite), $h = 750$ km, $i = 6°$. The French–Indian satellite belonging to the Megha-Tropiques project, $h = 866$ km, will be in a slightly inclined LEO orbit ($i = 20°$), devoted to study of the tropical regions.

Figure 5.6 (upper) shows the ground track of ERBS, and Fig. 5.9 (upper), the ground track of TRMM. For Megha-Tropiques, as well as the ground track shown in Fig. 5.6 (lower), the orbit is represented in Colour Plate X.

The joint project (United States, Japan, Europe) GPM (Global Precipitation Mission) continues and expands the TRMM mission to study rainfall. It includes a core satellite called GPM-core, $h = 450$ km, $i = 70°$ and a constellation of 6 to 8 Sun-synchronous satellites. Some of these, such as EGPM (European GPM), $h = 666$ km, are specialised in this field. Others, such as MetOp-1, the two GCOM and the two NPOESS, have a wider field of investigation.

Earth Resources, Remote-Sensing, and Environment

This category contains satellites carrying instruments whose resolution at ground level is between 50 and 5 m. Colour Plates IV and V show images obtained by the MISR and MODIS imagers aboard Terra. These satellites are all LEO and, apart from those in the Resurs-F series and a few special cases, they are all Sun-synchronous. Recurrent and frozen orbits are required for these satellites.

The first programme, Landsat, dates from 1972, and its first three satellites had the same orbit characteristics: $h = 910$ km, $i = 99.1°$. From Landsat-4, the altitude was reduced to $h = 700$ km, $i = 98.2°$, and this orbit has been used ever since, not only for all the Landsat satellites,[28] but by other NASA satellites, such as EO-1 (Earth Observing) in the NMP programme (New Millenium Program) and the majority of the EOS satellites (Earth Observation Satellite) in the ESE programme (Earth Science Enterprise), formerly MTPE (Mission To Planet Earth). Several of these satellites are already on this orbit, and others are planned: EOS-AM-1 and EOS-AM-2 (EOS Morning, AM = *ante meridiem*), EOS-PM-1 and EOS-PM-2 (EOS Afternoon, PM = *post meridiem*), EOS-Chem-1 (to study atmospheric chemistry), OCO (Orbiting Carbon Observatory), LDCM (Landsat Data Continuity Mission). Three of

[28] Launch dates: ERTS-1 (Earth Resources Technology Satellite) on 23 July 1972, renamed Landsat-1 on 13 January 1975, Landsat-2 on 22 January 1975, Landsat-3 on 5 March 1978, Landsat-4 on 16 July 1982, Landsat-5 on 1 March 1985, Landsat-6 on 5 October 1993 (lost in a launch failure), Landsat-7 on 15 April 1999, Terra (EOS-AM-1) on 18 December 1999, MTI on 12 March 2000, EO-1 and SAC-C on 21 November 2000, EO-2 (lidar aboard the Shuttle) cancelled, Aqua (EOS-PM-1) on 4 May 2002, Aura (EOS-Chem-1) on 15 July 2004.

them have been attributed new and less technical names: Terra for EOS-AM-1, Aqua for EOS-PM-1, Aura for EOS-Chem-1.

The satellite Aqua should be followed on the same orbit, from a few tens of kilometres, by the two satellites[29] CloudSat and Calipso (Cloud Aerosol Lidar and Infrared Pathfinder Satellite Observation), which should observe the same fields of view. The French microsatellite PARASOL (Polarization and Anisotropy of Reflectances for Atmospheric Science coupled with Observations from a Lidar) is also planned as part of the convoy, which will be brought to a close by Aura. This sequence of five satellites on the same orbit, called the A-train, with Aqua at the head and Aura bringing up the rear, is a novel project. A sixth satellite, OCO, is now planned for this space train (see Fig. 6.8).

The satellite EO-1 follows Landsat-7 at an interval of just one minute (of time). In the following, we shall call this orbit, first used by Landsat-4, the Terra orbit. It can be defined to great accuracy by its recurrence.

As part of the ESE programme, the Sun-synchronous satellite Aquarius, $h = 600$ km, will measure the salinity of the sea surface.

The satellite MTI (Multispectral Thermal Imager or P97-3) is on a lower Sun-synchronous orbit, $h = 585$ km, to observe both night and day, like the two satellites, currently under development, NEMO (Navy Earth Map Observation), $h = 606$ km, for observations in hyperspectral mode, and HYDROS (Hydrosphere State Mission), $h = 670$ km.

The SSTI mission (Small Spacecraft Technology Initiative), built around the Sun-synchronous satellites[30] Lewis and Clark, did not live up to expectations.

The French programme of commercial remote-sensing has been carried out by the SPOT family of satellites[31] (*Satellites Pour l'Observation de la Terre*), all on strictly the same orbit ($h = 822$ km), from SPOT-1 to -5. One may therefore speak of the SPOT orbit. The images produced by these satellites are used by the military (during the Gulf War, for example), who also have

[29] These two satellites, also called ESSP-4 and ESSP-3, respectively, belong to NASA's ESSP programme (Earth System Science Pathfinder) which also includes the two satellites -A and -B of the GRACE mission (ESSP-2), for geodesy, and VCL (Vegetation Canopy Lidar, ESSP-1), $h = 400$ km, $i = 67°$, for environmental studies. The US satellite with French collaboration ESSP-3 was originally called Picasso–Cena (Pathfinder Instruments for Cloud and Aerosol Spaceborne Observations – *Climatologie Etendue des Nuages et des Aérosols*). However, the artist's family was opposed to free use of the name and it was renamed Calipso.

[30] Lewis, $h = 523$ km, and Clark, $h = 479$ km: the first launched on 23 June 1997, whilst the second was cancelled in February 1998. Meriwether Lewis and William Clark were two American officers who explored Louisiana just after the French ceded it to the United States in 1803, eventually descending the Columbia river to the Pacific.

[31] Launch dates: SPOT-1 on 22 February 1986, SPOT-2 on 11 January 1990, SPOT-3 on 26 September 1993, SPOT-4 on 24 March 1998, SPOT-5 on 4 May 2002.

their own specific satellites of SPOT type, namely, the Hélios satellites,[32] which are Sun-synchronous but at lower altitude ($h = 680$ km). The spatial resolution of the SPOT satellites (5 m for SPOT-4, 2.5 m for SPOT-5) will be further improved (1 m) with the next generation of satellites known as Pléiades (Pléiades-1 and -2), at even lower altitude, $h = 695$ km. Linked with Pléiades, Italy has launched the COSMO-SkyMed project (Constellation of Small Satellites for Mediterranean Basin Observation), a constellation of 4 satellites equipped with radar, $h = 620$ km. The same orbit is planned for HypSEO (HyperSpectral Earth Observer).

The two German projects in this field also used Sun-synchronous satellites: RapidEye, a constellation of 4 satellites, $h = 600$ km, and the satellites Diamant-1, -2 and -3, $h = 670$ km. The TerraSAR mission arose from a common project between ESA and a private organisation InfoTerra. It involves two satellites equipped with SAR radar, TerraSAR-X1 (X-band) and TerraSAR-L1 (L band).

The Soviet then Russian programme began in 1979 with the series Resurs-F1 then -F2, using 6 tonne satellites in very low near-polar orbits, which operated for 14 days, then 30 days for the later version. Dozens of these were launched,[33] in near-polar orbit, $i = 82.3°$, with altitude $h = 275$ km for Resurs-F1, $h = 240$ km for Resurs-F2. Satellites in the series[34] Resurs-O most resemble other remote-sensing satellites: they are in Sun-synchronous orbits, $h = 600$ km, $i = 97.9°$, for Resurs-O1-1 to -O1-3, $h = 820$ km, $i = 98.8°$ for Resurs-O1-4 ('resurs' means 'resource', while F stands for film and O pour operational). The Resurs programme is the follow-on of the Meteor–Priroda programme.

Large remote-sensing and environmental satellites weighing several tonnes require powerful launch vehicles which may be able to offer several piggyback positions for very light passenger satellites. Such satellites, with various missions (although usually technological) also follow Sun-synchronous orbits, very close to the orbit of the main satellite. These grouped launches[35] provide

[32] Launch dates: Hélios-1A on 7 July 1995, Hélios-1B on 3 December 1999. Hélios-2A and -2B are currently under development.

[33] The first 39 are recorded as Kosmos, from Kosmos-1127 in 1979 to Kosmos-1990 in 1989. There were then 20 more under the name of Resurs-F, Resurs-F-1 (type F1) in 1989 to Resurs-F-20 (type F2) in 1995, followed by the modified version, Resurs-F1M-1 in 1997 and Resurs-F1M-2 in 1999 (type F1M).

[34] Launch dates: Resurs-O1-1 (Kosmos-1689) on 3 October 1985, Resurs-O1-2 (Kosmos-1939) on 20 November 1988, Resurs-O1-3 on 4 November 1994, Resurs-O1-4 on 10 July 1998.

[35] Here, in chronological order, are four examples of grouped launches where the main satellite is a large Sun-synchronous remote-sensing satellite. For the first two, ERS-1 and SPOT-3, launched by Ariane, the passenger satellites were called ASAP (Ariane Structure for Auxiliary Payload). With ERS-1 (Europe): UoSAT-5 (or OSCAR-22) (UK), Orbcomm-X (USA), Tubsat-A (Germany), SARA (France). With SPOT-3 (France): Kitsat-2 (South Korea), PoSAT-1 (Por-

an opportunity for countries with little experience in space to get their own satellite into orbit.

Countries occupying a very large territory use Sun-synchronous remote-sensing satellites. For India, in its IRS programme[36] (Indian Remote Sensing), the first satellites, IRS-1A and -1B, are on a rather high orbit, $h = 910$ km, whilst the rest, IRS-P2, -P3 and -P6 (Resourcesat-1) are on a lower orbit, $h = 817$ km. The future satellites Cartosat-1 (IRS-P5) and Cartosat-2 are planned for orbits at $h = 617$ km and $h = 630$ km, respectively. The experimental satellite TES (Technology Experiment Satellite) was launched on an even lower orbit, at $h = 565$ km. The IRS-2 series (Oceansat-2, Climatsat-1, Atmos-1) will be integrated mission that will cater to global observations of climate, ocean and atmosphere.

China and Brazil have a joint programme[37] called CBERS (China–Brazil Earth Resources Satellite), or Zi Yuan (meaning 'resources' in Chinese), with the satellites CBERS-1 and -2, $h = 774$ km and the following ZY-1 series. Independently, China has also launched[38] two satellites ZY-2 and -2B (series ZY-2) in a low orbit, $h = 495$ km and $h = 476$ km, then Tan Suo-1, $h = 610$ km. Australia intends to launch its satellite ARIES-1 (Australian Resource Information and Environment Satellite). The US private company Resource21 (21 indicates the 21st century) should launch five satellites, RS21-1 to RS21-5, around $h = 480$ km, with a resolution of 10 m.

Japan has launched JERS-1, already mentioned, and plans to send ALOS (Advanced Land Observation Satellite), which should be followed by GCOM-

tugal), Stella (France), HealthSat-2 (UK), ItamSat (Italy), EyeSat-1 (USA). With Resurs-O1-4 (Russia): FaSat-1 (Chile), TMSat (or Thai-Phutt) (Thailand), TechSat-1B (Israel), Westpac-1 (Australia), Safir-2 (Germany). With Meteor-3M-1 (Russia): Badr-B (Pakistan), Maroc-Tubsat (Morocco/Germany), Kompass and Reflektor (Russia).
To complete this note on passenger satellites, we give a few examples of grouped launches with oceanographic or technological satellites. Orbits are Sun-synchronous, except for TOPEX/Poseidon and its passengers. With TOPEX/Poseidon (USA/France): Uribyol ('our star' in Korean, Kitsat-1), S80/T (France). With ARGOS (USA): Ørsted (Denmark), SunSat (South Africa). The orbit remained circular for the main satellite, but was made elliptical ($e = 0.01545$) for Ørsted and its companion. With Oceansat-1 (India): Kitsat-3, DLR-Tubsat. With TES (India): BIRD (Germany), PROBA (Belgium/Europe).

[36] Launch dates: IRS-1A on 17 March 1988, IRS-1B on 29 August 1991, IRS-1C on 28 December 1995, IRS-1D on 4 June 1997, IRS-1E (IRS-P1) on 20 September 1993 (before IRS-1C), failed, IRS-P2 on 15 October 1994, IRS-P3 on 21 March 1996, TES on 22 October 2001, IRS-P6 on 17 October 2003.

[37] Launch dates: CBERS-1 (ZY-1, Zi Yuan-1) on 14 October 1999, CBERS-2 (ZY-1B, Zi Yuan-1B) on 21 October 2003, CBERS-3 and -4 are planned to follow.

[38] Launch dates: ZY-2 (Zi Yuan-2, DFH-50, Jian Bing-3, JB-3) on 1 September 2000, ZY-2B (Zi Yuan-2B, DFH-55, Jian Bing-3B, JB-3B) on 27 October 2002, (the satellites ZY-2 use a CBERS platform), Tan Suo-1 (ExperimentSat-1) on 18 April 2004.

A1 and -B1 (Global Change Observing Mission), then later by GCOM-A2 and -B2. The European Space Agency has many projects in this field.[39]

Satellite-based environmental studies are now very varied. Amongst these, we mention the detection of forest fires, where instruments have a ground resolution of about 100 m, as in the case of the Sun-synchronous German satellite BIRD (Bi-spectral InfraRed Detection), $h = 575$ km. The projected Spanish satellite FuegoSat, $h = 700$ km, $i = 47.5°$, will be the precursor of a constellation of 12 satellites, FuegoFOC (Fire Observation Constellation). For surveillance of the Amazonian forest, Brazil is developing a project for two satellites in equatorial orbit, $h = 900$ km, $i = 0°$, SSR-1 and -2 (*Satelete de Sensoriamento Remoto*).

Satellites designed for general environmental studies are rather large,[40] equipped with radar, at altitudes $h \approx 780$ km: for Canada, Radarsat-1, for Europe, ERS-1, -2 (European Remote Sensing Satellite) and Envisat (Environmental Satellite) [see Fig. 5.26 (upper)].

To study the polar ice caps and make precise measurements of variations in their thickness, a novel orbit (near-polar non-Sun-synchronous LEO) has been chosen for two missions,[41] respectively American and European: ICESat (Ice, Clouds, and Land Elevation, formerly EOS-LAM), $h = 592$ km, $i = 94.0°$ and CryoSat (Cryosphere Satellite), $h = 716$ km, $i = 92.0°$. At these altitudes, the Sun-synchronous inclinations would be 97.8° and 98.3°, respectively.

The British project DMC (Disaster Monitoring Constellation), with international cooperation, is currently underway.[42] It comprises a constellation of Sun-synchronous microsatellites, $h = 686$ km.

---

[39] The satellite ADM (Atmospheric Dynamics Mission), renamed Aeolus-ADM, will carry a lidar to study the winds. The satellite SMOS (Soil Moisture and Ocean Salinity), comprising a large Y-shaped antenna, will analyse water emissions in the centimetre band. In the more distant future, ESA has opted for six missions: the Sun-synchronous satellites SPECTRA (Surface Processes and Ecosystem Changes Through Response Analysis) which takes over from LSPIM (Land-Surface Processes and Interactions Mission), EarthCARE, which takes over from ERM and the Japanese mission Atmos-B1, WALES (Water vapour and Lidar Experiment in Space), EGPM (European contribution to the Global Precipitation Monitoring mission), and the constellations ACE+ (Atmosphere and Climate Explorer), and Swarm, a constellation of small satellites to study the dynamics of the Earth's magnetic field.

[40] Launch dates: Radarsat-1 on 4 November 1995, ERS-1 on 17 July 1991, ERS-2 on 21 April 1995, Envisat on 1 March 2002.

[41] Launch date: ICESat on 13 January 2003.

[42] Launch dates: AlSat-1 (Algeria) on 28 November 2002, BilSat-1 (Turkey), NigeriaSat-1 (Nigeria) and BNSCSat (UK) with a grouped launch on 27 September 2003.

Demeter (Detection of Electro-Magnetic Emissions Transmitted from Earthquake Regions)[43] is a French scientific microsatellite, $h = 695$ km, Sun-synchronous, which will measure electrical effects generated by seismic events.

We end this category of Earth-observation satellites with Triana,[44] a US project with an unusual orbit for this type of mission. After a 3.5 month journey, this satellite will be placed in a halo orbit around the Lagrange point L_1 (orbit type L1LO, period 6 months). Its instruments will have a view of the Earth which is permanently illuminated, but from a very great distance (234 Earth radii, or four times the distance from the Earth to the Moon). The projected pixel size (resolution) is 8 km (1 arcsec). Due to the large dimensions of the halo orbit, it will be possible to observe alternately the North and South Poles of the Earth, with a special concern for the stratospheric ozone. The project has been resumed under the name DSCO (Deep Space Climate Observatory) or DSCOVR.

Remote-Sensing, Surveillance

Satellites in this category have a resolution of the order of 1 m in the visible frequency range (and a few metres if they carry out infrared observations), which was a level reserved for military satellites until 1994. Unless otherwise specified, these are US commercial satellites. (OrbView-1 was the first remote-sensing satellite to belong to a private organisation, in 1995.)

Almost all of these satellites, launched[45] or under development, are Sun-synchronous. We give the satellite series (and their resolutions[46]) in order of decreasing altitude:

- Ikonos[47] (resolution 0.8 m), $h = 680$ km.

---

[43] Launch date: Demeter on 29 June 2004, with eight other microsatellites.

[44] Rodrigo Triana was the first person to see the New World, in 1492, among the sailors aboard Christopher Columbus' caravels.

[45] Launch dates: OrbView-1 (Microlab-1) on 3 April 1995 (with Orbcomm-FM-1 and -2, non-Sun-synchronous), EarlyBird/EarthWatch-1 on 24 December 1997, Ikonos-1 on 27 April 1999, failed, Ikonos-2 on 24 September 1999, EROS-A1 on 5 December 2000, OrbView-4 (before OrbView-3) on 21 September 2001, failed, QuickBird-1 on 20 November 2000, failed, QuickBird-2 on 18 October 2001, OrbView-3 on 26 June 2003.

[46] The resolution in panchromatic mode corresponds to black and white images, and in multispectral mode, to colour images, generally composed of blue, green, red and near-infrared.

[47] The satellite Ikonos-1, lost in launch, was quickly replaced by Ikonos-2, launched five months later, and renamed Ikonos to exorcise the failure of the first launch. The resolution planned for Ikonos-3 is 0.5 m. The Greek name ὁ εἰκών,όνος means 'image'. But why choose the genitive, *ikonos*?

- EarlyBird[48] (resolution 3 m), $h = 528$ km.
- EROS[49] (resolution 1.8 m for EROS-A, 0.8 m for -B), $h = 474$ km.
- OrbView (resolution 1 m in black and white, 4 m in colour), $h = 451$ km.
- QuickBird[50] (resolution 0.6 m in black and white, 2.5 m in colour), $h = 443$ km.

The SPIN-2 missions[51] are carried out on a non-Sun-synchronous orbit with $h = 260$ km, $i = 70°$.

For military applications, the main category of US surveillance satellites (or spy satellites, depending on one's point of view) carry the suggestive name Key Hole (KH). They fall into several series, from KH-1 in 1959 to the current KH-12. For the first few series up to KH-9, the basic principle was always the same: a camera took photos, the film was placed in a capsule, and as astonishing as it may seem, the capsule was then thrown back to Earth. A parachute opened at an altitude of 20 km and, all being well, an aircraft equipped with a net intercepted the prize in flight (although it could also be picked up by ship in the ocean). Now that the results of the first few series have been declassified, as they say in military circles, we may observe that the success rate was actually very low, with only two successful missions, Discoverer-14 and Discoverer-18, amongst the 27 satellites of the KH-1 and -2 series (satellites Discoverer-1 to -27). The subsequent series met with more success. Concerning series KH-11 and -12, the results are transmitted via SDS military satellites. It is claimed that the images provided by the latest satellites have a resolution of 15 cm. (However, one is led to raise several questions, such as the influence of atmospheric turbulence, the problem of data accumulation, etc.) Satellites in the future series KH-13 are KH-12 satellites will be made undetectable to radar and IR sensors (stealthy satellites), and those in the series 8X will apparently be gigantic telescope

---

[48] To avoid confusion with the satellite Early Bird in the Intelsat series, the name of the company operating the satellite is added to the name EarlyBird-1 or 2. This private company, founded under the name World-View Imaging, was renamed EarthWatch, and then in 2000, DigitalGlobe. The satellite EarlyBird/EarthWatch-1 never functioned correctly and the EarlyBird/EarthWatch-2 mission was cancelled. The company is now concentrating its efforts on the QuickBird satellites.

[49] The first part of the EROS programme (Earth Resources Observation Satellite, Israel), with EROS-A1, was considered to be so successful that EROS-A2 was cancelled. The second part includes six satellites, from EROS-B1 to -B6.

[50] The QuickBird satellites were initially planned for the orbit $h = 610$ km, $i = 52°$. The orbit of QB-1, never reached, should have been $h = 590$ km, $i = 66°$. It was eventually decided, from QB-2, to adopt a Sun-synchronous orbit.

[51] Launch date: SPIN-2 (Kosmos-2349) on 18 February 1998. A private US–Russian venture developed the SPIN-2 project (Space Information 2-meter), for commercial use of high-resolution images (2 m in panchromatic mode) obtained by satellites in the Resurs-F series.

satellites called Monstersats, with centimetre resolution,[52] in Molniya orbit, refuelled by the Space Shuttle.

Missions were very short (a few days) for the first series, then longer (19 days for KH-4B, 50 days and two capsules for KH-8). Orbits were generally low, with h between 200 and 400 km, near-polar (e.g., Discoverer-35, h = 260 km, $i = 82°$) up to KH-3, then with lower inclination (e.g., KH-4A-14, or Orbis, OPS/3360, h_p = 117 km, h_a = 329 km, $i = 70°$) up to KH-6, and subsequently Sun-synchronous or with high inclination (e.g., KH-7-27, h_p = 139 km, h_a = 375 km, $i = 117°$) up to KH-11, where missions become much longer (which can justify use of a Sun-synchronous orbit, as we have seen). The first eleven series involved a total of 262 satellites.

For KH-12, the satellites, at 20 tonnes, are sent up for long periods. The first, in 1990, USA-53, was launched by STS-36, with h = 200 km, i = 62°. The next two, in 1992 and 1996, USA-86 and USA-129, were placed in elliptical Sun-synchronous orbits with h_p = 256 km, h_a = 911 km, $i = 97.7°$ and h_p = 153 km, h_a = 949 km, $i = 97.9°$, respectively.

'All weather' military observation is carried out by the Lacrosse radar satellites, each with a mass of 20 tonnes. They have circular orbits with medium inclination: Lacrosse-1 (USA-34), h = 440 km, $i = 57°$, launched by STS-27 in 1988; the three others, Lacrosse-2, -3 and -4 (USA-69, -133 and 152), launched in 1991, 1997 and 2000, $h \approx 680$ km with inclinations $i = 68°$, $i = 57°$, and $i = 68°$, respectively.

Soviet military surveillance was carried out by a multitude of Kosmos satellites. The first, in the Zenit series, had very low altitude, $h \approx 150$ km and characteristic inclinations of $i = 63°, 73°, 82°$. Missions lasted a few days and the film was recovered with the satellite. The technique of recovering the capsule in flight appeared with satellites in the Yantar series, in 1975. The Arkon series is the equivalent of KH-12. Radar observation is carried out by the Almaz series ('almaz' meaning 'diamond' in Russian, from the Arabic word *al mās* with the same meaning), 19 tonne satellites in low circular orbits h = 300 km, $i = 72°$, with Kosmos-1870 and Almaz-1, launched in 1987 and 1991, and the Oblik series, equivalent to Lacrosse.

Chinese military surveillance and remote-sensing satellites belong to the FSW-2 and -3 series (Fanhui Shi Weixing), such as FSW-2-3, launched in 1996, h = 125 km, $i = 63°$, FSW-3 and FSW-3-2, launched in 2004, $h_p \approx 170$ km, $i = 63°$. They return to Earth after two weeks (as their name suggests: 'fanhui shi' means 'return' and 'weixing' means 'satellite').

In parallel with the EROS programme, Israel has developed its military programme Ofeq ('horizon' in Hebrew), with two operational satellites, Ofeq-

[52] Resolutions announced by the US Air Force, which is in charge of the programme: KH-1 (begun in 1959) 12 m, KH-2 (1960) 9 m, KH-3 (1961) 7.6 m, KH-4A (1963) 2.7 m, KH-6 (1963) 1.8 m, KH-8 (1966) 0.5 m. For KH-11 (1976) and KH-12 (1992) 0.15 m with a telescope similar to the Hubble. Projects: KH-13 at 0.10 m and 8X at 0.05 m.

3 and -5, launched in 1995 and 2002, on a highly inclined elliptical orbit: $h_p = 370$ km, $h_a = 750$ km, $i = 143.5°$. These satellites thus cover latitudes below 37°. Although there is no precise information available about this programme, we may imagine that one motivation for the retrograde orbit is to increase the synodic frequency of the satellite (with $\nu \approx 15$, we have $\nu' \approx 16$ rather than $\nu' \approx 14$ for a prograde orbit), and hence the viewing frequency.

5.4.3 Oceanographic Satellites

The first oceanographic satellites had highly inclined orbits: GEOS-3 (Geodynamics Experimental Ocean Satellite), $h = 847$ km, $i = 115.0°$, and SeaSat, $h = 780$ km, $i = 108.1°$. The latter orbit was then taken over, to within a few kilometres, by Geosat and GFO-1 (Geosat Follow On). The orbit of TOPEX/Poseidon is rather high, $h = 1330$ km, to avoid atmospheric drag as far as possible, and has a rather high inclination, $i = 66°$, so as to overfly almost the whole expanse of the oceans.[53] To avoid any bias due to the influence of the Sun on the tides, it was essential that the orbit should not be Sun-synchronous. The satellite Jason-1 is placed on exactly the same orbit as TOPEX/Poseidon to guarantee the continuity of the French–US mission, for which Jason-2 is then planned (see the lower part of Fig. 5.26).

These satellites, together with the two Sun-synchronous, have carried out altimetric measurements on the oceans (precise measurements of the sea level and its evolution), with very good results.[54]

The satellites[55] in the Okean series of Soviet, then Russian/Ukrainian, then Russian satellites (Ukraine having opted for the Sich series) are devoted to study of the polar region and oceans. The Okean-O1 series comprises four satellites. The first three, Okean-O1-1, -2 and -3, have altitude $h \approx 650$ km and inclination $i = 82.5°$, typical of many Meteor and hundreds of Kosmos satellites. The last in the series, Okean-O1-4, follows a Sun-synchronous orbit at the same altitude. It is on this orbit, $h = 663$ km, $i = 98°$, that Okean-O-1 was placed, the first satellite in the Okean-O series. The Ukrainian satellite Sich-1 follows a similar orbit to Okean-O1-3.

The Chinese programme is based on the Haiyang satellites[56] (meaning 'ocean'), or HY, $h = 798$ km, $i = 98.8°$.

---

[53] Launch dates: GEOS-3 on 9 April 1975, Seasat on 28 June 1978, Geosat on 13 March 1985, GFO-1 on 10 February 1998, TOPEX/Poseidon on 10 August 1992, Jason-1 on 7 December 2001.

[54] Estimated measurement accuracy: GEOS-3: 25 cm; Seasat: 5 cm; Geosat: 4 cm; ERS-1 and -2: 3 cm; TOPEX/Poseidon: 2 cm.

[55] Launch dates: Okean-O1-1 on 5 July 1988, Okean-O1-2 on 28 February 1990, Okean-O1-3 (generally called Okean-3) on 4 June 1991, Okean-O1-4 on 11 October 1994, Sich-1 on 31 August 1995, Okean-O-1 (generally called Okean-O) on 17 July 1999.

[56] Launch date: HY-1 (Ocean-1 or DFH-54) on 12 May 2002 (with FY-1D). In preparation: HY-2 (Ocean-2).

When their main mission is not altimetry, oceanographic satellites[57] are Sun-synchronous: the Japanese satellites MOS-1 and MOS-1B (Marine Observation Satellite, also called Momo and Momo-1B, where 'momo' means 'peach flower'), h = 908 km, the Indian Oceansat-1 (IRS-P4), h = 720 km, the US SeaStar (OrbView-2), h = 700 km, the South Korean Kompsat-1 (Korea Multi-purpose Satellite), to study the oceans and land masses. Satellites equipped with scatterometers to study the winds over the sea are also Sun-synchronous, e.g., QuikScat (Quick Scatterometer), h = 805 km and Coriolis[58] (also called WindSat or P98-2), h = 830 km.

5.4.4 Navigation Satellites

The first US navigation system was provided by the Transit satellites,[59] in polar (often strictly polar) LEO orbit, as discussed in Chap. 4. They played a very important part at the inception of space geodesy.[60] Several of these satellites were equipped with nuclear generators.[61] A comparable Soviet then

[57] Launch dates: MOS-1 on 19 February 1987, MOS-1B on 7 February 1990, SeaStar on 1 August 1997, Oceansat-1 on 26 May 1999, QuikScat on 20 June 1999, Kompsat-1 (with ACRIMSAT) on 21 December 1999, Coriolis on 6 January 2003.

[58] Gustave Gaspard Coriolis (1792–1843) was a French mathematician and engineer. In his first work, *Du calcul de l'effet des machines* (1829), he introduced the ideas of work done by a force (force times displacement) and kinetic energy. In his paper *Sur le principe des forces vives dans le mouvement relatif des machines* (1831), he examined the various accelerations: absolute, relative, transport and complementary. The last was subsequently given the name of Coriolis acceleration. This is today a basic feature in the study of geophysical fluids in motion, such as marine and atmospheric currents on the surface of the globe.

[59] This system, an integral part of the Polaris nuclear submarine programme, began with the successful launch of Transit-1B and Transit-2A in 1960, followed by the doublet Transit-3B with Lofti-1.

[60] In the geodesy literature, the satellites Transit-5B-1 and -5B-2 are often designated by the simplified notation VBN-1 and VBN-2.

[61] In 1961, Transit-4A was the first satellite to be equipped with a nuclear generator for its electricity supply, the so-called SNAP (System for Nuclear Auxiliary Power). These generators are also referred to by the acronym RTG (radioisotope thermoelectric generator). Other satellites in this series were equipped with RTG: Transit-4B in 1961, Transit-5B-1, -5B-2 and -5B-3 in 1963, Triad-1 in 1972. The fuel was polonium-210 for the Transit-4 satellites, and plutonium-238 for the Transit-5 satellites. In the other US series, satellites with RTG (plutonium-238) were OPS/4682 (or Snapshot, a pun on SNAP), Nimbus-B (launch failure) and Nimbus-3 in LEO orbit, together with the two satellites LES-8 and -9 in GEO orbit. Concerning the Soviet satellites equipped with RTG, it is known that there were accidents with Kosmos-954 and Kosmos-1402. Probes travelling far out into the Solar System are also equipped with nuclear generators. (Cassini is carrying 35 kg of plutonium-238 to produce a power output of 750 W.)

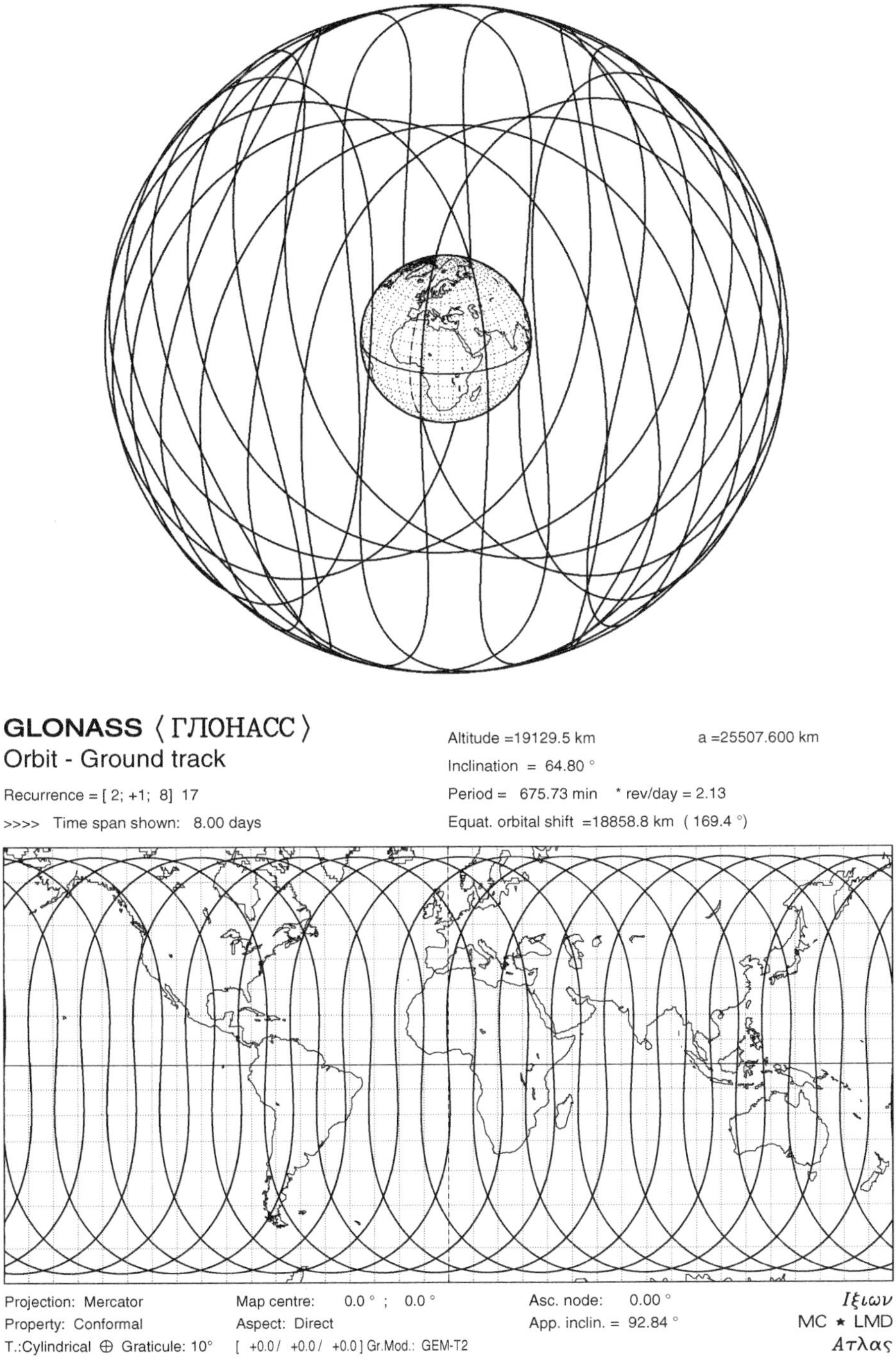

Figure 5.11. Orbit (relative to the Earth) and ground track of the orbit of a satellite in the GLONASS constellation, over a time span of eight days (duration of its recurrence cycle)

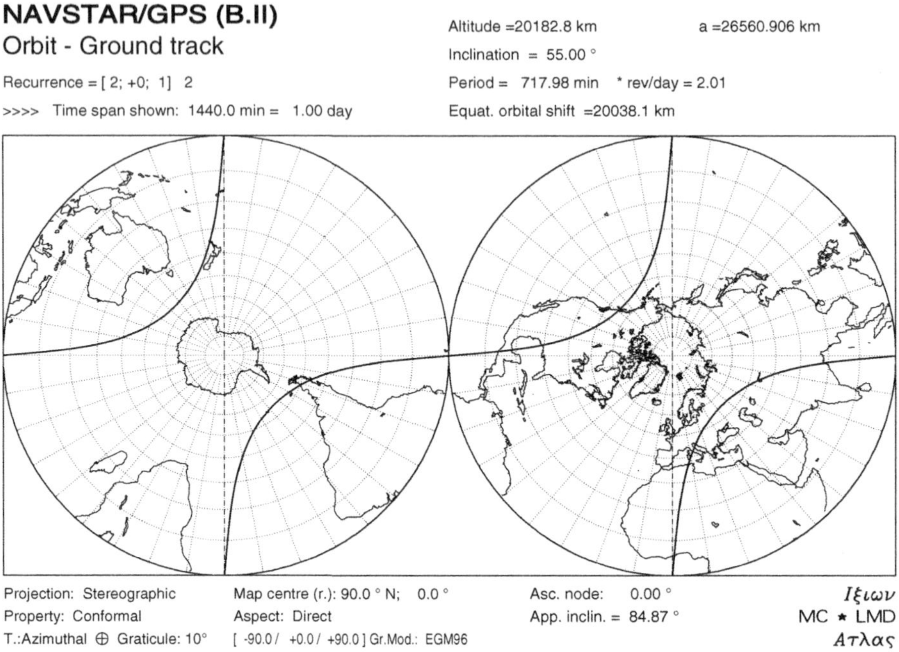

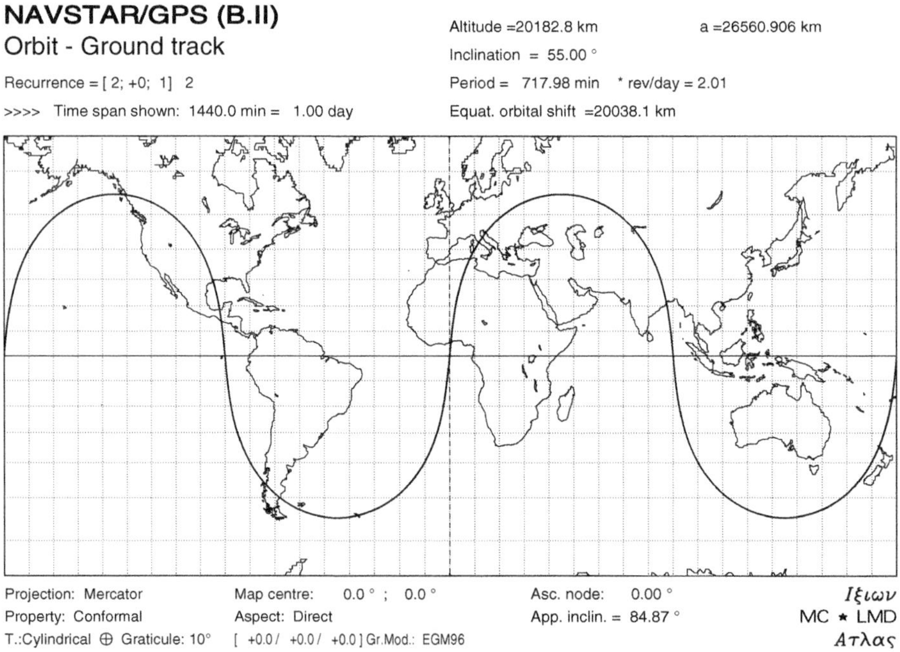

Figure 5.12. Ground track of the orbit of a satellite in the NAVSTAR/GPS constellation, over one day (duration of its recurrence cycle)

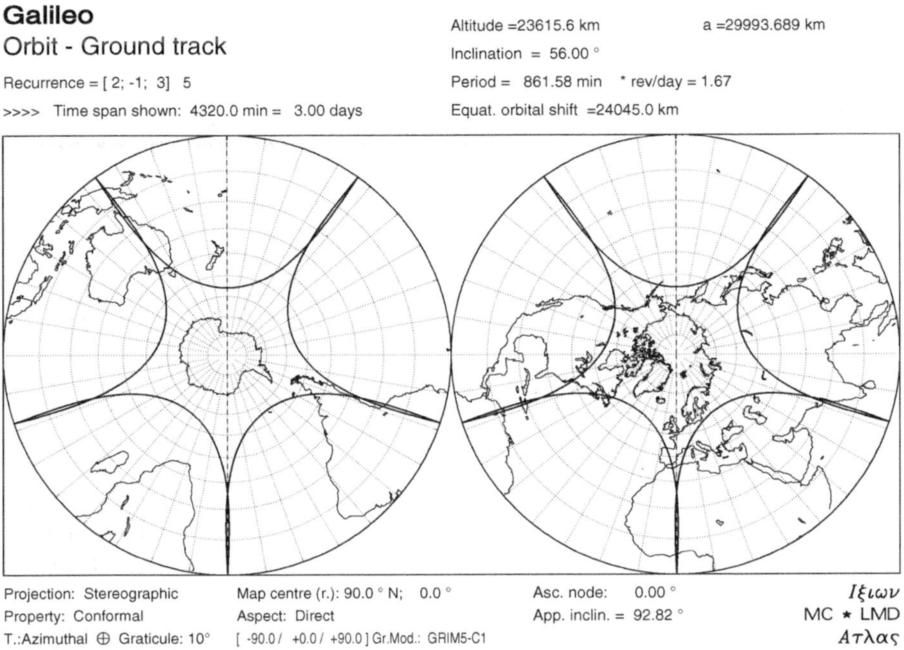

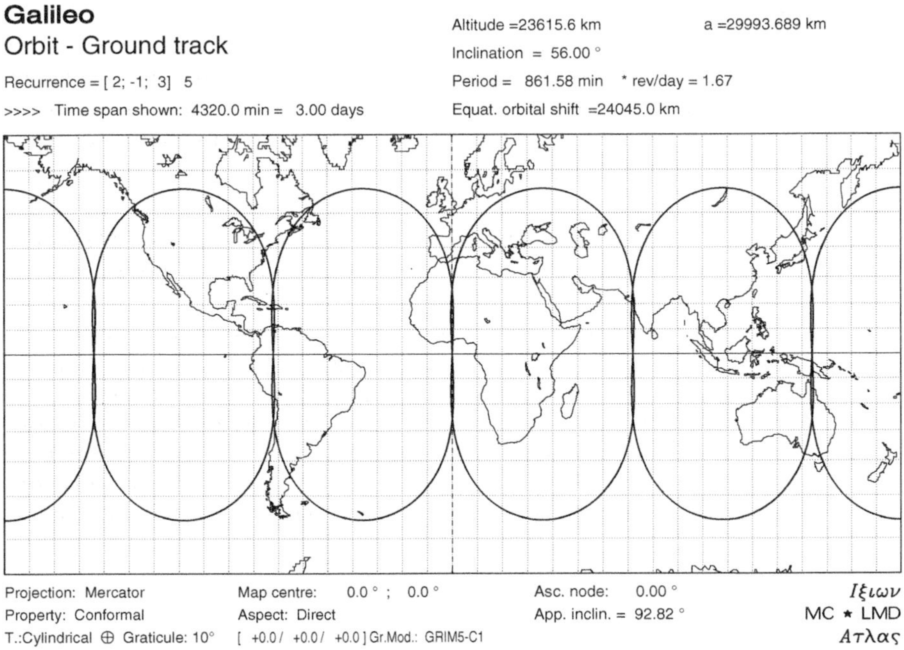

Figure 5.13. Ground track of the orbit of a satellite in the Galileo constellation, over a time span of three days (duration of its recurrence cycle)

Russian system was constructed using the Tsikada and Nadezhda constellations of satellites in polar orbit $h = 1\,000$ km, $i = 82.9°$, such as Kosmos-1383, launched in 1982 or Nadezhda-5 in 1998.

Results of astonishing accuracy (positioning to within a few metres) are currrently obtained with MEO satellite constellations. For the US system known as NAVSTAR/GPS (Navigation System with Timing and Ranging/Global Positioning System), the circular orbit at $h = 20\,200$ km is such that the period is 12 hr, or more exactly, half a sidereal day. Satellites in Block I were experimental, with $i = 63°$, from NAVSTAR-1 (OPS/5111) in 1978 to NAVSTAR-11 (USA-10) in 1985. Block II comprises 24 satellites in 6 planes of 4 operational satellites, with $i = 55°$ (see Figure 5.12), from NAVSTAR-13 (USA-35) in 1989 to NAVSTAR-36 (USA-100), launched on 10 March 1994, at which date the system became operational. To replace defective satellites, so-called Block IIR satellites have been launched on a regular basis since 1996, such as NAVSTAR-37 (USA-117). By the end of 2004, NAVSTAR-55 (USA-178 or GPS-2R-12)) had been reached.

The underlying principle of GPS is simple. Each of the 24 satellites is equipped with a very accurate clock,[62] synchronised from a ground station, and a transmitter which sends the clock signal and satellite position, calculated to great accuracy and regularly adjusted. An arbitrary point, which can be on the Earth, an aircraft or a satellite, equipped with a receiver can determine its distance from the emitting satellite by measuring the journey time of the signal. Since the position of the emitting satellite is known, this therefore determines a sphere, centred on the satellite, on which the given point must be located. The intersection of this sphere with a second, centred on another satellite, specifies a circle, and the intersection of this circle with yet another sphere, centred on a third satellite, specifies two points. One of these two points is an impossible position. A fourth intersecting sphere can confirm this. But the real need for this fourth satellite is to correct the time measurement made by the receiver clock. Indeed, this clock, paid for by the user, is a much cheaper one and hence also much less accurate than the transmitter clock.

The selective availability policy, introduced by the military to limit the accuracy of the system for civilian activities, was abolished in May 2000. Position determinations are constantly being refined by technological advances, such as more powerful signals and better correction for atmospheric pertur-

[62] Block II satellites each have caesium and rubidium clocks. Those in Block IIR should each carry a hydrogen maser atomic clock, with deviation less than 10^{-14} s every three hours.

bations, and improved statistical processing of the signal.[63] The displacement velocity of the receiver is obtained by measuring the Doppler shift.

The Russian system GLONASS (Global Navigation Satellite System) is on a slightly lower orbit, at $h = 19\,100$ km, $i = 64.8°$ (see Fig. 5.11). It comprises 21 satellites, distributed in 7 planes of 3 satellites. The first satellite in the constellation was Kosmos-1413, launched in 1982. The others were generally launched in groups of three, about once a year. The last to be placed in orbit were GLONASS-791, -792, -793 (Kosmos-2394, -2395, -2396), launched together in 2002, and GLONASS-794, -795, -701 (Kosmos-2402, -2403, -2404), launched together in 2003.

The European navigation system of MEO satellites was agreed in 2002 (financed in equal parts by Europe and ESA): this is the Galileo system. It is made up of 30 satellites, in 3 planes of 9 satellites, with 3 back-up satellites. The chosen orbit is circular with $h = 23\,616$ km, $i = 56°$ (see Fig. 5.13).

The European system EGNOS (European Geostationary Navigation Overlay Service) is based on three GEO satellites and uses GPS/NAVSTAR and GLONASS. It will be operational from 2004 and aims primarily at aviation in Europe.

The Chinese system BNS (Beidou Navigation System) resembles EGNOS, with a constellation GEO satellites. The first two, Beidou-1 and Beidou-1B (DFH-51 and -52), were launched in 2000, followed by Beidou-1C (DFH-56) in 2003. ('Beidou' is the Chinese name for the the constellation of Ursa Major.)

5.4.5 Communications Satellites

Telecommunications

The principle of communication by relay is to send a signal, e.g., telephone, television, telecommunications, from a given point on the Earth to another by relaying it through a satellite which detects, amplifies and retransmits it. A GEO satellite can of course do this, provided it is visible from the two points and in suitable conditions. For high latitudes, a group of HEO satellites can guarantee the link. With LEO satellites, the time for which the satellite is visible is rather short and a constellation is required (see Sect. 5.6).

GEO Satellites. Hundreds of GEO communications satellites are currently operating. The geostationary orbit, which is by definition one-dimensional, is

[63] We know that light travels 30 cm in 1 ns. If the receiver clock has an accuracy of 200 ns, the accuracy on the ground is of the order of 60 m. The result is improved by taking measurements over a period of a few minutes and then using statistical analysis.
An accuracy better than one metre is achieved with the DGPS system (Differential GPS), which uses the well established positions of reference points, such as airports. Averaging a DGPS signal can yield a position to within one centimetre.

beginning to get congested today.[64] Generally speaking, a country or group of countries sets up its satellite at the longitude of one of its meridians.[65] A country like Indonesia can use a geostationary satellite to set up a telecommunications network between the hundreds of islands composing it, much more easily than using a network at ground level. Moreover, for many countries, a geostationary communications satellite has an important symbolic value.[66]

---

[64] For example, even in 1989, there were four television and telecommunications satellites operating at the position $\lambda_S = 19°$W, only a few tens of kilometres apart: TDF-1 and TDF-2 (France), TVSat-2 (Germany), and Olympus-1 (UK). In 1999, there were eight Astra satellites, from Astra-1A to Astra-1H, operating between 19.2°E and 19.3°E, dispensing eight bouquets of television programmes over the same regions. The same company duplicates this achievement with the three satellites Astra-2A, -2B, -2D, operating since 2001 at longitude 28.2°E.

[65] As an example, we mention several satellites with their parking position on 30 June 1999. We give one satellite per country or organisation, operating at this date, going round the equator in the positive direction: Sirius-2 (Sweden) 4.83°E, Eutelsat-W3 (European organisation) 7.01°E, Italsat-F2 (Italy) 16.24°E, AfriStar (international foundation) 20.98°E, Kopernikus-3 [DFS-3] (Germany) 23.50°E, Inmarsat-3-F5 (international organisation) 25.07°E, Arabsat-3A (Arab League) 27.20°E, Astra-2A (Luxembourg) 28.23°E, Gorizont-31 (Russia) 39.91°E, Turksat-1C (Turkey) 41.99°E, Intelsat-704 (international organisation) 66.04°E, Thaicom-3 (Thailand) 78.50°E, Zhongwei-1 [Chinastar-1] (China) 87.53°E, Measat-1 (Malaysia) 91.49°E, INSAT-2C (India) 93.49°E, AsiaSat-3S (Hong-Kong/China) 105.50°E, Cakrawarta-1 [Indostar-1] (Indonesia) 107.65°E, Mugunghwa-2 [Koreasat-2] (South Korea) 115.89°E, N-Star-B (Japan) 133.98°E, Optus-A3 [AusSat-3] (Australia) 164.05°E, Solidaridad-2 (Mexico) 113.00°W, Galaxy-8 (United States) 94.94°W, Nimiq-1 [Telesat-DTH-1] (Canada) 91.16°W, Brasilsat-B3 (Brazil) 84.05°W, Nahuel-1A (Argentina) 71.87°W, Hispasat-1B (Spain) 29.99°W, Telecom-2D (France) 5.03°W, Amos-1 (Israel) 4.03°W, Thor-3 (Norway) 0.84°W. The first commercial satellite in this category, Anik-A1 (Canada), was launched in 1972.

[66] The names chosen for these satellites serve to demonstrate this. The multiethnic country Indonesia, sometimes torn by internal conflict, chose the name Palapa, which means 'unity' in Bahasa Indonesia (the official language).
Developed countries in the New World seek the names of their satellites in the Amerindian languages, perhaps as a way of finding their roots. Canada named its satellites Anik (ᐊᓂᒃ, in Inuit writing) and Nimiq (ᓂᒥᖅ), which means 'brother' (for a sister) and 'union' (or 'bond that unites') in Inuktitut (the Inuit or Eskimo language). Argentina uses the Araucanian word 'Nahuel' (Araucan, Mapuche language), meaning 'tiger'.

These GEO communications satellites, which are becoming increasingly common, bigger, and more expensive,[67] currently represent the largest part of the commercial market for space activities.

GEO satellites are also widely used for military communications. Examples are the US series DSCS (Defense Satellite Communications System),[68] LES (Lincoln Experimental Satellite),[69] TDRS (Tracking and Data Relay Satellite)[70] and Milstar (Military Strategic and Tactical Relay System),[71] the Soviet series Luch (with Kosmos-2054 and Luch-1, where 'luch' means 'light beam' in Russian) and Raduga (up to Raduga-32, where 'raduga' means 'rainbow'), and the Chinese series STTW (with China-26).

The satellite Telecom-2C is shared by French civilian and military organisations. The first high-speed data transmission by laser was carried out in

[67] The two satellites Westar-6 (Western Union Communications Satellite, US) and Palapa-B2 (Indonesian) were launched by STS-11 (STS-41-B) on 3 February 1984, but they never reached the geostationary orbit. The insurers, the new owners, paid for recovery and return of the satellites by the shuttle flight STS-19 (STS-51-A) on 16 November 1984. China bought Westar-6, and turned it into AsiaSat-1, launching the satellite itself on 7 April 1990. The other satellite became Palapa-B2R and was launched on 13 April 1990. The adventures of AsiaSat did not end there. The satellite AsiaSat-3, launched by a Russian rocket on 24 December 1997, was placed on the wrong orbit, too highly inclined (GTO orbit, $i = 56°$). After purchasing it, the new owner (Hughes) attempted a novel maneuver: the satellite was sent on two revolutions around the Moon ($r_a = 488\,000$ km, $T = 15$ day, see the Luna-3 satellite). It then returned to a geostationary orbit and became HGS-1 (Hughes Global Services), before being renamed PAS-22. The replacement, AsiaSat-3S, was inserted into the wrong orbit on 21 March 1999: $h_p \approx 10\,000$ km, $h_a = h_{GS}$, $i = 13°$. Using its thrust motors, it was then moved into GEO orbit. Another example of successful recovery, although less spectacular, is the following. The satellite Palapa-C1, launched in 1996, broke down in 1998. The manufacturer bought it back, repaired it under the name of HGS-3 whilst it remained in GEO orbit, then resold it to Turkey under the name of Anatolia-1, having displaced it in longitude.

[68] Launched since 1971, they are stationed over the American continent. The super secure communication satellites DSCS-3A3 (USA-167) and DSCS-3B6 (USA-170) were launched in 2003.

[69] In GEO orbit, from LES-5 to LES-9.

[70] The TDRS satellites in the TDRSS series (TDRS System) are launched at intervals varying between two and five years. The first were launched from the US Space Shuttle, such as TDRS-1 in 1983 (STS-6), or TDRS-2, lost in the Challenger explosion of 1986, up to TDRS-7 in 1995 (STS-70). In 2000, TDRS-8 was launched directly, and in 2002, TDRS-9 and -10 likewise.

[71] The Milstar-1 satellites such as Milstar-1-1 (USA-99), launched in 1994, provide low data rate (LDR) communications. Those in the Milstar-2 series, such as Milstar-2-3 (Milstar-DFS-5 or USA-164) launched in 2002 and Milstar-2-4 (Milstar-DFS-6 or USA-169), launched in 2003 provide medium data rate (MDR) communications. The next generation, Milstar-AEHF, will operate at very high speed (high data rate or HDR).

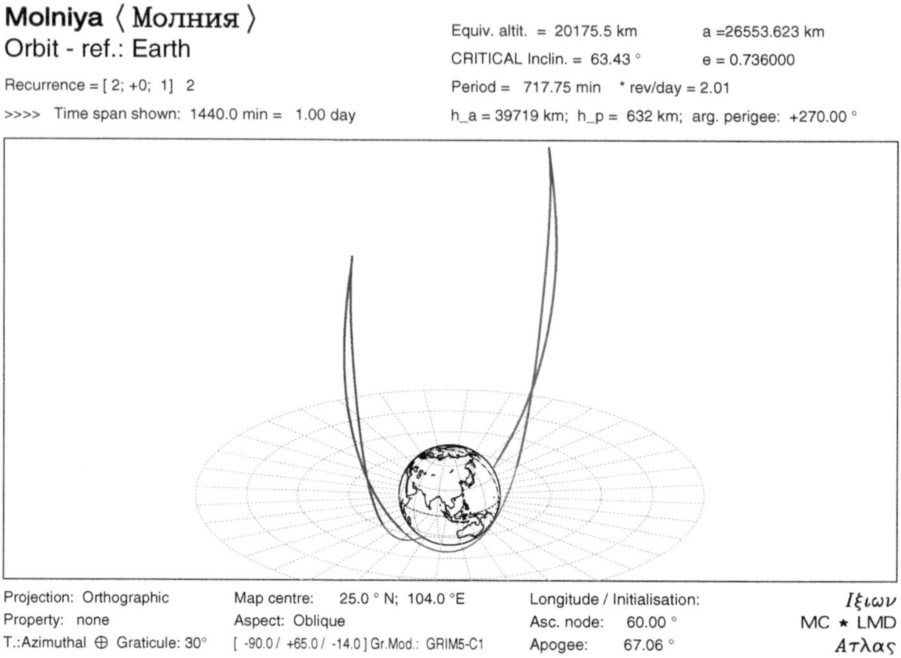

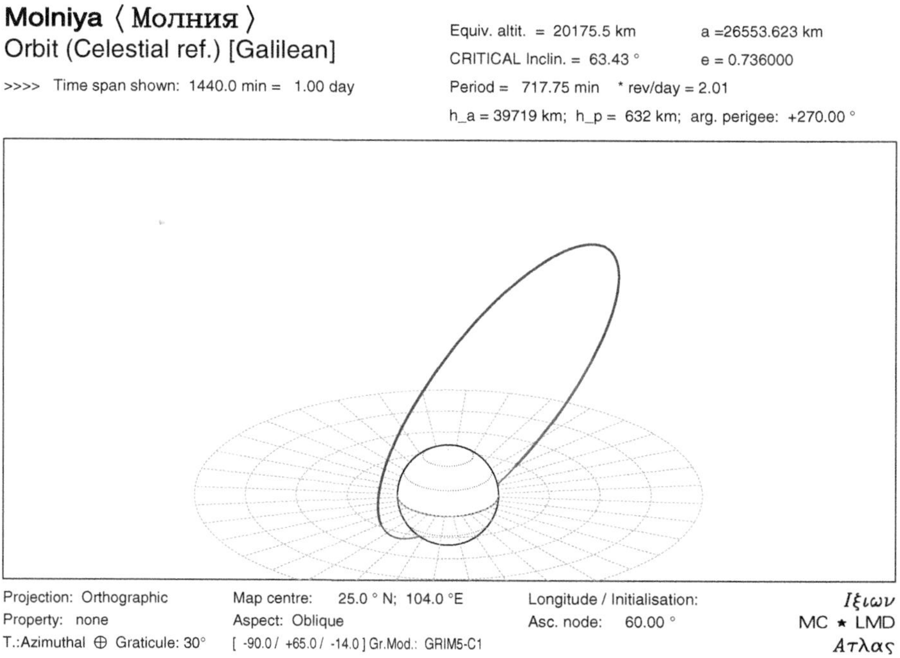

Figure 5.14. Orbit of a Molniya satellite, over 1 day (2 revolutions). *Upper*: Orbit, relative to the Earth (terrestrial reference frame). *Lower*: Orbit, relative to the stars (celestial reference frame)

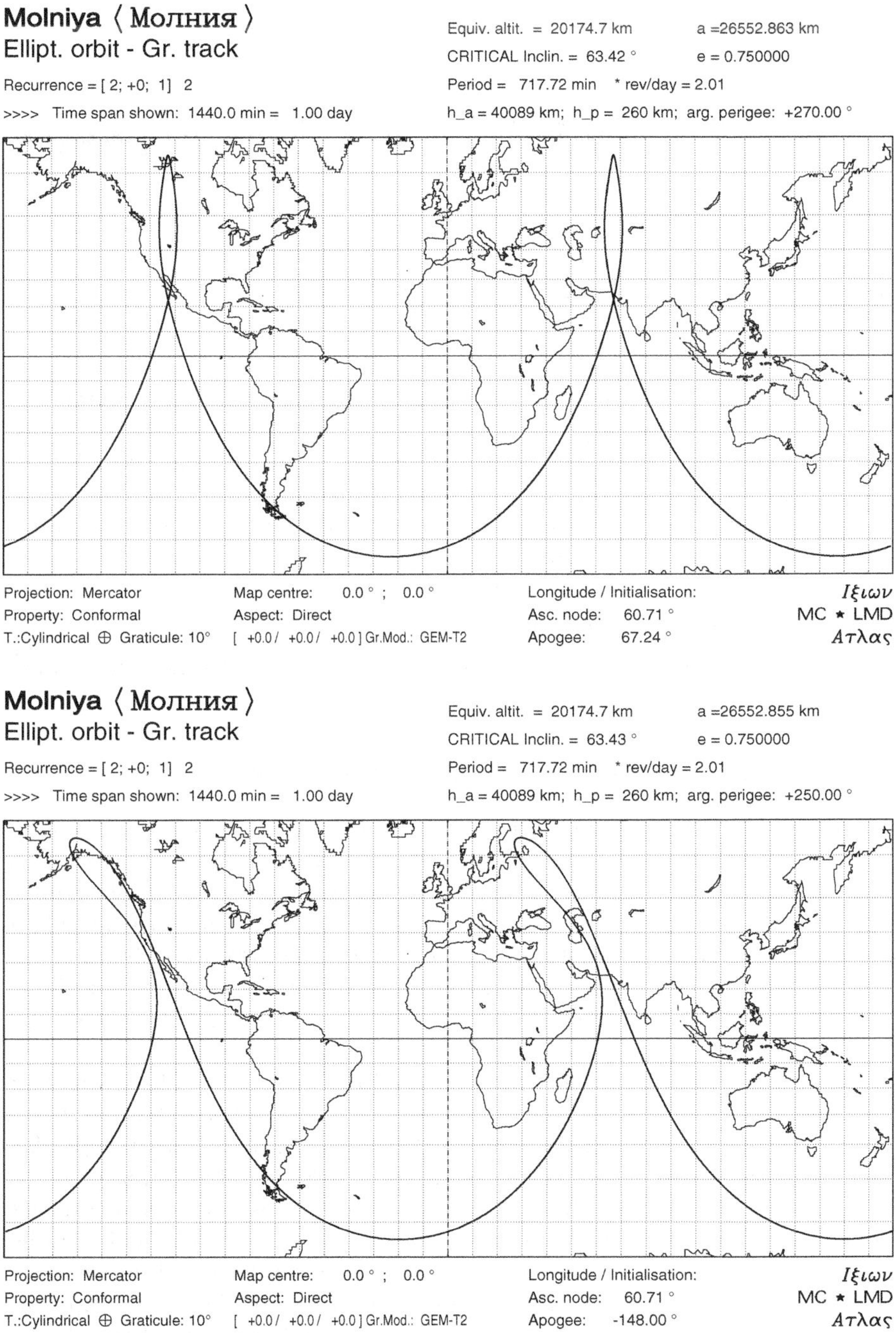

Figure 5.15. Ground tracks of the (HEO) orbit of a Molniya satellite, with two different perigee positions, over a time span of one day

Tundra
Ellipt. orbit - Gr. track

Recurrence = [1; +0; 1] 1
>>>> Time span shown: 1440.0 min = 1.00 day

Equiv. altit. = 35785.2 km a =42163.383 km
CRITICAL Inclin. = 63.43 ° e = 0.266800
Period = 1436.04 min * rev/day = 1.00
h_a = 47034 km; h_p =24536 km; arg. perigee: +270.00 °

Projection: Orthographic Map centre: 15.0 ° N; 95.0 ° W Longitude / Initialisation: Ιξιων
Property: none Aspect: Oblique Asc. node: -69.80 ° MC ★ LMD
T.:Azimuthal ⊕ Graticule: 10° [-90.0 / +75.0 / -175.0] Gr.Mod.: GEM-T2 Apogee: -100.00 ° Ατλας

Supertundra
Ellipt. orbit - Gr. track

Recurrence = [1; +0; 1] 1
>>>> Time span shown: 1440.0 min = 1.00 day

Equiv. altit. = 35785.1 km a =42163.191 km
CRITICAL Inclin. = 63.43 ° e = 0.423000
Period = 1436.03 min * rev/day = 1.00
h_a = 53620 km; h_p =17950 km; arg. perigee: +270.00 °

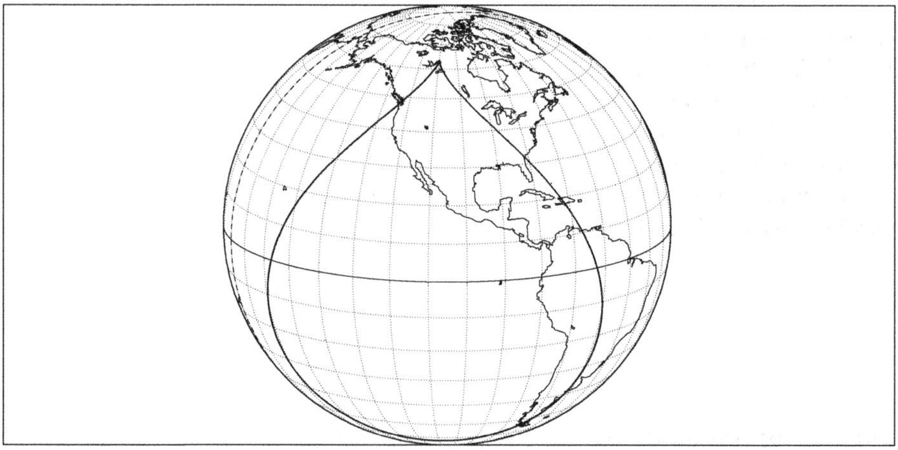

Projection: Orthographic Map centre: 15.0 ° N; 105.0 ° W Longitude / Initialisation: Ιξιων
Property: none Aspect: Oblique Asc. node: -63.02 ° MC ★ LMD
T.:Azimuthal ⊕ Graticule: 10° [-90.0 / +75.0 / -165.0] Gr.Mod.: GEM-T2 Apogee: -110.00 ° Ατλας

Figure 5.16. Ground tracks of satellites in (HEO) Tundra and Supertundra orbits, over a time span of one day

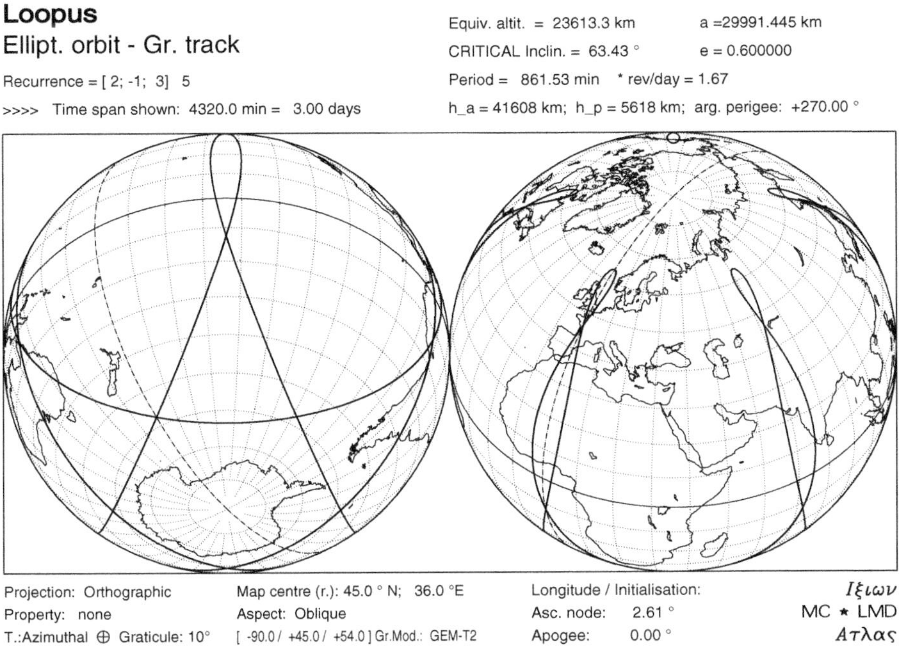

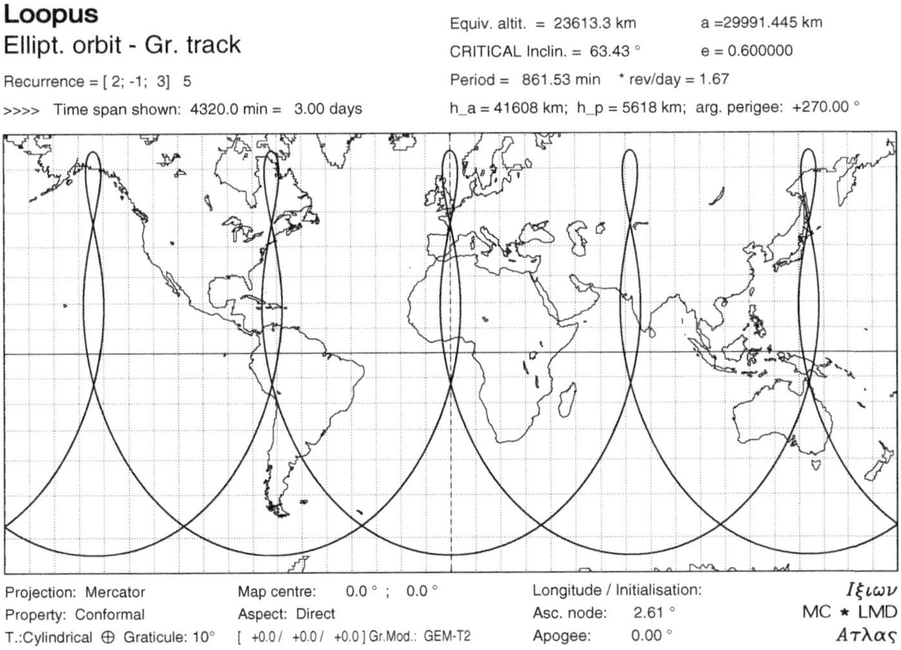

Figure 5.17. Ground track of the (HEO) orbit of a satellite in the Loopus constellation, over a time span of three days

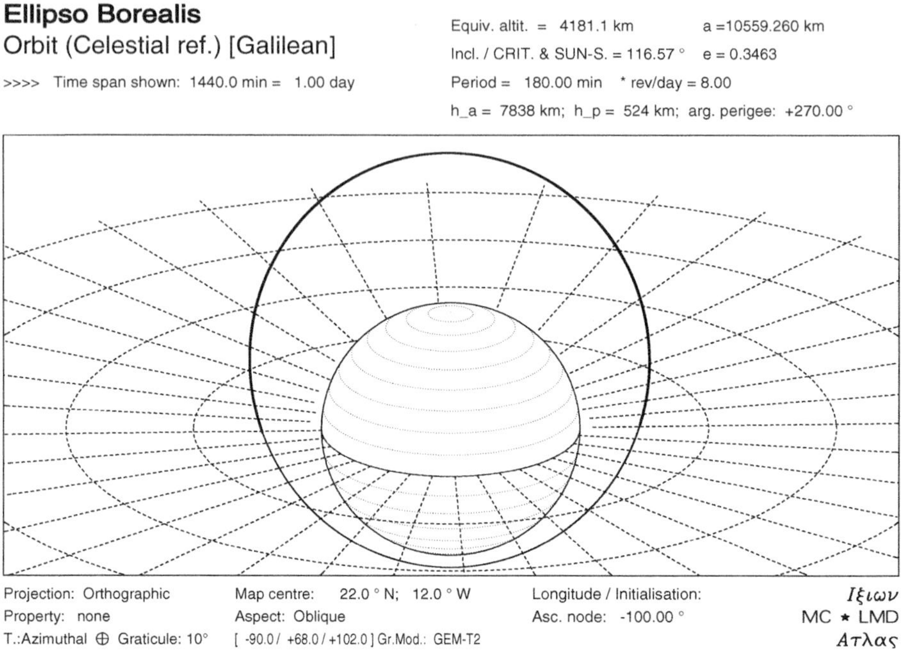

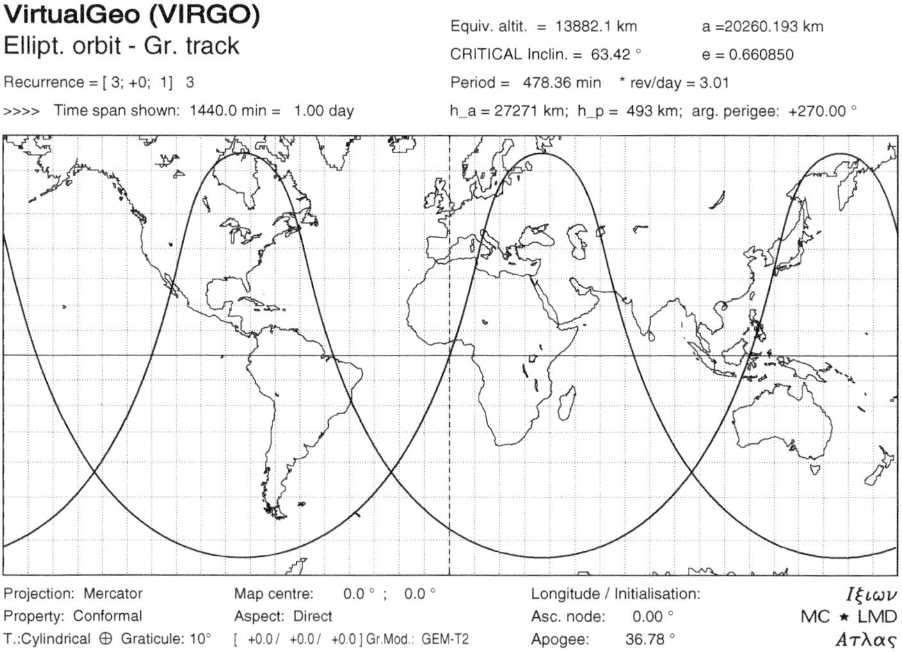

Figure 5.18. *Upper* Orbit and ground track of the (HEO) orbit of a satellite in the Ellipso Borealis constellation. *Upper* Ground track of a satellite in the VIRGO constellation, over a time span of one day

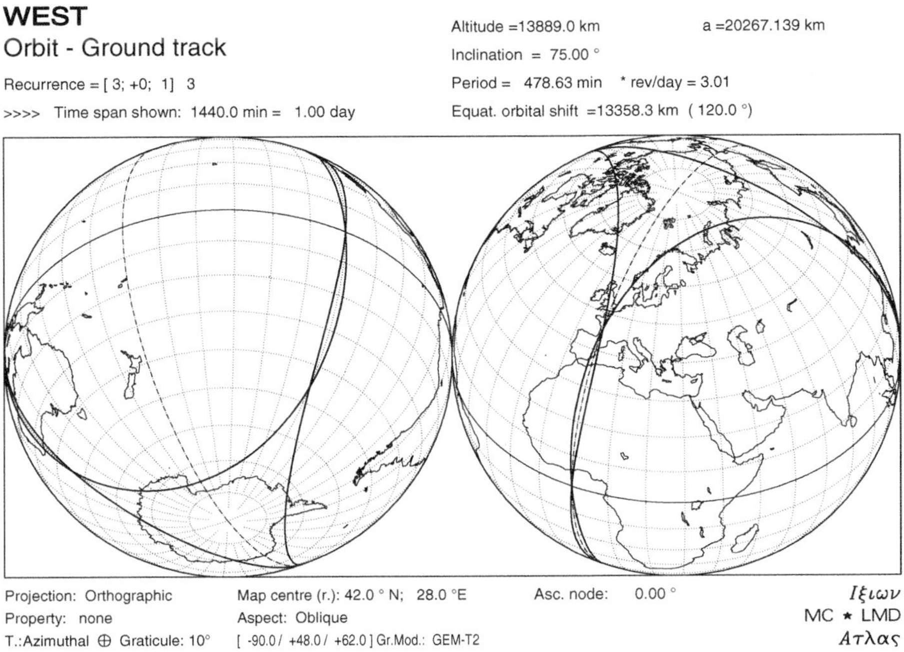

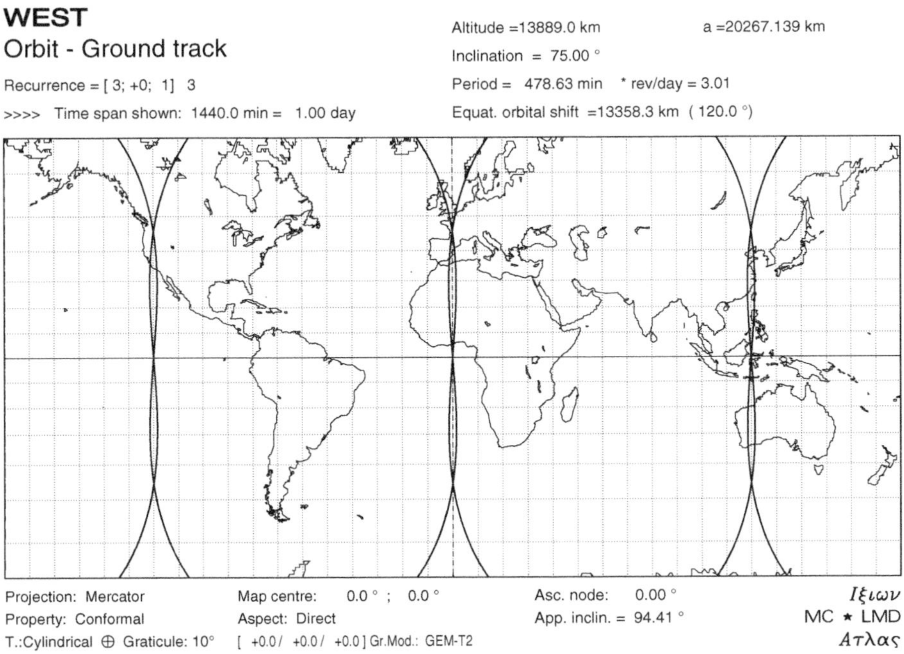

Figure 5.19. Ground track of the (MEO) orbit of a satellite in the WEST constellation, over a time span of one day

2001 between the European GEO satellite Artemis, launched just before for this purpose, and the French LEO satellite SPOT-4. This prefigured transmissions between LEO satellites via a GEO satellite as intermediary.

HEO Satellites. The most common HEO orbit is the one used by Soviet then Russian satellites, the Molniya orbit (meaning 'lightning'), since Kosmos-41 in 1964. The number of satellites launched in these orbits is staggering: 93 for Molniya-1, from Molniya-1-01 in 1965 to Molniya-1-91 in 1998, then Molniya-1-92 in 2003 and Molniya-1-93 (Kosmos-2405) in 2004, 17 for Molniya-2, from Molniya-2-01 in 1971 to Molniya-2-17 in 1977, 53 for the present series Molniya-3, from Molniya-3-01 in 1974 to Molniya-3-53 in 2003.

The satellites in these three series have the same orbit, to within a few tens of kilometres:[72] $h_p \approx 500$ km, $h_a \approx 40\,000$ km, $i = 63°$ (critical inclination). The period is 12 hr (half a sidereal day): $T = 717.7$ min, $a = 26\,553$ km, e between 0.72 and 0.75. The argument of the perigee is $\omega = -90°$, which means that the perigee is located in the southern hemisphere, region of the Earth that the satellite overflies very quickly (see Example 1.3). On the other hand, at the apogee, the satellite is almost stationary for 8 hr, when above Russia (see Example 5.7). The orbit and the ground track of this orbit are shown in Fig. 5.14 and 5.15.

This orbit is at the critical inclination, which fixes the position of the perigee (and the apogee). Furthermore, the orbit is recurrent with a cycle of one day: the ground track passes through the same point every day. With three regularly spaced satellites on the same orbit, one almost achieves the equivalent of a geostationary satellite for the regions close to the ground track at apogee. This is a judicious method for solving the problem of geosynchronicity at high latitudes.

The US military satellites SDS (Space Defense System), from SDS-1 in 1976 to SDS-7 (USA-21) in 1987, are in Molniya orbit.

The Tundra and Supertundra orbits, shown in Fig. 5.16 and discussed in Chap. 4, are far less commonly used. The constellation SD-Radio, which provides radio transmissions for American road users, is in a Tundra orbit. The perigee of these two orbits is high, well above the Van Allen radiation belt. This is not so for the Molniya orbits, where satellites cross this belt on every revolution, an ordeal for electronic equipment.

---

[72] As an example, here are the characteristics of several Molniya orbits with the launch date of the satellite, using the notation $[h_p/h_a/i]$ (altitudes in km, angle in degrees): Molniya-1-01 (23 April 1965) [538/39300/65.5], Molniya-2-01 (25 November 1971) [516/39553/65.0], Molniya-3-01 (21 March 1974) [250/40095/64.1], Molniya-3-50 (8 July 1999) [464/39889/62.8], Molniya-3-51 (20 July 2001) [407/40831/62.9], Molniya-3-52 (25 October 2001) [615/40658/62.9], Molniya-3-53 (19 June 2003) [637/39709/62.8], Molniya-1-93 (18 February 2004) [634/39729/62.8].

The Loopus orbit, $a = 30\,000$ km, $e = 0.6$, Fig. 5.17, lies outside the radiation belts. The period of the satellite is 3/5 of a day ($T = 861.526$ min, see Galileo orbital characteristics).

The Ellipso Borealis orbit is highly original, as we have already seen: it is Sun-synchronous and has the critical inclination. The parameters are: $h_\mathrm{p} = 524$ km, $h_\mathrm{a} = 7\,838$ km, $i = 116.57°$, $\omega = -90°$, with a period of 3 hr [see Fig. 5.18 (upper) and Colour Plate VII]. The Ellipso project includes a HEO constellation, Borealis, with 10 satellites in 2 planes, and a constellation, Concordia, with 4 (or perhaps 7) satellites in equatorial orbit, $h = 8\,050$ km, $i = 0.0°$.

The US projects COBRA (Communications Orbiting Broadband Repeating Arrays) and VIRGO (or VirtualGEO, Virtual Geostationary) use HEO orbits with a period of 8 hr, precisely one third of a sidereal day: $a = 20\,261$ km, $e = 0.6458$, $i = 63.4°$ (see the lower part of Fig. 5.18). As in the Molniya case, the VIRGO constellation is effectively playing the role of a geostationary satellite.

MEO Satellites. The WEST orbit is circular with $h = 13\,900$ km, $i = 75°$ and a period of 8 hr (one third of a sidereal day), Fig. 5.19. The WEST constellation project (Wideband European Satellite Telecommunication) includes 9 satellites on this MEO orbit, together with 2 GEO satellites.

Satellites Between LEO and MEO. The Odyssey project, abandoned in 2000, consisted of a constellation of 3 planes of 4 satellites, $h = 10\,354$ km, $i = 50°$. The ICO constellation (Intermediate Circular Orbit)[73] comprises 2 planes of 5 satellites, $h = 10\,390$ km, $i = 45°$.

LEO Satellites. Communications using LEO satellites always require a constellation. For telephone communications, the advantage of LEO constellations is the very short response time: the journey time of the signal transiting via a GEO satellite is about 250 ms, and this quarter of a second is sometimes considered to be a nuisance. However, this is not a big enough advantage to ensure the success of a commercial venture. The failure of Iridium[74] in 2000 and GlobalStar in 2002 clearly demonstrates this. The following satellite constellations were operating or under construction (or had failed) in 2004:

- Orbcomm, 35 satellites (3 planes of 8), $h = 810$ km, $i = 45.0°$,
- Iridium, 88 satellites (6 planes of 11), $h = 780$ km, $i = 86.4°$,
- GlobalStar, 52 satellites (8 planes of 6), $h = 1\,410$ km, $i = 52.0°$,

[73] Launch date: ICO-F2 on 19 June 2001.

[74] Faced with commercial failure, the first reaction was to remove all the satellites from orbit. However, in the end, the system was taken over by the US Department of Defense. Originally, the constellation was to include 77 satellites, whence the name Iridium, which is the chemical element (Ir) with atomic number 77.

- Teledesic,[75] 288 satellites (12 planes of 24), $h = 550$ km, $i = 97.7°$.

The total number of satellites in each constellation[76] includes all spare satellites.

The Russian system Gonets-D1 ('gonets' means 'messenger') is the commercial version of the military system Strela-3 ('strela' means 'arrow'). The constellation was set up[77] with the following characteristics:

- Gonets-D1, 48 satellites (6 planes of 8), $h = 1400$ km, $i = 82.5°$.

The OSCAR satellites (Orbiting Satellite Carrying Amateur Radio), not to be confused with Transit-Oscar, are launched on a regular basis for amateur radio transmissions. Orbits are often near-polar and LEO, as for OSCAR-1 in 1961 or OSCAR-30 in 1996.

Passive Communication

As their name suggests, the satellites Echo-1 and -2 were launched as passive telecommunications relays. As one might imagine, the results were not very convincing and this experimental system was dropped. However, these two satellites were originally intended for space geodesy.

We should also mention another attempt to create a passive space communications system. The idea was to place a ring in orbit around the Earth to reflect radio waves. This was the US experiment called Westford Needles. Two packages (Westford-1 and -2) were inserted into orbit. Once opened, they distributed 350 then 475 million small copper needles (1.7 cm long, 0.1 mm in diameter) along their trajectory. These military experiments were carried

[75] The original project, in 1994, was to launch 840 active satellites (21 planes of 40).

[76] Launch dates of the first and last satellite in each constellation: Orbcomm constellation: Orbcomm-FM-1 and -FM-2 on 3 April 1995, launched with Microlab-1, $h = 740$ km, $i = 69.9°$, Orbcomm-FM-3 and -FM-4 on 10 February 1998, launched with GFO-1, $h = 830$ km, $i = 108.0°$, the other Orbcomm satellites were launched with inclination $i = 45.0°$, in clusters of 8, (beginning) Orbcomm-FM-5 to -FM-12 on 23 December 1997, (end) Orbcomm-FM-30 to -FM-36 on 4 December 1999. Iridium constellation: (beginning) Iridium-4 to -8 on 5 May 1997, (end of initial programme) Iridium-83 to -86 on 6 November 1998, (restart) Iridium-90 to -96 on 11 February 2002, Iridium-97 and -98 on 20 June 2002. GlobalStar constellation: (beginning) GlobalStar-M001 to -M004 on 14 February 1998, (end) GlobalStar-M061 to -M064 on 4 December 1999. Teledesic constellation: (demonstration satellite) Teledesic-1 on 26 February 1998.

[77] The Strela-3 satellites were launched in clusters of 6. With the establishment of the Gonets-D1 constellation, each cluster of 6 satellites has included 3 military and 3 civilian. Launch dates: Gonets-D1-1, -2 and -3 (with the Strela-3 satellites, Kosmos-2328, -2329, -2330) on 19 February 1996, Gonets-D1-4 and the following on 14 February 1997, Gonets-D1-7 and the following on 17 December 2001.

out between 1961 and 1963, from the satellites MIDAS-4, -5 and -7, in circular near-polar orbit $h = 3\,600$ km. They were called Westford-1, Westford Drag, and Westford-2, respectively. The experiment was severely criticised by astronomers who saw in this a source of optical and radio pollution. In the end, the needles constituted neither a reflector nor a pollutant.

5.4.6 Satellites for Fundamental Physics

The satellite Gravity Probe-B (GP-B), launched on 20 April 2004, has a strictly polar LEO orbit, already mentioned in the last chapter. Its ground track is illustrated in the lower part of Fig. 5.5. The aim of the mission is to measure, using gyroscopes, the extent to which space and time are distorted by the presence of the Earth, within the framework of the general theory of relativity. The preliminary experiment GP-A, in June 1976, sent a hydrogen maser clock into space in a sub-orbital flight up to an altitude of 10 000 km.

Two missions are being planned to provide experimental corroboration for the principle of equivalence.[78] The French project µSCOPE (*Micro-Satellite à Compensation de Traînée pour l'Observation du Principe d'Equivalence*), or Microscope, is designed to check the universality of free fall. The satellite will have a circular, Sun-synchronous LEO orbit at $h \approx 700$ km. Drag effects on the satellite will be compensated by electric thrusters to ensure that the two test masses carried aboard undergo perfect free-fall conditions over thousands of kilometers (rather than just a few tens of meters, and without compensation for air resistance, if one drops the objects from the leaning tower of Pisa). The US satellite STEP (Satellite Test of the Equivalence Principle) will be much bigger, with cooled instruments to increase measurement accuracy.[79] It will also follow a circular LEO orbit, with $h = 400$ km.

The LISA project (Laser Interferometer Space Antenna), a joint venture between NASA and ESA, will attempt to detect gravitational waves using three satellites in formation on a heliocentric orbit.[80]

[78] The principle of equivalence postulated by Einstein is based on the observation that all bodies, independently of their mass, have the same acceleration in a gravitational field for identical initial conditions. This is the universal principle of free fall: the passive gravitational mass of a body ($m = m_{\mathrm{g}}$ in the expression for the gravitational force, involving $\mathcal{G}$) is equal to the inertial mass ($m = m_{\mathrm{i}}$ in Newton's second law).

[79] The relation $m_{\mathrm{g}}/m_{\mathrm{i}} = 1$ has been checked on Earth to an accuracy of 10^{-12}. The µSCOPE experiment aims to achieve 10^{-14} and STEP 10^{-17}.

[80] The three satellites will be placed at the corners of an equilateral triangle with side 5 million kilometres. The centre of the triangle will lie in the plane of the ecliptic at a distance of 1 a.u. from the Sun. The angle between the centre of the triangle, the Sun and the Earth will be equal to 20° (measured positively). The three satellites, linked by laser, will constitute a gigantic Michelson interferometer.

5.4.7 Satellites for Astronomy and Astrophysics

Astronomy and Astrometry

In most cases, astronomical missions are one-off experiments and the satellites are not systematically placed on the same orbits, as they might be for remote-sensing, for example. We shall discuss here several satellites for each region of the electromagnetic spectrum used in astronomy. Most of them are American and the remainder are mainly European. NASA launched a programme of large orbiting observatories (Great Observatories Program): GRO (Gamma Ray Observer), renamed Compton,[81] for γ rays, AXAF (Advanced X-ray Astrophysics Facility), renamed Chandra,[82] for X rays, HST (Hubble Space Telescope) often called Hubble[83] for the visible, and SIRTF (Space InfraRed Telescope Facility) renamed[84] Spitzer or SST (Spitzer Space Telescope) after launch, for the infrared.

Concerning more modest missions, the Explorer programme is NASA's continuing programme for orbital astronomy and space physics packages. Explorers come in three sizes defined by budget considerations: Medium Explorer (MIDEX), Small Explorer (SMEX), and University-class Explorer (UNEX).

Cosmic Rays, Gamma Rays. Gamma astronomy began in 1961 with Explorer-11, $h_{\mathrm{p}} = 480$ km, $h_{\mathrm{a}} = 1\,460$ km, $i = 29°$, which provided the first detection of gamma rays from space. This was followed by OSO-3 and OSO-7, both on the orbit $h = 550$ km, $i = 33°$, then Explorer-48 (SAS-2), $h = 526$ km, $i = 1°$, not forgetting the Vela satellites mentioned again below.

In this field of investigation, orbits are generally low and not highly inclined, as for the HEAO satellites (High Energy Astrophysical Observatory), HEAO-1, -2 (renamed Einstein), -3, $h \approx 500$ km, $i = 23°$, $23°$ and $44°$,

[81] Arthur Holly Compton (1892–1962) was an American physicist. His work on X rays led him to discover in 1923 the effect which now carries his name (interaction between matter and X rays). He also studied cosmic rays.

[82] Subrahmanyan Chandrasekhar (1910–1995) was an American astrophysicist of Indian birth. He carried out a great many theoretical studies on the internal structure of stars. He studied radiative transfer in stellar atmospheres. The root 'chand' means 'Moon' or 'bright' in Sanskrit.

[83] Edwin Powell Hubble (1889–1953) was an American astronomer. He produced a classification of extragalactic nebulas and, in 1928, established the law of spectral shifts now known as the Hubble law, which says that the spectral shift of a galaxy (redshift) is proportional to its distance, thus confirming the hypothesis that the Universe is expanding. The constant of proportionality H_0, called the Hubble constant, is not precisely known and may even vary in time. It is measured in $\mathrm{km\,s^{-1}}$ per megaparsec. Its reciprocal, which has units of time, gives the age of the Universe to within an order of magnitude: $1/H_0 \approx 10$ billion years.

[84] In honour of Lyman Spitzer Jr. (1914–1957), the American astrophysicist who was the first to suggest placing a large telescope in space.

or the HETE satellites (High Energy Transient Experiment), HETE-1, -2, $h \approx 500$ km, $i = 38°$ and 2°. The Compton satellite (also called CGRO), launched on 5 April 1991 by STS-37, $h = 450$ km, $i = 28.5°$, had a mass of 15 tonnes, of which 7 tonnes represented instrumentation. Great precautions were taken when it was de-orbited on 3 June 2000. The international project GLAST (Gamma-ray Large Area Space Telescope) will use a similar orbit: $h = 550$ km, $i = 28.5°$, and the Italian Agile satellite will be in equatorial orbit, $h = 550$ km, $i \approx 0°$. The US project Swift (MIDEX-3) for detection of gamma-ray bursts will follow a similar orbit, $h = 600$ km, $i = 19°$. The European satellite Integral (International Gamma-Ray Astrophysics Laboratory), launched on 17 October 2002, is on an HEO orbit: $a = 87\,698.656$ km, $e = 0.8204$ (i.e., $h_{\mathrm{p}} = 9\,400$ km, $h_{\mathrm{a}} = 153\,300$ km), $i = 57.1°$, $T = 71.8$ hr $= 3J_{\mathrm{sid}}$,see Fig.5.20 and 5.21 (upper).

X Rays. Certain orbits are like those seen above, i.e., low, with $h \approx 550$ km, and with low inclination: $i = 53°$ for ROSAT, $i = 4°$ for BeppoSAX (*Satellite per Astronomia a raggi X*), $i = 23°$ for XTE (X-ray Timing Explorer), $i = 38°$ for HESSI (High-Energy Solar Spectroscopic Imager, SMEX-6), renamed RHESSI (Reuven Ramaty HESSI).

The rest are very high and highly eccentric, as in the case of the Soviet satellites Astron and Granat, or the European satellite Exosat, launched in 1983 (one revolution is shown in the upper part of Fig. 5.22). In 1999, two large satellites, one American and the other European, used this type of orbit: Chandra (also called CXO, Chandra X-ray Observatory), launched on 23 July by STS-93, $h_{\mathrm{p}} = 10\,157$ km, $h_{\mathrm{a}} = 138\,672$ km, $i = 29.0°$, $T = 63.5$ hr, and XMM (ESA's X-ray Multi-Mirror Space Observatory, renamed XMM-Newton), launched on 10 December, $h_{\mathrm{p}} = 7\,417$ km, $h_{\mathrm{a}} = 113\,678$ km, $i = 38.8°$, $T = 47.9$ hr.

Ultraviolet (UV). These satellites use slightly inclined LEO orbits. The satellites OAO-1, -2, -3 (Orbiting Astronomical Observatory), launched between 1966 and 1972, were on the orbit $h = 750$ km, $i = 35°$. OAO-3 was renamed Copernicus to mark the 500th anniversary of the birth of the great man. The US satellites EUVE (Extreme UV Explorer), $h = 515$ km, $i = 28.4°$, FUSE (Far UV Spectroscopic Explorer, MIDEX-0), $h = 760$ km, $i = 25.0°$, and GALEX (Galaxy Evolution Experiment, SMEX-7), $h = 690$ km, $i = 28.0°$ were launched in 1992, 1999 and 2003, respectively. The orbit of the Spanish Minisat-01, launched in 1997, $h = 570$ km, is also tilted by about thirty degrees to the equator, since $i = 151°$ (which is the highest value yet encountered, to our knowledge).

The satellite CHIPSat (Cosmic Hot Interstellar Plasma Spectrometer Satellite, UNEX-1) is on a near-polar orbit since it was launched as a passenger of ICESat. CHIPSat was the first US mission to use end-to-end satellite operations with TCP/IP and FTP.

Visible. In the visible region of the electromagnetic spectrum, the two largest missions have been the European Hipparcos and the American Hubble (with ESA participation). Both have provided excellent results, despite a difficult start.

The Hipparcos satellite (High Precision Parallax Collecting Satellite, combined with the name of the great Greek astronomer Hipparchos) was devoted to astrometry[85] (accurate measurement of stellar positions). Launched on 8 August 1989, the satellite did not reach the planned geostationary orbit but instead remained in the highly eccentric transfer orbit (GTO): $h_\mathrm{p} = 542$ km, $h_\mathrm{a} = 35\,840$ km, $i = 6.7°$. Rewriting the computer programs and redistributing the ground receiving stations, the mission was nevertheless carried successfully to completion.

The Hubble observatory was correctly launched by STS-31 on 24 April 1990. The problems began as soon as the first image came through, clearly showing that the telescope was shortsighted due to an inexcusable mirror defect. The intention had been to carry out maintenance via the Shuttle, and this possibility was used primarily to correct the optics. The results did indeed come up to expectations and astronomy was revolutionised in the process.

In contrast, Hubble's successor must be beyond reproach from the moment of launch: NGST (New Generation Space Telescope), renamed[86] JWST (James Webb Space Telescope), should be placed at the Lagrange point L_2 of the Sun–Earth system (see Sect. 3.14), and it will be quite impossible to go out there and repair it! In addition to the visible region, JWST will also observe in the infrared and for this reason will be equipped with cryogenic systems. The successor to Hipparcos, named GAIA (Global Astrometric Interferometer for Astrophysics, with reference to ἡ Γαῖα, Gæe or Gaia, the personification of the Earth according to the Ancient Greeks) will also be placed there.[87] The main advantage of the halo orbit (L2LO) around the point L_2 for stellar observation is that the Sun, the Earth and the Moon are all located behind the line of sight of the telescope. The whole of the celestial sphere can then be observed as the year goes by, without blind spots. This region is also very stable as regards the thermal and particle environment.

---

[85] The satellite determined the position, luminosity, and distance of 118 218 stars. The accuracy of the measurements (2 milliarcsec), was 100 times better than ground-based measurements of the day.

[86] In honour of James E. Webb (1906–1992), NASA's second administrator. He directed the Apollo programme and was one of the instigators of the first interplanetary exploration.

[87] GAIA's mission is to observe and record a billion stars with an accuracy of 10 μarcsec. For stars within 500 light-years, the distance will then be calculable to within a few light-years.

Infrared (IR). The detection of IR radiation involves cooling the optical system and associated instrumentation in the satellite. The satellite ceases to function once its reserves of cryogenic fluid have been exhausted.

Certain satellites have used Sun-synchronous orbits, such as the helium-cooled IRAS (IR All-sky Survey), $h = 890$ km, launched in 1983, and WIRE (Wide field IR Explorer, SMEX-5), launched in 1999 with $h = 560$ km, whose optical system is cooled by 3 kg of solid hydrogen. The satellite WISE (Wide field IR Survey Explorer, MIDEX-5), $h = 600$ km, will also be in this type of orbit.

The European satellite ISO (Infrared Space Observatory), with American and Japanese participation, was equipped with a cryostat filled with superfluid helium. It operated between November 1995 and May 1998 on a highly eccentric orbit, with a period of revolution of one day.[88] Its successor Herschel[89] (or HSO, Herschel Space Observatory), taking over from the FIRST project (Far IR and Submillimeter Telescope), will be placed in orbit around the Lagrange point L_2, for three years. JASMINE (Japan Astrometry Satellite Mission for Infrared Exploration) will measure parallaxes, positions with the accuracy of 10μarcsec (Lissajous Orbit around the Sun-Earth L_2 point) in 2014.

The US satellite Spitzer, launched on 25 August 2003 under the name of SIRTF, is on a different orbit. It follows the Earth at a distance of 0.1 a.u. or 15 million kilometres. This heliocentric orbit, known as ETHO (Earth-trailing heliocentric orbit), maintains the satellite in an extremely cold environment, propitious for technological innovation.[90]

---

[88] Launched on 17 November 1995. Orbital characteristics: $h_p = 1\,110$ km, $h_a = 70\,504$ km, $i = 5.1°$, $a = 42\,185$ km, $e = 0.822$, $T_d = 1436$ min = 1 sidereal day. Cryostat: 2 200 litres of superfluid helium. Temperatures of the various components: detector 2 K, optical system 3–4 K, instruments 8 K.

[89] The life of William Herschel (1738–1822), British astronomer of German origin, attests to an unusual intellectual development, revealing an exceptionally curious nature. He was led to mathematics by music, and from there moved on to astronomy. He made his own telescopes and they were the best of his day. He discovered Uranus in 1781 and then later, two of its moons, followed by two of the moons of Saturn. He demonstrated the displacement of the Solar System through the Galaxy and gave the coordinates of the apparent convergence point (the so-called apex) in 1783. In 1801, he discovered infrared radiation.

[90] Unlike IRAS and ISO, however, Spitzer adopts an innovative 'warm-launch' cryogenic architecture. The observatory is launched at ambient temperature and radiatively (or passively) cooled in the deep recesses of space. Only the focal-plane instruments and the liquid helium cryostat are enclosed in a vacuum shell containing the cryostat. It is important to recognize that the warm-launch architecture is enabled by the choice of orbit. This innovative launch architecture, combined with 360 liters of liquid helium, yields an estimated mission lifetime of about five years. For the sake of comparison, IRAS used 520 liters of cryogen during its 10-month mission.

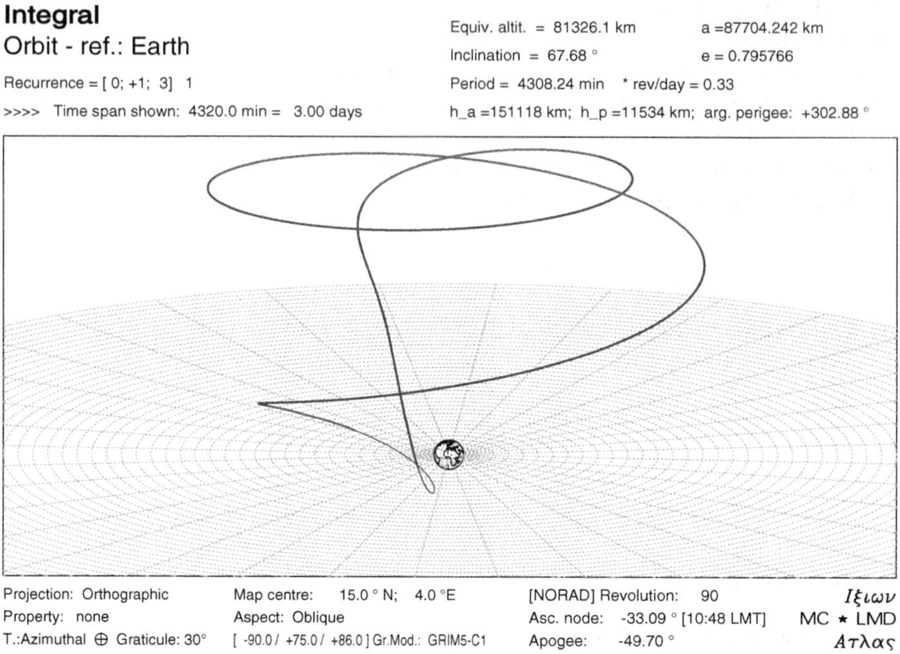

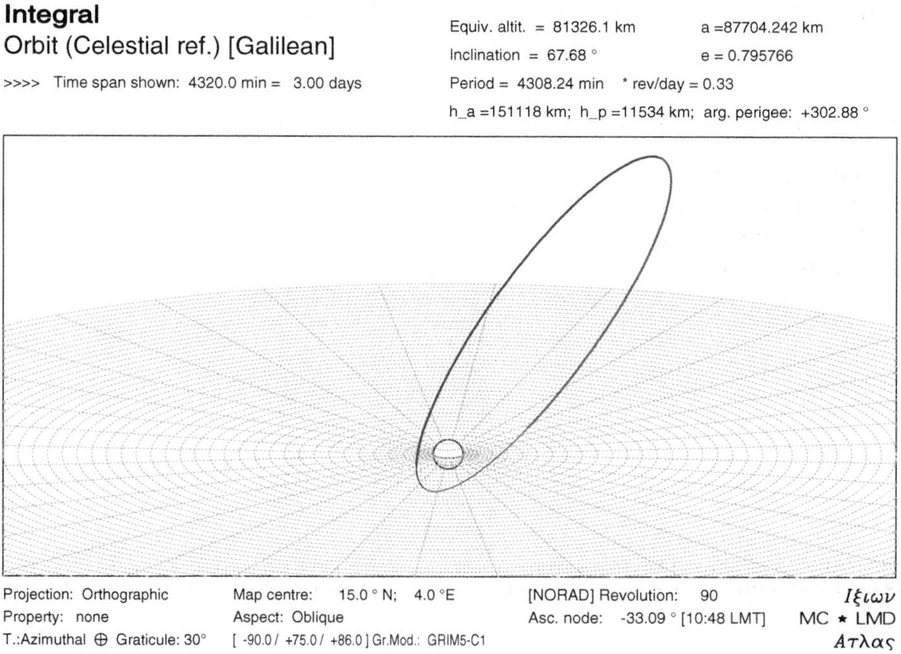

Figure 5.20. Orbit of the satellite Integral, over 3 days (1 revolution). *Upper*: Orbit, relative to the Earth (terrestrial reference frame). *Lower*: Orbit, relative to the stars (celestial reference frame)

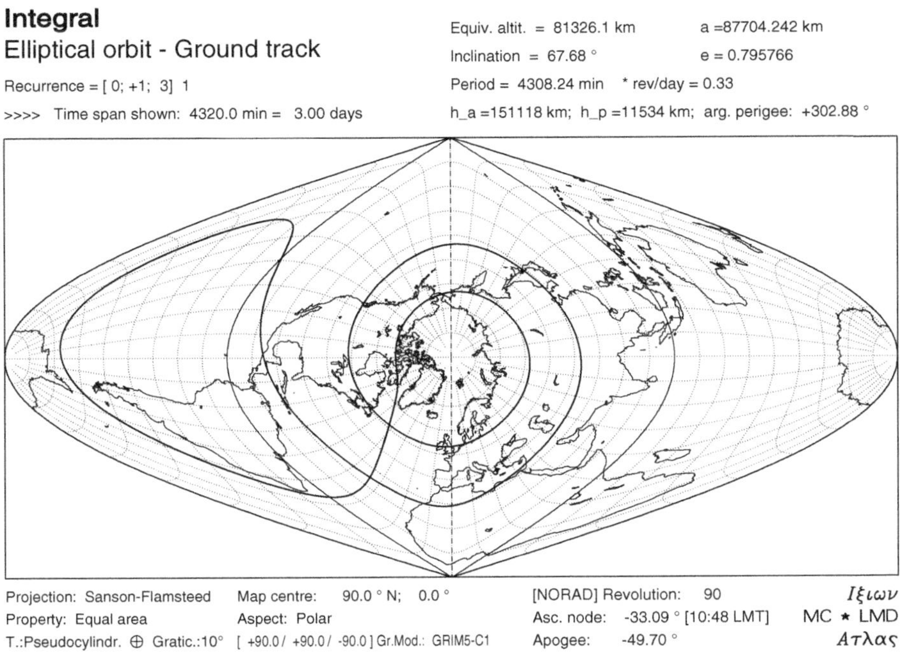

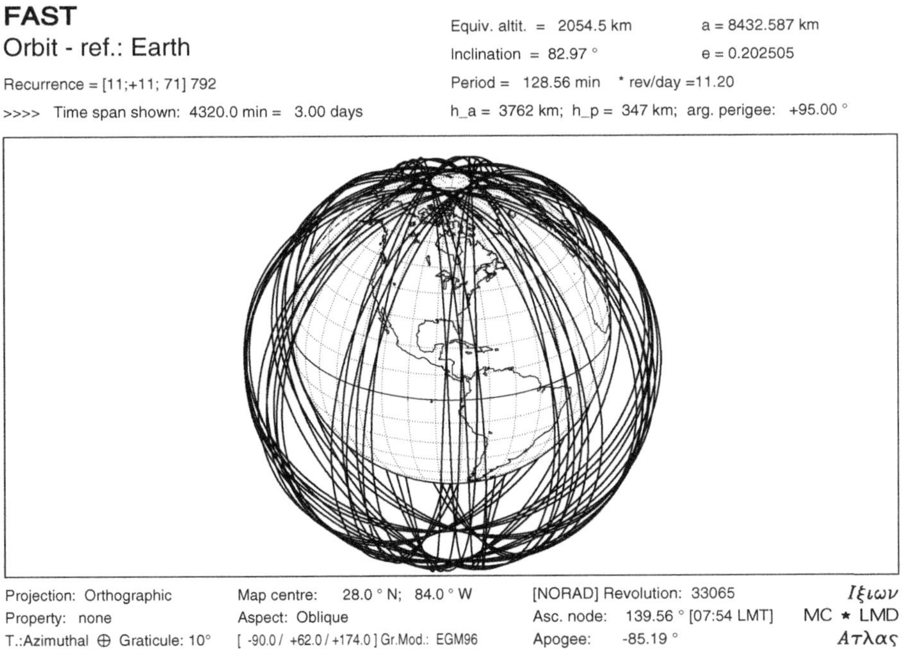

Figure 5.21. *Upper*: Ground track of the satellite Integral, over 3 days (1 revolution, recurrence cycle). *Lower*: Orbit of the satellite FAST, relative to the Earth

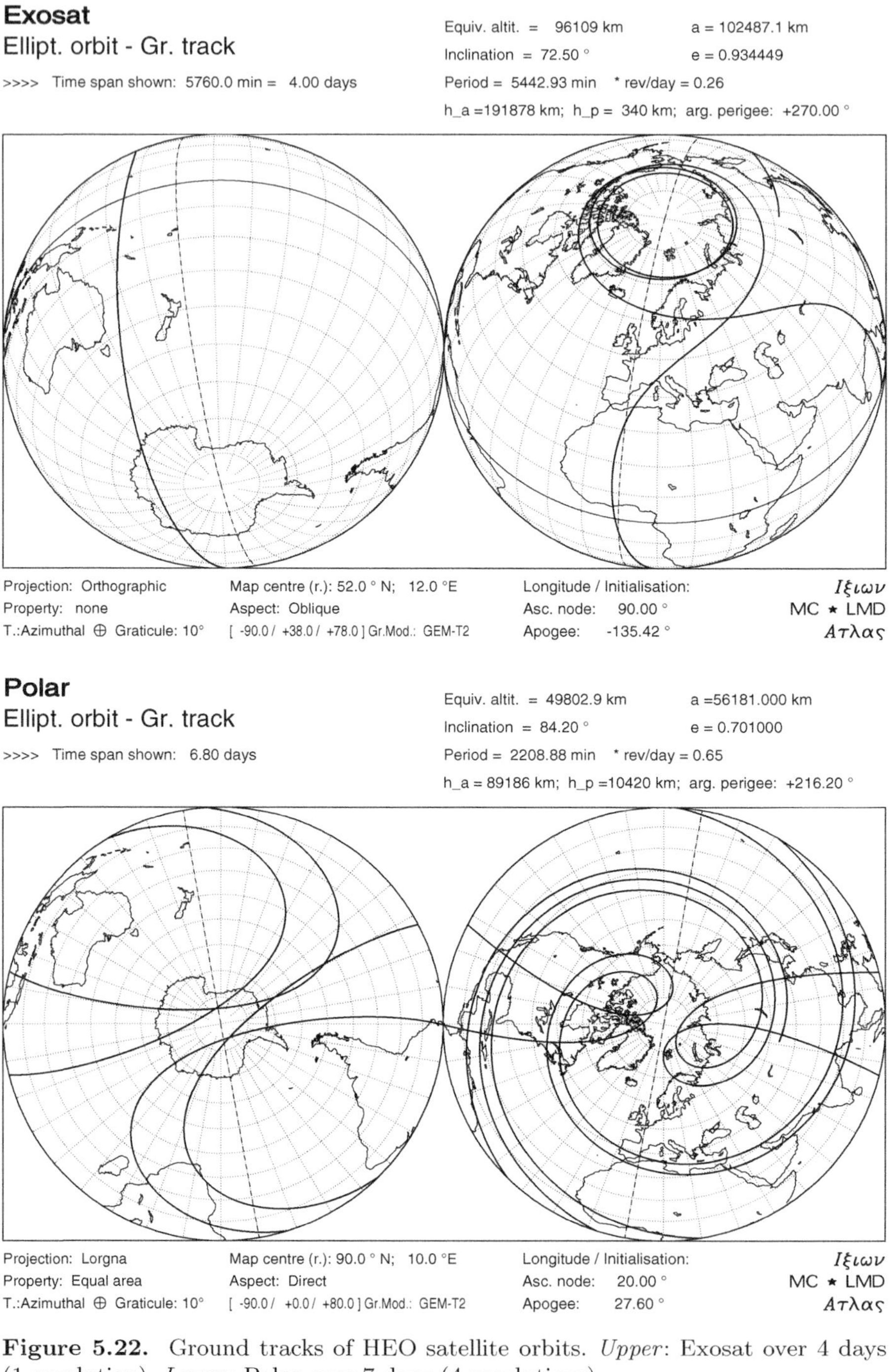

Figure 5.22. Ground tracks of HEO satellite orbits. *Upper*: Exosat over 4 days (1 revolution). *Lower*: Polar over 7 days (4 revolutions)

Microwave. The US satellite COBE (Cosmic Background Explorer), launched in 1989 on a Sun-synchronous orbit at $h = 880$ km, $i = 99.0°$, explored the millimetre radiation in space in order to study temperature fluctuations in the diffuse cosmological background via extremely precise measurements. (The temperature varies from 2.722 49 K to 2.722 51 K depending on the observed region.) This informs us about fluctuations in the matter density of the early universe. Its successor MAP (Microwave Anisotropy Probe, MIDEX-2), with angular resolution 0.2° as compared with 7° for COBE, is placed[91] in L2LO orbit. The European satellite Planck[92] (formerly COBRAS/SAMBA) launched slightly later, should further improve these results. It will also be placed at the point L_2.

For the study of galactic molecular clouds, SWAS (Submillimeter Wave Astronomy Satellite, SMEX-3) was launched in 1998, and now follows a low prograde orbit, with $h = 640$ km, $i = 69.9°$.

Radio. The first satellites to study radio sources were RAE-A and -B (Radio Astronomy Explorer, Explorer-38 and -49), launched in 1968 and 1973 on a highly inclined circular orbit with $h = 5850$ km and $i = 120.9°$. The Japanese satellite Haruka (Muses-B, 'haruka' means 'remote'), launched in 1997, is on an HEO orbit, with $h_\mathrm{p} = 569$ km, $h_\mathrm{a} = 21\,415$ km, $i = 31.4°$. The Russian project with European participation KRT-25 (a 25 m radio telescope) will be placed in an orbit which, over seven years, will become more and more eccentric: $h_\mathrm{p} \approx 5000$ km, h_a from 20 000 to 150 000 km, $i = 63°$

Solar Astrophysics

Solar radiation is studied across the whole range of wavelengths. The Sun has been the subject of investigation since the very beginning of the Space Age, from 1962 to 1976, with LEO satellites at altitudes between 500 and 600 km, such as the eight observatories OSO-1 to -8 (Orbiting Solar Observatory), $i = 33°$ and three Explorers, IQSY (International Quiet Sun Year, Explorer-

[91] Launched on 30 June 2001, the MAP probe made four revolutions around the Earth on ever more eccentric orbits, reaching the vicinity of the Moon a month later. Using a lunar swing-by, it took another two months to arrive at the point L_2 of the Sun–Earth system on 1 October 2001 and go into the halo orbit. MAP was re-christened Wilkinson MAP (or WMAP) in February 2003, in honour of David T. Wilkinson of Princeton University, a world-famous cosmologist and MAP team member who died in September 2002 at the age of 67.

[92] Max Planck (1858–1947) was a German physicist. He studied blackbody radiation and found an expression for the blackbody spectrum as a function of temperature and frequency. This problem had stumped many physicists before him. Planck solved it in 1900 by introducing the idea of the energy quantum. The theory of these quanta then became the basis for much of modern physics.

30), Solrad-9 and -10 (Solar Radiation,[93] Explorer-37 and -44), $i \approx 55°$, or with slightly inclined HEO orbits, such as the EPE programme (Energetic Particles Explorer), with EPE-A, -B, -C and -D (Explorer-12, -14, -15, -26). The German satellites Helios-1 and -2 are in heliocentric orbit (perihelion 0.309 a.u., aphelion 0.985 a.u., $i = 0°$, ecliptic plane, $T = 190$ day).

Launched in 1980, the US satellite SMM (Solar Maximum Mission) was the first satellite to be repaired in flight by the Shuttle STS-13 (STS-41-C), in 1984. It operated until 1989 on the orbit $h = 405$ km, $i = 26.5°$.

Recent LEO satellites are in near-polar orbits. Some are Sun-synchronous like the US satellites TRACE (Transition Region and Coronal Explorer, SMEX-4), with altitude $h = 620$ km, ACRIMSAT (Active Cavity Radiometer Irradiance Monitor Satellite), $h = 700$ km (launched in 1998 and 1999, respectively), or the French project Picard (see note on Picard), $h = 700$ km. Others are prograde, such as the Russian–Ukrainian satellites Koronas-I and Koronas-F (AUDS-SM-KI and -KF), $h = 520$ km, $i = 82.5°$ (launched in 1994 and 2001).[94] The US satellite SORCE (Solar Radiation and Climate Experiment) was launched in 2003 on the orbit $h = 641$ km, $i = 40.0°$. The satellite Solar-B, currently in the planning stages for a Sun-synchronous orbit, should take over from the Japanese Solar-A (or Yohkoh, 'Sun'), which was launched in 1991 with $h_{\mathrm{p}} = 526$ km, $h_{\mathrm{a}} = 795$ km, $i = 31.3°$.

Two large projects have inspired a collaboration between ESA and NASA in this area: Ulysses and SOHO. The Ulysses probe, launched on 6 October 1990 from STS-41, set off first in the direction of Jupiter, to use it for a gravity-assist maneuver (see Chap. 11) that would take it out of the plane of the ecliptic. On 1 November 1994 it overflew the south pole of the Sun and on 1 October 1995, the north pole. The satellite SOHO (Solar and Heliospheric Observatory), launched on 2 December 1995, went to Lagrange point L_1 to acquire a halo orbit. It subsequently discovered a great many comets.

Two projects should use the stable Lagrange points. The Japanese satellite L5-Mission will observe the Sun in order to provide a regular space 'weather' forecast, and ensure better protection for space activities. The two US satellites STEREO (Solar-Terrestrial Relation Observatory) are planned for the Lagrange points, with STEREO-Ahead at L_4, and STEREO-Behind at L_5, thus providing a 3D observation of the Sun.

---

[93] It later transpired that the first Solrad missions, from Solrad-1 to -7B, between 1960 and 1965, were primarily spy satellites (ELINT). To give the Solrad programme a more scientific countenance, it was also referred to as GREB (Galactic Radiation Experimental Background) or GRAB (Galactic Radiation And Background).

[94] The letters I and F are the initials of IZMIRAN and FIRAS, the names of the institutes which designed the project.

Stellar Seismology and the Search for Exoplanets

Three satellites, each carrying a small telescope, have been designed to study the seismology of stars. They follow three very different orbits. The Canadian satellite MOST (Microvariability and Oscillations of Stars, jokingly nicknamed the Humble Space Telescope by its developers), launched on 30 June 2003, follows a Sun-synchronous orbit at $h = 625$ km, $i = 98.7°$. The Danish satellite MONS (Measuring Oscillations in Nearby Stars) is on a Molniya orbit. The French satellite Corot (*Convection, rotation et transits planétaires*) is looking for planets in orbit around stars other than the Sun by detecting the transit (when the planet passes in front of its star and thereby diminishes its luminosity), as well as studying stellar seismology. It has a circular, strictly polar orbit with $h = 827$ km. These three satellites are a preparatory stage for the European project Eddington,[95] which will study the internal structure of stars and seek to detect extra-solar planets. This satellite will be in an L2LO orbit. In November 2003, ESA postponed the mission indefinitely. Another project, a forerunner of Eddington that has since been abandoned, would have sent the satellite STARS to the Lagrange point L_5 of the Earth–Moon system.

The European mission Darwin,[96] will be devoted to astrobiology, but it is a long-term project. (One also uses the term 'exobiology' to remove all possible confusion with 'astrology'.) The Darwin 'flotilla'[97] will also be in

[95] Arthur Stanley Eddington (1882–1944) was a British astronomer and physicist. He did much to promote the theory of relativity (see the note on Einstein), through the publication of his book *Space, Time and Gravitation* (1920), which is still being reprinted. He also laid the foundations for a new discipline, stellar dynamics, with *The Internal Constitution of the Stars* (1926), in which he shows that a star is subject to two opposing effects: it tends to contract under the effect of gravity, whilst the release of energy tends to push it apart.

[96] Charles Darwin (1809–1882) was an English naturalist. From 1831 to 1836, he took part in an expedition to South America (and in particular, to the Galapagos islands) and Oceanica, aboard the HMS Beagle. As a geologist and botanist, he elaborated his theory of evolution on the basis of notes taken and collections brought back from this expedition. He concluded that the variability of the species is due to the effects of their environment and to sudden variations. These variations are only favoured by natural selection if they give the individual organism an advantage in its struggle for survival (subsistence and reproduction). His famous book *The Origin of Species* was published in 1859. Darwin's theory, supported and developed by a great many intellectuals, was attacked without scientific argument by the conservative-minded and religious classes. Here was another problem of divine order!

[97] The Darwin flotilla comprises six satellites in formation. They lie strictly in the same plane and each is equipped with a telescope in such a way as to form an infrared interferometer. One master satellite, a short distance from the others, oversees the satellite positions and provides the link with the Earth. The aim is

an L2LO orbit. The US project Kepler will follow a heliocentric orbit with $a = 1.013\,19$ a.u., $e = 0.03188$, and $T = 372.5$ day.

Space Exploration

Probes sent out to observe other planets in the Solar System or their moons will be studied in Chaps. 10 and 11, along with satellites in orbit around these celestial bodies.

We restrict ourselves here to probes involved in the study of comets. Halley's comet,[98] the most famous of all comets, was approached by six probes in March 1986: ISEE-3 renamed ICE for the mission, Vega-1 and -2 (Soviet probes, 'Ve' for 'Venera', 'Ga' for 'Galleia', 'Venus' and 'Halley' in Russian), Giotto (a European probe, which flew by the cometary nucleus at only 600 km), Sakigake and Suisei (Japanese probes, 'scout' and 'comet', respectively, in Japanese).

The CONTOUR probe (Comet Nucleus Tour) was launched on 3 July 2002 and failed shortly afterwards. It was to investigate short period comets such as P/Encke, whose orbit never extends beyond that of Jupiter.

The Rosetta mission[99] is one of ESA's cornerstone missions. The probe, launched on 2 March 2004, should follow a complex path[100] in order to acquire an orbit around comet[101] 67P/Churyumov-Gerasimenko in May 2114. In November 2114, the Philae module should land on the cometary nucleus and ride with it to its perihelion.

NASA's Stardust and Genesis missions involve in situ exploration with sample-return.[102]

---

to detect planets orbiting other stars and to spy out possible signatures of life beyond our own Solar System.

[98] Edmond Halley (1656–1742) was a British astronomer. In 1682, he observed a very bright comet near its perihelion. He calculated the orbital elements and predicted that the comet would return in 1758. Once the period of 76 years had been established, it was realised that this comet had been known since ancient times. The first pictorial representation was due to the Italian painter Giotto in 1304. In 1716, Halley published a method using the transit of Venus, predicted for 1761, to measure the distance from the Earth to the Sun.

[99] The Rosetta stone carries a bilingual inscription in three scripts, which allowed the French linguist Champollion to decipher the hieroglyphs of Ancient Egypt in 1822. The name Rosetta comes from 'Rachid', the Arabic name for the village in the Nile delta where the stone was found. ESA chose this appellation for its cornerstone project because it expects the results to throw light upon some of the mysteries of the formation of the Solar System.

[100] EMEEGA: Earth (March 2005) Mars (March 2007) Earth (November 2007) Earth (November 2009) Gravity Assist.

[101] For the initially planned launch in January 2003, Rosetta was to meet up with comet 46P/Wirtanen ten years on.

[102] Stardust, launched on 7 February 1999, collected interstellar dust in 2002 and particles from comet Wild-2 on 2 January 2004. The capsule should be returned

5.4.8 Technological Satellites

This is the category of satellites whose mission it is to assist in various kinds of technological development. The very first satellites fit well in this class. Technological satellites are used to test instruments, orbital maneuvers, communication techniques, and reentry vehicles (RV technology). They are also used to test the design of electric or ion propulsion motors.

The satellite DODGE (Department Of Defense Gravity Experiment), $h = 33\,400$ km, $i = 12°$, was sent into space in 1967 to further the development of geostationary satellites, although the first had already been launched some years previously. A camera aboard this satellite took the very first colour pictures of the Earth's disk.

The military satellite ARGOS (Advanced Research Global Observation Satellite or P91-1) was launched into a Sun-synchronous LEO orbit in 1999 to test electric propulsion systems and detect X rays.

To study the effect of radiation on various types of equipment, several small technological satellites have been placed in highly elliptical orbits. These are in fact the geostationary transfer orbits (GTO) of the main satellites with which they were launched (see the lower part of Fig. 5.10). Such GTO orbits expose the satellites to an extremely testing environment, since they cross the Van Allen radiation belts on each revolution. With this mission, we find the British microsatellites STRV (Space Technology Research Vehicle), $h_{\mathrm{p}} = 300$ km, $h_{\mathrm{a}} = 36\,000$ km, $i = 7°$, STRV-1A and -1B launched in 1994 as passengers with Intelsat-702, and STRV-1C and -1D in 2000, passengers with PAS-1R (PanAmSat-1R). Likewise for Teamsat, launched with the geostationary satellite Maqsat. The Japanese satellite MDS-1 (Mission Demonstration Satellite) will go into a more highly inclined GTO orbit, with $i = 28.5°$.

This category also covers calibration satellites, such as Radcal, $h \approx 800$ km, $i = 89.5°$, for radar calibrations, Reflektor, launched with Meteor-3M-1, for laser calibrations, and the twelve nanosatellites (a few kg) launched by STS-60 and -63, ODERACS-A to -F, and ODERACS-2A to -2F (Orbital Debris Radar Calibration Sphere), to calibrate spacecraft debris.

One type of experiment, called a tether experiment, involves connecting the satellite to a secondary body once it is already in orbit, using cables of different lengths. These cables may or may not be conducting. The first tests were carried out on the manned flights Gemini-11 and -12 (with a cable of

---

to the Utah desert on 15 January 2006. Stardust's heliocentric orbit involves one trip beyond the orbit of Mars.
Genesis, launched on 8 August 2001, reached LOI (Lissajous Orbit Insertion) on 16 November 2001. After two years at Lagrange point L_1 (halo orbit), it collected particles from the solar wind and returned the sample. The capsule was to have been ejected and recovered by helicopter in the Utah desert, on 8 September 2004. However, the parachutes failed to deploy, resulting in a hard-landing, and the samples are considered to be virtally unusable.

30 m). Also worth mentioning, launched into LEO orbit between 1993 and 1996, are the satellites SEDS-1 then -2 (Small Expendable Deployer System), with cables 20 km long, and TiPS (Tether Physics and Survivability), with a 4 km cable. The orbit of the TiPS central satellite had the characteristics: $h = 1\,022$ km, $i = 63.4°$, $e = 0.000$.

Among planned missions, we note the BOLAS experiment (Bistatic Observations with Low Altitude Satellites), devoted to the study of the ionosphere with a view to improving communications for GPS. It consists of two satellites, BOLAS-1 and -2, connected by a non-conducting cable 100 m long and rotating about their centre of gravity. They will be launched at the same time as Radarsat-2, on a Sun-synchronous orbit, and will then acquire an elliptical orbit with $h_{\mathrm{p}} = 350$ km, $h_{\mathrm{a}} = 800$ km, $i = 102.81°$.

In the United States, the universities carry out missions with small technological satellites, such as TERRIERS or SNOE (Student Nitric Oxide Explorer). In Europe, this policy is mainly applied in the United Kingdom, with the UoSAT (University of Surrey Satellites), and in Germany with the Tubsat (Technische Universität Berlin Satellites). These satellites, which can generally be subsumed under the heading of technological satellites, are almost always placed in Sun-synchronous orbits as passengers.

In this category we may also include satellites with biology missions. The satellite Sputnik-2 carried a female dog aboard. Animals were sent in the Russian Bion satellites to study the effects of radiation. For example, in 1996, two monkeys flew in Bion-11, $h = 300$ km, $i = 63°$. We may also mention the US LEO satellite OFO-1 (Orbital Frog Otolith), occupied by two toads. The aim was to study their inner ear, seat of the vestibular organ (balance system).

5.4.9 Satellites with Specific Military Missions

Many programmes have been developed by the military and civilian sectors, either jointly or in parallel, with similar areas of interest, such as remote-sensing, surveillance and communications. However, certain programmes are specifically military, such as the detection of nuclear explosions. Here is a brief review.

Early Warning

These satellites are designed to detect enemy missiles as soon as possible during or after their launch. In the United States, the first programme was MIDAS (Missile Defense Alarm System), with near-polar LEO satellites, $h \approx 3000$ km, from MIDAS-3 in 1962 to MIDAS-12 in 1966. Satellites in the following programme, IMEWS (Integrated Missile Early Warning Satellites), extended by DSP (Defense Support Program), have been geostationary, $i \approx 0°$, from IMEWS-2 in 1971 to DSP-F-21 (USA-159) in 2001, and DSP-F-22

(USA-176) in 2004. The NMD project (National Missile Defense), currently under discussion, is based on the SBIRS satellites (Space-Based IR System). As soon as a missile is fired, it is detected by SBIRS-High (a system of 4 GEO satellites and 2 HEO satellites). The missile is then tracked by SBIRS-Low (a constellation of 24 satellites in 6 orbital planes).

The Soviet programme SPRN was equivalent to IMEWS, with a fleet of GEO satellites. Satellites in the Oko series ('eye') are in Molniya orbit, from Kosmos-520 in 1972 to Kosmos-2368 in 1999.

Nuclear Surveillance

In order to keep an eye on nuclear testing, the various space powers have launched satellites that can detect explosion by their gamma-ray emissions. The US satellites Vela had novel circular orbits, at very high altitudes $h \approx 110\,000$ km, and with inclinations varying between 34° and 61°. They were launched in pairs, diametrically opposed with respect to the centre of the Earth, from Vela-1 and -2 in 1963 to Vela-11 and -12 in 1970, and remained operational until 1984. ('Vela' means 'lookout' in Spanish. These satellites were also called Watchdogs or Vela Hotel.) They fullfilled a scientific mission that had not been planned at all at the outset: between 1969 and 1979, the satellites Vela-9 and -10 (OPS/6909 and OPS/6911), Vela-11 and -12 (OPS/7033 and OPS/7044) mapped gamma-ray sources in space. They were the first to observe what have since become known as gamma-ray bursts (GRB).

Destruction of Satellites, Star Wars

It is of little interest in this context to dwell upon satellite destruction programmes like ASAT (Air-Launched Anti-Satellite Missile), FOBS (Fractional Orbital Bombardment System), or what is popularly referred to as 'star wars', i.e., the US programme SDI (Space Defense Initiative).

Intelligence

Satellite interception of all kinds of electronic signal is considered to be of great importance by the military. This is SIGINT (signal intelligence) or ELINT (electronic intelligence). Military photographic surveillance is IMINT (image intelligence). The US–British (and Commonwealth) project Echelon is planning a constellation of 120 satellites.

The US SIGINT missions began with the GREB (or GRAB) series and Ferret (beginning in 1962 with Ferret-2), on LEO orbits. Subsequently, the whole range of different orbits was used. Each programme lasted for roughly a decade. We note the following programmes, from the second generation (the 1970s) to the fifth (the 2000s):

- LEO: SSF (Subsatellite Ferrets), NOSS (Navy Ocean Surveillance Satellite, also called White Cloud) and NOSS-Sub-sats (satellites in formation), SB-WASS (Space-Based Wide Area Surveillance System), SB-WASS Next Generation (with planned orbits inclined at 155°).
- HEO: Jumpseat, Trumpet,[103] Prowler.
- GEO: Canyon, Rhyolite/Aquacade, Chalet/Vortex, Magnum/Orion, Mercury,[104] Mentor, Intruder.

There have been similar Soviet then Russian programmes, beginning in 1970 with Kosmos-389. The Tselina programme, in LEO orbit, is the equivalent of NOSS, with $i = 82.6°$ for the Tselina-D series and $i = 71.0°$ for the most recent Tselina-2 series.

5.4.10 Satellites with Human Occupation

We give here a few dates marking out the history of manned spaceflight, i.e., satellites in orbit with humans aboard: Vostok-1 ('Orient' in Russian), launched on 12 April 1961, for the first man in orbit (one revolution); Apollo-11, launched on 16 July 1969, for the first man on the Moon.

The idea of an orbital space station was first put into practice with the Soviet Salyut ('salvation') from 1971 to 1986, followed by Mir ('mir' means both 'world' and 'peace' in Russian) from 1986 to 2000, on a near-circular orbit, $h \sim 300$ km, $i = 51.6°$. The United States used Skylab, in 1973, on an equivalent orbit, $h \sim 400$ km, $i = 50.0°$. They then began to develop the ISS (International Space Station), from 1998, in collaboration with Russia and other nations. The orbit is circular, with h between 355 and 400 km, $i = 51.6°$.

The space shuttle idea is based on the possibility of a reusable spacecraft, a satellite becoming a plane in the landing stage. The five American Space Shuttles are Columbia (1981–2003), Challenger (1983–1986), Discovery (since 1984), Atlantis (since 1985), Endeavour (since 1992). Their flights are numbered with the prefix STS (Space Transportation System). At the end of 2003, there had been 113 shuttle flights: 111 successes, 2 failures, with Challenger STS-33 (STS-51-L) and Columbia STS-107.

The shuttle has two configurations: a payload of 24.4 tonnes for a low orbit, $h = 204$ km, $i = 28.5°$, and a payload of 12.5 tonnes for a higher orbit, $h = 407$ km, $i = 51.6°$. It is the latter configuration that is used

[103] The three satellites in the series Trumpet-1 (USA-103), Trumpet-2 (USA-112) and Trumpet-3 (USA-136) were launched between 1994 and 1997. They follow Molniya orbits with apogee over Russia. Their antennas have diameters of several tens of metres, and between 100 and 150 m apparently for Trumpet-3. Data is collected by the SDS satellites, also in Molniya orbit.

[104] Mercury-1 (USA-105 or Jeroboam) and Mercury-2 (USA-118) were launched in 1994 and 1996. This programme is sometimes called Mercury ELINT to distinguish it from the Mercury programme of manned flights.

for rendezvous with ISS. For maintenance of Hubble, the shuttle goes into a higher orbit. A great many satellites have been placed in orbit by the shuttle. Once in orbit, the satellite leaves the cargo bay of the shuttle and moves to its destination under its own power (a nearby orbit, either geostationary or heliocentric).

5.4.11 Non-Scientific Satellites

The Celestis satellites are placed under the 'others' heading of the miscellaneous category in the otherwise rather detailed sample listed in the Celestrak directory. Furthermore, they are the only representatives of this category. Other US organisations classify them as burial satellites.[105] Without wishing to carry out unnecessary advertising for the Celestis company, we merely note that the orbit of Celestis-1 has the high inclination of 151° (because launched with Minisat-01). Celestis-2 has an inclination of 108° (because launched with GFO-1) and Celestis-3 is Sun-synchronous (launched with ACRIMSAT and Kompsat). They have altitude $h \sim 700$ km. Celestis-4 burnt up in space following launch failure (grouped with OrbView-4 and QuikTOMS). Note that these satellites have been renamed Earthview-01, -02, -03 and -04, which might lead to some surprises in a remote-sensing bibliography.

5.5 Appendix: Velocity of Satellite and Ground Track in Circular Orbit

Accurate calculation of the satellite velocity can be achieved using the equations of motion discussed in Chap. 3. Here we shall calculate the velocity of the satellite and its ground track to very high accuracy, considering the case of circular and Keplerian motion. This discussion could well have been included in Chap. 2. However, we prefer to present it here, now that we have defined the angular velocity of the Earth's rotation and the different types of satellite, especially geosynchronous satellites.

5.5.1 Definitions of the Different Velocities

The velocity of the satellite S and that of its ground track S_0, in $\Re$, can be expressed in a simple manner in terms of the mean motion n:

[105] Taking a closer look, one may read in the informative note of the Celestis company: 'Celestis offers to launch a symbolic portion of the cremated remains of the individuals into space.' Business is clearly booming: several satellites have been placed in orbit since 1997, not to mention a lunar impact with Lunar-01 (in fact, a capsule carried by Lunar Prospector) in 1998, following the same idea. There is a project to send such spacecraft into deep space and out of the Solar System.

Table 5.2. Velocity of satellite and ground track, and relative velocity of the ground track for various satellites, in circular (Keplerian) orbit. For each satellite, we give the altitude h (in km) and the length of the semi-major axis a, or distance from the centre of the Earth (in km), the daily frequency ν (in revs per day), the Keplerian period T_0 (in hours and minutes), the speeds V, V_0, w_E (for the three values of the angle i, 0°, 90°, 180°), defined in the present section (in $\mathrm{km\,s^{-1}}$). Also given is the corresponding type of satellite $\mathcal{T}$: GL (ground level), E (espionage, surveillance), SS (space shuttle, manned flights and Earth observation), O (Earth observation from LEO orbit), G (geodesy), I (communications with ICO-type orbit, between LEO and MEO), P (positioning by GPS from MEO orbit), S (geostationary, GEO), V (Vela-type), M (Moon)

h [km]	a [km]	ν [rev/day]	T_0 [d h m]	V	V_0	w_E 0	w_E 90	w_E 180	$\mathcal{T}$
0	6378	17.04	1h24	7.91	7.91	7.44	7.92	8.37	GL
100	6478	16.65	1h26	7.84	7.72	7.26	7.74	8.19	E
200	6578	16.27	1h28	7.78	7.55	7.08	7.56	8.01	E
300	6678	15.91	1h31	7.73	7.38	6.91	7.39	7.84	SS
400	6778	15.56	1h33	7.67	7.22	6.75	7.23	7.68	SS
500	6878	15.22	1h35	7.61	7.06	6.60	7.07	7.52	O
600	6978	14.89	1h37	7.56	6.91	6.44	6.92	7.37	O
700	7078	14.58	1h39	7.50	6.76	6.30	6.78	7.23	O
800	7178	14.28	1h41	7.45	6.62	6.16	6.64	7.09	O
900	7278	13.98	1h43	7.40	6.49	6.02	6.50	6.95	O
1000	7378	13.70	1h45	7.35	6.35	5.89	6.37	6.82	O
1100	7478	13.42	1h47	7.30	6.23	5.76	6.24	6.69	O
1200	7578	13.16	1h49	7.25	6.10	5.64	6.12	6.57	O
1300	7678	12.90	1h52	7.21	5.99	5.52	6.00	6.45	O
1400	7778	12.66	1h54	7.16	5.87	5.41	5.89	6.33	O
1500	7878	12.42	1h56	7.11	5.76	5.29	5.78	6.22	O
1600	7978	12.18	1h58	7.07	5.65	5.19	5.67	6.11	
2000	8378	11.32	2h07	6.90	5.25	4.79	5.27	5.71	
3000	9378	9.56	2h31	6.52	4.43	3.97	4.46	4.90	
4000	10378	8.21	2h55	6.20	3.81	3.34	3.84	4.27	
5000	11378	7.15	3h21	5.92	3.32	2.85	3.35	3.78	G
6000	12378	6.30	3h48	5.67	2.92	2.46	2.96	3.39	G
10390	16768	4.00	6h00	4.88	1.85	1.39	1.91		I
20183	26561	2.01	11h58	3.87	0.93	0.47	1.04		P
35786	42164	1.00	23h56	3.07	0.47	0.00			S
110000	116378	0.22	4d13h45	1.85	0.10	−0.36	0.47		V
376805	383183	0.04	27d07h43	1.02	0.02				M

$$\boldsymbol{V} = \frac{\mathrm{d}(\boldsymbol{OS})}{\mathrm{d}t} \,, \qquad V = an = \sqrt{\frac{\mu}{a}} \,, \tag{5.33}$$

$$\boldsymbol{V}_0 = \frac{\mathrm{d}(\boldsymbol{OS}_0)}{\mathrm{d}t} \,, \qquad V_0 = Rn = \frac{R}{a} V = \frac{R}{a}\sqrt{\frac{\mu}{a}} \,. \tag{5.34}$$

In the terrestrial frame $\Re_{\mathrm{T}}$, consider a point on the surface of the Earth, e.g., S_0, and the right-handed triple of unit vectors associated with spherical coordinates $(\boldsymbol{e}_r, \boldsymbol{e}_\lambda, \boldsymbol{e}_\phi)$. In a plane tangent to the Earth (local horizontal plane), the plane $(\boldsymbol{e}_\lambda, \boldsymbol{e}_\phi)$, the vector $\boldsymbol{e}_\lambda$ lies along a line of latitude and the vector $\boldsymbol{e}_\phi$ lies along a meridian. In this frame, the velocity of the ground track, which we shall denote by $\boldsymbol{w}$, is equal to

$$\boldsymbol{w} = \boldsymbol{V}_0 - R\cos\phi\,\dot{\Omega}_{\mathrm{T}}\boldsymbol{e}_\lambda \,.$$

If $\boldsymbol{e}_{\mathrm{u}}$ is the unit vector along $\boldsymbol{V}_0$, we obtain the expression for $\boldsymbol{w}$, which can be called the relative velocity of the ground track, or the velocity of the ground track relative to the ground:

$$\boldsymbol{w} = R\left(n\boldsymbol{e}_{\mathrm{u}} - \dot{\Omega}_{\mathrm{T}}\cos\phi\,\boldsymbol{e}_\lambda\right) \,. \tag{5.35}$$

5.5.2 Velocity at the Equator

To compare values obtained for various satellites, we consider the relative velocity $\boldsymbol{w}$ and the ascending (or descending) node, i.e., at the equator, where it will be written $\boldsymbol{w}_{\mathrm{E}}$:

$$\frac{\boldsymbol{w}_{\mathrm{E}}}{R} = n\boldsymbol{e}_{\mathrm{u}} - \dot{\Omega}_{\mathrm{T}}\boldsymbol{e}_\lambda \,.$$

It can be written in terms of the inclination i of the satellite as

$$\frac{\boldsymbol{w}_{\mathrm{E}}}{R} = \left(n\cos i - \dot{\Omega}_{\mathrm{T}}\right)\boldsymbol{e}_\lambda + n\sin i\,\boldsymbol{e}_\phi \,. \tag{5.36}$$

Let w_{E} be the magnitude of $\boldsymbol{w}_{\mathrm{E}}$ taken in the direction of the velocity. We obtain the results for three values of the inclination, namely, $i = 0°$, $i = 90°$ and $i = 180°$, respectively:

$$\frac{w_{\mathrm{E}}}{R} = n - \dot{\Omega}_{\mathrm{T}} \,, \qquad \frac{w_{\mathrm{E}}}{R} = \sqrt{n^2 + \dot{\Omega}_{\mathrm{T}}^2} \,, \qquad \frac{w_{\mathrm{E}}}{R} = n + \dot{\Omega}_{\mathrm{T}} \,. \tag{5.37}$$

Table 5.2 shows for various satellites the velocity of the satellite and its ground track, V and V_0, in $\Re$, and the relative velocities of the ground track w_{E}, in $\Re_{\mathrm{T}}$ for these three values of i. Satellites are represented for a whole range of altitudes between 0 and 1600 km, in steps of 100 km, and then for a few altitudes characteristic of various types of mission.

The maximal velocity of the ground track relative to the ground is 8.4 $\mathrm{km\,s^{-1}}$, obtained for a retrograde satellite moving at ground level. More realistically, the velocity of the ground track at the equator for an operational Sun-synchronous satellite is 6.6 $\mathrm{km\,s^{-1}}$. For a geosynchronous satellite with $i = 0$, hence geostationary, we may check that $w_\mathrm{E} = w_\mathrm{E}(i = 0) = 0.00$.

The last satellite in the table is not as artificial as it may look. It is in fact a simplified model of the Moon, in circular, Keplerian orbit at a distance of 380 000 km, with period about 27 day. Note that, in $\Re$, the sidereal period is equal to 27.32 day, whilst in $\Re_\mathrm{T}$, the synodic period[106] which takes the Earth's motion into account, is equal to 29.53 day, a lunar month.[107] The synodic period T' (in days) is calculated using the relation (2.23), with $T = 27.32$ and $T_1 = N_\mathrm{sid}$.

5.6 Appendix: Satellite Visibility Time

The span of time over which a satellite S is visible from a given point P on the Earth is called the visibility time of S from P. This idea arises mainly in the study of satellite constellations.

5.6.1 Satellite in Circular Orbit

Consider an arbitrary point P on the (spherical) surface of the Earth and a satellite S whose orbital plane passes through P at a given time. This is shown schematically in Fig. 5.23 (upper). The satellite S can be seen from P as long as it remains above the local horizon for P, represented by the straight line S_1PS_2, i.e., on the circular arc S_1AS_2. The angle $\alpha = (\boldsymbol{OP}, \boldsymbol{OS}_1)$ can be found immediately (reduced distance η):

$$\cos\alpha = \frac{R}{R+h} = \frac{R}{a} = \frac{1}{\eta}\,. \tag{5.38}$$

The period of the satellite is taken equal to the Keplerian period T_0. The visibility time Δt_v is therefore

---

[106] The noun ἡ σύνοδος, ου, 'synod', is composed from σύν, 'with, together' and ἡ ὁδός, ου, 'way, journey'. In Ancient Greek, it already had the two meanings of 'meeting' and 'conjunction of heavenly bodies', both of which illustrate the idea of arriving at the same time.

[107] In English it is no accident that the word 'month' should be so similar to the word 'Moon'. This similarity can be found in German and other related languages. The Indo-European root **men, *mes* refers to the Moon, to lunation (= month), and to measurement (of time). Many languages in this family have held on to this proximity of meaning, although this has not happened in Greek or Latin. These two languages referred to the Moon as 'the shining one' (ἡ σελήνη, ης and *luna, æ*). See also the note on Chandrasekhar.

$$\Delta t_{\mathrm{v}} = \frac{\alpha}{\pi} T_0 \,. \tag{5.39}$$

Expressing α and T_0 as functions of the semi-major axis a (here the radius) of the orbit, we obtain

$$\Delta t_{\mathrm{v}} = 2\sqrt{\frac{a^3}{\mu}} \arccos \frac{R}{a} \,. \tag{5.40}$$

Using the altitude h and the value of the period $T_{0(h=0)}$ defined by (2.17), we have (for time in minutes and angles in radians)

$$\Delta t_{\mathrm{v}} \text{ (min)} = 84.5 \left(1 + \frac{h}{R}\right)^{3/2} \frac{1}{\pi} \arccos \frac{R}{R+h} \,. \tag{5.41}$$

One may also consider the question by fixing a minimal zenithal angle of sight ζ, with $\zeta = (\boldsymbol{PA}, \boldsymbol{PS}_1')$. The visibility time then corresponds to the time taken by the satellite to travel along the arc $S_1'AS_2'$. In this case, the angle $\alpha = (\boldsymbol{OP}, \boldsymbol{OS}_1')$ depends on η and ζ. Simple trigonometric considerations (projecting S_1' on PA) show that this angle is the solution of the equation

$$\cos\alpha - \frac{1}{\tan\zeta} \sin\alpha = \frac{1}{\eta} \,.$$

Hence,

$$\alpha = 2 \arctan \frac{\sqrt{1 + Z^2 - H^2} - Z}{1 + H} \,. \tag{5.42}$$

$$\text{with} \quad H = 1/\eta \,, \quad Z = 1/\tan\zeta \,,$$

and we obtain the visibility time from (5.39).

When the satellite does not pass through the vertical at the relevant point, the visibility time is obviously shorter.

Example 5.6. *Calculate the visibility time for LEO and MEO satellites.*

For a (SPOT-type) LEO satellite with $h = 800$ km, we find $\eta = 1.125$ and $T_0 = 101$ min. Using (5.38), we obtain $\alpha = 27°$ and hence, by (5.39),

$$\Delta t_{\mathrm{v}} = (27/180) \times 101 = 15 \text{ min} \,.$$

If the visibility condition consists in requiring that the satellite should be at least 15° above the horizon, i.e., $\zeta = 75°$, the calculation with (5.42) gives $\alpha = 16°$ and we deduce that $\Delta t_{\mathrm{v}} = 9$ min. These visibility times are maximal values, assuming that the satellite passes through the local vertical.
For a (NAVSTAR/GPS-type) MEO satellite with $h = 20\,200$ km, we find $\eta = 4.167$ and $T_0 = 718$ min. It follows that $\alpha = 76°$ and hence,

$$\Delta t_{\mathrm{v}} = (76/180) \times 718 = 304 \text{ min} \approx 5 \text{ hr} \,.$$

With $\zeta = 60°$ (visibility if the satellite is more than 30° above the horizon), we have $\alpha = 48°$ and $\Delta t_{\mathrm{v}} = 192$ min ≈ 3 hr.

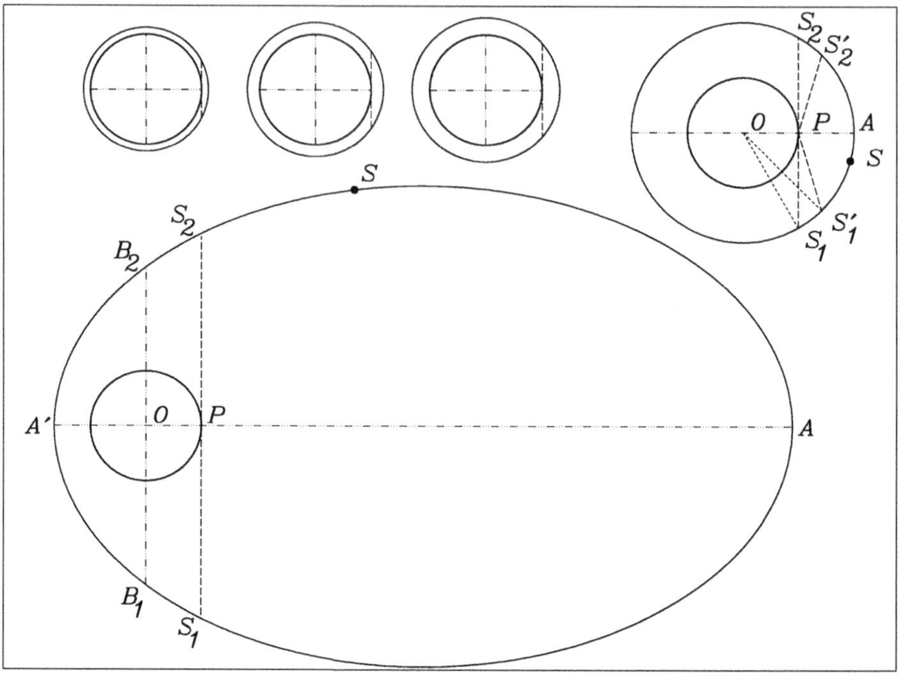

Figure 5.23. Schematic representation of the Earth and satellite trajectories. The Earth and the orbits are drawn on the same scale. *Upper*: circular orbits. LEO orbits are shown with $h = 800$, 1400 and 2000 km, together with the orbit $h = R$, indicating the points mentioned in the text. *Lower*: HEO orbit with period $T \approx$ 24 hr, $e = 0.75$, indicating the points mentioned in the text

5.6.2 Satellite in Highly Eccentric Orbit

For a highly eccentric elliptical orbit, with eccentricity e and type HEO, we consider once again the most favourable situation: the relevant point P is the subsatellite point when the satellite goes through its apogee A, as shown in Fig. 5.23 (lower).

The satellite S can be seen from P as long as it remains on the elliptical arc S_1AS_2. Given the approximations made here, one may replace the local horizon S_1PS_2 by the parallel B_1OB_2 which goes through the centre of the Earth O. We thus find the visibility time as the time taken by the satellite S to travel along the elliptical arc B_1AB_2.

At a given time, the position of S is specified relative to the perigee A' by the true anomaly $v = (\boldsymbol{OA'}, \boldsymbol{OS})$. The mean anomaly M of the point B_1 is calculated from (1.54) with $v = \pi/2$. We obtain

$$M(B_1) = 2\arctan\sqrt{\frac{1-e}{1+e}} - e\sqrt{1-e^2}\,. \tag{5.43}$$

Molniya ⟨ Молния ⟩
Elliptical orbit - Ground track

Recurrence = [2; +0; 1] 2

>>>> Time span shown: 1440.0 min = 1.00 day

TIME MARKER

Equiv. altit. = 20175.5 km a =26553.629 km

CRITICAL Inclin. = 63.42 ° e = 0.736000

Period = 717.75 min * rev/day = 2.01

h_a = 39719 km; h_p = 632 km; arg. perigee: +270.00 °

Time marker: one point every 60.0 minutes

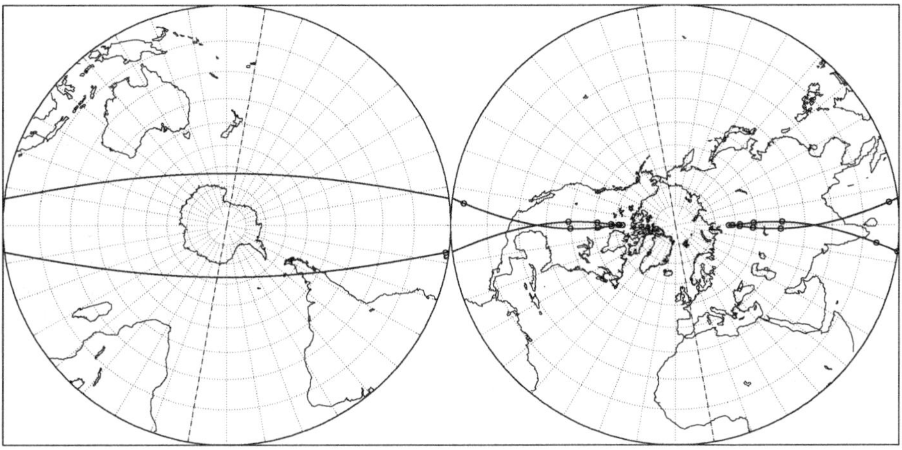

Projection: Stereographic
Property: Conformal
T.:Azimuthal ⊕ Graticule: 10°

Map centre (r.): 90.0 ° N; 10.0 ° W
Aspect: Direct
[-90.0 / +0.0 / +100.0] Gr.Mod.: GEM-T2

Longitude / Initialisation:
Asc. node: 72.94 °
Apogee: 80.00 °

Iξιων
MC ★ LMD
Aτλας

Supertundra
Elliptical orbit - Ground track

Recurrence = [1; +0; 1] 1

>>>> Time span shown: 1440.0 min = 1.00 day

TIME MARKER

Equiv. altit. = 35785.1 km a =42163.191 km

CRITICAL Inclin. = 63.43 ° e = 0.423000

Period = 1436.03 min * rev/day = 1.00

h_a = 53620 km; h_p =17950 km; arg. perigee: +270.00 °

Time marker: one point every 60.0 minutes

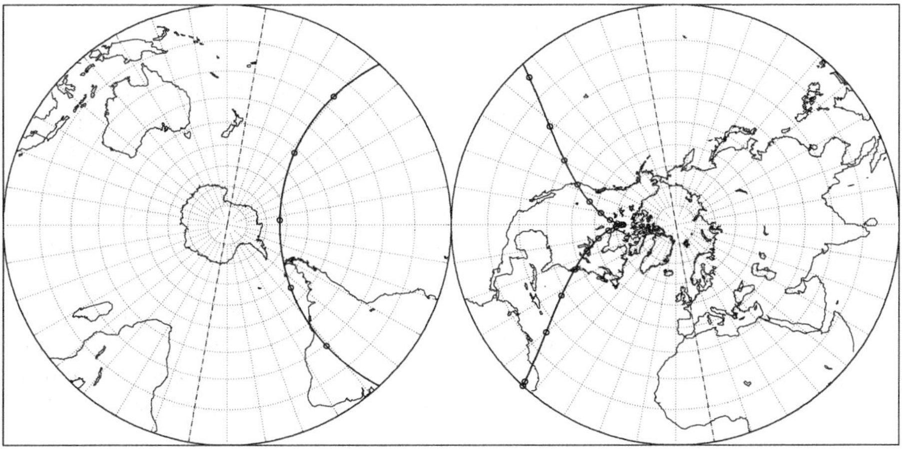

Projection: Stereographic
Property: Conformal
T.:Azimuthal ⊕ Graticule: 10°

Map centre (r.): 90.0 ° N; 10.0 ° W
Aspect: Direct
[-90.0 / +0.0 / +100.0] Gr.Mod.: GEM-T2

Longitude / Initialisation:
Asc. node: -53.02 °
Apogee: -100.00 °

Iξιων
MC ★ LMD
Aτλας

Figure 5.24. Ground tracks. *Upper*: Molniya orbit. *Lower*: Supertundra orbit

We deduce the visibility time as a function of the period:

$$\Delta t_v = \left[1 - \frac{M(B_1)}{\pi}\right] T_0 . \tag{5.44}$$

Example 5.7. *Calculate the visibility time for HEO satellites.*

For a (Molniya-type) HEO satellite, we have $e = 0.736$ and $T_0 = 718$ min. From (5.43) and (5.44), we obtain

$$M(B_1) = 0.744 - 0.498 = 0.245 \text{ rad} = 14° , \quad 1 - 14/180 = 0.92 .$$

During one revolution, the satellite thus spends 92% of its time in going from B_1 to B_2 via A (and 8% in going from B_2 to B_1 via A'), which corresponds to $\Delta t_v = 11$ hr for a period of 12 hr.[108] This result appears clearly in Fig. 5.24 (upper), where the time interval between two consecutive points marked on the ground track of the orbit is one hour. If we impose a minimal zenithal angle of sight $\zeta \sim 70°$ and if P is not exactly at the subsatellite point of A, we obtain a visibility span of about 8 hr. For a (Supertundra-type) HEO satellite, we have $e = 0.423$ and $T_0 = 1436$ min. We obtain

$$M(B_1) = 0.751 \text{ rad} = 43° , \quad 1 - 43/180 = 0.76 ,$$

which represents a visibility time of 18 hr, illustrated in Fig. 5.24 (lower). Imposing the above restrictions, a time of more than 12 hr is nevertheless obtained.

5.7 Appendix: NORAD Orbital Elements

The organisation known as NORAD (North American Aerospace Defence Command), founded in 1962, provides an inventory of all orbiting objects (satellites, whether active or otherwise, rocket debris, etc.). The position of the object on its orbit and the position of the orbit relative to the ECI Galilean frame[109] are defined by the orbital elements. The calculated positions are adjusted by radar measurements.

This data, referred to as the NORAD Two-Line Element Set Format, is updated twice a day and can be accessed by Internet. Data for each satellite consists of three lines in the following format:

```
AAAAAAAAAAAAAAAAAAAAAAAA
1 NNNNNU NNNNNAAA NNNNN.NNNNNNNN +.NNNNNNNN +NNNNN-N +NNNNN-N N NNNNN
2 NNNNN NNN.NNNN NNN.NNNN NNNNNNN NNN.NNNN NNN.NNNN NN.NNNNNNNNNNNNNN
```

[108] We observe that, although the name Molniya, lightning, suggests speed, it is when the satellite is at its slowest that it is actually used.

[109] Earth-Centered Inertial (ECI) coordinate system: the z axis runs along the Earth's rotational axis pointing North, the x axis points in the direction of the vernal equinox, and the y axis completes the direct orthogonal system.

Line 0 is a twenty-four character name. Lines 1 and 2 are the standard Two-Line Orbital Element Set Format identical to that used by NORAD and NASA. The format description is:

Line	*Column*	*Description*	
1	01	Line Number of Element Data	
1	03-07	Satellite Number	
1	08	Classification (U=Unclassified)	
1	10-11	International Designator (Last two digits of launch year)	
1	12-14	International Designator (Launch number of the year)	
1	15-17	International Designator (Piece of the launch)	
1	19-20	Epoch Year (Last two digits of year)	
1	21-32	Epoch (Day of the year and fractional portion of the day)	
1	34-43	First Time Derivative of the Mean Motion	
1	45-52	Second Time Deriv. of Mean Motion (decimal point assumed)	
1	54-61	Drag term (decimal point assumed), 'B*' model	
1	63	Ephemeris type	
1	65-68	Element number	
1	69	Checksum (Modulo 10)	
2	01	Line Number of Element Data	
2	03-07	Satellite Number	
2	09-16	i	Inclination (Degrees)
2	18-25	Ω	Right Ascension of the Ascending Node (Degrees)
2	27-33	e	Eccentricity (decimal point assumed)
2	35-42	ω	Argument of Perigee (Degrees)
2	44-51	M	Mean Anomaly (Degrees)
2	53-63	n	Mean Motion (Revolutions per day)
2	64-68	Revolution number at epoch	
2	69	Checksum (Modulo 10)	

The correspondence between the six standard Keplerian elements (as discussed in Chap. 2) and the six NORAD elements is immediate for the metric elements i and e and the angle elements Ω, ω, and M. The semi-major axis a is obtained from the mean motion n.

In order to use these orbital elements in a practical context, two of them require preliminary calculations:

- As we have just seen, the semi-major axis a is not given directly by the NORAD elements. The number of revolutions per day gives the anomalistic period T_{a} (because the period, in this orbital study, is defined as the time elapsed between two successive transits at perigee). By an iterative method like the one used in Example 4.2, we obtain the value of a.
- The angle Ω, the right ascension of the ascending node, is measured in the ECI frame from the direction of the vernal equinox. But in practice, one needs to know λ_{AN}, the longitude of the ascending node, i.e., the angular elongation of this point in a terrestrial frame (measured from the Greenwich meridian). With the usual notation, we can say that Ω is measured in $\Re$, and λ_{AN} in $\Re_{\mathrm{T}}$.

We first calculate the angle Ω_{G00} between the Greenwich meridian and the vernal equinox at 0 h UTC, on the relevant day. This angle corresponds to the mean sidereal time GMST (Greenwich Mean Sidereal Time), at 0 h, written q_{G00} and measured in seconds. It is obtained from the relation

$$q_{\mathrm{G00}} = 24110.54841 + 8640184.812866 T_{\mathrm{u}} + 9.3 \times 10^{-2} T_{\mathrm{u}}^2 - 6.2 \times 10^{-6} T_{\mathrm{u}}^3 , \tag{5.45}$$

where $T_{\mathrm{u}} = d_{\mathrm{u}}/36525$ and d_{u} is the number of days elapsed since 1 January 2000 at 12 h, which corresponds to the Julian date JD 2 451 545.0.

With the equivalence between days and round trips (1 day – 360°), we obtain Ω_{G00} in degrees from q_{G00} in seconds:

$$\Omega_{\mathrm{G00}} = \frac{q_{\mathrm{G00}}}{240} . \tag{5.46}$$

We then calculate $\Omega_{\mathrm{G}t}$, the angle between the Greenwich meridian and the vernal equinox at the relevant UT time:

$$\Omega_{\mathrm{G}t} = \Omega_{\mathrm{G00}} + \dot{\Omega}_{\mathrm{T}} \Delta t , \tag{5.47}$$

where $\dot{\Omega}_{\mathrm{T}}$ is the angular speed of rotation of the Earth, in degrees/day, given by (4.22), and Δt is the fraction of the day elapsed at the relevant UT time since 0 hr UT.

The positions of the ascending node and the Greenwich meridian, Ω and $\Omega_{\mathrm{G}t}$, respectively, are measured from the same origin at the same time. We thus obtain the longitude λ_{AN} of the ascending node in a terrestrial frame:

$$\lambda_{\mathrm{AN}} = \Omega - \Omega_{\mathrm{G}t} . \tag{5.48}$$

The longitude and the UT time then give the LMT crossing time at the ascending node.

Example 5.8. *Calculate the orbital elements of the ICESat satellite from the NORAD elements.*

During the first few months of its mission, the satellite ICESat followed a so-called calibration orbit. Its ground track had to repeat every 8 days (we return to the notion of recurrence in Chap. 7). The calibration orbit overflew the White Sands ground station calibration site every 8 days. Once the satellite and its instruments had been commissioned, burns were performed to attain ICESat's 183-day repeat ground track, mission orbit. The NORAD elements for the given day during this calibration phase were as follows:

```
ICESAT
1 27642U 03002A   03175.25018279  .00000722  00000-0  75456-4 0  1631
2 27642  94.0031 263.4514 0002250  85.5696 274.5785 14.90462832 24163
```

We obtain the date from

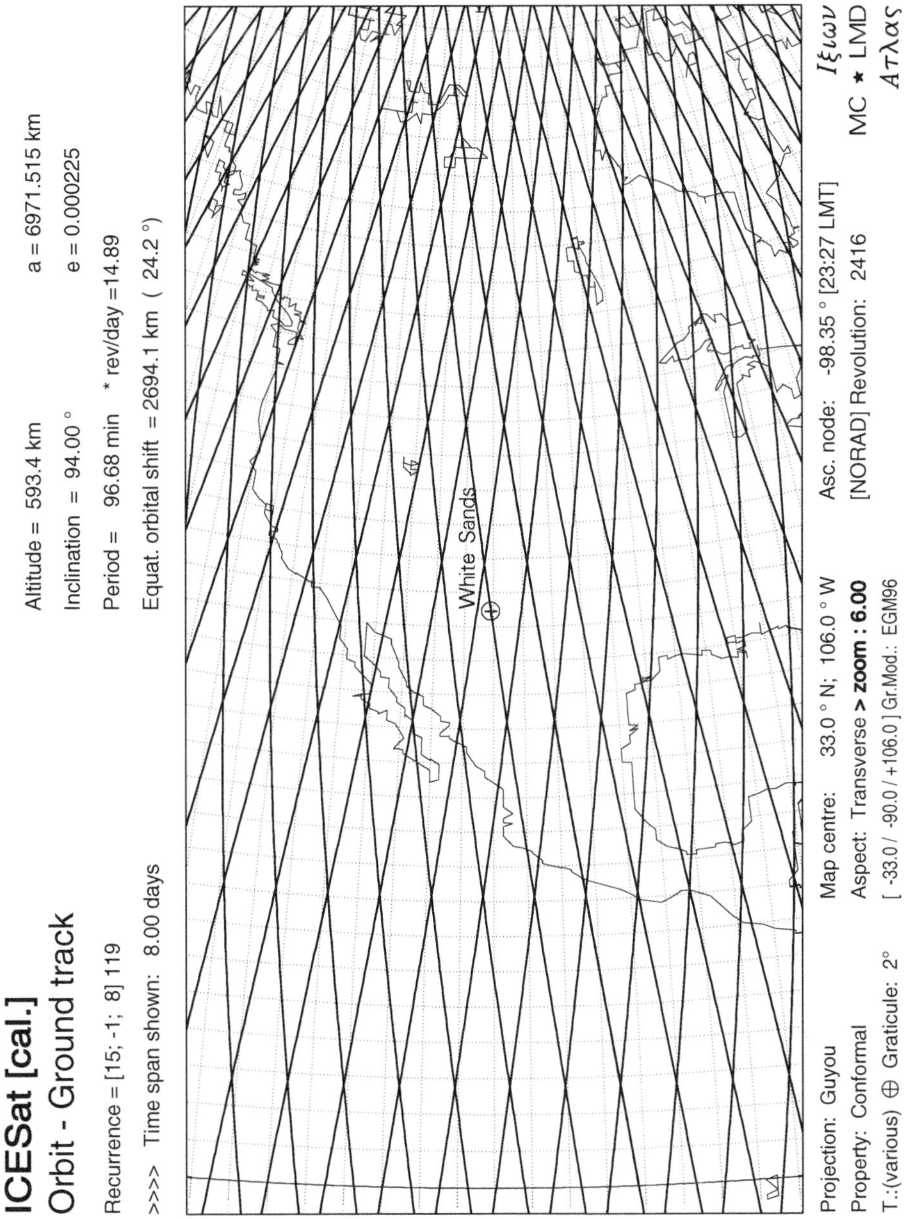

Figure 5.25. Transit of the ICESat satellite through the vertical of the calibration site at White Sands (Arizona, USA). Calibration phase

$$\texttt{03175.25018279}\text{: year} = 03,\ \text{day} = 175,\ \text{hour} = 0.25018279/24\ ,$$

which gives 24 June 2003, at 06:00:15.79 UTC. Just for information, this was revolution number 2416, counting from the satellite's first transit at the ascending node. The following elements are obtained immediately:

$$n = 14.904\,628\,32\ \text{rev/day}\ , \quad e = 0.000\,225\,0\ , \quad i = 94.003\,1^\circ\ ,$$

$$\Omega = 263.451\,4^\circ\ , \quad \omega = 85.569\,6^\circ\ , \quad M = 274.578\,5^\circ\ .$$

With the mean anomaly M, we can calculate the eccentric and true anomalies, very close to M since the eccentricity is extremely small: $E = 274.566^\circ$ and $v = 274.553^\circ$. With n, we find the anomalistic period T_a (min) $= 1440/n$, or $T_a = 96.614\,28$ min. To begin with, we set $T_0 = T_a$, and with this Keplerian period, we obtain the value of the semi-major axis of the Keplerian orbit: $a_0 = 6974.6$ km. We calculate the secular variation related to the mean motion, $\Delta n/n = -0.6695 \times 10^{-3}$, then recalculate a. After several iterations, this yields $a = 6971.515$ km, that is, an altitude of $h = a - R = 593$ km. With the values of a and i, we can calculate the precession rates and periods. This yields:

nodal precession	$\dot{\Omega} = +0.5079^\circ/\text{day}$,
apsidal precession	$\dot{\omega} = -3.5508^\circ/\text{day}$,
anomalistic period	$T_a = 96.6143$ min ,
draconitic period	$T_d = 96.6782$ min.

To calculate the longitude of the ascending node, we determine the Julian date of the relevant time, JD 2 452 814.750, which gives JD-JD2000 $= 1\,269.750$ and hence $d_u = 1\,269.5$. Using (5.45), we obtain $q_{G00} = 65\,217.588$ (mod 86 400), or $\Omega_{G00} = 271.740^\circ$. Given the fraction of the day elapsed, we obtain from (5.47), $\Omega_{Gt} = 1.806^\circ$ whence, by (5.48),

$$\lambda_{AN} = 263.451 - 1.806 = 261.646\ .$$

For a transit at the ascending node at longitude 261.646° (i.e., 98.354° W), the time 06:00 UT corresponds to 23:27 LMT. These elements suffice to represent the ground track of the ICESat satellite over 8 days. During this time, the satellite should pass through the vertical at the calibration site in White Sands (Arizona). Figure 5.25 shows that this constraint was respected.

5.8 Appendix: Cartographic Projections

A cartographic projection is a transformation mapping a point of the sphere (or an ellipsoid), specified by its spherical coordinates λ, ϕ (longitude, latitude), in a bijective manner onto its coordinates x, y on the map:

$$\text{cartographic projection}\quad f = \{f_1, f_2\} \qquad \begin{cases} x = & f_1(\lambda, \phi)\ , \\ y = & f_2(\lambda, \phi)\ . \end{cases}$$

There exist infinitely many such projections.

The main problem to be solved by a cartographic projection can be put succinctly as follows: the sphere is not developable. This means that the surface of a sphere cannot be mapped onto a plane without distorting or tearing it.[110] Theoretical study of this question was carried out contemporaneously by Lambert[111] (1772), Euler (1777) and Lagrange (1779). It was definitively solved by Gauss (1822), who studied the conditions under which an arbitrary surface could be mapped onto another arbitrary surface.

A cartographic projection can have (exclusively) one or other of the following two properties:

- angles are conserved, and the projection is said to be conformal,
- areas are conserved, and the projection is said to be equal area or equivalent.

It may be that it has neither of these properties,[112] but it can never have both. In fact, the map can conserve the angles or the area of a figure, but it can never conserve the perimeter. No map can conserve distances in all directions. In other words, no projection can have constant scale over the whole field of projection.

In a conformal projection, the parallels and the meridians on the map intersect at right-angles, since the same is true on the sphere, where they form two sets of orthogonal curves. In an equal area projection, a country twice as big as another is represented on the map by an area twice as big. When the whole Earth is represented, it can be considered as spherical since projecting the terrestrial sphere or ellipsoid on a plane leads to quite imperceptible differences in the resulting maps, whatever projection may be used. The same is not true for accurate regional maps, however.

Projections can be classified by type or by aspect. The type tells us how the sphere appears when projected onto the map, i.e., cylindrical, conical,

[110] Unlike the sphere, a cylinder is developable. If the body of a big cat could be assimilated to a cylinder, one could understand how the tiger might change into a bedside rug without deformation.

[111] Jean Henri Lambert (1728–1777) was a Swiss and German astronomer, mathematician and physicist, with French ancestry. In astronomy, he calculated the trajectories of comets and understood that the Milky Way was just a modest galaxy in the Universe. In physics, he discovered the fundamental law of photometry. In his many mathematical works, among which he demonstrated the irrationality of π (1766), he assigned great importance to problems of perspective and cartographic projections. He defined a great many projections, several of which bear his name today. The best known is the conformal conical projection, used in France for the map of France since 1922 and the cadastral survey since 1938.

[112] The older literature is full of different adjectives describing these properties: autogonal and orthomorphic for conformal; authalic, homolographic for equivalent; aphylactic, in the absence of these properties. Such terms have now gone into disuse.

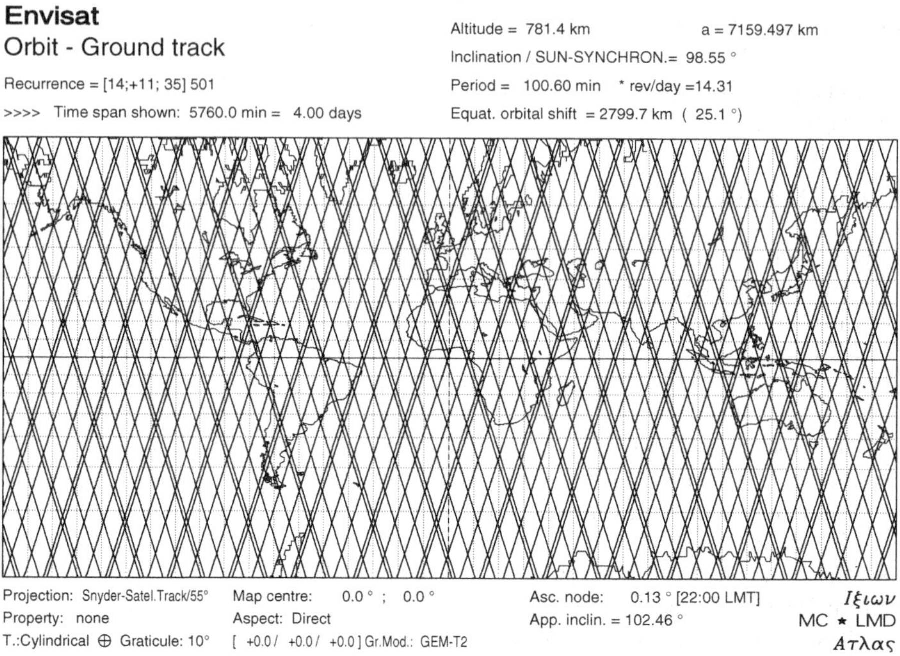

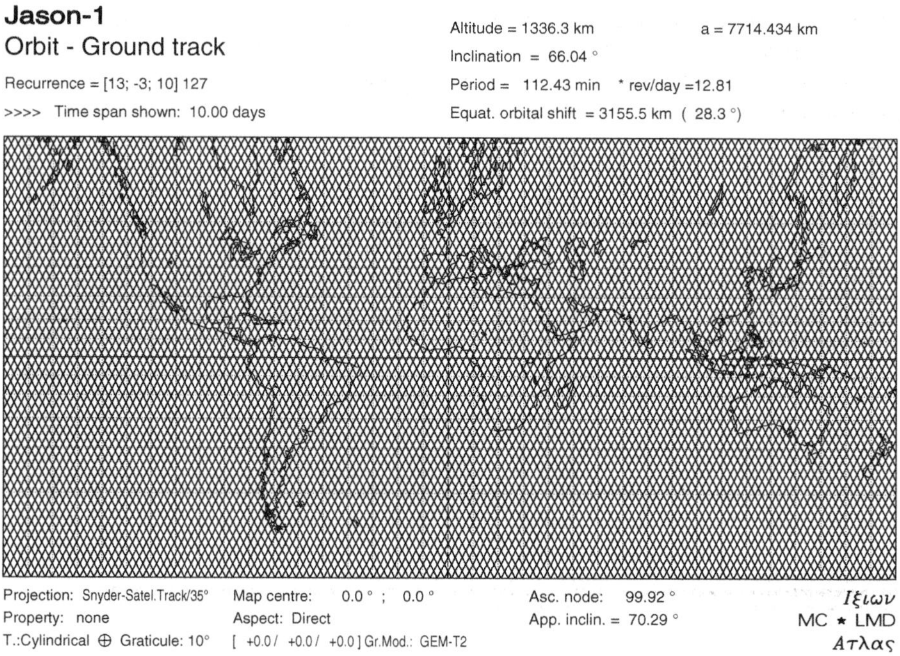

Figure 5.26. Ground tracks of two Sun-synchronous and prograde satellites. Snyder projection, with different standard parallels. *Upper*: 55°. *Lower*: 35°

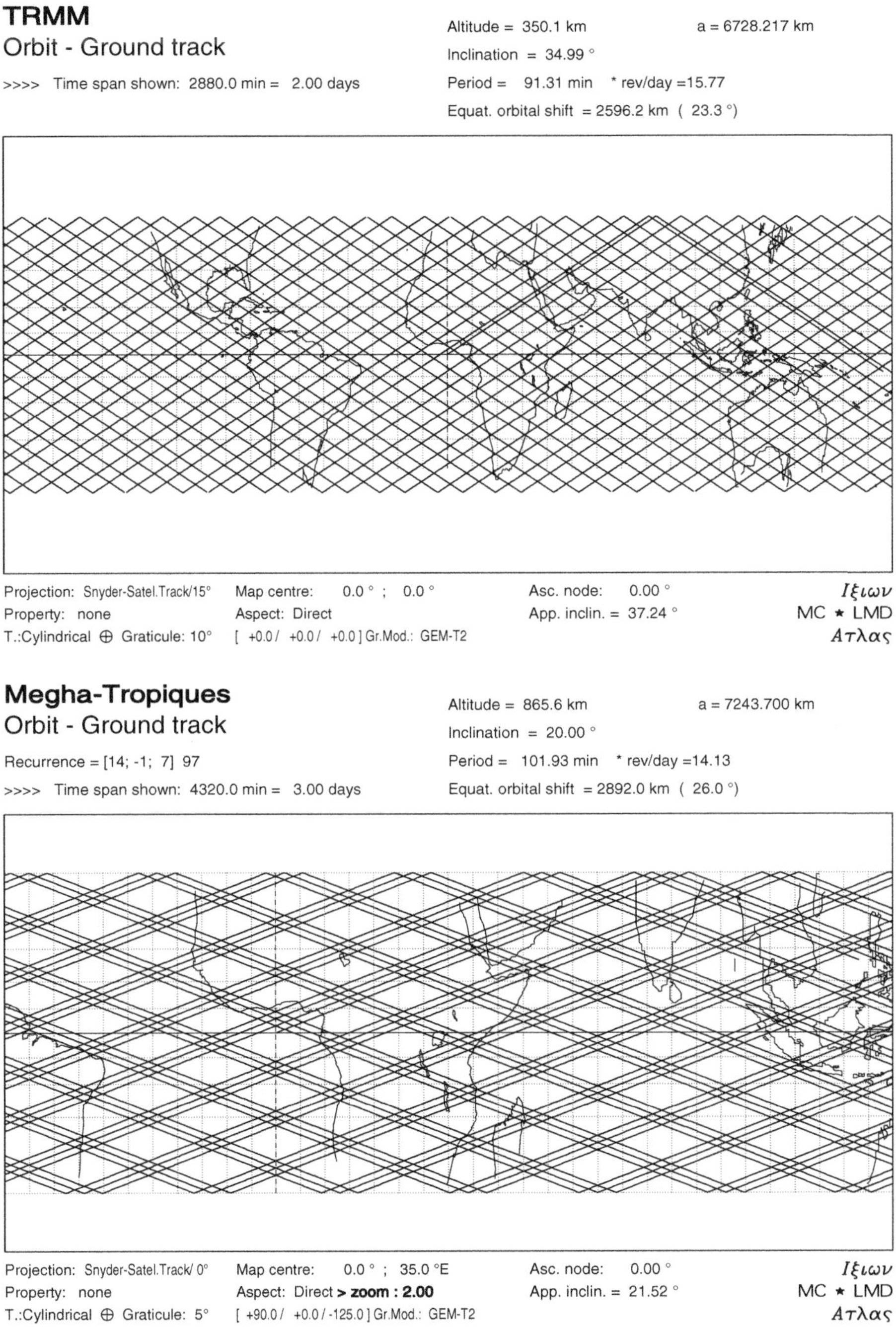

Figure 5.27. Ground tracks of two prograde (low inclination) satellites. Snyder projection, with different standard parallels. *Upper*: 15°. *Lower*: 0°

azimuthal, and so on. We use the word 'appears' because a cartographic projection is not usually (and the exceptions are very rare) a projection in the sense of an intersection between a straight line and a plane. For example, the Mercator projection is said to be cylindrical, but it is not the 'projection' of this sphere from its centre onto a cylinder that is tangent to it at the equator (as one often reads).

The aspect of a projection can be direct (or normal), transverse, or oblique. For example, for a stereographic projection (of azimuthal type) with direct aspect (also called polar in this case), the point of contact of the plane of projection with the sphere occurs at the pole, whilst it occurs on the equator for a transverse projection (also called equatorial in this case). For an arbitrary point of contact, the projection is said to be oblique.

The computer software Atlas, which we have devised, is coupled with the orbitography part of our program Ixion. Any satellite ground track can thus be mapped out with the chosen projection. In each representation, every effort is made to apply the most suitable cartographic projection. On all the maps presented here, plotted using Atlas, we indicate the main features of the projection: name, properties, type and aspect. Also given in the key are the coordinates (longitude and latitude) of the centre of the map, together with the three Euler angles which define the rotation of the globe for this projection from the standard initial position.

The projections used in the present book can be grouped as follows:

- **Conformal Projections.** The angle between the ground track (of the satellite or its swath) and the given meridian is conserved. The main projections here are:
  - the Mercator projection,
  - the stereographic projection,
  - the projections due to Guyou and Adams, based on elliptic integrals of the first kind.
- **Equal Area Projections.** These projections are used when it is important to respect surface areas. The main projections used in this book are:
  - the Behrmann projection (dilated Lambert equal area cylindrical projection) and the Lorgna projection,
  - the Mollweide and Sanson projections,
  - the Goode homolosine projection (in interrupted form),
  - the Hammer–Aitoff projection.
- **Perspective Projections.** Although they have no special properties, these projections are rather visual, representing the planet as if it were seen from space, viewed from various distances. The main projections used are:
  - the perspective view projection, where the viewing point is at a finite distance (expressed as a number of planetary radii),
  - the orthographic projection, where the viewing point is at infinity,

  – the Armadillo projection, due to Raisz, which represents the sphere in a rather picturesque manner projected on a torus.
- **Specific Projections.** In 1977, the American cartographer John P. Snyder invented a specific projection to represent satellite ground tracks for the satellite ERTS-1 (Landsat-1) and the following satellites in the Landsat programme. This projection, Satellite Tracking, keeps the meridians regularly spaced and modifies the spacing of the parallels in such a way that the satellite ground track is a straight line. We have adapted this to any type of satellite. It is presented in the two parts of Fig. 5.26 and Fig. 5.27, with four different dilatations of the latitudes.
- **Archaic Projection.** The so-called *plate-carrée* projection is very frequently used (if not exclusively) in books and documents concerned with satellite ground tracks. It represents longitudes and latitudes linearly along the abscissa and ordinate, respectively. This projection is somewhat simple-minded ($x = \lambda$, $y = \phi$), even primitive (it was fashionable ... in the Middle Ages). It has no particular properties (it is neither conformal, nor equivalent) and its only mathematical value is its simplicity. This is no longer an argument with the advent of computer programming. It has not been used here, except for the first three maps in the book, at the beginning of the chapter. (We did not wish to upset habits too early on.)

6 Orbit Relative to the Sun. Crossing Times

We begin by studying the position of the orbit and ground track of an arbitrary satellite relative to the direction of the Sun. We then turn more specifically to Sun-synchronous satellites for which this relative position provides the very definition of their orbit.

6.1 Cycle with Respect to the Sun

6.1.1 Crossing Time

At a given time, it is useful to know the local time on the ground track, i.e., the LMT, deduced in a straightforward manner from the UT once the longitude of the place is given, using (4.49). The local mean time (LMT) on the ground track at this given time is called the crossing time or local crossing time.

To obtain the local apparent time (LAT), one must know the day of the year to specify the equation of time E_{T}. In all matters involving the position of the Sun (elevation and azimuth) relative to a local frame, this is the time that should be used.

The ground track of the satellite can be represented by giving the crossing time. We have chosen to represent the LMT using colour on Colour Plates IIb, IIIb and VII to XI.

For Sun-synchronous satellites such as MetOp-1, it is clear that the crossing time (in the ascending or descending direction) depends only on the latitude. At a given place (except near the poles), one crossing occurs during the day, the other at night. For the HEO of the Ellipso Borealis satellite, the stability of the crossing time also shows up clearly.

In the case of non-Sun-synchronous satellites such as Meteor-3-07, the time difference shows up through a shift in the time from one revolution to the next. For a low-inclination satellite such as Megha-Tropiques, we see that, if the ascending node crossing occurs at 06:00, the northern hemisphere will be viewed during the day, and the southern hemisphere during the night. After a few days the crossing time at the equator will have changed.

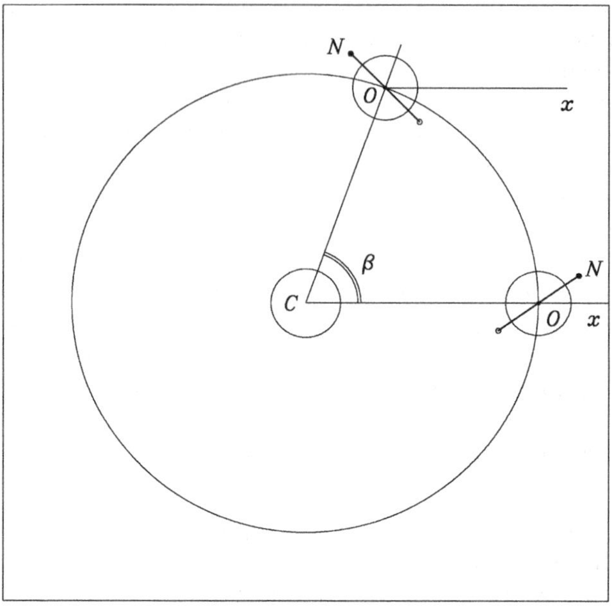

Figure 6.1. Nodal precession in the ecliptic. In the plane of the ecliptic, the centre of the Sun has been marked C, and the centre of the Earth O at two given times. The direction Ox is fixed relative to the stars. A satellite orbits the Earth. Its ascending node N is shown by a *small black dot*, whilst the descending node is shown by a *small circle*. The line segment joining the two nodes is the projection of the line of nodes on the ecliptic, not on the satellite orbit

6.1.2 Calculating the Cycle C_S

We consider the orbit of the Earth around the Sun, treating it as circular, since in this calculation of the cycle, we identify LAT and LMT. In Fig. 6.1, the centre of the Sun, and of the Earth's orbit, is denoted by C, whilst the centre of the Earth is O. The ascending node of a satellite in orbit around the Earth is denoted by N. The dihedral angle between the meridian plane of the Earth containing N and that containing C gives H, the hour angle of the ascending node. This angle is represented in Fig. 6.1 by $H = (\boldsymbol{OC}, \boldsymbol{ON})$. The diagram is schematic. To be precise, N should represent the projection of the ascending node on the plane of the ecliptic. However, this will not affect the following argument.

At the time $t = t_0$, the direction $\boldsymbol{CO}$ defines an axis $\boldsymbol{Ox}$, with fixed direction in the Galilean frame $\mathfrak{R}$. The hour angle of N is thus written $H(t_0) = H_0$. At another time $t = t_1$, the plane of the satellite orbit will have changed due to the phenomenon of nodal precession by an angle Ω relative to the frame $\mathfrak{R}$, i.e., relative to the direction $\boldsymbol{Ox}$. The hour angle of N is then given by

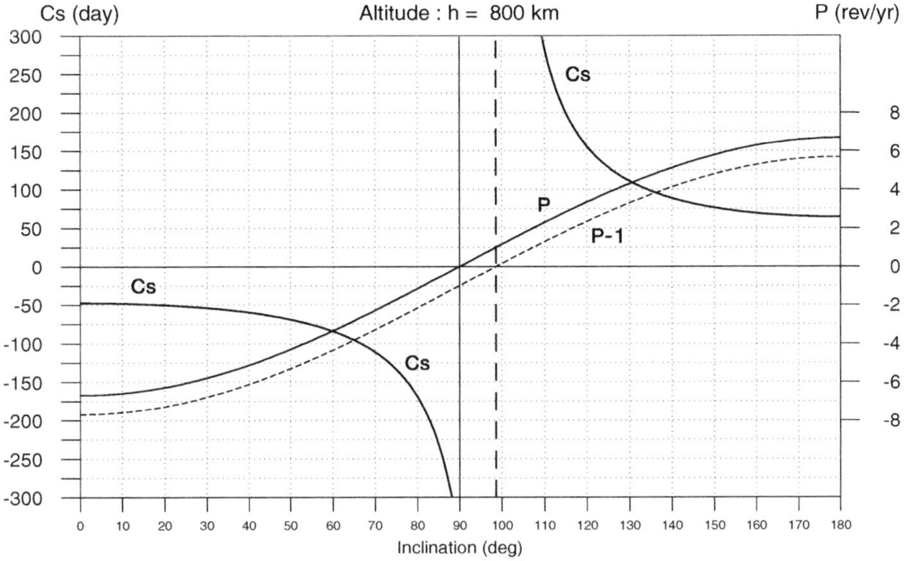

Figure 6.2. Variation of the cycle C_S relative to the Sun as a function of the inclination for a satellite at altitude $h = 800$ km. The cycle C_S is given in days on the *left ordinate*, and the nodal precession rate P is given in rev/yr on the *right ordinate*

$$H(t_1) = H_1 = H_0 + \Omega - \beta \; ,$$

where β is the angle through which the Earth has moved on its orbit around the Sun, viz.,

$$\beta = \big(\boldsymbol{CO}(t_0), \boldsymbol{CO}(t_1)\big) \; .$$

This angle β is equal to the difference in ecliptic longitude l of the Sun at the two given times. Hence,

$$\Delta H = H_1 - H_0 = \Omega - \beta \; ,$$

which represents the variation of the orbital plane relative to the direction of the Sun.

Setting $m = t_1 - t_0$ for the time interval, the angles can be written in terms of the angular speeds:

$$\Omega = m\dot{\Omega} \; , \quad \beta = m\dot{\Omega}_S \; ,$$

whence,

$$\Delta H = m(\dot{\Omega} - \dot{\Omega}_S) \; . \tag{6.1}$$

The time interval m_0 needed for the hour angle of the ascending node to vary by 24 hr, or one round trip, is called the cycle relative to the Sun. Hence,

$$H(t + m_0) = H(t) \quad [2\pi] \, ,$$

which implies that

$$m_0 = \frac{2\pi}{\dot{\Omega} - \dot{\Omega}_S} \, .$$

Bringing in the nodal precession rate P in round trips per year as defined by (4.30) and using (4.28), m_0 becomes

$$m_0 = -J_M \frac{N_{tro}}{1 - P} \, .$$

The cycle relative to the Sun is usually given in days and we shall denote it by C_S (with C for 'cycle' and S for 'Sun'). Since m_0 is expressed in SI units, i.e., in seconds, we obtain C_S from the very simple expression

$$C_S = \frac{N_{tro}}{P - 1} \, . \tag{6.2}$$

The quantity P can be expressed in terms of the constant k_h defined by (4.63). This rate P is given by

$$P = -k_h \left(\frac{R}{a} \right)^{7/2} \cos i \, . \tag{6.3}$$

One can check that for a Sun-synchronous satellite we do indeed have $P = 1$.

In this way we obtain the cycle relative to the Sun as a function of the orbital characteristics, taking care to note the signs:

$$C_S = C_S(a, i) = -\frac{N_{tro}}{k_h \left(\dfrac{R}{a} \right)^{7/2} \cos i + 1} \, , \tag{6.4}$$

or with approximate numerical values (C_S in days),

$$C_S = -\frac{365.25}{10.11 \left(\dfrac{R}{a} \right)^{7/2} \cos i + 1} \, . \tag{6.5}$$

The cycle relative to the Sun, $C_S = C_S(a, i)$, is a very important feature of any satellite, but especially Earth-observation satellites.

6.1.3 Cycle C_S and Orbital Characteristics

Cycle C_S as a Function of Altitude and Inclination

The cycle C_S is a function of a and i. Figure 6.2 shows the variation $C_S(i)$ for a fixed altitude $h = 800$ km. The cycle $C_S(i)$ is given in days, with the sign indicating the direction of rotation. We have also plotted the nodal precession $P(i)$ in rev/yr, which is a sinusoidal curve, and $P - 1$ which determines the vertical asymptote of $C_S(i)$ by its intersection with the horizontal axis through the origin.

For the altitude represented here, typical of LEO satellites, we see that the cycle remains in the vicinity of two months ($C_S \sim -60$ day) for inclinations below 45°. When i increases, the length of the cycle also increases. Above the Sun-synchronous inclination, the cycle decreases (but there are very few satellites in this configuration).

Specific Cases of the Cycle C_S

We note here certain specific values of the cycle C_S for different orbits.

- **Polar Satellites.** We see immediately from (6.4) or (6.5) that, if the satellite is strictly polar, $C_S = -365.25$ day. The cycle is annual. One year goes by before we return to the same orbital configuration relative to the Sun, since the plane of the orbit does not rotate with respect to $\Re$. The negative value of C_S shows that the line of nodes moves in the retrograde direction relative to $\Re_T$.
- **Sun-Synchronous Satellites.** Equation (6.2) shows that if $\dot{\Omega} = \dot{\Omega}_S$, the cycle is infinite. This happens for Sun-synchronous satellites and we may indeed treat C_S as infinite, since after a very great number of days, the angle H will not have changed. For Sun-synchronous satellites, the hour angle of the ascending node, and hence the crossing time[1] of the satellite at the ascending node, is constant. For a given altitude, the cycle C_S is negative provided that i is less than the Sun-synchronous inclination given by (4.68). Beyond this value, C_S is positive.
- **Shortest Cycle.** The smallest value for the cycle is given by the minimum of $|C_S(a,i)|$. According to (6.5), it is obtained for $i = 0$ and $a = R$ and the value is

$$|C_S|_{\min} = \frac{N_{\mathrm{tro}}}{k_{\mathrm{h}} + 1} = \frac{365.25}{11.11} = 32.9 \text{ day} . \tag{6.6}$$

The cycle relative to the Sun C_S can never be less than 33 days.

---

[1] The time related to the hour angle is LAT. A Sun-synchronous satellite transits at the ascending node at the same LMT. If there is no difference between LAT and LMT here, it is because we have used a simplified scenario for the Earth orbit. However, for the calculation of the cycle C_S, this could not be otherwise: we only want to know how many days it will be before the next crossing (to within a few minutes), whatever time of year it is. To treat an elliptical Earth orbit, we would have to specify the day we choose to begin the cycle.

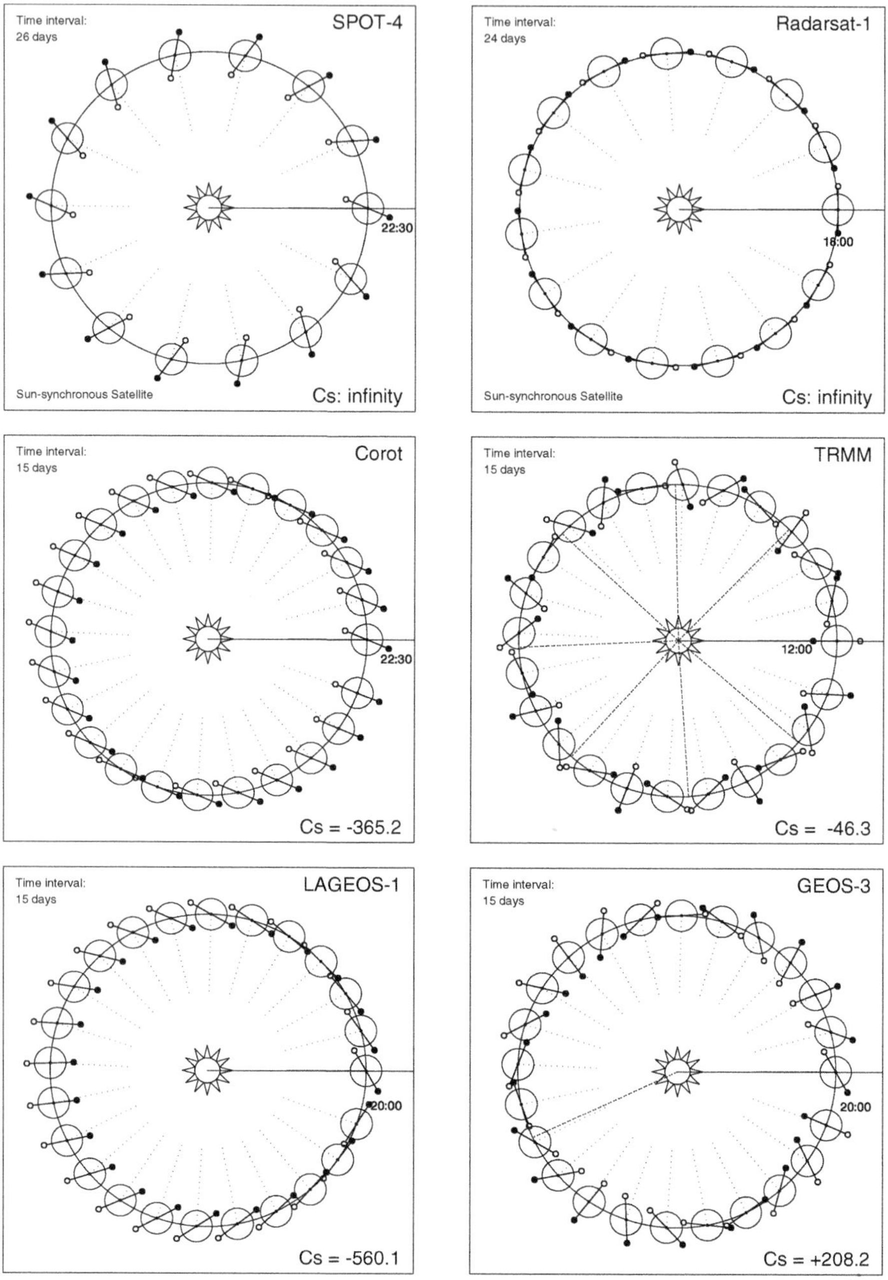

Figure 6.3. Cycle relative to the Sun for various satellites. The time given is the crossing time at the first ascending node

Example 6.1. *Calculate the cycle relative to the Sun for the satellites Meteor-3-07, TOPEX/Poseidon, ICESat, ERBS and UARS.*

These satellites have near-circular orbits. For Meteor-3-07, we have $h = 1\,194$ km and $i = 82.56°$. Using (6.5), we obtain

$$C_S = -\frac{365.25}{10.11\left(\dfrac{6378}{7572}\right)^{7/2}\cos(82.56)+1} = -\frac{365.25}{10.11\times 0.5477\times 0.1295+1}$$

$$= -\frac{365.25}{0.7169+1} = -\frac{365.25}{1.7169} = -212.73\ ,$$

which gives a cycle of 213 days (advance of crossing time). In this case it is easier to use (6.2) because the value of P has already been calculated in Example 4.2:

$$P = -0.716 \qquad \Longrightarrow \qquad C_S = \frac{365.25}{P-1} = -212.73\ .$$

For TOPEX/Poseidon, with $h = 1\,336$ km and $i = 66.04°$, we obtain $P = -2.107$, which gives a cycle $C_S = -117.47$, or 117 days (advance of crossing time).
ICESat is at low altitude, $h = 592$ km, with inclination $i = 94°$ between the polar inclination for which the cycle is one year ($C_S = -365.25$) and the Sun-synchronous inclination ($i_{HS} = 97.8°$ at this altitude) for which the cycle is infinite. The calculation gives $P = 0.515$, whence $C_S = -752.7$, which corresponds to a very long cycle of more than two years.
ERBS and UARS, both launched by the space shuttle, have the same inclination and the same altitude to within a few kilometres. The calculation gives $P = -3.986$ for ERBS, whence $C_S = -73.2$, and $P = -4.090$ for UARS, whence $C_S = -72.0$. One often reads for these satellites that their cycle relative to the Sun is 36 days. However, this is the half-cycle.

Nodal Precession and Cycle C_S

In order to visualise the nodal precession and bring out the significance of the cycle C_S as clearly as possible, let us return to the graph in Fig. 6.1 and apply it to a few satellites in the following example.

Example 6.2. *Visualising the cycle C_S for various satellites in prograde, polar, retrograde and Sun-synchronous orbit.*

Figure 6.3 shows the position of the Earth on its orbit around the Sun and the position of the nodes (ascending in black, descending in white) of the satellite orbit.
For the two Sun-synchronous satellites, SPOT-4 and Radarsat-1, it is clear that the shift of the orbital plane compensates the Earth's annual motion. For Radarsat-1, the normal to the orbit lies in the meridian plane passing through the Sun.
For a strictly polar satellite like Nova-1 or Corot, the orbital plane is fixed in $\Re$.

For Corot, which has this inertial orbit, stars are observed perpendicularly to the orbit, six months in one direction, and six months in the opposite direction, in such a way as to avoid viewing the Sun.
Let us consider now several retrograde (negative) cycles, one very short, for TRMM (prograde orbit), one very long, for LAGEOS-1 (retrograde orbit). The satellite GEOS-3 (retrograde orbit) provides a rare case of precession in the prograde direction.

6.1.4 Cycle and Ascending Node Crossing Time

Knowing the initial conditions, it is a simple matter to obtain the crossing times at the ascending node at an arbitrary date, provided that we also know the cycle relative to the Sun C_S. Indeed, since the crossing time increases or decreases by 24 hours every C_S days, it is easy to calculate the increase or decrease per day. Here is an example of this calculation.

Example 6.3. *Calculate the dates during the year 1999 for which the LMT of the ascending node crossing is the same for the satellites TRMM and Resurs-O1-4.*

In order to study the Earth's radiation budget, TRMM and Resurs-O1-4 were equipped with the CERES and ScaRaB instruments, respectively. A joint measurement campaign was organised in January and February 1999. The aim was to compare the measurements obtained for the same region viewed by the two instruments at roughly the same times (with a leeway of ± 15 min). The Sun-synchronous satellite Resurs-O1-4 crosses the ascending node at 22:20 LMT. The initial conditions for TRMM are given by an ascending node crossing (t_{AN} given in month day hr min s):

$$t_{AN} = 1999\ 01\ 21\ 20{:}43{:}47\ (\mathrm{UT})\ , \quad \lambda = +5.157^\circ\ .$$

We calculate the value of τ_{AN}, LMT crossing time:

$$\tau_{AN} = t_{AN} + \frac{\lambda}{15} = 20{:}43{:}47 + 00{:}20{:}38 = 21{:}04{:}25\ .$$

In Example 4.1, we found $P = -6.89$, which gives the cycle

$$C_S = -\frac{365.25}{7.89} = -46.29\ \mathrm{day}\ .$$

We thus obtain the daily drift as

$$\frac{1440}{C_S} = -\frac{1440}{46.42} = -31.02\ \mathrm{min}\ .$$

The difference between $\tau_{AN} = 21{:}04$ on 21 January 1999 ($J = 21$) and the chosen time of 22:20 is 76 min. The passage of TRMM at the chosen time thus occurs with a shift of $-76/31 = -2.45$ days, or 2 days earlier, i.e., on 19 January 1999 ($J = 19$). The ascending node crossing around 22:20 thus occurs on the days J_k given by

$$J_k = 19 + k|C_S|\ ,$$

where the integer k takes 8 values over one year (since $|365/C_S| = 7.9$). Here, with $J_0 = 19$ and the values $k = 0, \dots, 7$, we obtain all the dates required for the year 1999. If we need to know the dates of passage of TRMM at 22:20 at the descending node, we merely add a half-cycle to the values of J_k, which gives dates shifted by 23 days with respect to the first series.

6.2 Crossing Time for a Sun-Synchronous Satellite

6.2.1 Passage at a Given Latitude

The time in LMT at which a Sun-synchronous satellite crosses the ascending node is constant in time (provided that the orbit is suitably maintained, of course), because in the frame $\Re$, the nodal precession balances the motion of the Earth's axis about the Sun. This is the defining feature of Sun-synchronous orbits, brought out in the next example.

Example 6.4. *Calculate the crossing time at two consecutive ascending nodes for a Sun-synchronous satellite.*

Consider the first crossing at the ascending node at longitude λ_1 and time $t = t_0$ in UT. Let τ_1 be the corresponding LMT, so that, according to (4.50),

$$\tau_1 = t_0 + \frac{\lambda_1}{15} ,$$

with time in hours and longitude in degrees.

The next passage (nodal period T) will occur at longitude λ_2 and at time $t = t_0 + T$. The corresponding LMT at the second crossing, denoted by τ_2, is therefore

$$\tau_2 = t_0 + T + \frac{\lambda_2}{15} .$$

The longitude λ_2 is obtained simply by considering the equatorial shift given by (5.22):

$$\lambda_2 = \lambda_1 + \Delta_E \lambda = \lambda_1 - 15T .$$

We thus have

$$\tau_2 = t_0 + T + \frac{\lambda_1 - 15T}{15} = t_0 + \frac{\lambda_1}{15} = \tau_1 ,$$

which shows that the LMT remains constant.

Since the mean motion is constant, the time taken to reach a given latitude from the equator will be the same for each revolution. We may thus say that, for a Sun-synchronous satellite:

- the LMT crossing time at a given latitude is constant,
- the LMT crossing time at a given meridian depends only on the latitude.

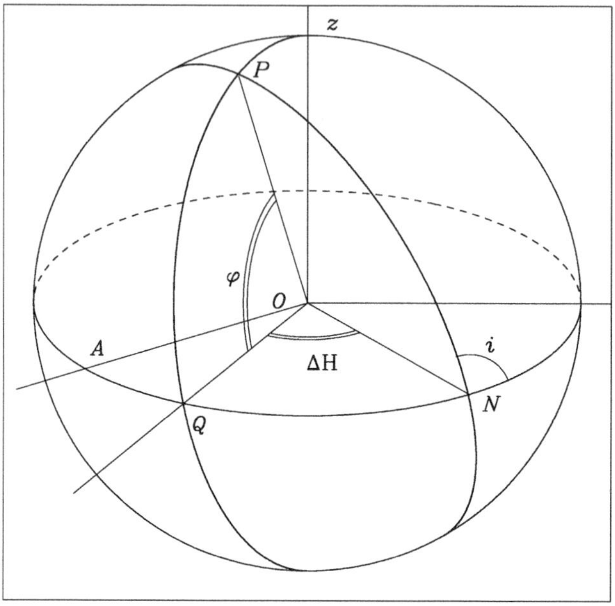

Figure 6.4. Intersection of the ground track of a Sun-synchronous satellite orbit (ascending node N) with a given meridian plane, defined by the point Q on the equator

Establishing the Relation Between ϕ and $\Delta\tau$

The relation between τ (the crossing time at the meridian in LMT) and ϕ (latitude) is found using the equations for the ground track and calculating the longitude corresponding to each latitude, whereupon the time can be found in LMT. But there is a simpler way to obtain this relation from geometric considerations.

Consider the Earth in the Galilean frame, as shown in Fig. 6.4. At a given time, let A be the intersection of the meridian plane of the direction of the Sun with the Earth's equator. We consider the orbital plane of a Sun-synchronous satellite. Its ground track cuts the equator at N, the projection of the ascending node on the Earth's surface. This plane makes an angle $i = i_{\text{HS}}$ with the equatorial plane (this is indeed i since we are working in $\Re$, rather than the apparent inclination).

The angle $H_{\text{AN}} = (\boldsymbol{OA}, \boldsymbol{ON})$ remains constant by the Sun-synchronicity condition, since H_{AN} measures the hour angle, and hence the time in LMT, of the ascending node.

Consider a meridian defined by a point Q on the equator. The ground track of the orbit cuts this meridian at a point P of latitude ϕ. The hour angle of P and of Q is $H = (\boldsymbol{OA}, \boldsymbol{OQ})$. We define

$$\Delta H = H - H_{\text{AN}} = (\boldsymbol{ON}, \boldsymbol{OQ}) \,.$$

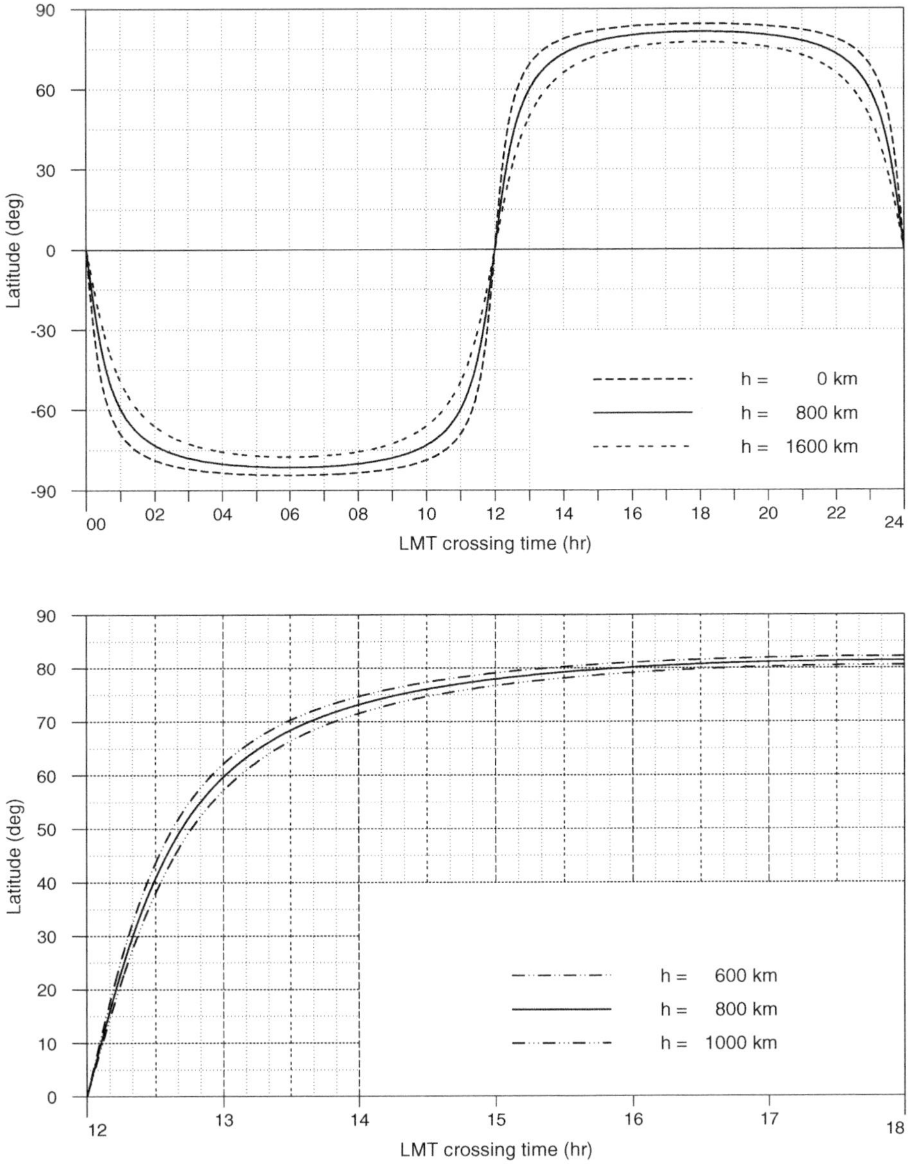

Figure 6.5. Graph of $\phi(\Delta\tau)$, the relation between the latitude of the point under consideration and the LMT time difference between transit at the ascending node and transit at this latitude, for a Sun-synchronous satellite. *Upper*: for three values of the altitude, $h = 800$ km and $h = (800 \pm 800)$ km. *Lower*: for three values of the altitude, $h = 800$ km and $h = (800 \pm 200)$ km. This is a magnified view of part of the upper figure

This angle thus measures the difference in hour angle between N and P (or Q).

In the spherical triangle PQN, with a right-angle at Q, we know the side PQ, $(\boldsymbol{OQ}, \boldsymbol{OP}) = \phi$ and the angle at N, representing the inclination of the orbital plane. We obtain ΔH from the standard relation of spherical trigonometry, corresponding to the relation (ST XII), identifying PQN with CAB:

$$\sin \Delta H = \frac{\tan \phi}{\tan i_{\mathrm{HS}}} . \tag{6.7}$$

Naturally, this formula is valid whether the satellite orbit is prograde or retrograde. In the prograde case, $\tan N$ and $\sin \Delta H$ are positive. In the retrograde case, as here, $\tan N = \tan(\pi - i_{\mathrm{HS}})$ and ΔH are negative.

Let τ_{AN} and τ be the local crossing times at the ascending node and P, respectively. Then,

$$\Delta \tau = \tau - \tau_{\mathrm{AN}} = \frac{1}{K} \Delta H , \tag{6.8}$$

where K is a constant depending on the units, so that if time is in hours and angles in degrees, then $K = 15$ (since 1 hr corresponds to 15°).

We thus have the following relations between the latitude ϕ and the difference in crossing times $\Delta \tau$:

$$\Delta \tau = \frac{1}{K} \arcsin \left(\frac{\tan \phi}{\tan i_{\mathrm{HS}}} \right) , \tag{6.9}$$

or

$$\phi = \arctan \left(\tan i_{\mathrm{HS}} \sin K \Delta \tau \right) . \tag{6.10}$$

Crossing Time at an Arbitrary Latitude

Let τ_{AN} and τ_{DN} be the crossing times at the ascending and descending nodes, respectively. Then,

$$\tau_{\mathrm{AN}} = 12 + \tau_{\mathrm{DN}} \quad [\mathrm{mod}\ 24] .$$

For $\Delta \tau$, we take the value defined by (6.9), i.e., between -6 hr and $+6$ hr. We thereby obtain the two daily crossing times $\tau_{(\mathrm{A})}$ and $\tau_{(\mathrm{D})}$ in the ascending and descending parts of the ground track, respectively:

$$\begin{cases} \tau_{(\mathrm{A})} = \tau_{\mathrm{AN}} + \Delta \tau , \\ \tau_{(\mathrm{D})} = \tau_{\mathrm{DN}} - \Delta \tau = \tau_{\mathrm{AN}} + 12 - \Delta \tau . \end{cases} \tag{6.11}$$

The time difference $\delta(\phi)$ between two crossings, one in the ascending part and the other in the descending part, at a given latitude is given by

$$\delta(\phi) = \tau_{(A)} - \tau_{(D)} = 12 + 2\Delta\tau \ . \tag{6.12}$$

We now give some examples of this calculation.

Example 6.5. *Calculate the LMT crossing time at latitude 15° N for a Sun-synchronous satellite at altitude h = 800 km, when the crossing time at the ascending node is 00:00 LMT.*

We have seen that the inclination of the satellite is $i = 98.6°$ for this altitude. Equation (6.9) yields

$$\Delta\tau = \frac{1}{K} \arcsin\left(\frac{\tan 15}{\tan 98.6}\right) = \frac{1}{15} \arcsin(-0.04052) = \frac{-2.32}{15} \text{ hr} = -9.3 \text{ min} \ .$$

We thus take $\Delta\tau = -9$ min, and inserting $\tau_{\rm AN} = 00{:}00$ in (6.11), this implies that

$$\tau_{(A)} = \tau_{\rm AN} + \Delta\tau = 24 \text{ h } 0 \text{ min } -9 \text{ min} = 23{:}51 \ ,$$

$$\tau_{(D)} = \tau_{\rm AN} + 12 - \Delta\tau = 12 \text{ h } 0 \text{ min } +9 \text{ min} = 12{:}09 \ .$$

The two passages at this latitude thus occur at 23:51 LMT and 12:09 LMT, as can be checked on the upper part of Fig. 6.5.

Example 6.6. *Calculate the LMT crossing time at latitude 50° for the Sun-synchronous satellite SPOT-5, which transits the ascending node at 22:30 LMT.*

For this satellite and latitude 50°, (6.9) gives $\Delta\tau = -42$ min. With (6.11) and $\tau_{\rm AN} = 22{:}30$, we will thus have

$$\phi = 50°\text{N} \quad \longrightarrow \quad 21{:}48 \text{ and } 11{:}12 \ ,$$

$$\phi = 50°\text{S} \quad \longrightarrow \quad 23{:}12 \text{ and } 09{:}48 \ .$$

The daytime crossing will occur, in the northern hemisphere, well after 10:30, in fact, close to midday, with good solar lighting conditions. On the other hand, in the southern hemisphere, the crossing occurs rather early in the morning and the lighting conditions are not so good. The choice of node, e.g., descending at 10:30 rather than ascending) favours observation of the high latitudes of one hemisphere at the expense of the other. We shall return to this point.

6.2.2 Choice of Local Time at the Ascending Node

Restrictions on the Choice of Crossing Time

The local crossing time at the ascending node is determined by the aims of the mission. It is chosen as a compromise between various constraints which we shall number here from C1 to C6 (where C stands for 'constraint'):

(C1) to obtain the best solar lighting conditions for the regions observed,
(C2) to reduce the risks of antisolar or specular reflection,[2]
(C3) to take meteorological factors into account, e.g., a certain region may be under cloud cover every day in the middle of the morning,
(C4) to take into account the crossing time of another Sun-synchronous satellite carrying out the same type of mission,
(C5) to limit periods of solar eclipse,
(C6) to limit thermal variations during each revolution.

We shall now discuss the various times chosen according to the type of mission.

Different Choices Depending on the Constraints

Satellites with High Energy Requirements. It is important to avoid long breaks in the power supply when satellites carry a radar or other instrument with high energy requirements. The solar panels must be almost continuously illuminated. To achieve this, the best-suited orbit has normal in the meridian plane (the normal at the centre of the orbit and the Earth–Sun direction are coplanar), because eclipses are then kept to a minimum (see Sect. 6.3). This Sun-synchronous orbit is such that $\tau_{\mathrm{AN}} = 06{:}00$ or 18:00 and it is called the dawn–dusk orbit.

Radarsat-1 is such a satellite ($\tau_{\mathrm{AN}} = 18{:}00$), as can be seen from Fig. 6.3: the constraint (C5) is given precedence. This orbit has been chosen for the future satellite Radarsat-2 ($\tau_{\mathrm{AN}} = 06{:}00$), for the Indian RISat-1 (Radar Imaging Satellite), and the Argentinian SAOCOM-1A. The same goes for oceanographic satellites using scatterometers, i.e., instruments measuring wind speeds at the sea surface, such as QuikScat ($\tau_{\mathrm{AN}} = 17{:}55$) and Coriolis ($\tau_{\mathrm{AN}} = 18{:}00$). It is also the orbit of the satellite Odin ($\tau_{\mathrm{AN}} = 18{:}00$).

This orbit is planned for the European projects GOCE, at very low altitude ($h \simeq 250$ km), Aeolus-ADM (Atmospheric Dynamics Mission) and WALES, at low altitude ($h \simeq 400$ km), and SMOS ($h = 755$ km). Other planned radar satellites will also be in dawn–dusk orbits: TerraSAR-X1 and TerraSAR-L1, and the COSMO-SkyMed constellation.

Satellites with Orbits Requiring a Specific Configuration Relative to the Sun. Solar observing satellites, if placed near the Earth, must gain

[2] Specular reflection occurs when the normal at the point P, viewed by the satellite, and the two directions P–satellite and P–Sun lie in the same plane to within a few degrees and, in addition, the normal is close to the bisector of these two directions. In this case, the satellite sensor may be blinded by the Sun, with the Earth's surface playing the role of mirror. This kind of reflection can be very efficient, in the case of a calm sea, for example, or quite imperceptible. All intermediate cases are possible, too. Antisolar reflection can occur when the Sun, the satellite and the point P being viewed are collinear. This can only happen between the two tropics.

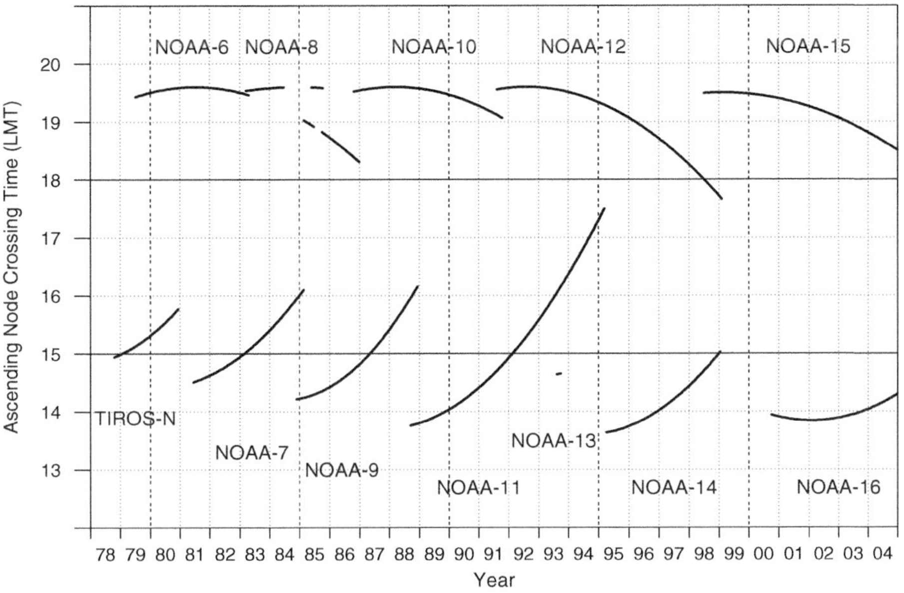

Figure 6.6. Drift of the ascending node crossing time τ_{AN} for Sun-synchronous meteorological satellites in the POES programme. The time τ_{AN} is given for the operating period of each satellite. From NOAA data

maximum advantage of their view of the day star. In its response to the constraint (C5), only the dawn–dusk orbit can allow such continuous observation. The satellite TRACE (τ_{AN} = 06:00) is on this type of orbit, also expected for Picard.

Satellites Subject to Limited Temperature Variation. It is of the utmost importance for satellites carrying out fundamental physics experiments on the equivalence principle that temperature variations should be kept to a minimum. The dawn–dusk orbit satisfies constraint (C6). This will be the orbit for µSCOPE and STEP.

Oceanographic Satellites. When they are not specialised in altimetry, oceanographic satellites are Sun-synchronous. If they do not carry scatterometers, the equatorial crossing is often chosen around midday and midnight, to satisfy constraint (C1): τ_{AN} = 00:00 for Oceansat-1, τ_{AN} = 00:20 for SeaStar, and τ_{N} around midday for Ocean-1 and -2 (also called HY-1 and -2).

Meteorological Satellites. For these satellites which observe meteorological phenomena, the crossing time is not critical. The ascending node crossing times of the various satellites are therefore rather varied, as can be seen from Table 6.1. Moreover, in most cases, these satellites are not kept at their station, the crossing time being allowed to drift. This drift is quadratic in time, as shown by (4.77). For the NOAA satellites, the drift, which can become

Table 6.1. Ascending node crossing time τ_{AN} for various Sun-synchronous satellites. The value of τ_{AN} is that of the first orbits for satellites actually launched or the planned value for satellites still under development. Meteorological satellites. For NOAA satellites, see also Fig. 6.6

Sun-synchronous satellite	τ_{AN}	Sun-synchronous satellite	τ_{AN}
NOAA-2	20:30	Nimbus-6	11:45
NOAA-3	20:30	Nimbus-7	23:50
NOAA-4	20:30	HCMM	14:00
NOAA-5	20:30	DMSP-5D2 F-8	06:15
NOAA-17	22:20	DMSP-5D2 F-10	19:30
FY-1A	15:30	DMSP-5D2 F-11	18:11
FY-1B	19:50	DMSP-5D2 F-12	21:22
FY-1C	18:20	DMSP-5D2 F-13	17:42
FY-1D	20:15	DMSP-5D2 F-14	20:29
FY-3A	21:30	DMSP-5D3 F-15	21:15
FY-3B	21:30	DMSP-5D3 F-16	19:58
Meteor-3M-1	09:15	NPP	22:30
MetOp-1	21:30	NPOESS-1, -4	21:30
MetOp-2	21:30	NPOESS-2, -5	13:30
MetOp-3	21:30	NPOESS-3, -6	17:30

quite significant, is shown in Fig. 6.6. The same goes for the DMSP satellites. For example, for the satellite DMSP-5D2 F-10, the drift was 47 min during 1991.

For the NOAA satellites from TIROS-N and NOAA-6 onwards, the constraint (C4) has been taken into account: for a given region, and with solar illumination, one satellite overflies in the morning and the other in the afternoon.

Satellites for Remote-Sensing of Earth Resources. A satellite may carry instruments pertaining to different types of mission. For example, the Russian satellite Resurs-O1-4 carries the Russian imaging device MSU for remote-sensing and the French instrument ScaRaB to study the Earth radiation budget (which can be classified as meteorological). But it is the remote-sensing aspect that determined the choice of crossing time.

As already mentioned, satellites of this type are Sun-synchronous, with very few exceptions. For satellites devoted to remote sensing of Earth resources, constraints (C1) and (C2) are given priority. The local crossing time at the node must be close to midday for (C1), but not too close because of (C2). Moreover, considering the curve $\phi(\Delta\tau)$, a shift away from midday yields good solar lighting conditions for high latitudes. Mission designers generally consider that the optimal time slot for viewing lasts for three hours centered on noon, i.e., from 10:30 to 13:30 LMT for the crossing at the relevant place, although these limits do not have to be strictly observed.

Table 6.2. Ascending node crossing time τ_{AN} for various Sun-synchronous satellites. The value of τ_{AN} is that of the first orbits for satellites actually launched or the planned value for satellites still under development. Remote-sensing and resource management satellites

Sun-synchronous satellite	τ_{AN}	Sun-synchronous satellite	τ_{AN}
Landsat-1	21:30	IRS-1A	22:25
Landsat-2	21:30	IRS-1B	22:25
Landsat-3	21:30	IRS-1C	22:30
Landsat-4	21:45	IRS-1D	22:30
Landsat-5	21:45	IRS-P2	22:40
Landsat-7	22:00	IRS-P3	22:30
EO-1	22:01	IRS-P6	22:30
SAC-C	22:15	TES	22:30
SPOT-1	22:30	Cartosat-1	22:30
SPOT-2	22:30	Cartosat-2	22:30
SPOT-3	22:15	CBERS-1	22:30
SPOT-4	22:30	CBERS-2	22:30
SPOT-5	22:30	TMSat	22:20
Hélios-1A	13:17	OrbView-3	22:30
Hélios-1B	13:16	OrbView-4	22:30
Pléiades-1	22:15	QuickBird-2	22:20
Pléiades-2	22:15	Ikonos-2	22:30
ERS-1	22:15	EarlyBird-1	22:30
ERS-2	22:30	QuikTOMS	22:30
Envisat	22:00	BIRD	22:30
EarthCARE	22:30	Terra (EOS-AM-1)	22:30
MOS-1	22:25	Aqua (EOS-PM-1)	13:30
MOS-1B	22:30	CloudSat	13:31
JERS-1	22:30	Calipso	13:31
ADEOS-1	22:30	PARASOL	13:33
ADEOS-2	22:30	Aura (EOS-Chem-1)	13:38
ALOS	22:30	OCO	13:15
EROS-A1	21:45	Rocsat-2	21:45
Kompsat-1	22:50	Tan Suo-1	23:00
Resource21-01	22:30	Diamant-1	23:30
Resource21-02	22:30	Diamant-2	23:30
Resurs-O1-4	22:15	RapidEye-1	12:00
TechSat-1B	22:15	NEMO	10:30
FaSat-2	22:20	SSR-1/ss	09:30

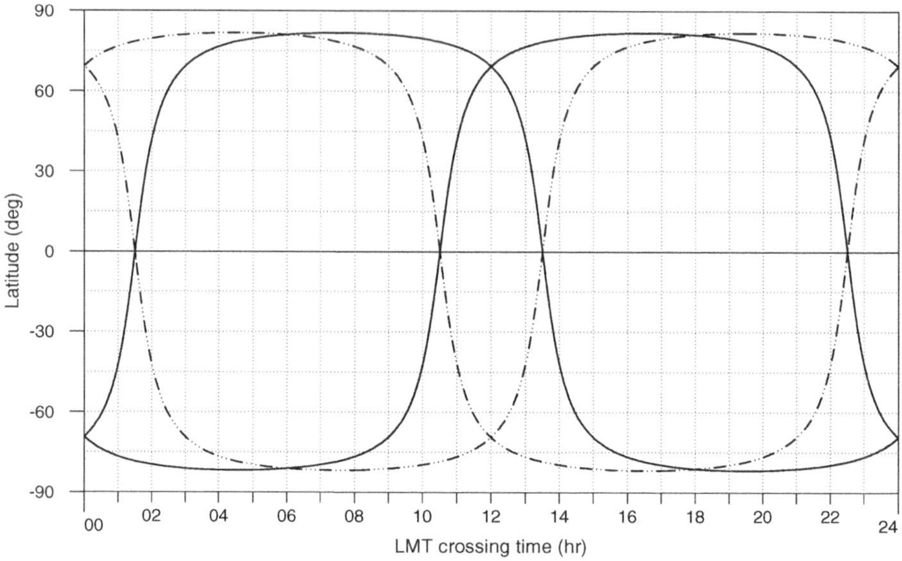

Figure 6.7. Complementarity of Terra and Aqua. LMT crossing time as a function of latitude for the Sun-synchronous EOS satellites. For various values of the LMT ascending node crossing time: 10:30 and 22:30 for EOS-AM-1, 01:30 and 13:30 for EOS-PM-1. The *continuous curve* shows the graph for values corresponding to the crossing time retained in the final project, i.e., 22:30 for EOS-AM-1 and 13:30 for EOS-PM-1 (Terra and Aqua, respectively)

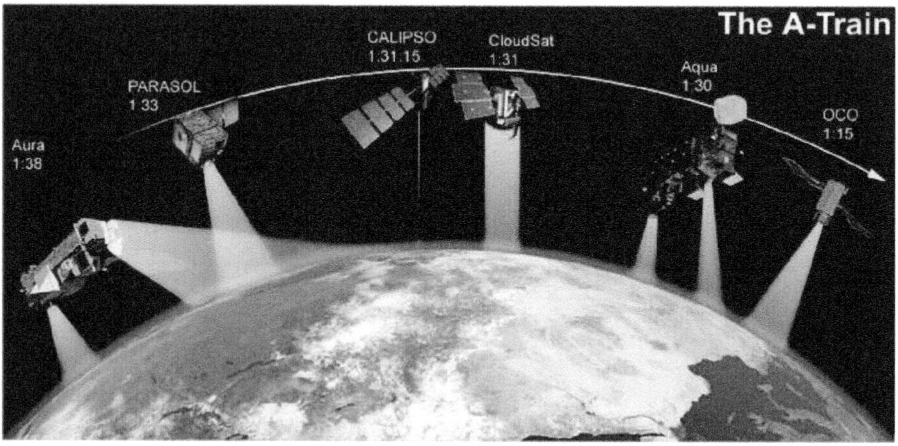

Figure 6.8. A-Train mission spacing (with notation of descending node crossing time). Credit: NASA, ESMO Project

One can thus envisage the following cases, calculated for a satellite at altitude $h = 800$ km:

- Equatorial crossing at the lower time limit. If the ascending node is at 10:30, $\tau_{\text{AN}} = 10{:}30$, latitudes viewed between $\Delta\tau = 0$ and $\Delta\tau = 03{:}00$ are obtained using (6.10). With $K = 15$, the calculation for $\Delta\tau = 3$ gives

$$\phi = \arctan\left[(\tan 98.6) \times (\tan 45)\right] = -78^\circ ,$$

  which corresponds to latitudes lying between 0° (at 10:30) and 78°S (at 13:30). If the descending node occurs at 10:30, $\tau_{\text{AN}} = 22{:}30$, latitudes viewed during this time interval lie between 0° (at 10:30) and 78°N (at 13:30).
- Equatorial crossing at the upper time limit. If the ascending node is at 13:30, $\tau_{\text{AN}} = 13{:}30$, latitudes viewed between $\Delta\tau = 0$ and $\Delta\tau = -3{:}00$ then lie between 78°N (at 10:30) and 0° (at 13:30). If the descending node is at 13:30, $\tau_{\text{AN}} = 01{:}30$, latitudes lie between 78°S (at 10:30) and 0° (at 13:30).
- Equatorial crossing at midday. If the ascending node occurs at 12:00, $\tau_{\text{AN}} = 12{:}00$, latitudes are viewed between $\Delta\tau = -1{:}30$ and $\Delta\tau = 1{:}30$. The calculation for $\Delta\tau = 1.5$ gives

$$\phi = \arctan\left[(\tan 98.6) \times (\tan 22.5)\right] = -68^\circ ,$$

  which corresponds to latitudes lying between 68°N (at 10:30) and 68°S (at 13:30). If the descending node is at 12:00, $\tau_{\text{AN}} = 00{:}00$, latitudes lie between 68°S (at 10:30) and 68°N (at 13:30).
- Choice of time. As the midday crossing time at the node is not chosen, to avoid specular reflection, the choice of the equatorial crossing time at 10:30 or 13:30 is guided by the choice between the northern and southern hemispheres. Naturally, the northern hemisphere is generally favoured, since it encompasses more visible land mass than the other hemisphere, but also because it comprises more nations financing satellite launches.

For satellites observing Earth resources, the choice is between the two equatorial crossing times:

$$\tau_{\text{AN}} = 22{:}30 \quad \Longrightarrow \quad \text{descending node 10:30} ,$$

$$\tau_{\text{AN}} = 13{:}30 \quad \Longrightarrow \quad \text{ascending node 13:30} .$$

The graphs in Fig. 6.7 clearly explain these choices for the satellites EOS-AM-1 and EOS-PM-1 (renamed Terra and Aqua, respectively).

The A-Train refers to the constellation of satellites that plan to fly together with EOS Aqua to enable coordinated science observation. These satellites have an afternoon crossing time close to the local mean time of the lead satellite, Aqua, which is 1:30 p.m. This explains the name: A is short for 'afternoon' and 'Train' is self-explanatory (see Fig. 6.8).

The EROS satellites should form a constellation of six satellites for which the choice of crossing times corresponds to the same strategy. The crossing times retained for this project are τ_{AN} = 22:00, 22:30, and 23:00 for EROS-B1, -B2 and -B3, and τ_{AN} = 13:00, 13:30, and 14:00 for EROS-B4, B-5 and -B6.

The choice between the two possibilities τ_{AN} = 22:30 or τ_{AN} = 13:30 is generally decided in response to the constraint (C3). In this way, one avoids the rather systematic formation of cloud cover at certain times of the day in certain well-defined regions. For example, the descending node was chosen at the end of the morning for the seven satellites in the Landsat series and the five SPOT satellites.

Table 6.2 shows the supremacy of the 22:30 crossing time for the ascending node with this type of satellite.

We may lay stakes that, if Australia sends up a satellite to study Earth resources across its territory, the ascending node will be at 10:30! Remaining for a moment in the southern hemisphere, note that Brazil had a project for a Sun-synchronous satellite, SSR-1 (here called SSR-1/ss), with ascending node at 09:30. This project has been transformed into another, for surveillance of the Amazon, requiring an equatorial orbit, although the satellite will still be called SSR-1.

The crossing times of remote-sensing satellites are generally maintained quite accurately, to within a few minutes.

Other Types of Mission. Other types of mission not mentioned above use Sun-synchronous orbits. Here are a few examples of ascending node crossing times: τ_{AN} = 12:00 for TOMS-EP, τ_{AN} = 14:00 for ARGOS, τ_{AN} = 22:50 for ACRIMSAT. Note that τ_{AN} = 08:40 was planned for TERRIERS.

6.3 Appendix: Duration of Solar Eclipse

The satellite undergoes solar eclipse when the Sun is hidden from it by the Earth. During the eclipse, the satellite cools down and its solar panels no longer produce electricity. For some satellites, an eclipse is a critical phenomenon, and in this case, the Earth–Sun–satellite geometry is examined in detail. We shall discuss here two types of orbit: dawn–dusk LEO and GEO.

6.3.1 Dawn–Dusk LEO Orbit

Consider a Sun-synchronous satellite in low circular orbit. If the LMT crossing time at the equator is around midday and midnight, the satellite is illuminated by the Sun for roughly a little more than half the period. The rest of the time, it moves in the shadow of the Earth.

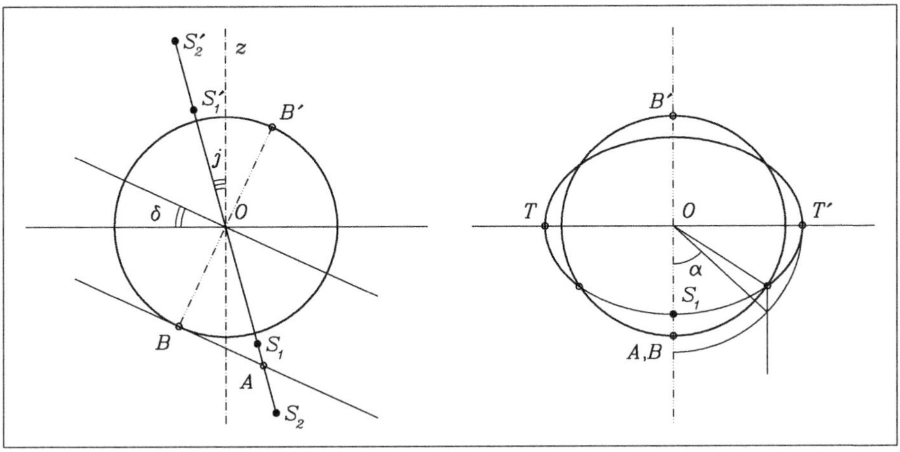

Figure 6.9. Schematic diagram of the Earth and the orbit of a Sun-synchronous satellite in a dawn–dusk configuration. *Left*: Meridian plane. Intersection of orbit with this plane: S_i and S_i'. *Right*: Plane perpendicular to the meridian plane and perpendicular to the direction of the Sun. The projection of the circular orbit on this plane is an ellipse

On the other hand, if the equatorial crossing times are around 06:00 and 18:00, the satellite is rarely in the Earth's shadow. This Sun-synchronous LEO orbit, with $\tau_{\rm AN} = 06{:}00$ or 18:00 is called a dawn–dusk orbit, as we have seen. In this configuration, which limits the length of the eclipse, one finds satellites that cannot tolerate long breaks in their power supply, or are sensitive to the sudden temperature change between day and night.

Eclipse Conditions

A Sun-synchronous satellite at altitude h (reduced distance η) has inclination $i = i_{\rm HS}$ given by (4.69). We set

$$j = i_{\rm HS} - \frac{\pi}{2} \,. \tag{6.13}$$

Figure 6.9 (left) shows the Earth (polar axis Oz, radius R) in the meridian plane containing the Sun (hour angle zero). Light rays from the Sun make an angle δ (declination) with the equatorial plane. The satellite orbit, which is perpendicular to the meridian plane because it is a dawn–dusk orbit, cuts this plane at S_i and S_i'. One of the two points is illuminated, e.g., S_i', and so is the other if it is not in the Earth's shadow, i.e., if $OS_i > OA$, where the point A is the intersection of the edge of the Earth's shadow with the plane of the orbit, in the meridian plane (the plane of the figure). In the example given in the figure, if the satellite is in position S_1, it undergoes solar eclipse, whereas if it is at S_2, there is no eclipse.

We have immediately

$$OS = R + h = a \;, \qquad OA = \frac{R}{\cos(\delta + j)} \;.$$

The condition for there to be no eclipse is therefore

$$K > H \quad \text{(for a given declination)} \tag{6.14}$$

$$\text{with:} \quad H = 1/\eta \;, \qquad K = \cos(\delta + j) \;.$$

The strongest constraint obtains at the two solstices, with $|\delta| = \varepsilon = 23.44°$. In these conditions, when η is varied between 1 and 1.9367, the maximal value for a Sun-synchronous satellite, given by (4.72), the condition ($K > H$) is satisfied when η lies between 1.2181 and 1.5221. Using the altitude, we obtain

$$\text{no eclipse} \quad \Longleftrightarrow \quad 1\,391 < h < 3\,330 \text{ km} \;.$$

If the altitude of the satellite is less than 1 391 km, there is eclipse, because the satellite is not high enough to escape from the Earth's shadow (at least, at the solstice). If the altitude is greater than 3 330 km, the orbit is close enough to the equatorial plane (i tends to 180°) and the ecliptic to mean that, despite its high altitude, the satellite moves into the shadow.

These observations are rather theoretical. In practice, most satellites in dawn–dusk orbit are equipped with radar – with the constraint (C5) considered earlier – and an altitude less than 800 km is thus the norm. The eclipse phenomenon is then inevitable at some point during the year.

Calculating the Duration of Eclipse

We calculate the duration of eclipse when the satellite has altitude less than the limiting value $h = 1\,391$ km. Looking along the direction of the Sun's rays, the Earth appears as a circle $(\mathcal{C}_1)$ of radius R and the dawn–dusk orbit appears as an ellipse $(\mathcal{C}_2)$ with semi-major axis a, the actual radius of the circular orbit, and semi-minor axis b, the projection of a on a plane perpendicular to the direction of the Sun. Figure 6.9 (right) shows, for $(\mathcal{C}_1)$, $R = OB$, and for $(\mathcal{C}_2)$, $a = OT$, $b = OS_1$, so that

$$a = R + h \;, \qquad b = a\cos(\delta + j) = aK \;.$$

With the axes (O, x, y), the equations defining curves $(\mathcal{C}_1)$ and $(\mathcal{C}_2)$ can be written

$$(\mathcal{C}_1) : \quad x^2 + y^2 = R^2 \;, \qquad (\mathcal{C}_2) : \quad x^2 + \frac{y^2}{K^2} = a^2 \;.$$

We calculate the intersection (x_1, y_1) of these two curves, which yields

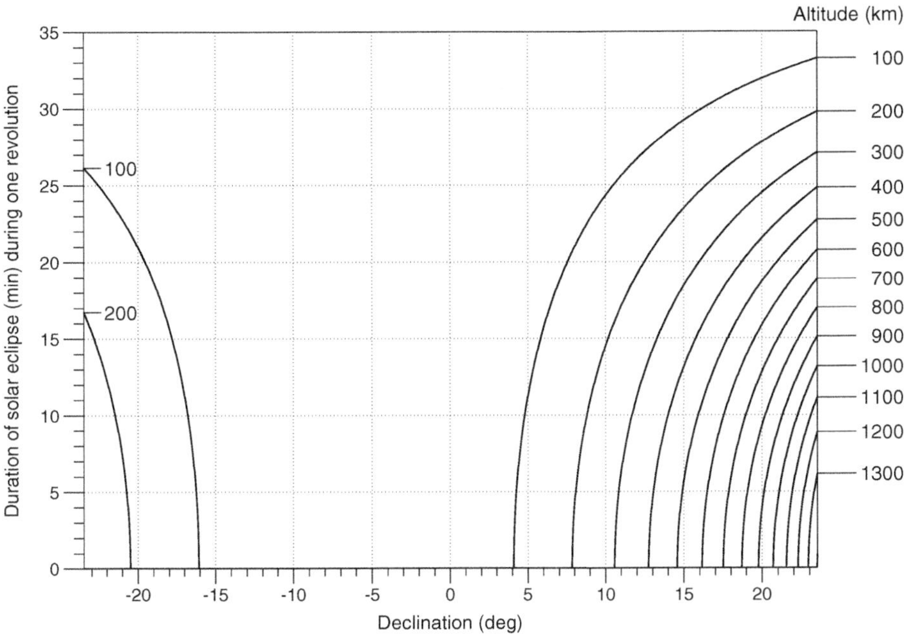

Figure 6.10. Sun-synchronous satellite in dawn–dusk orbit. Duration of solar eclipse in minutes during one revolution, for the altitude indicated, as a function of the declination. Graphs are drawn for $\tau_{\mathrm{AN}} = $ 18:00. For $\tau_{\mathrm{AN}} = $ 06:00, take the opposite value of the declination

$$x_1^2 = \frac{1 - \eta^2 K^2}{1 - K^2} R^2 \,.$$

Rotating the orbital plane onto the plane of the figure, we obtain the actual value of the angle α which determines the duration of the eclipse (see Fig. 6.9, right). Hence,

$$\sin\alpha = \frac{x_1}{a} = \sqrt{\frac{H^2 - K^2}{1 - K^2}} \,. \tag{6.15}$$

The duration Δt_{e} of the eclipse is

$$\Delta t_{\mathrm{e}} = \frac{\alpha}{\pi} T_0 \,, \tag{6.16}$$

since the orbit is circular, with uniform motion of period T_0.

In the case $K > H$, there is no eclipse, as explained above, and we put $\alpha = 0$.

Using the altitude h, the angle i_{HS} and the value of the period $T_{0(h=0)}$ defined by (2.17), we have (time in minutes, angles in radians), for declination δ,

$$\Delta t_e \text{ (min)} = 84.5 \left(1 + \frac{h}{R}\right)^{3/2} \frac{1}{\pi} \arcsin \frac{\sqrt{[R/(R+h)]^2 - \sin^2(\delta + i_{HS})}}{|\cos(\delta + i_{HS})|} . \tag{6.17}$$

Figure 6.10 plots representative graphs of the duration of solar eclipse over one revolution for various altitudes, as a function of the declination. For easier understanding, Fig. 6.11 (upper) shows the same as a function of the day of the year.

Ascending Node Crossing Time and Dates of Eclipse

Figure 6.9 (left) shows the situation in a northern summer ($\delta > 0$, Sun at the zenith in the northern hemisphere) with a satellite orbit crossing the ascending node at 18:00 (taking into account the direction of rotation of the Earth). The maximal eclipse occurs at the summer solstice when the satellite passes close to the South Pole. The most favorable situation with regard to the question of eclipse, even at very low orbit, occurs for $\delta = -j$, i.e., during the northern winter, when the direction of the Sun is exactly perpendicular to the orbit.

For a satellite crossing the ascending node at 06:00, the straight line $S_i S_i'$ occupies a symmetric position with respect to the polar axis Oz. All the above calculations remain the same, provided that we take the opposite value of the declination. For example, in this case, the value $\delta = -23.44°$ corresponds to the northern summer solstice.

Example 6.7. *Calculate the eclipse dates and duration of eclipse for Radarsat-1 and SMOS.*

Radarsat-1 has a near-circular Sun-synchronous dawn–dusk orbit with characteristics: $a = 7\,167.064$ km, $i_{HS} = 98.58°$, $T_d = 100.76$ min, $\tau_{AN} = $ 18:00. Using the above notation, we obtain

$$\eta = 1.1237 , \quad H = 0.8899 , \quad j = 8.58° .$$

To find the date of eclipse, we apply (6.14). With $\arccos(0.8899) = 27.14°$, we obtain

$$\delta + j = \pm 27.14 , \quad \text{hence} \quad \delta = \pm 27.14 - 8.58 .$$

The solution $\delta = 18.56°$ is the only possible value, since for the other, $|\delta| > \varepsilon$. As we have $\tau_{AN} = $ 18:00, the dates are given directly by the values of δ. In the northern summer, there is eclipse for days when the declination is greater than 18.56°, i.e., in the interval from 15 May to 20 July. To calculate the duration of the longest eclipse, at the summer solstice, we use (6.15). With $K = \cos(\varepsilon + j) = \cos(32.02) = 0.8479$, we obtain $\sin \alpha = 0.5096$, whence $\alpha = 30.6°$. Then, with (6.16),

$$\Delta t_e = 0.170 T_0 \approx 17 \text{ min} .$$

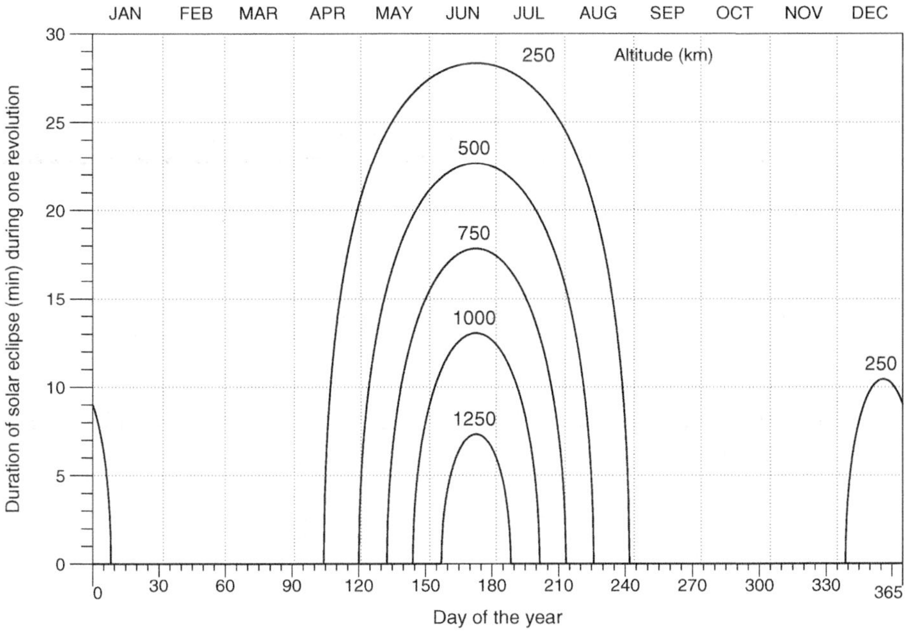

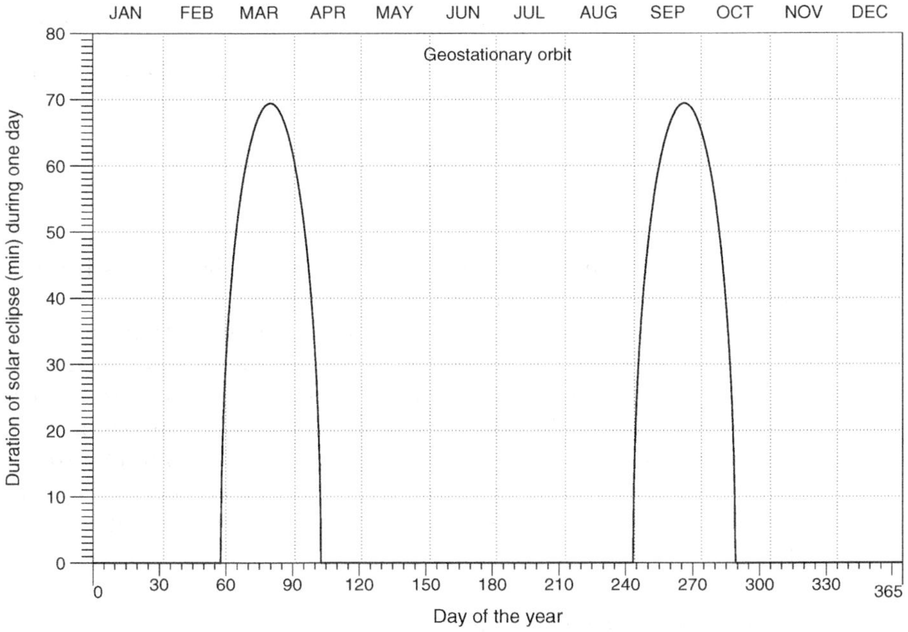

Figure 6.11. Duration of solar eclipse in minutes as a function of the day of the year. *Upper*: Sun-synchronous satellite in dawn–dusk orbit. Duration of the eclipse during one revolution at the given altitude. Graphs are drawn for $\tau_{AN} = 18{:}00$. For $\tau_{AN} = 06{:}00$, take the opposite value of the declination. *Lower*: geostationary satellite. Duration of eclipse over one day

We can also obtain these results directly using (6.17).

The satellite SMOS should fly at an altitude of 755 km, so that $i_{HS} = 98.44°$. We deduce that $H = 0.8942$ and $j = 8.44°$. For the eclipse dates, we find once again that there is only one possible value for the declination, viz., $\delta = 26.60 - 8.44 = 18.16$. If we choose $\tau_{AN} = 18{:}00$, eclipse will occur for declinations greater than $+18.16°$, or in the interval between 13 May and 31 July, around the summer solstice. If we choose $\tau_{AN} = 06{:}00$, eclipse will occur for declinations less than $-18.16°$, that is, in the interval from 15 November to 28 January, around the winter solstice. Concerning the duration of eclipse at the solstice, the calculation gives $\Delta t_e = 18$ min per revolution.

Example 6.8. *Constraints imposed by eclipse on the satellites STEP and GOCE, in very low orbit.*

For STEP, the requirements of temperature stability forbid any period of solar eclipse during its operating time. Moreover, the orbit must be low, the altitude being fixed at 400 km. A Sun-synchronous dawn–dusk orbit is the only suitable orbit. With $i_{HS} = 97.05°$ and $\eta = 1.0627$, we obtain $j = 7.05°$ and $H = 0.9410$ and hence, $\delta + j = \pm 19.78$. We thus have just one value for the declination, viz., $\delta = 19.78 - 7.05 = 12.73°$. Depending on the value of τ_{AN}, this corresponds to the interval 25 April to 21 August or the interval 28 October to 15 February. Note that, in the first case, the eclipse lasts for 118 days, whilst in the second case, it lasts for 110 days, a consequence of the eccentricity of the Earth's orbit. Finally, with this orbit, there is a period of 8 months without eclipse. The accelerometers of the STEP experiment are maintained at a temperature of 2 K, using a superfluid helium cryostat, which limits the time over which the experiment can operate to around 6 months. The satellite must be launched shortly after 21 August, if $\tau_{AN} = 18{:}00$ is chosen, or shortly after 15 February, if $\tau_{AN} = 06{:}00$ is chosen.

GOCE flies at the very low altitude $h = 250$ km. With $i_{HS} = 96.52°$ and $\eta = 1.0392$, we obtain $j = 6.52°$ and $H = 0.9623$, whence $\delta + j = \pm 15.79$ and $\delta = \pm 15.79 - 6.52$. There are now two values of the declination:

$$\delta = -22.31° \;, \quad \delta = +9.27° \;.$$

We can say that there are two eclipse 'seasons', one short, with $|\delta|$ close to ε, the other long, as can be seen very clearly in Fig. 6.11 (upper). The mission is planned to last 20 months (limited by the amount of fuel needed to compensate for atmospheric drag) and the satellite should be launched at the end of the long eclipse season. There are therefore only two possible launch windows: either in July, taking $\tau_{AN} = 18{:}00$, or in January, taking $\tau_{AN} = 06{:}00$.

6.3.2 GEO Orbit

For a geostationary satellite, no shadow is cast by the Earth on the circular orbit as long as the direction of the Sun has an inclination (declination δ), with respect to the equatorial plane, greater than the angle with which the satellite views the Earth. Let f_0 be this angle, which is the half-angle at the apex of the observation cone with which the satellite views the Earth, and to which we shall return in Chap. 8 [see (8.24)]. With η_{GS} defined by (4.58), the relation $\sin f_0 = 1/\eta_{GS}$ gives $f_0 = 8.7°$.

There is therefore an eclipse if $|\delta| < f_0$. This happens twice a year, around the equinoxes:

$$\text{eclipse for GEO} \quad \Longleftrightarrow \quad [27\ \text{Feb–12 Apr}]\ , \quad [01\ \text{Sep–16 Oct}]\ .$$

During these periods, each lasting 45 days (from $J = 58$ to $J = 102$ and from $J = 244$ to $J = 289$, although dates may vary by one day from year to year), the longest eclipse occurs at the equinox itself. On this day, it lasts Δt_{e0} given by

$$\Delta t_{e0} = \frac{f_0}{\pi} T_0 = \frac{8.7}{180} J_{sid} = 69.5\ \text{min} \approx 1\ \text{hr}\ 10\ \text{min}\ . \tag{6.18}$$

On the other days, the duration of the eclipse is found by considering the Earth's disk, viewed by the satellite, occulting the Sun. The 'ground track' of the Sun cuts the disk along parallel chords, passing through the centre of the disk for $\delta = f_0$. This gives, for the duration Δt_e of the eclipse as a function of δ,

$$\Delta t_e = \sqrt{1 - \left(\frac{\delta}{f_0}\right)^2}\, \Delta t_{e0}\ . \tag{6.19}$$

The value of Δt_e as a function of the day of the year is shown in Fig. 6.11 (lower).

7 Orbit Relative to the Earth. Recurrence and Altitude

In this chapter, we discuss the position of the satellite orbit relative to the Earth. There are two distinct parts. The first concerns the position of the satellite ground track relative to the Earth, and the second the altitude of the satellite measured from the terrestrial ellipsoid.

7.1 The Recurrence Constraint

7.1.1 Definition of Recurrence

In observation missions using non-geosynchronous satellites, one often requires repeated coverage of the Earth, in the sense that the satellite must periodically overfly the same points of the Earth's surface. This means that, for a given point, one is sure to recover geometrically identical observing conditions with this periodicity. This recurrence constraint on the mission imposes specific characteristics on the orbital elements.

The recurrence period after which the satellite ground track repeats itself exactly on the Earth's surface is called the recurrence cycle, but also the repetitivity or repeat cycle.[1] This corresponds to the cycle relative to the Earth and we denote its value by C_{T}, where the subscript T stands for 'terrestrial', just as we denoted the cycle relative to the Sun by C_{S} in the last chapter.

When recurrence has been achieved, the ground track of the satellite forms a fixed grid with respect to the Earth, which covers the globe between the highest attained latitudes. One point on the grid (the ascending node is generally chosen) fixes the whole thing. We shall study this recurrence grid below.

The calculational methods developed here to analyse recurrence remain valid even if the satellite has an eccentric orbit. However, in practice, all

[1] The term 'resonance' is sometimes found. In the present context, this is inappropriate. Indeed, in mechanics, resonance is a phenomenon in which a mutual interaction modifies the motion of each body involved, e.g., Laplace resonance for the Galilean moons of Jupiter. When we consider an artificial satellite in orbit around the Earth, the motion of the Earth is not affected by the motion of the satellite.

recurrent LEO satellites are in near-circular orbits. On the other hand, recurrence is not restricted to LEO satellites, but is used for MEO satellites and HEO communications satellites.

Daily Recurrence Frequency

The angle precession, the Euler angle ψ, plays a key role in the study of recurrence. Using (4.27) and (5.14), we obtain

$$\dot{\psi} = -(\dot{\Omega}_\mathrm{T} - \dot{\Omega}) = -\frac{2\pi}{J_\mathrm{M}}\left(1 + \frac{1-P}{N_\mathrm{sid}}\right) , \tag{7.1}$$

$$\dot{\psi} = -\frac{n}{\kappa} , \tag{7.2}$$

with the notation N_sid, J_M, P and κ defined in Chap. 4.

The daily recurrence frequency defined by (4.32), viz.,

$$\kappa = \frac{\nu}{1 + \dfrac{1-P}{N_\mathrm{sid}}} , \tag{7.3}$$

allows one to compare the Earth's rotation, the satellite motion, and its nodal precession via the angular speeds. It is close to ν, the daily orbital frequency, but it is not the same, except for Sun-synchronous satellites. Indeed only in the latter case do we have $\kappa = \nu$, since $P = 1$.

7.1.2 Calculating the Recurrence Cycle C_T

Consider a non-geosynchronous satellite and the intersection of its ascending ground track with the equator, which defines an ascending node of longitude λ_0. If the satellite is recurrent, its ground track will pass precisely through this point λ_0 on the equator, C_T days later. The satellite will have made a whole number of round trips between these two crossings. The number of round trips is denoted by $N_{\mathrm{T}_\mathrm{o}}$, whereas C_T is an arbitrary real number, a priori non-integer. For the rest of this chapter, we shall attach the subscript 'o' to whole numbers (integers) entering our calculations.

From the above discussion, we obtain the following relation which gives the length L of the time interval between the two crossings at the same ascending node λ_0 :

$$L = C_\mathrm{T} J_\mathrm{M} = N_{\mathrm{T}_\mathrm{o}} T_\mathrm{d} . \tag{7.4}$$

During this time L, the plane of the orbit makes a whole number of round trips, denoted by k_o, relative to the frame $\Re_\mathrm{T}$, since the ground track returns exactly to an earlier position. This yields

$$L(\dot{\Omega}_{\mathrm{T}} - \dot{\Omega}) = 2\pi k_{\mathrm{o}} \,. \tag{7.5}$$

Equations (7.1), (7.2) and (7.4) now imply

$$N_{\mathrm{T_o}} T_{\mathrm{d}} \frac{n}{\kappa} = 2\pi k_{\mathrm{o}} \,,$$

which gives (since $T_{\mathrm{d}} = 2\pi/n$)

$$\kappa = \frac{N_{\mathrm{T_o}}}{k_{\mathrm{o}}} \,.$$

This relation shows that, for a recurrent satellite, the parameter κ which we have called the daily recurrence frequency is a rational number:

$$\text{recurrent satellite} \quad \Longleftrightarrow \quad \kappa \text{ rational} \,. \tag{7.6}$$

In terms of the daily orbital frequency ν, we have

$$C_{\mathrm{T}} J_{\mathrm{M}} = N_{\mathrm{T_o}} \frac{J_{\mathrm{M}}}{\nu} \,,$$

which implies for the value of the cycle C_{T},

$$C_{\mathrm{T}} = \frac{N_{\mathrm{T_o}}}{\nu} \,. \tag{7.7}$$

The whole number k_{o} defined above, which represents a whole number of days, will be denoted by $C_{\mathrm{T_o}}$. Hence,

$$C_{\mathrm{T_o}} = \frac{N_{\mathrm{T_o}}}{\kappa} \,. \tag{7.8}$$

In the general case, we distinguish the recurrence cycle C_{T} from the integer recurrence cycle $C_{\mathrm{T_o}}$. In the special case of Sun-synchronous satellites (and it should be noted that this type of satellite covers most cases of recurrence), C_{T} and $C_{\mathrm{T_o}}$ coincide, since in this case $\kappa = \nu$. This means that, for a recurrent satellite:

- if it is Sun-synchronous, its ground track always returns to the same point at the same time, i.e., at the end of a whole number of days, and C_{T} is an integer;
- if it is not Sun-synchronous, its ground track returns to a given point at different times, and C_{T} is not an integer.

Connection with the Cycle Relative to the Sun

The cycle C_{S} relative to the Sun and the cycle C_{T} relative to the Earth both depend on the orbital characteristics, but not in a one-to-one manner. The orbital parameters of a satellite can be varied in such a way that, for example, C_{S} remains constant whilst C_{T} takes any value we wish.

However, a useful relation can be brought out concerning the difference $C_{\mathrm{T}} - C_{\mathrm{T_o}}$ and the cycle C_{S}. From the definitions of C_{S}, C_{T}, $C_{\mathrm{T_o}}$, P and κ, we may write

$$\frac{\nu}{\kappa} = 1 - \frac{1}{C_{\mathrm{S}}} ,$$

$$\frac{C_{\mathrm{T}} - C_{\mathrm{T_o}}}{C_{\mathrm{T}}} = 1 - \frac{C_{\mathrm{T_o}}}{C_{\mathrm{T}}} = 1 - \frac{\nu}{\kappa} = \frac{1}{C_{\mathrm{S}}} .$$

We thus obtain C_{T}, given $C_{\mathrm{T_o}}$ and C_{S}:

$$C_{\mathrm{T}} = \frac{C_{\mathrm{T_o}}}{1 - 1/C_{\mathrm{S}}} . \tag{7.9}$$

For a Sun-synchronous satellite, $C_{\mathrm{T}} = C_{\mathrm{T_o}}$, since C_{S} is infinite.

7.1.3 Recurrence Triple

The rational number κ, the daily recurrence frequency, can thus be expressed in the form

$$\kappa = \frac{N_{\mathrm{T_o}}}{C_{\mathrm{T_o}}} . \tag{7.10}$$

It can be written as the sum of an integer and a positive or negative fractional part with modulus less than 1/2:

$$\kappa = \nu_{\mathrm{o}} + \frac{D_{\mathrm{T_o}}}{C_{\mathrm{T_o}}} . \tag{7.11}$$

In this expression, ν_{o} is the whole number closest to κ and $D_{\mathrm{T_o}}$ the unique integer such that

$$D_{\mathrm{T_o}} = N_{\mathrm{T_o}} - \nu_{\mathrm{o}} C_{\mathrm{T_o}} . \tag{7.12}$$

Hence,

$$\begin{cases} |D_{\mathrm{T_o}}| < \dfrac{1}{2} C_{\mathrm{T_o}} , \\ |D_{\mathrm{T_o}}| \text{ and } C_{\mathrm{T_o}} \text{ coprime} . \end{cases}$$

We shall call the triple of numbers ν_{o}, $D_{\mathrm{T_o}}$ and $C_{\mathrm{T_o}}$ the recurrence triple of the satellite, written

$$[\nu_{\mathrm{o}}, D_{\mathrm{T_o}}, C_{\mathrm{T_o}}] .$$

The recurrence of a satellite orbit can thus be defined equivalently via the recurrence triple or the pair of whole numbers $N_{\mathrm{T_o}}$, $C_{\mathrm{T_o}}$. The value of κ obtained in this way from (7.10) or from (7.11) thus yields ν via (7.3), and hence the period or mean motion, after an iterative calculation on P. The period in minutes is given as a function of $N_{\mathrm{T_o}}$, $C_{\mathrm{T_o}}$ and P by

$$T_{\mathrm{d}}\ (\mathrm{min}) = 1440 \frac{C_{\mathrm{T_o}}}{N_{\mathrm{T_o}}} \left(1 + \frac{1-P}{N_{\mathrm{sid}}} \right) . \tag{7.13}$$

This iterative calculation involves making a first estimate of T, which gives a, and using a with i to obtain P. This in turn gives a new value of T. This iteration converges rapidly to give a final value for the period.

We provide example calculations below. Naturally, all these calculations are much simpler for a Sun-synchronous satellite, since $P = 1$. It is for this reason that we separate the following discussion into two parts, depending on whether the satellite is Sun-synchronous or not.

7.2 Recurrence for a Sun-Synchronous Satellite

7.2.1 Method for Obtaining Recurrence

We have seen that the altitude of a Sun-synchronous satellite in near-circular orbit lies between the theoretical bounds $h = 0$ and $h = 5\,964$ km, which corresponds to values of the daily orbital frequency of $\nu = 17.03$ and $\nu = 6.34$, respectively. In current practice, when h is situated between 400 and 1 000 km, ν varies between 15.5 and 13.8 round trips per day.

For a Sun-synchronous satellite, it is a simple matter to obtain recurrence conditions since $\nu = \kappa$. The daily orbital frequency $\nu = \nu(a)$, which only depends on a here, since i and a are related, is a rational number which can be written in the form

$$\nu = \nu_{\mathrm{o}} + \frac{D_{\mathrm{T_o}}}{C_{\mathrm{T_o}}} . \tag{7.14}$$

The satellite ground track repeats every $C_{\mathrm{T_o}}$ days, after $N_{\mathrm{T_o}} = \nu C_{\mathrm{T_o}}$ revolutions.

7.2.2 Recurrence Diagram

The recurrence diagram is designed as an aid to visualising the altitudes leading to different recurrence situations. It is basically a graph in which the altitudes, from lowest to highest, are marked on the ordinate axis and the recurrence cycles (in days) on the abscissa.

For each value of ν_{o}, for each cycle $C_{\mathrm{T_o}}$, the quantity $D_{\mathrm{T_o}}$ is varied over its range of possible values and ν is obtained from (7.14). This in turn gives the mean motion n and draconitic period $T = T_{\text{d}}$. We thus obtain the altitude and inclination, calculating a and i by an iterative method, as in Example 4.2, or Examples 7.1 and 7.2 below.

The value obtained is then marked on the diagram. In Fig. 7.1, these values are indicated by small squares. In Figs. 7.2 and 7.3, the value of D_{T_o} is noted explicitly in each case. The small dots on these diagrams indicate a recurrence for which $|D_{\text{T}_o}|$ and C_{T_o} are not coprime. Strictly speaking, this is not a recurrence with cycle C_{T_o}. For example, for a given altitude, if the recurrence is over three days, the satellite is sure to repeat its ground track every 6 days, every 9 days, and so on. But the cycle considered is 3 days. In its relationship with the prime numbers, this diagram is reminiscent of the sieve of Eratosthenes.[2]

The diagram gives an overview of the possibilities for recurrence. For short cycles, we see that these possibilities are limited to a handful of values. Between 450 and 1 000 km, there is only one possible altitude for a 2 day recurrence cycle ($h = 720$ km for Oceansat-1), and only three possible altitudes for a 3 day cycle. On the other hand, for long cycles, there are many more opportunities, especially if C_{T_o} is a prime number. For $C_{\text{T}_o} = 31$, there are about 120 possibilities between 0 and 1 200 km, or roughly one available altitude for recurrence every 10 km.

---

[2] Eratosthenes of Cyrene (284–192 BC), ὁ Ἐρατ οσθένης,ους, was a Greek astronomer, mathematician and geographer. He discovered a systematic method for obtaining the sequence of prime numbers up to any desired value. One writes down the sequence of positive integers, then crosses out the multiples of 2, of 3, of 5, and so on. This method, which sifts the positive integers, keeping only the primes, is known as the sieve of Eratosthenes. His abilities as an astronomer and geographer are revealed by a scientific and relatively accurate measurement of the Earth's radius, in which he measured the shadow cast by a column in Alexandria at midday on a day when he knew that the Sun's rays reached the bottom of the wells in ancient Syene (Assouan), the day of the summer solstice at Syene, under the Tropic of Cancer. He determined the obliquity of the ecliptic and estimated at 47°42′ the arc of the meridian between the two tropics.

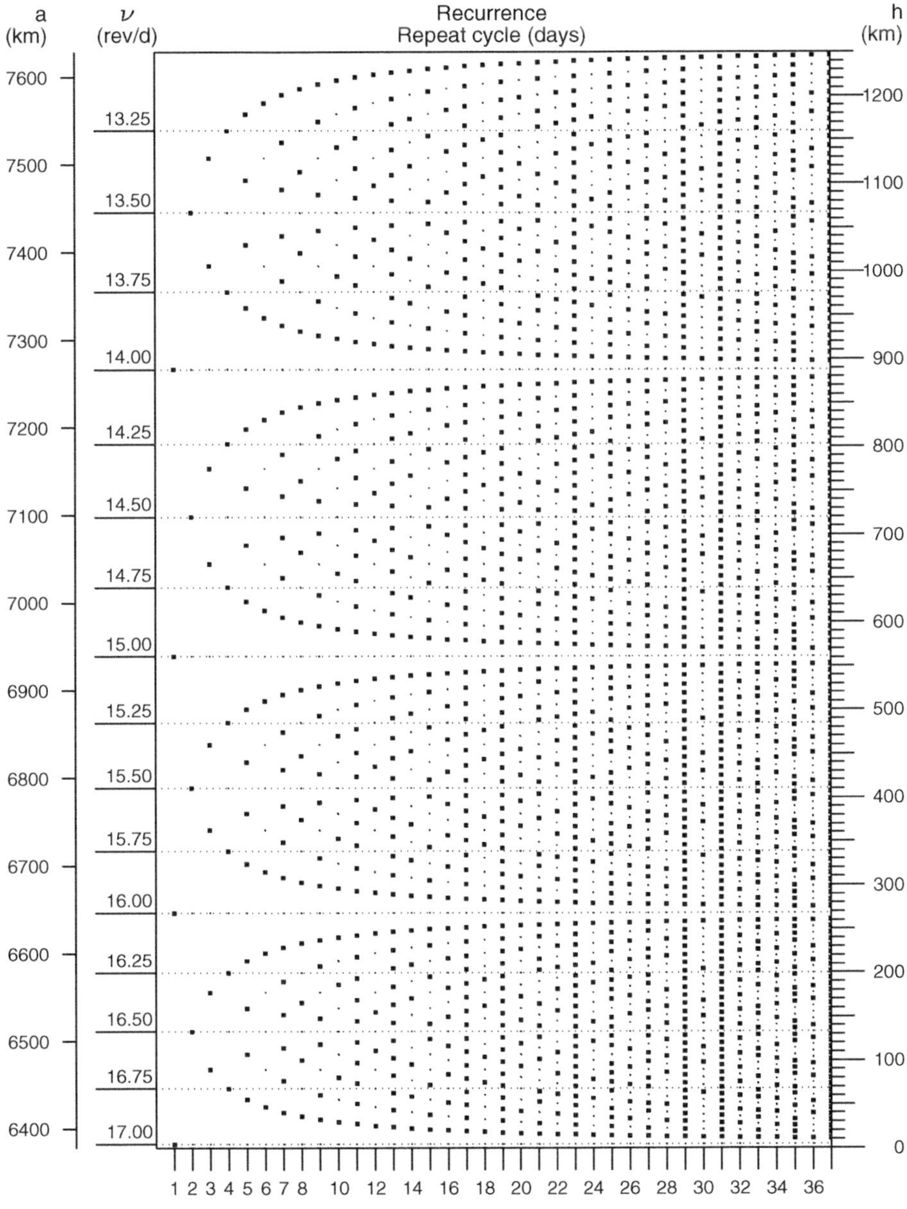

Figure 7.1. Recurrence diagram for Sun-synchronous satellites. For altitudes between 0 and 1 250 km, the *small squares* denote values of the altitude h (and the semi-major axis a) for which recurrence is possible. *Abscissa*: value of the recurrence cycle in days. The daily frequency ν is in round trips per day

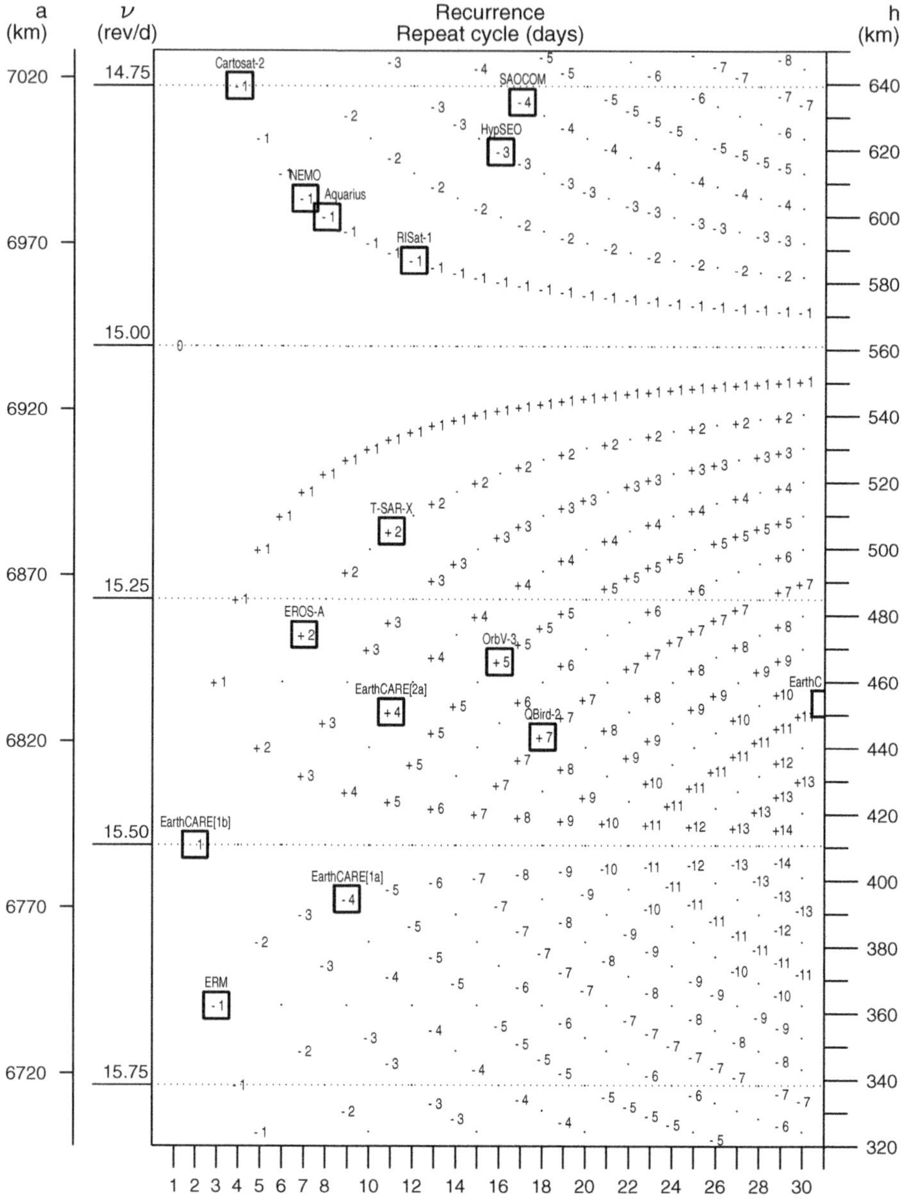

Figure 7.2. Recurrence diagram for Sun-synchronous satellites. For altitudes between 320 and 650 km, the possible recurrences are indicated by the value of D_{T_o}. *Boxed values* correspond to the satellites appearing in Table 7.1. For example, for OrbView-3, we find the triple $[15, +5, 16]$, i.e., $\nu_o = 15$ (the integer closest to ν, as ordinate), $D_{T_o} = +5$ (indicated on the diagram) and $C_{T_o} = 16$ (abscissa)

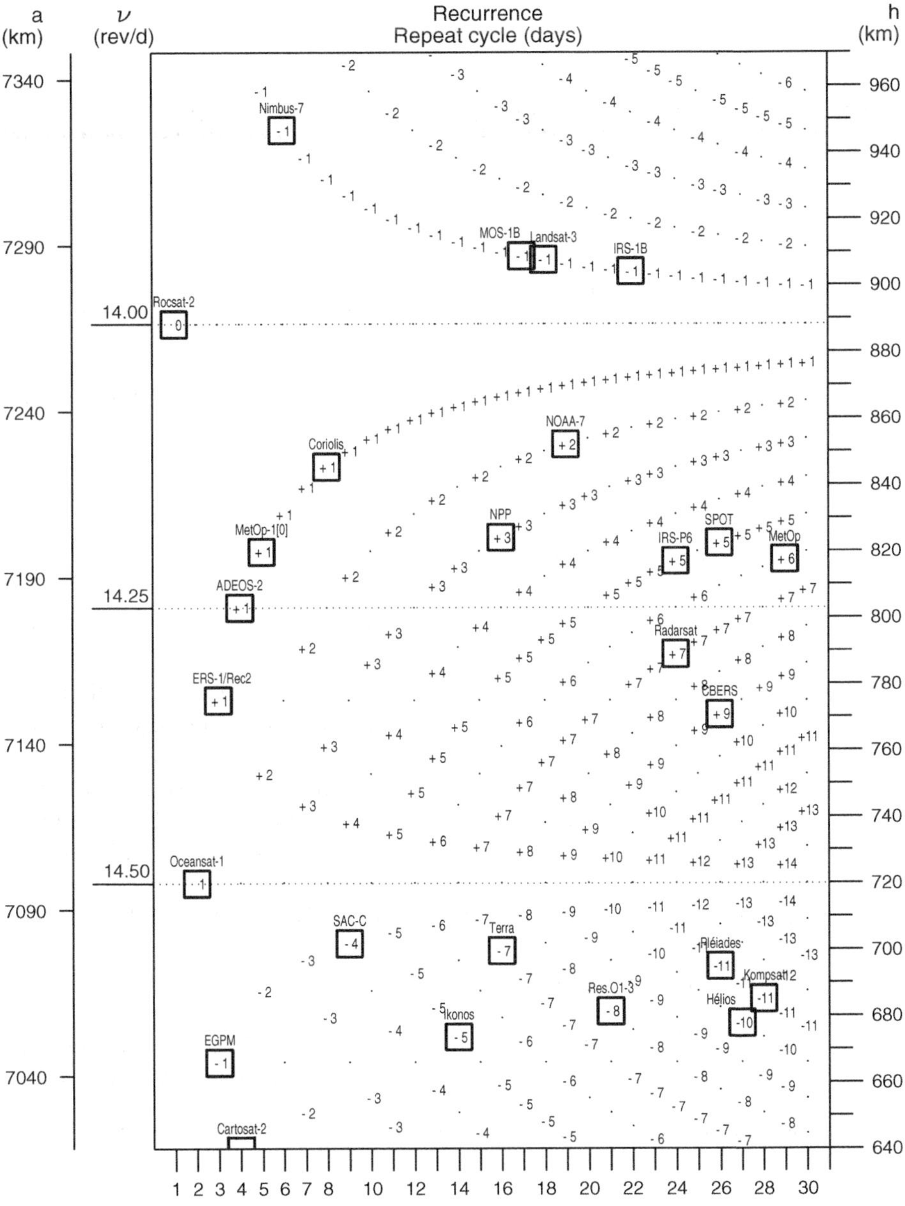

Figure 7.3. Recurrence diagram for Sun-synchronous satellites. For altitudes between 640 and 970 km, the possible recurrences are indicated by the value of $D_{\mathrm{T_o}}$. *Boxed values* correspond to the satellites appearing in Table 7.1. For example, for SPOT-5, we find the triple [14, +5, 26], i.e., $\nu_{\mathrm{o}} = 14$ (the integer closest to ν, as ordinate), $D_{\mathrm{T_o}} = +5$ (indicated on the diagram) and $C_{\mathrm{T_o}} = 26$ (abscissa)

Sun-syn. sat.	ν_o	D_{T_o}	C_{T_o}	N_{T_o}	T_d	a	h	i_{HS}
⋆ Landsat-3	14	−1	18	251	103.27	7 285.799	908	99.09
⋆ Terra	15	−7	16	233	98.88	7 077.738	700	98.21
⋆ NOAA-7	14	+2	19	268	102.09	7 230.200	852	98.85
Nimbus-7	14	−1	6	83	104.10	7 324.842	947	99.27
⋆ HCMM	15	−3	16	237	97.22	6 997.706	620	97.89
OrbView-3	15	+5	16	245	94.04	6 844.207	466	97.30
⋆ Ikonos-2	15	−5	14	205	98.34	7 051.765	674	98.11
QuickBird-2	15	+7	18	277	93.57	6 821.490	443	97.21
Coriolis	14	+1	8	113	101.89	7 223.450	845	98.82
Aquarius	15	−1	8	119	96.81	6 978.050	600	97.81
NEMO	15	−1	7	104	96.92	6 983.652	606	97.83
NPP	14	+3	16	227	101.50	7 202.173	824	98.73
⋆ SPOT-5	14	+5	26	369	101.46	7 200.546	822	98.72
⋆ Hélios-1B	15	−10	27	395	98.43	7 056.025	678	98.12
⋆ Pléiades-1	15	−11	26	379	98.79	7 073.059	695	98.19
⋆ Envisat	14	+11	35	501	100.60	7 159.496	781	98.55
ERS-1 (Rec. 2)	14	+1	3	43	100.46	7 153.138	775	98.52
ERS-1 (Rec. 3)	14	+59	168	2 411	100.34	7 147.192	769	98.50
⋆ MetOp-1	14	+6	29	412	101.36	7 195.606	817	98.70
MetOp-1 [0]	14	+1	5	71	101.41	7 197.940	820	98.71
TerraSAR-X1	15	+2	11	167	94.85	6 883.512	505	97.45
EGPM	15	−1	3	44	98.18	7 044.115	666	98.07
⋆ MOS-1B	14	−1	17	237	103.29	7 286.941	909	99.10
JERS-1	15	−1	44	659	96.15	6 946.179	568	97.69
ADEOS-1	14	+11	41	585	100.92	7 174.906	797	98.61
⋆ ADEOS-2	14	+1	4	57	101.05	7 181.058	803	98.64
ALOS	15	−19	46	671	98.66	7 069.809	692	98.18
⋆ IRS-1B	14	−1	22	307	103.19	7 282.277	904	99.08
⋆ IRS-P6	14	+5	24	341	101.35	7 195.119	817	98.70
Oceansat-1	14	+1	2	29	99.31	7 098.105	720	98.29
RISat-1	15	−1	12	179	96.54	6 965.021	587	97.76
Cartosat-1	15	−21	116	1 719	97.17	6 995.667	618	97.88
Cartosat-2	15	−1	4	59	97.63	7 017.502	639	97.97
Resurs-O1-3	15	−8	21	307	98.50	7 059.437	681	98.14
⋆ Radarsat-1	14	+7	24	343	100.76	7 167.064	789	98.58
⋆ CBERS-2	14	+9	26	373	100.38	7 148.868	771	98.50
SAC-C	15	−4	9	131	98.93	7 079.991	702	98.22
SAOCOM-1A	15	−4	17	251	97.53	7 012.831	635	97.95
Kompsat-1	15	−11	28	409	98.58	7 063.280	685	98.15
EROS-A1	15	+2	7	107	94.21	6 852.218	474	97.33
Rocsat-2	14	0	1	14	102.74	7 266.473	888	99.00

Table 7.1. Opposite page. Orbital characteristics of Sun-synchronous satellites obtained from the recurrence triple $[\nu_o, D_{T_o}, C_{T_o}]$, where ν_o is the whole number closest to the number of revolutions per day, D_{T_o} is the whole number equal to $N_{T_o} - \nu_o C_{T_o}$, C_{T_o} is the recurrence cycle (whole number of days), N_{T_o} is the number of revolutions per cycle. The numbers in the triple give the draconitic period T_d (min) and hence the orbital characteristics: the semi-major axis a (km), the altitude h (km) obtained from $h = a - R$, and the Sun-synchronous inclination i_{HS} (degrees). For satellites marked $\star$, see Table 7.3

Table 7.2. As in Table 7.1, but for the ERM then EarthCARE project. The altitude increases as the launch is delayed, since we are moving from the years of minimal solar activity towards the years of maximal solar activity

Sun-syn. sat.	ν_o	D_{T_o}	C_{T_o}	N_{T_o}	T_d	a	h	i_{HS}
ERM	16	−1	3	47	91.91	6 740.440	362	96.92
EarthCARE [1a]	16	−4	9	140	92.57	6 772.571	394	97.03
EarthCARE [1b]	15	+1	2	31	92.90	6 788.780	411	97.09
EarthCARE [2a]	15	+4	11	169	93.73	6 828.979	451	97.24
EarthCARE [2b]	15	+11	31	476	93.78	6 831.593	453	97.25

Table 7.3. List of orbits common to different Sun-synchronous satellites. The B satellites are on the same orbit as the corresponding A satellites

Satellite A	Satellite B	Satellite A	Satellite B
Landsat-3	Landsat-1, -2	SPOT-5	SPOT-1, -2, -3, -4
Terra	Landsat-4, -5, -7	Hélios-1B	Hélios-1A
Terra	SeaStar, EO-1	Pléiades-1	Pléiades-2
Terra	Aqua, Aura	Envisat	ERS-1, ERS-2
Terra	CloudSat, Calipso	MetOp-1	MetOp-2, -3
Terra	PARASOL, OCO	MOS-1B	MOS-1
NOAA-7	NOAA-6	ADEOS-2	QuikScat
HCMM	COSMO-SkyMed	IRS-1B	IRS-1A
HCMM	HypSEO	IRS-P6	IRS-1C, -1D
Ikonos-2	LSPIM	IRS-P6	IRS-P2, -P3
Radarsat-1	Radarsat-2	CBERS-2	CBERS-1

7.2.3 Recurrence Defined by the Recurrence Triple

Recurrent Sun-synchronous satellites are defined by the recurrence triple. All Sun-synchronous satellites with the same recurrence triple have the same period, and hence the same values of a and i. (They also share the same value of e, because the orbit is frozen, as we shall see at the end of the chapter.) The recurrence triple thus defines the orbit. We may speak of the SPOT orbit for all satellites having the same recurrence triple as SPOT-1.

Table 7.1 shows the recurrence triple for various satellites, some no longer in service, such as Landsat-3, some still operating, such as Terra, and some under development, such as Aquarius. The recurrence originally planned for MetOp-1, but later changed, is indicated by [0]. Regarding the EarthCARE project, there are several candidates for recurrence, given separately in Table 7.2. We have also mentioned the recurrence intended for several projects such as ERM and LSPIM which have now been abandoned.

Figures 7.2 and 7.3 give the recurrence diagram over the range of altitudes h from 320 to 970 km and cycles $C_{\mathrm{T_o}}$ from 1 to 30 days. Recurrence possibilities are indicated in each case with $D_{\mathrm{T_o}}$, $C_{\mathrm{T_o}}$ and ν_{o}, deduced from ν. Values in use by satellites in Tables 7.1 and 7.2 are boxed.

Note that the draconitic period in minutes is given by

$$T_{\mathrm{d}}\ (\mathrm{min}) = 1440 \frac{C_{\mathrm{T_o}}}{N_{\mathrm{T_o}}} , \tag{7.15}$$

which is the adaptation of (7.13) to Sun-synchronous satellites.

Example 7.1. *Calculate the characteristics of the SPOT orbit.*

Concerning the orbit of the SPOT satellites, the recurrence triple is [14, 5, 26]. We observe that 5 and 26 are coprime and that 5 is less than 13. For strict recurrence, the draconitic period is held at its exact value given by

$$T_{\mathrm{d}}\ (\mathrm{min}) = 1440 \frac{26}{369} = 101.463 .$$

To begin with, following the same lines as Example 4.2, we set $T_{\mathrm{O}} = T_{\mathrm{d}}$, which gives the corresponding Keplerian values:

$$a_{\mathrm{O}} = 7\,206.1\ \mathrm{km} , \quad i_{\mathrm{O}} = i_{\mathrm{HS}} = 98.7^\circ .$$

Using a_{O} and i_{O}, we then calculate the relative values of the secular variations:

$$\frac{\Delta n}{n_{\mathrm{O}}} = -0.593 \times 10^{-3} , \quad \frac{\dot{\omega}}{n_{\mathrm{O}}} = -0.564 \times 10^{-3} ,$$

which yields a new value a_1 for the semi-major axis:

$$\frac{\Delta a}{a_{\mathrm{O}}} = -\frac{2}{3} \times 1.157 \times 10^{-3} , \quad \Delta a = -5.6\ \mathrm{km} ,$$

$$a_1 = a_{\mathrm{O}} + \Delta a = 7\,200.5\ \mathrm{km} .$$

The iterative calculation (one step is actually enough) delivers the semi-major axis and the inclination as

$$a = 7\,200.546\,4\ \mathrm{km} , \quad i = i_{\mathrm{HS}} = 98.723^\circ .$$

Values given in the CNES documentation concerning SPOT are

$$a = 7\,200.547 \text{ km} , \quad i = i_{\text{HS}} = 98.723° ,$$

implying an altitude of $h = a - R = 822$ km.

If the secular variations are calculated with the expansion cut off at the term in J_2, this yields the following values after iteration:

$$a' = 7\,200.537 \text{ km} , \quad i = i_{\text{HS}} = 98.70° .$$

Comparing the various values found for the semi-major axis, we thus observe that:

- the term $a_0 = 7\,206.1$ km is obtained from the central term of the Newtonian acceleration (degree 0 of the potential),
- the difference $|a' - a_0| = 5.6$ km arises from the terms of degree 2 (the J_2 term) in the geopotential,
- the difference $|a - a'| = 10$ m arises from the terms of degree 4 (the J_2^2 and J_4 terms) and higher order in the geopotential.

Concerning the relative orders of magnitude of the values found for the semi-major axis, viz.,

$$|a - a_0| \sim 10^{-3} a_0 , \quad |a - a'| \sim 10^{-3} |a - a_0| ,$$

we obtain the same values as in Chap. 3 when comparing orders of magnitude of J_2 and 1, then J_4 and J_2.

These theoretical values can also be compared with the actual values as obtained from the NORAD elements. We choose three dates, the second and third being separated by 26 days and one year, respectively, from the first:

```
SPOT 5
1 27421U 02021A   03040.18015505  .00000155  00000-0  93359-4 0  9661
2 27421  98.7244 116.8304 0000554  58.9354 301.1883 14.20029420 39902
SPOT 5
1 27421U 02021A   03066.18009415  .00000138  00000-0  85210-4 0   469
2 27421  98.7212 142.4627 0000619  93.4275 266.6981 14.20038040 43590
SPOT 5
1 27421U 02021A   04040.23558731  .00000015  00000-0  27564-4 0  1157
2 27421  98.7426 116.1385 0000814  76.9405 283.1877 14.20014756 91717
```

The orbital elements obtained are as follows:

- First set, 9 February 2003, revolution 3 990: $a = 7\,200.542$ km, $e = 5.54 \times 10^{-5}$, $i = 98.7244°$, $\omega = 58.9354°$, $\lambda_{\text{AN}} = 273.297°$, $\tau_{\text{AN}} = 22{:}33$.
- Second set, 6 March 2003, revolution 4 359: $a = 7\,200.513$ km, $e = 6.19 \times 10^{-5}$, $i = 98.7212°$, $\omega = 93.4275°$, $\lambda_{\text{AN}} = 273.324°$, $\tau_{\text{AN}} = 22{:}33$.
- Third set, 9 February 2004, revolution 9 171: $a = 7\,200.593$ km, $e = 8.14 \times 10^{-5}$, $i = 98.7426°$, $\omega = 76.9405°$, $\lambda_{\text{AN}} = 252.888°$, $\tau_{\text{AN}} = 22{:}32$.

Between the first two sets, separated by 369 revolutions, we note a discrepancy in λ_{AN} of 0.027°, or 3.0 km, less than the tolerated maximum (5 km) for recurrence. The elements a and i remain very close to the theoretical values. However, e is much lower than the value needed to freeze the orbit (see the end of the chapter). Consequently, there is a large variation in ω around 90°. The crossing time τ_{AN} is assured to within 2 minutes, which is well below the variations in solar time due to the equation of time.

Example 7.2. *Calculate the characteristics of the Terra orbit.*

For the orbit of satellites in the Terra family (see Table 7.3), the recurrence triple is $[15, -7, 16]$. The draconitic period is maintained at

$$T_{\rm d}\ (\text{min}) = 1440\frac{16}{233} = 98.884\ .$$

As before, we obtain

$$a_0 = 7\,083.4\ \text{km}\ , \quad i = i_{\rm HS} = 98.2°\ .$$

We then calculate the relative values of the secular variations:

$$\frac{\Delta n}{n_0} = -0.619 \times 10^{-3}\ , \quad \frac{\dot{\omega}}{n_0} = -0.591 \times 10^{-3}\ ,$$

which gives the new value of the semi-major axis as $a_1 = 7077.7$ km. By iteration,

$$a = 7\,077.737\,8\ \text{km}\ , \quad i = i_{\rm HS} = 98.211°\ .$$

We have seen that, although the value of a can be accurately determined, the same cannot be said of the altitude, which varies due to the ellipticity of the orbit and the Earth. (This question will be tackled at the end of the chapter.) If we define the altitude h as the difference between a and the equatorial radius R, the value of h is well-determined. In the case of Terra, we obtain $h = 700$ km. The NASA documentation gives the value $h = 705$ km for all satellites in the Terra orbit, the altitude having been defined in a slightly different way.
The Argentinian satellite SAC-C (*Satélite de Aplicaciones Científicas*) was launched with EO-1. The latter, coupled with Landsat-7, is on the Terra orbit. The satellite SAC-C was placed on an orbit 2.2 km higher, leading to a recurrence cycle of 9 days. The value of ν for this satellite is $\nu = 15 - 4/9 = 14.5555$, whilst for EO-1, we have $\nu = 15 - 7/16 = 14.5625$.

Example 7.3. *Calculate the orbital characteristics of the satellite ERS-1 which has had three different recurrence cycles: 35 days, 3 days and 168 days.*

The ESA satellite ERS-1 was launched on 17 July 1991. It flies at an altitude of about 780 km and its orbit is Sun-synchronous. It initially had a recurrence cycle of 35 days. From 20 December 1993, a slight change of orbit brought the cycle to 3 days, better suited to the study of ice flows during the arctic winter, according to the mission controllers. A further maneuver brought the cycle to 168 days so that it could carry out geodetic observations. The satellite ERS-2, launched on 21 April 1995, carried out a mission in tandem with ERS-1 until 10 March 2000.
Let us now calculate the orbital characteristics for each recurrence condition:

- **Recurrence cycle 1.** 35 day cycle over 501 revolutions:

$$\nu_1 = \frac{501}{35} = 14 + \frac{11}{35} \quad \Longrightarrow \quad \text{recurrence triple } [14, 11, 35] \ .$$

From T_d, we calculate a and i by the above method. The results are given in Table 7.1: ERS-1 (Rec. 1) as Envisat, ERS-1 (Rec. 2) and ERS-1 (Rec. 3) (see Fig. 5.26, upper).
- **Recurrence cycle 2.** 3 day cycle over 43 revolutions:

$$\nu_2 = \frac{43}{3} = 14 + \frac{1}{3} \quad \Longrightarrow \quad \text{recurrence triple } [14, 1, 3] \ .$$

The change of cycle was brought about by going from ν_1 to ν_2, and this in turn was done by reducing the altitude by 6.358 km (and the inclination by 0.027°).
- **Recurrence cycle 3.** 168 day cycle over 2411 revolutions:

$$\nu_3 = \frac{2411}{168} = 14 + \frac{59}{168} \quad \Longrightarrow \quad \text{recurrence triple } [14, 59, 168] \ .$$

The change of cycle was brought about by going from ν_2 to ν_3, and this in turn was done by reducing the altitude by 5.948 km (and the inclination by 0.025°).

Very small variations in altitude lead to totally different recurrence cycles, as can be seen very clearly from Fig. 7.1. These variations in altitude are small enough to make them easy to calculate (without carrying out the above type of iteration), applying the relation for finite increments already used:

$$\frac{\Delta\nu}{\nu} = -\frac{3}{2}\frac{\Delta a}{a} \ .$$

Setting

$$a_{\text{rec1}} = 7\,159.496 \text{ km} \ ,$$

we obtain

$$\Delta\nu = \nu_2 - \nu_1 = \frac{1}{3} - \frac{11}{35} = \frac{2}{3 \times 35} \ , \qquad \frac{\Delta\nu}{\nu_1} = \frac{2}{3 \times 35} \times \frac{35}{501} = \frac{2}{3 \times 501} \ ,$$

$$\Delta a_{\text{rec1}\to\text{rec2}} = -6.358 \text{ km} \ ,$$

$$\Delta\nu = \nu_3 - \nu_2 = \frac{59}{168} - \frac{1}{3} = \frac{1}{56} \ , \qquad \frac{\Delta\nu}{\nu_2} = \frac{1}{56} \times \frac{3}{43} = \frac{3}{56 \times 43} \ ,$$

$$\Delta a_{\text{rec2}\to\text{rec3}} = -5.946 \text{ km} \ .$$

Note that a recurrence condition based on such a long time scale as $C_{T_o} = 168$ days is quite exceptional. Values of C_{T_o} do not generally exceed 45 days.

When the orbit of a Sun-synchronous satellite is defined by its recurrence cycle C_{T_o} and its approximate altitude, one determines the recurrence triple, then returns to the previous analysis. From the altitude, one can calculate ν, which gives ν_o. Once the cycle C_{T_o} is known, one can deduce N_{T_o} by an iterative method, and then find D_{T_o}.

Table 7.4. Recurrence triple $[\nu_o, D_{T_o}, C_{T_o}]$ and number of revolutions N_{T_o} per cycle for various non-Sun-synchronous satellites, giving the draconitic period T_d (min) and orbital characteristics: semi-major axis a, altitude h (km), and inclination i_{HS} (degrees). ⋆ T/P is the common orbit of TOPEX/Poseidon, Jason-1 and -2. For ICESat and CryoSat, we have indicated the recurrence conditions for the mission orbit and for the calibration orbit [cal]

Non-Sun-sync.	ν_o	D_{T_o}	C_{T_o}	N_{T_o}	T_d	a	h	i
Seasat	14	+8	25	358	100.85	7 173.367	795	108.00
Geosat	14	+6	17	244	100.62	7 162.520	784	108.00
⋆ T/P	13	−3	10	127	112.42	7 714.428	1 336	66.04
ICESat	15	−22	183	2 723	96.65	6 970.030	592	94.00
ICESat [cal]	15	−1	8	119	96.68	6 971.522	593	94.00
CryoSat	14	+178	369	5 344	99.25	7 094.555	716	92.00
CryoSat [cal]	14	+1	2	29	99.10	7 087.812	710	92.00

7.2.4 One-Day Recurrence Cycle

For LEO satellites (at least, for those which observe the Earth), one-day recurrence cycles are avoided. Indeed, if such a satellite views exactly the same regions every day, many other places must necessarily be missed out, unless the instruments have a very wide field of view. Those altitudes between 0 and 1250 km giving a one-day recurrence cycle appear in the recurrence diagram of Fig. 7.1 and are also indicated in Table 7.5. The corresponding recurrence triples are of the form $[\nu_o, 0, 1]$.

To our knowledge, the only LEO satellite with a one-day recurrence cycle is Taiwan's Rocsat-2. The triple $[14, 0, 1]$ induces a recurrence grid that has been put to very judicious use in connection with the geographic position of Taiwan. We shall return to this point later, when discussing recurrence grids.

7.3 Recurrence for a Non-Sun-Synchronous LEO Satellite

7.3.1 Obtaining the Recurrence Triple

For a non-Sun-synchronous satellite, recurrence is defined by the recurrence cycle C_{T_o}, in a whole number of days, differing here from the cycle C_T. To determine the recurrence triple, the altitude and inclination must also be known, with a certain interval of possible variation.

The daily recurrence frequency is given by (7.3), where $\nu = \nu(a, i)$ and $P = P(a, i)$. We calculate the product (κC_{T_o}), allowing a and i to vary until we obtain a whole number, which then represents N_{T_o}. With this value, we obtain ν_o, the integer closest to κ, and by $N_{T_o} = \nu_o C_{T_o} + D_{T_o}$, we obtain

$D_{\mathrm{T_o}}$. This gives the recurrence triple $[\nu_o, D_{\mathrm{T_o}}, C_{\mathrm{T_o}}]$ and the value of C_{T}. Recall that C_{T} and $C_{\mathrm{T_o}}$ are given by (7.7) and (7.8).

Table 7.4 gives the main recurrent, non-Sun-synchronous LEO satellites. These are oceanographic satellites, except for ICESat and CryoSat. The satellite ICESat, which repeats over a very long period (2 723 revolutions in 183 days), carries just one instrument, a laser altimeter. This is why it is important that the satellite does not repeat its ground track before half a year.

Example 7.4. *Calculate the recurrence triple and exact altitude for the satellite TOPEX/Poseidon, which has a 10 day cycle and inclination $i = 66.04°$. Its altitude is around 1 335 km.*

The French–US satellite TOPEX/Poseidon comprises the US platform TOPEX (Topography Experiment for Ocean Circulation) and the French altimetric instrument Poseidon, equipped with the DORIS system (*Détermination d'Orbite et Radiopositionnement Intégrés par Satellite*). Poseidon is used to calculate the orbit to within a radial accuracy of 2 cm. Consequently the mean altitude of the oceans can be measured to the same accuracy. The TOPEX/Poseidon experiment is being continued by Jason-1, which uses exactly the same orbit. (The T/P orbit is shifted by a half-interval on the ground track.)
To obtain the recurrence characteristics, treating the inclination as fixed at $i = i_0 = 66.04°$, we vary h, allowing a margin of 10 km on either side of the central value, doing the calculations with a. With the usual units, we obtain:

$$h = 1\,325 \Longrightarrow P = -2.12\ ,\ \nu = 12.838\ ,\ \kappa = 12.728 \Longrightarrow \kappa C_{\mathrm{T_o}} = 127.276\ ,$$
$$h = 1\,335 \Longrightarrow P = -2.11\ ,\ \nu = 12.811\ ,\ \kappa = 12.703 \Longrightarrow \kappa C_{\mathrm{T_o}} = 127.032\ ,$$
$$h = 1\,345 \Longrightarrow P = -2.10\ ,\ \nu = 12.786\ ,\ \kappa = 12.679 \Longrightarrow \kappa C_{\mathrm{T_o}} = 126.788\ .$$

By interpolation, we seek the value of a leading to the integer value $\kappa C_{\mathrm{T_o}} = 127$. With a further iteration, we obtain the result:

$$h = 1\,336.297 \quad \Longrightarrow \quad P = -2.106898\ ,$$

$$\nu = 12.8080\ , \quad \kappa = 12.7000 \quad \Longrightarrow \quad \kappa C_{\mathrm{T_o}} = 127.00000\ .$$

Finally, for this satellite, with $C_{\mathrm{T_o}} = 10$, we thus have

$$a = 7\,714.434\ \mathrm{km}\ , \quad i = 66.04°\ ,$$

$$T = 1440 \frac{10}{270} \left(1 + \frac{1-P}{N_{\mathrm{sid}}}\right) = 1440 \frac{10}{270} \left(1 + \frac{3.1068}{365.25}\right) = 112.4295\ \mathrm{min}\ ,$$

$$T_{\mathrm{d}} = 112.4295\ \mathrm{min}\ , \qquad T_{\mathrm{a}} = 112.4185\ \mathrm{min}\ ,$$

and the recurrence triple is

$$\kappa = \frac{127}{10} = 13 + \frac{-3}{10} \quad \Longrightarrow \quad [13, -3, 10]\ .$$

The reference values communicated by the TOPEX/Poseidon team are

$$a = 7\,714.429\,38 \text{ km} , \quad i = 66.0408^\circ .$$

Every 127 revolutions, the satellite repeats its ground track precisely. However, the time elapsed between two passages is not exactly 10 days, but 9.92 days. Indeed,

$$C_\text{T} = \frac{127}{\nu} = \frac{127}{12.8080} = 9.9156 \text{ day} = 9 \text{ d } 21 \text{ h } 58 \text{ m } 27 \text{ s} \approx 9 \text{ d } 22 \text{ h} .$$

In 10 days, the satellite thus gains 2 h 01 m 33 s = 2.043 h on its local crossing time. After 117.45 days, it has gained $117.45 \times 0.2043 = 24.00$, or a whole day, as we have seen in Example 6.1, where $C_\text{S} = -117.45$ corresponds to the cycle relative to the Sun. The relation between C_{T_o} and C_T can be obtained directly from (7.9). Let us now compare these theoretical values with those obtained using NORAD data:

```
TOPEX
1 22076U 92052A   04068.34164556 -.00000038  00000-0  10000-3 0  6642
2 22076  66.0420 112.9091 0008368 272.0846  87.9192 12.80929020541452
JASON
1 26997U 01055A   04067.94661617 -.00000061  00000-0  00000-0 0  3432
2 26997  66.0416 113.4703 0008092 269.2700  90.7389 12.80928744105199
```

For T/P, on 8 March 2004 at 08:12 UT, with $\lambda_\text{AN} = 183.880^\circ$, 20:27 LMT,

$$a = 7\,714.431 \text{ km} .$$

For Jason-1, on 7 March 2004 at 22:43 UT, with $\lambda_\text{AN} = 326.652^\circ$, 20:30 LMT,

$$a = 7\,714.433 \text{ km} .$$

7.3.2 Recurrence, Altitude, and Inclination

For a non-Sun-synchronous satellite, a given recurrence condition leaves a certain freedom to vary the altitude and inclination. For concreteness, consider a polar satellite, with $i = 90^\circ$ and recurrence $[14, -1, 17]$. Calculations then give its altitude as $h = 894.9$ km. With $i = 80^\circ$, the satellite must be brought down to an altitude $h = 880.9$ km in order to obtain the same recurrence triple, whereas with $i = 100^\circ$, it must be moved up to an altitude $h = 910.2$ km. We see that, in the neighbourhood of 90°, a change of one degree in the inclination implies a change of 1.5 km in the altitude, in order to maintain the same recurrence conditions.

For each triple, we can calculate the value of h as a function of i, whilst the inclination varies from 0 to 180°. The graphs $h(i)$ are shown in Fig. 7.4 (upper) for all triples $[14, \pm 1, C_{\text{T}_\text{o}}]$, where C_{T_o} varies between 5 and 17, and also for the triple $[14, 0, 1]$. A more restricted range of variations is also shown in the lower part of Fig. 7.4.

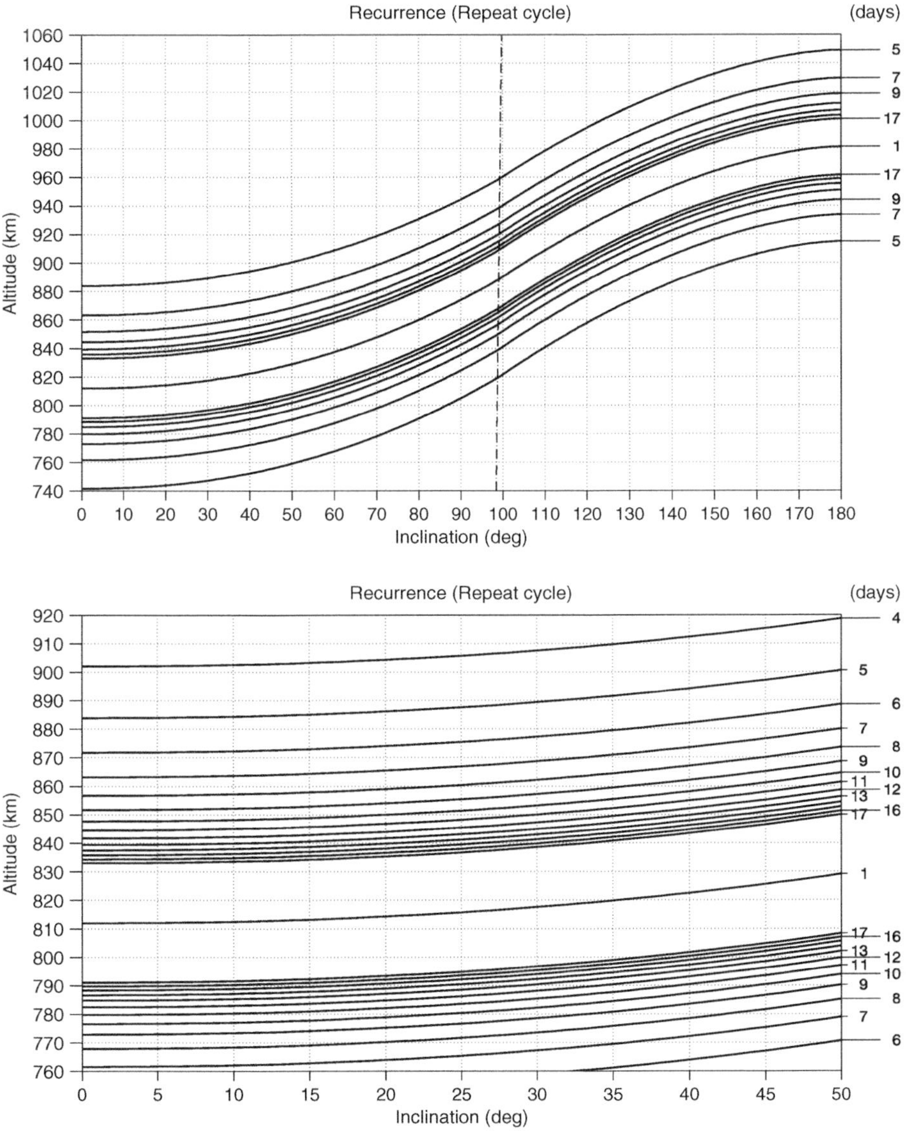

Figure 7.4. Altitude as a function of inclination for a satellite holding the same recurrence. *Upper*: Recurrence triples $[14, \pm 1, C_{\mathrm{T_o}}]$, for $C_{\mathrm{T_o}}$ varying between 5 and 17 (in steps of 2), and $[14, 0, 1]$. Values corresponding to $D_{\mathrm{T_o}} = -1$ are above the median, representing $[14, 0, 1]$, while those corresponding to $D_{\mathrm{T_o}} = +1$ are below it. The *dashed curve* denotes the Sun-synchronous inclination as a function of altitude. *Lower*: Detail from the upper diagram for a restricted range of inclinations, with $C_{\mathrm{T_o}}$ varying between 4 and 17

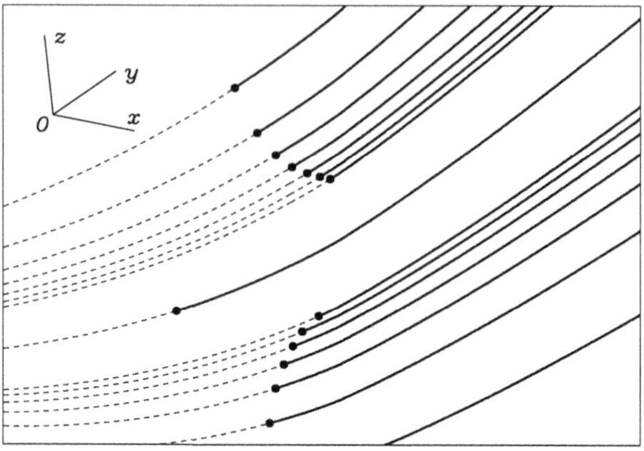

Figure 7.5. Schematic view of the recurrence space. Locus of points at which recurrence occurs. *Quantities along axes*: Ox for the inclination i, Oy for the recurrence cycle C_{T_o}, Oz for the altitude h

Table 7.5. Orbital and recurrence characteristics for satellites with a one-day cycle. Non-Sun-synchronous satellites with three different inclinations are compared with Sun-synchronous satellites. Altitudes h in kilometres and angles i in degrees. The daily orbital frequency ν is in rev/day

ν_o	$i = 20$	$i = 65$	$i = 110$	$i = i_{HS}$	
	h	h	h	h	i_{HS}
16	176.4	214.9	294.6	268.1	96.6
15	478.6	511.6	583.1	561.0	97.7
14	814.4	842.5	906.1	888.3	99.0
13	1 191.1	1 214.5	1 270.6	1 257.1	100.7

Let us now consider a three-dimensional space with the angle of inclination i along the Ox axis, the recurrence cycle C_{T_o} along the Oy axis, and the altitude h along the Oz axis. We then plot the locus of points for which recurrence occurs. This produces a graph like the one sketched in Fig. 7.5. Examination of the space in the diagram shows that:

- the projection of the curves onto the plane xOz gives Fig. 7.4 (upper),
- the intersection of the curves with the plane yOz, for the Sun-synchronous inclination, gives Fig. 7.1 (or Figs. 7.2 or 7.3), regarding the points for the recurrence triples $[14, \pm 1, C_{T_o}]$ and $[14, 0, 1]$.

As for Sun-synchronous satellites, one-day recurrence cycles are not generally favoured for non-Sun-synchronous satellites. The 'avoided' altitudes are noted in Table 7.5 for various inclinations.

Table 7.6. Orbital and recurrence characteristics for the (MEO) GPS navigation satellites and satellites in the constellation WEST. Distances a and h in km, angle i in degrees, period T in minutes

MEO satellite	a	h	i	T	Rec. triple	N_{T_o}
NAVSTAR (I)	26 559.967	20 182	63.0	717.97	$[2, 0, 1]$	2
NAVSTAR (II)	26 560.904	20 183	55.0	717.98	$[2, 0, 1]$	2
GLONASS	25 507.598	19 130	64.8	675.73	$[2, +1, 8]$	17
Galileo	29 993.689	23 616	56.0	861.58	$[2, -1, 3]$	5
WEST	20 267.139	13 889	75.0	478.63	$[3, 0, 1]$	3

Example 7.5. *Choice of orbit for surveillance satellites in the Resurs-F2 series.*

The Soviet then Russian surveillance satellites in the Resurs-F2 series are in very low orbit for one-month missions. With $i = 82.3°$ (the usual inclination for these satellites) and for an altitude $h < 300$ km, there is one possibility for a one-day recurrence cycle. Calculation gives

$$h = 241.850 \text{ km} , \quad \text{triple } [16, 0, 1] , \quad C_{\mathrm{T}} = 0.994 \text{ day} .$$

The orbit of the satellite Resurs-F-10, launched on 21 May 1991, had characteristics $h_{\mathrm{p}} = 196$ km and $h_{\mathrm{a}} = 277$ km, giving a mean altitude of $h = 237$ km. This value put the satellite into a one-day recurrence cycle.

Example 7.6. *Choice of orbit for the satellite Megha-Tropiques.*

To study meteorological phenomena in the inter-tropical zone, the French satellite Tropiques was intended to have a rather high altitude ($h \sim 1000$ km) and low inclination ($i \approx 15°$). After collaboration with India, its inclination was increased to $i = 20°$ (to cover the Himalayas) and its altitude reduced (a restriction imposed by new instruments). The new name of the satellite is Megha-Tropiques (*megha*, मेघ, means clouds in Sanskrit and *tropiques* is the French word for tropics). The altitude, then fixed at 817 km, put the satellite within 3 km of the altitude for a one-day recurrence cycle. It was modified to give a recurrence cycle of 7 days or more. Figure 7.4 (lower) can be used to choose this altitude. For the triple $[14, -1, 7]$, the altitude is $h = 866$ km. The calculation gives $a = 7243.700$ km and $C_{\mathrm{T}} = 6.87$ day.

7.4 Recurrence for MEO and HEO Satellites

The possible problems of one-day recurrence cycles arising for LEO satellites are quite irrelevant for MEO and HEO satellites, for which this type of recurrence is often desirable.

GPS navigation satellites have recurrence properties. For the US system NAVSTAR/GPS (Block I, Block II), the recurrence is based on one day (2

Table 7.7. Orbital and recurrence characteristics for the (HEO) Molniya and Tundra satellites, and also for other satellites under development. Semi-major axis a in km, period T in minutes. For all these orbits, with eccentricity e, the inclination is at the critical value $i = 63.43°$ (for Ellipso Borealis, $i = 180° - 63.43° = 116.57°$)

HEO satellite	a	e	T	Rec. triple	N_{T_o}
Supertundra	42 163.2	0.4230	1436.03	$[1, 0, 1]$	1
Tundra	42 163.4	0.2668	1436.04	$[1, 0, 1]$	1
Molniya	26 552.9	0.7500	717.72	$[2, 0, 1]$	2
Molniya	26 553.6	0.7360	717.75	$[2, 0, 1]$	2
Molniya	26 554.3	0.7222	717.77	$[2, 0, 1]$	2
VirtualGEO	20 260.2	0.6609	478.36	$[3, 0, 1]$	3
COBRA	20 260.9	0.6459	478.39	$[3, 0, 1]$	3
Loopus	29 991.4	0.6000	861.53	$[2, -1, 3]$	5
Ellipso Borealis	10 559.3	0.3463	180.00	$[8, 0, 1]$	8

revolutions per day). For the Russian system GLONASS, it is based on 8 days (17 revolutions over this period), while for the European system Galileo, it is based on 3 days (5 revolutions). For satellites in the constellation WEST, the recurrence cycle is one day (3 revolutions per day). Table 7.6 gives orbital and recurrence features of MEO satellites.

It is also extremely useful for HEO communications satellites to have one-day recurrence cycles so that they overfly ground stations (for transmission and reception) once or twice a day with the same viewing geometry. For these orbits, the apogee is fixed following the choice of the critical inclination. Table 7.7 gives orbital and recurrence characteristics of HEO satellites.

Although associated with a quite different type of mission, it is interesting to note the recurrence conditions of the satellite Integral, $i = 57.1°$, which makes one revolution in three days. This gives the unusual recurrence triple $[0, 1, 3]$, $N_{T_o} = 1$.

7.5 Recurrence Grid

7.5.1 Constructing the Recurrence Grid

As we saw in Chap. 6, the equatorial shift $\Delta\lambda_E$ represents the distance between two successive ground tracks (of the same kind, ascending or descending) at the equator. In the following, we consider near-circular satellites in low orbit (LEO), the only ones for which the recurrence grid is relevant. For the quantities already discussed, we have

$$\Delta\lambda_E = -(\dot{\Omega}_T - \dot{\Omega})T = -\frac{n}{\kappa}T = -\frac{2\pi}{\kappa} . \qquad (7.16)$$

The equatorial shift is thus very simply related to the daily recurrence frequency κ.

Consider a ground track defining one ascending node as origin, with longitude λ_0. If λ_1 is the longitude of the following ascending node, we thus have, by definition of the equatorial shift,

$$\lambda_1 - \lambda_0 = \Delta\lambda_{\rm E} = -2\pi \frac{C_{\rm T_o}}{N_{\rm T_o}} . \tag{7.17}$$

After $\nu_{\rm o}$ revolutions (or about one day), the ground track cuts the equator at the ascending node with longitude $\lambda_{\nu_{\rm o}}$ such that

$$\lambda_{\nu_{\rm o}} - \lambda_0 = \nu_{\rm o}\Delta\lambda_{\rm E} = -2\pi \frac{\nu_{\rm o} C_{\rm T_o}}{N_{\rm T_o}} .$$

According to (7.11), this can be written in the form

$$\lambda_{\nu_{\rm o}} - \lambda_0 = -2\pi \left(1 - \frac{D_{\rm T_o}}{N_{\rm T_o}}\right) = -2\pi + 2\pi \frac{D_{\rm T_o}}{N_{\rm T_o}} .$$

Since the longitudes are defined to within 2π, we thus have

$$\lambda_{\nu_{\rm o}} - \lambda_0 = 2\pi \frac{D_{\rm T_o}}{N_{\rm T_o}} . \tag{7.18}$$

If $D_{\rm T_o}$ is positive, i.e., if $\nu_{\rm o} < \kappa$, we have $\lambda_{\nu_{\rm o}} - \lambda_0 > 0$, i.e., $\lambda_{\nu_{\rm o}}$ to the east of λ_0. Indeed, if after $\nu_{\rm o}$ revolutions a whole day has not gone by since the crossing at λ_0, the ground track $\lambda_{\nu_{\rm o}}$ is indeed east of λ_0.

In the opposite situation, if $D_{\rm T_o}$ is negative, i.e., if $\nu_{\rm o} > \kappa$, a little more than one day has gone by and the ground track $\lambda_{\nu_{\rm o}}$ is situated west of λ_0, which shows that $\lambda_{\nu_{\rm o}} - \lambda_0 < 0$.

In the rest of this calculation, we shall denote the day of crossing by a superscript and the number of the crossing in this day by a subscript on the longitude of the ascending node, viz.,

$$\lambda^{\rm day}_{\rm crossing} .$$

The origin is denoted by λ^0_0. We consider the case where $D_{\rm T_o}$ is positive for concreteness. The other case would require us to use $\nu_{\rm o} + 1$ rather than $\nu_{\rm o}$ and change some of the signs. We thus have, with this notation,

$$\lambda^0_1 - \lambda^0_0 = \Delta\lambda_{\rm E} .$$

The last ascending node on day 0 is $\lambda^0_{\nu_{\rm o}}$, and the next, which is the first on day 1, is written λ^1_1. Hence,

$$\lambda^0_{\nu_{\rm o}} - \lambda^0_0 = \nu_{\rm o}\Delta\lambda_{\rm E} \qquad [2\pi] ,$$
$$\lambda^1_1 - \lambda^0_0 = (\nu_{\rm o} + 1)\Delta\lambda_{\rm E} \qquad [2\pi] ,$$

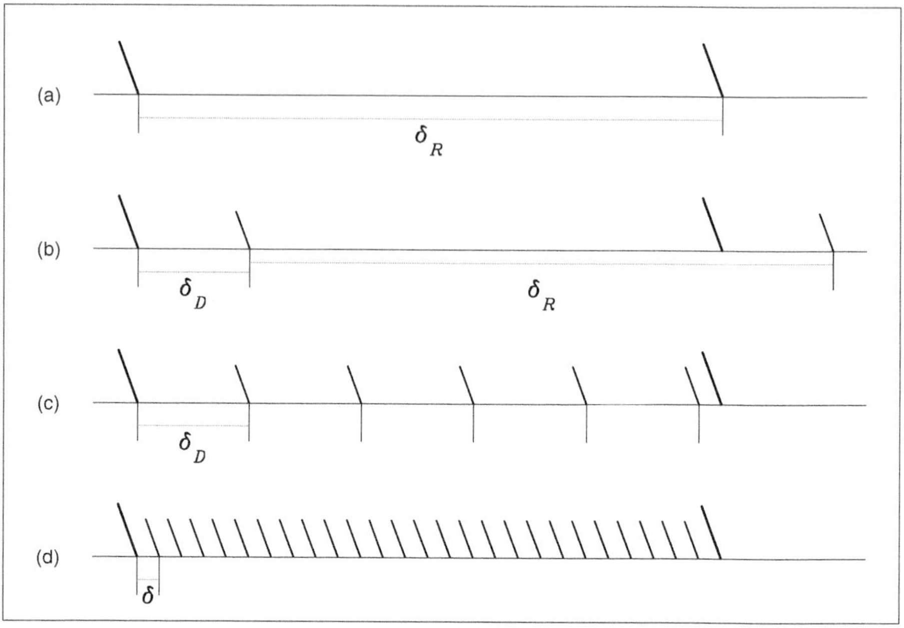

Figure 7.6. Constructing the recurrence grid. (**a**) Two consecutive ground tracks from day 0 determine the base interval – these ground tracks are plotted in *bold type*. (**b**) One ground track of day 1 passes through the base interval. (**c**) The ground tracks for the following days, $2, 3, \ldots, J$ pass through the base interval. (**d**) All ground tracks up to day $(C_{T_o} - 1)$ define the grid interval. Note: by 'ground track', we understand the ground track at the ascending node

where $[2\pi]$ indicates the congruence modulo 2π.

We now consider the interval $[\lambda_1^0, \lambda_0^0]$, which we shall call the base interval, counting positively towards the east. We take λ_1^0 as origin and set

$$\delta_R = \lambda_0^0 - \lambda_1^0 \,, \quad \delta_D = \lambda_1^1 - \lambda_1^0 \,,$$

where δ_R is the interval between ascending nodes for two consecutive revolutions, so that $\delta_R = -\Delta\lambda_E$, and δ_D the interval between ascending nodes for two consecutive days. The interval δ_R is shown in Figs. 7.6a and b, and the interval δ_D in Figs. 7.6b and c.

Calling δ the grid interval at the equator, defined by

$$\delta = \frac{2\pi}{N_{T_o}} \,, \tag{7.19}$$

we have the relations

$$\delta_R = C_{T_o} \delta \,, \quad \delta_D = D_{T_o} \delta \,.$$

The grid interval[3] δ, related in this way to δ_{R} and δ_{D}, is indicated in Fig. 7.6d.

In the base interval, we have for the different days:

$$\begin{array}{ll} \text{for day } 1 & \lambda_1^1 - \lambda_1^0 = \delta_{\mathrm{D}} \,, \\ \text{for day } 2 & \lambda_1^2 - \lambda_1^0 = 2\delta_{\mathrm{D}} \,, \\ \text{for day } J & \lambda_1^J - \lambda_1^0 = J\delta_{\mathrm{D}} \,. \end{array}$$

For day J, this relation holds if the point λ_1^J lies in the base interval. Otherwise, we subtract a whole number of values of δ_{R}, expressing this by congruence relations:

$$\begin{aligned} \lambda_1^J - \lambda_1^0 &= J\delta_{\mathrm{D}} & [\delta_{\mathrm{R}}] \,, \\ \lambda_1^J - \lambda_1^0 &= JD_{\mathrm{T_o}}\delta & [C_{\mathrm{T_o}}\delta] \,, \\ \frac{\lambda_1^J - \lambda_1^0}{\delta} &= JD_{\mathrm{T_o}} & [C_{\mathrm{T_o}}] \,. \end{aligned}$$

We note that the quantity $(\lambda_1^J - \lambda_1^0)/\delta$ is indeed a whole number. Hence, for a given day, we obtain the position of the ascending node λ_1^J in the base interval, hence also in the recurrence grid.

If $u(J)$ denotes the position of the ground track of day J in the base interval, in units of δ, i.e.,

$$u(J) = \frac{\lambda_1^J - \lambda_1^0}{\delta} \,, \tag{7.20}$$

we have the fundamental relation for the recurrence grid:

$$u(J) = JD_{\mathrm{T_o}} \qquad [C_{\mathrm{T_o}}] \,. \tag{7.21}$$

The whole number $u(J)$ can take $C_{\mathrm{T_o}}$ values between 0 and $C_{\mathrm{T_o}} - 1$.

When we do not wish to favour one bound of the interval rather than the other, we will consider the number $u^*(J)$ defined by

$$u^*(J) = \min\left\{u(J), C_{\mathrm{T_o}} - u(J)\right\} \,, \tag{7.22}$$

which is an integer taking values between 0 and $C_{\mathrm{T_o}}/2$.

[3] For a quick evaluation, δ can be obtained from an approximate relation for recurrent satellites of altitude $h = 900 \pm 300$ km. For these satellites, the daily orbital frequency ν lies between 13 and 15. We may thus take ν to be equal to 14 and $N_{\mathrm{T_o}}$ equal to $14C_{\mathrm{T_o}}$. Expressing δ in degrees, we thus find that

$$\delta \approx \frac{360}{14C_{\mathrm{T_o}}} \,, \qquad \text{i.e.,} \quad C_{\mathrm{T_o}}\delta \approx 25 \,.$$

We thus see that the product of the grid interval (in degrees) and the recurrence cycle (in days) is roughly equal to 25. This relation is often given in the literature. In certain texts, it is presented as a kind of magic formula, although we can see from the above that it is very easy to derive.

7.5.2 Using the Recurrence Grid

We now give several examples to show how the recurrence grid is determined and used.

Example 7.7. *Calculate the crossing order in the base interval for ADEOS-1.*

The recurrence triple for this satellite is $[14, 11, 41]$, whence $N_{T_o} = 585$. We thus have

$$\delta = \frac{2\pi}{585} = 0.010740 \text{ rad} = 0.6154^\circ = 68.504 \text{ km} ,$$

and for the equatorial shift, $\delta_R = 41\delta$, i.e., $\delta_R = 25.23^\circ$. Applying (7.21), we obtain the values $u(J)$ for each day of the cycle:

$$u(0) = 0 , \quad u(1) = 11 , \quad u(2) = 22 , \quad u(3) = 33 ,$$

$$u(4) = 3 , \quad u(5) = 14 , \quad u(6) = 25 , \quad \dots .$$

From these values $u(J)$, we obtain the grid in the base interval, from $u = 0$ to $u = 41$. The values of J are:

$$\begin{aligned}
u = 0\text{–}9 &\longmapsto 0, 15, 30, 4, 19, 34, 8, 23, 38, 12 ,\\
u = 10\text{–}19 &\longmapsto 27, 1, 16, 31, 5, 20, 35, 9, 24, 39 ,\\
u = 20\text{–}29 &\longmapsto 13, 28, 2, 17, 32, 6, 21, 36, 10, 25 ,\\
u = 30\text{–}39 &\longmapsto 40, 14, 29, 3, 18, 33, 7, 22, 37, 11 ,\\
u = 40\text{–}41 &\longmapsto 26, 41 = 0 .
\end{aligned}$$

The ascending node crossing $u = 1$ occurs on day $J = 15$. With (7.21), we can check that $u(15) = 15 \times 11 \ [41] = 165 \ [41] = 1$. Hence, $u^* = 1$ for $J = 15$ and $J = 26$.

Example 7.8. *Calculate the crossing order in the base interval for Landsat-3 and ADEOS-2.*

All recurrent satellites with $D_{T_o} = \pm 1$ have sequential recurrence grids: in the base interval, consecutive ground tracks are in the order of the days. Indeed, in this case, (7.21) implies the very simple relation

$$D_{T_o} = +1 \quad \Longrightarrow \quad u(J) = J , \qquad D_{T_o} = -1 \quad \Longrightarrow \quad u(J) = C_{T_o} - J .$$

Many satellites fall into this category – see Table 7.1. Later on, we shall discover the important consequences of this fact for the ground track during the recurrence cycle. For Landsat-3, with $D_{T_o} = -1$ for an 18 day cycle, the various ground tracks $u = 0, 1, 2, 3, \dots , 17, 18$, occur on days $J = 18\,(= 0), 17, 16, 15, \dots , 1, 0$. For ADEOS-2, with $D_{T_o} = +1$ for a 4 day cycle, the various ground tracks $u = 0, 1, 2, 3, 4$, occur on days $J = 0, 1, 2, 3, 4$.

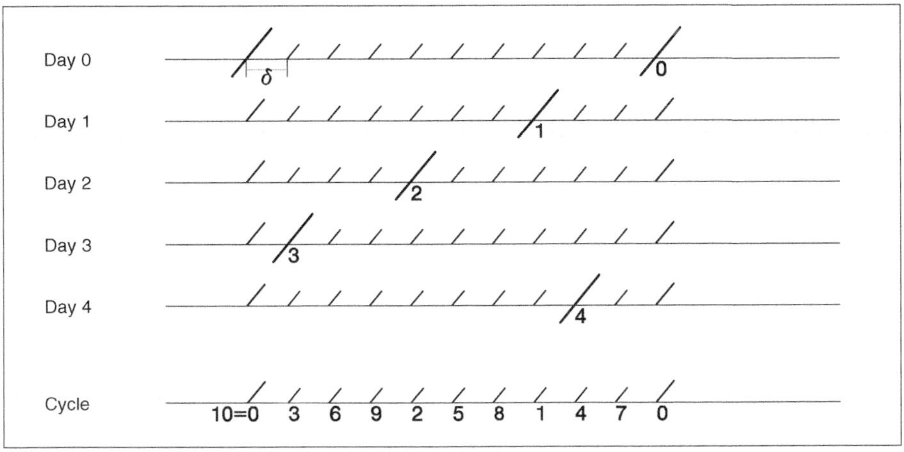

Figure 7.7. Construction of the recurrence grid for TOPEX/Poseidon (or Jason-1 and -2). For each cycle ($J = 0, 1, 2, 3, 4, \ldots$), the ground track at the equator is marked in the base interval

Example 7.9. *Calculate the crossing order in the base interval for TOPEX/Poseidon (and Jason-1, Jason-2).*

The recurrence triple for this satellite is $[13, -3, 10]$, which immediately gives $N_{\mathrm{T_o}} = 127$. We thus have

$$\delta = \frac{2\pi}{127} = 0.049474 \text{ rad} = 2.8346^\circ = 315.551 \text{ km} ,$$

and for the equatorial shift, $\delta_{\mathrm{R}} = 10\delta$, or $\delta_{\mathrm{R}} = 28.35^\circ$. Applying (7.21), we obtain $u(J)$ for each day of the cycle. We thus deduce the grid in the base interval, from $u = 0$ to $u = 10$, indicating the value of J for each u:

$$u = 0\text{–}10 \quad \longmapsto \quad 0, 3, 6, 9, 2, 5, 8, 1, 4, 7, 10 = 0 .$$

Note that $u^* = 1$ for $J = 3$ and $J = 7$. It is easy to obtain these values using a graph like the one shown in Fig. 7.7.

7.5.3 Reference Grids

For a recurrent satellite, a single point on the ground track completely fixes its ground track on the globe. Earth-observation satellites are maintained on their nominal orbits in order to guarantee the position of their ground tracks to within a few kilometres (generally ±5 km, and ±1 km for TOPEX/Poseidon and Jason-1, ±0.8 km for ICESat). All the SPOT satellites, for example, use the same grid, which is fixed by giving an ascending node longitude. Table 7.8 describes the main grids and their characteristics.[4]

[4] Abbreviations are as follows. WRS: Worldwide Reference System. GRS: *Grille de Référence SPOT*. ERS: ERS–SAR Reference System. MWRS: MOS-1 World

Table 7.8. Characteristics of reference grids for various satellites. N_{T_o} is the number of revolutions per recurrence cycle, δ the grid interval, and λ the longitude origin

Satellites	Grid	N_{T_o}	δ [deg]	δ [km]	λ origin [deg]
Landsat-1 to -3	WRS-1	251	1.4343	159.7	294.5200
Landsat-4 to -7, Terra	WRS-2	233	1.5451	182.0	295.4000
A-Train	WRS-2	233	1.5451	182.0	295.4000
SPOT-1 to -5	GRS	369	0.9756	108.6	330.4000
ERS-1, 2, Envisat	ERS	501	0.7186	80.0	0.1335
MOS-1, -1B	MWRS	235	1.5190	169.1	326.7500
Oceansat-1	IRSP4G	29	12.4138	1381.9	328.1900
TOPEX/P, Jason-1	T/P	127	2.8346	315.5	99.9242

The ground track of TOPEX/Poseidon over one recurrence cycle (see Fig. 7.10) represents the reference grid of this satellite and the satellites Jason-1 and -2, since the initial conditions agree with those given in Table 7.8 (see also Fig. 5.26, lower).

Example 7.10. *Using the recurrence grid for the first revolutions of ERS-2.*

We consider the initial conditions for ERS-2 given by ESA, at the first ascending node crossing of the measurement phase (revolution 147, the first 146 revolutions being used to check the instrumentation):

$$\text{day } 1995\ 04\ 21\ , \quad \text{time } 03{:}16{:}54\ \text{UT}\ , \quad \text{longitude } \lambda = +288.277\ .$$

We check that the satellite is indeed on the recurrence grid. To do so, we consider the value of the longitude origin given in Table 7.8, and denoted here by λ_1^0. The calculation gives

$$\frac{\lambda - \lambda_1^0}{\delta} = (288.2770 - 0.1335) \times \frac{501}{360} = 401.000\ ,$$

which is indeed an integer. We can also accurately calculate the ascending node crossing time:

$$\tau_{\mathrm{AN}} = \frac{288.277}{15} + t_{\mathrm{AN}} = 19{:}13{:}06 + 03{:}16{:}54 = 22{:}30{:}00\ .$$

We thus see that the satellite is on its nominal orbit.

Example 7.11. *Fitting a grid to geographic circumstances: the case of Rocsat-2.*

Reference System. IRSP4G: National Remote Sensing Agency (India) IRS-P4 Grid.

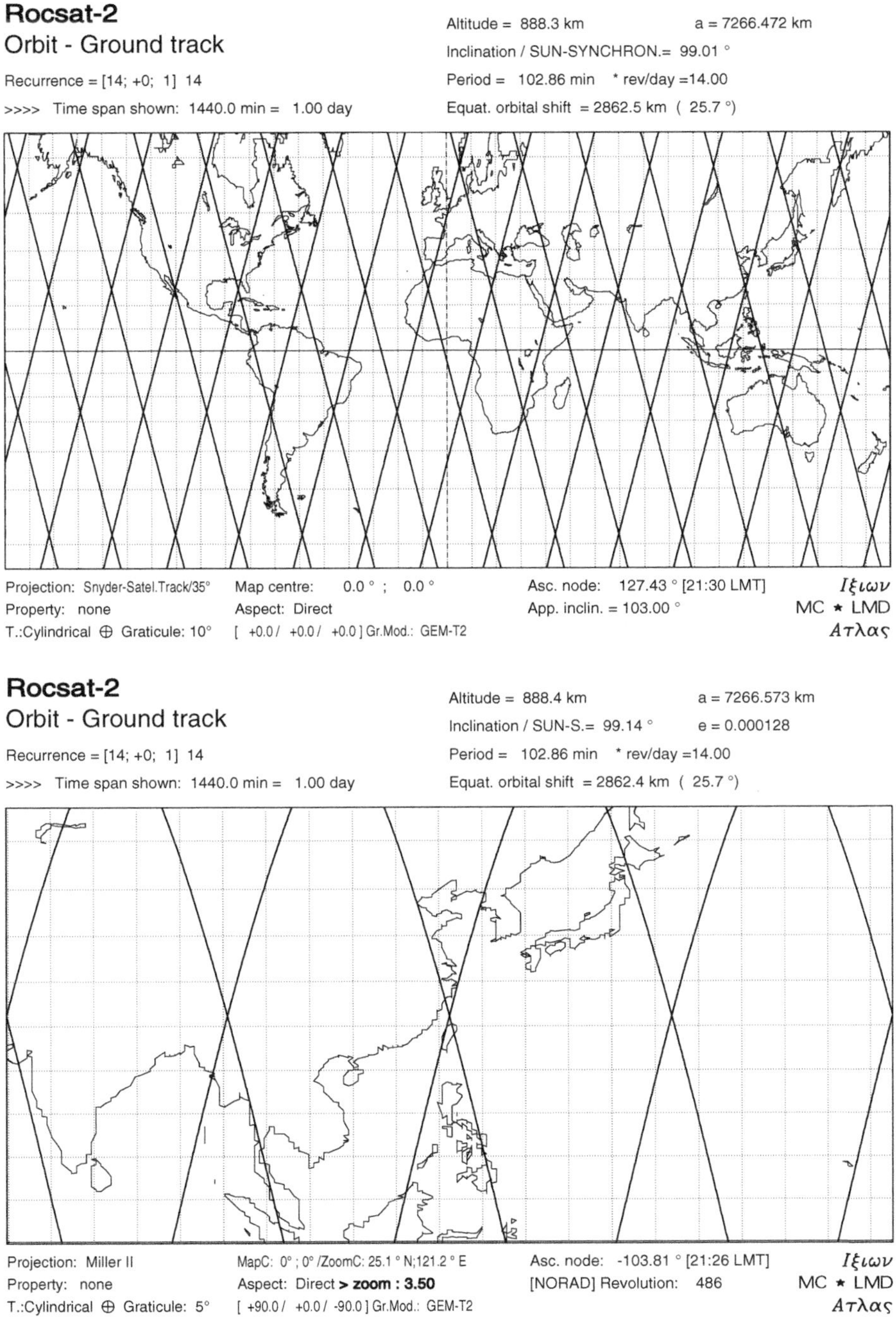

Figure 7.8. Ground track of the Taiwanese satellite Rocsat-2, with one-day recurrence cycle. *Upper*: Recurrence grid. *Lower*: Magnified view centred on the island of Taiwan, explaining this choice of grid

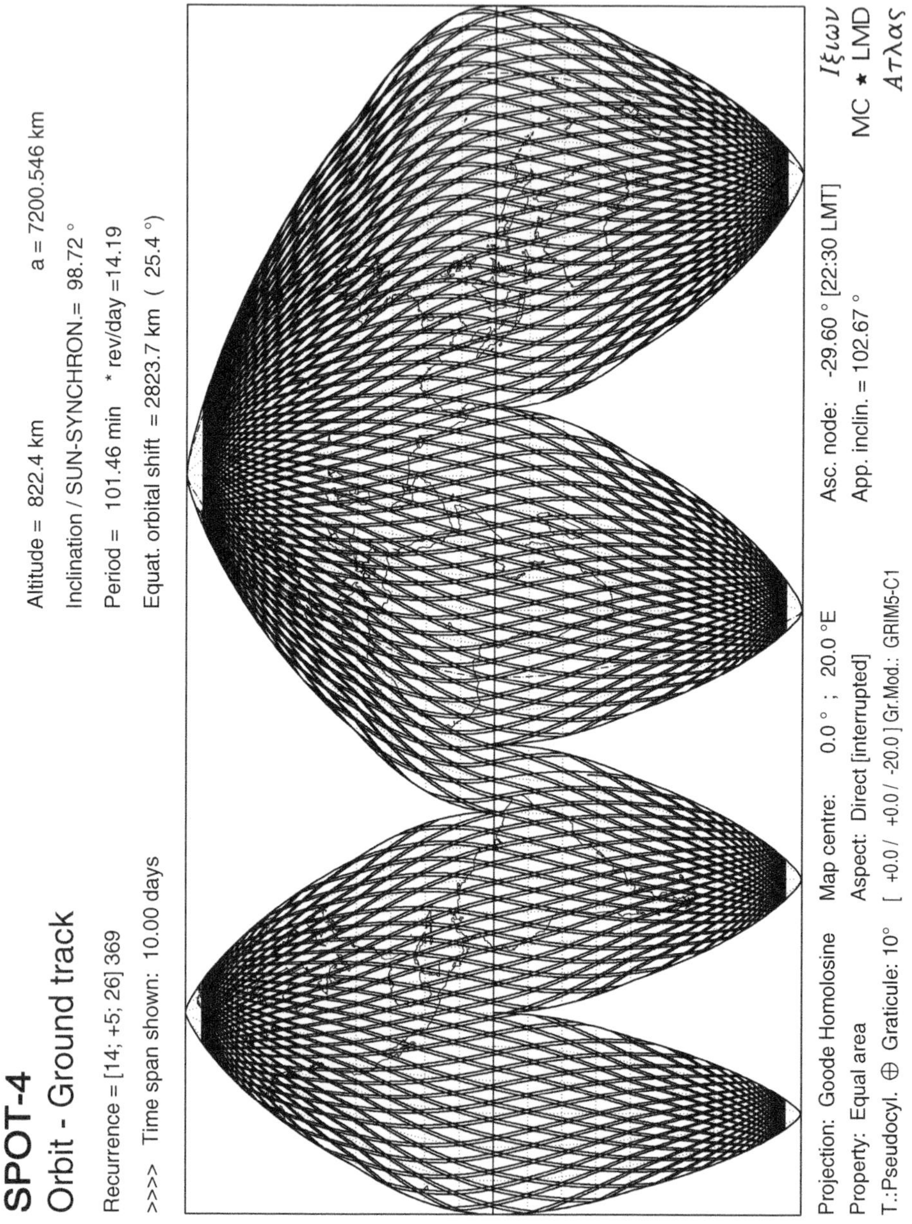

Figure 7.9. Ground track of the SPOT-4 orbit over two recurrence subcycles, i.e., 10 days

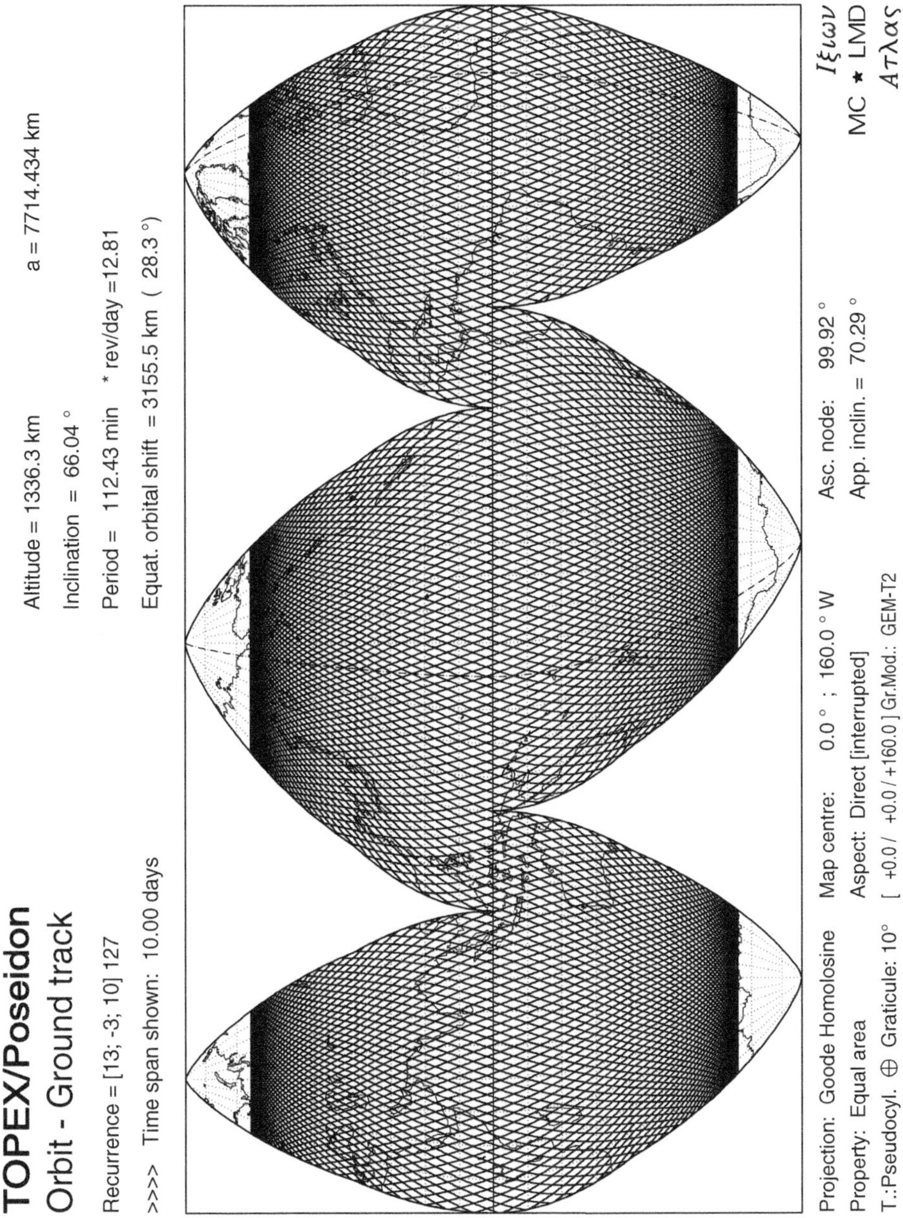

Figure 7.10. Ground track of the TOPEX/Poseidon orbit over a recurrence cycle, i.e., 10 days

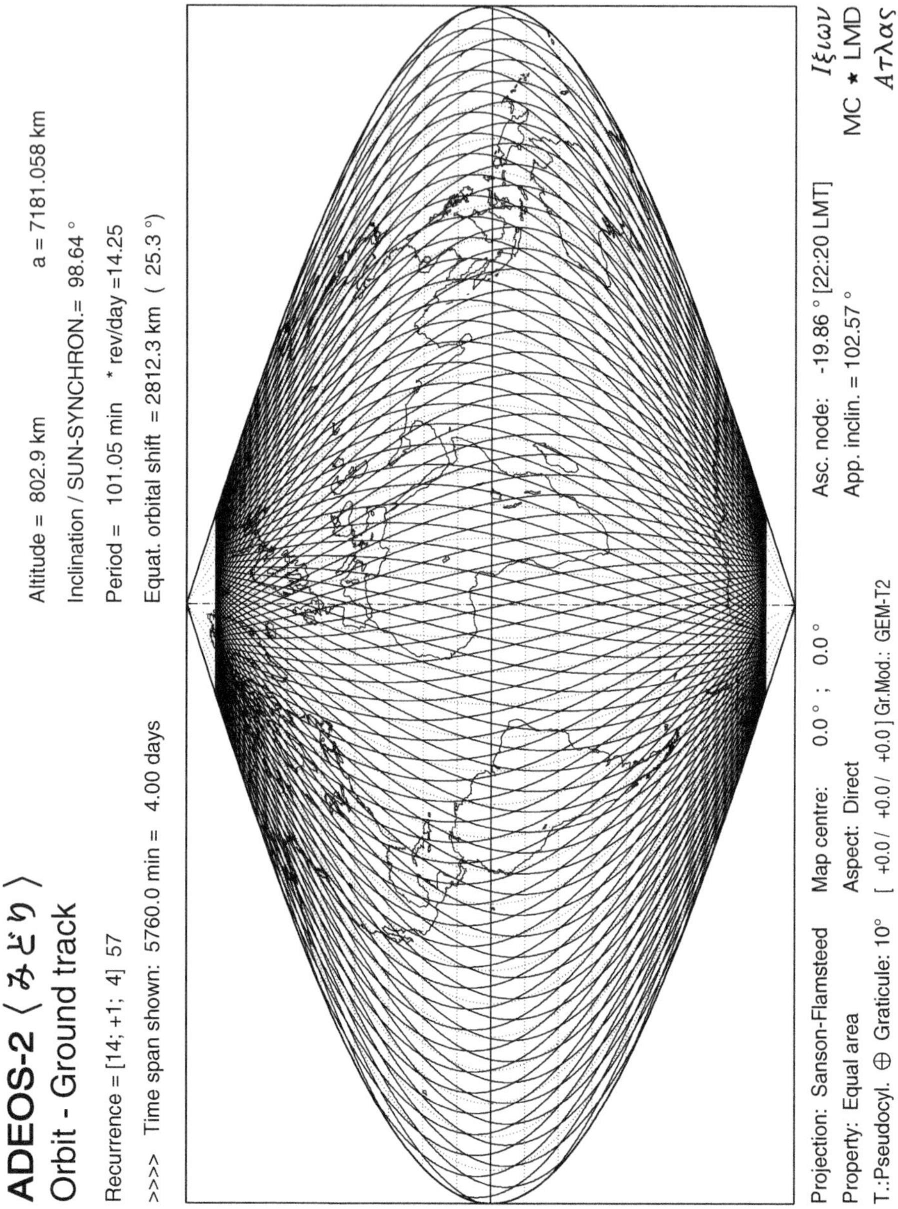

Figure 7.11. Ground track of the ADEOS-2 orbit over a recurrence cycle, i.e., 4 days

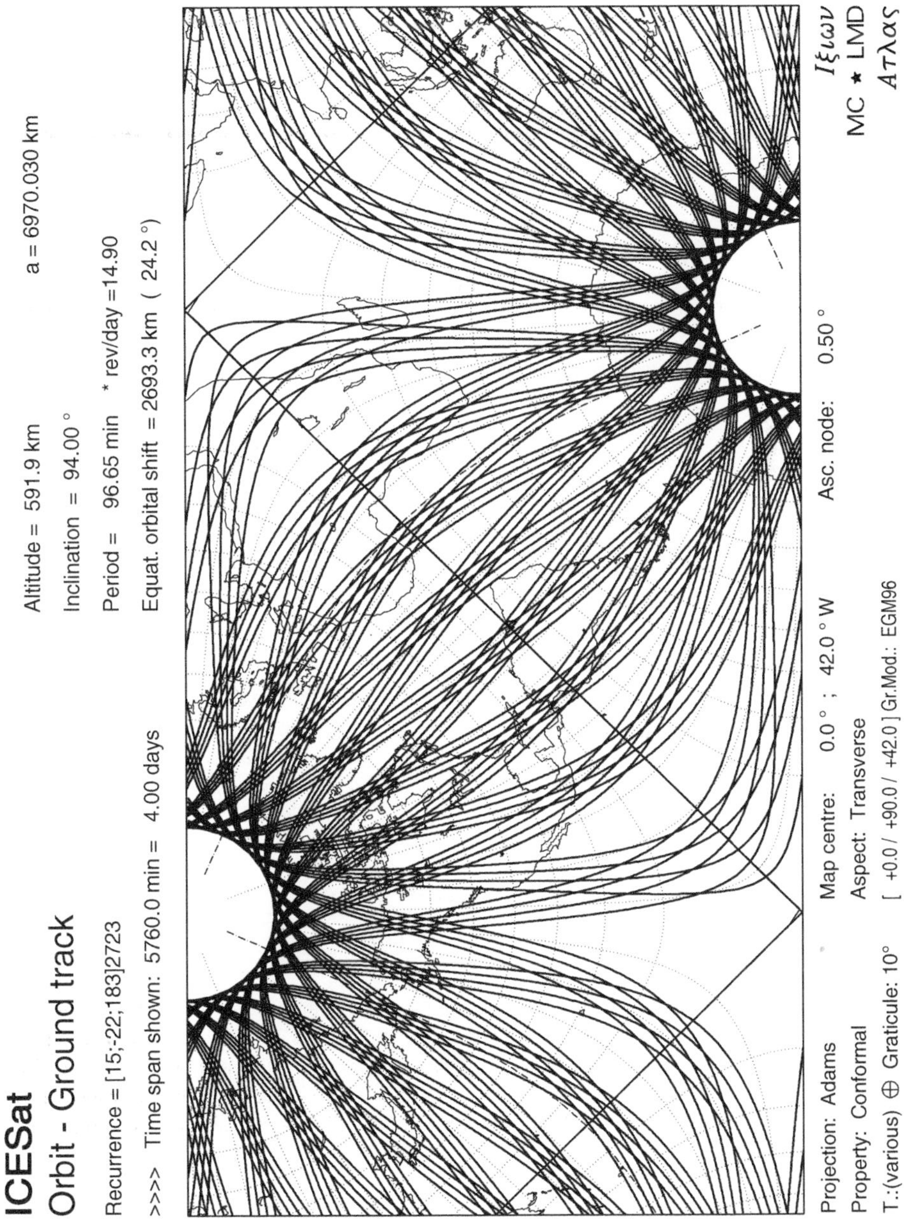

Figure 7.12. Ground track of the ICESat mission orbit over a recurrence half-sub-subcycle, i.e., 4 days

Rocsat-2, launched on 20 May 2004, is devoted to the study of Earth resources and meteorology, in particular, typhoon warnings. Its recurrence triple is [14, 0, 1]. This means that each day, after 14 round trips, it exactly repeats its ground track (see the upper part of Fig. 7.8). If we examine this grid, we find that the ascending and descending ground tracks cross at latitude 26.415° (see Sect. 7.9 on grid points). Now the island of Taiwan, formerly Formosa, lies between latitudes 22° N and 25° N. The grid can thus be fixed in such a way that the island is crossed twice a day by the (ascending and descending) ground track, as can be seen in Fig. 7.8 (lower). The equatorial shift is $\Delta\lambda_E = 360/14 = 25.71°$. By symmetry considerations, we understand that the longitude of the ground track intersection and the longitude of the ascending node must be separated by a distance equal to one quarter of the equatorial shift. The central longitude of Taiwan is $\lambda_C = 121.2°$E, which means that the ascending node must have longitude λ_O such that

$$\lambda_O = \lambda_C + \Delta\lambda_E/4 = 121.2 + 6.4 = 127.6°\text{E} .$$

7.5.4 Recurrence Subcycle

Definition of the Subcycle

Consider a day origin $J = 0$. A subcycle E_{T_o} is then the number of days required for the ground track to pass at a grid interval δ of the ascending node origin. This can be written

$$u(E_{T_o}) = \pm 1 \qquad [C_{T_o}] , \tag{7.23}$$

or

$$E_{T_o} D_{T_o} = \pm 1 \qquad [C_{T_o}] , \tag{7.24}$$

or bringing in the distance u^*,

$$u^*(E_{T_o}) = 1 , \tag{7.25}$$

which gives two value for E_{T_o}. If $E^*_{T_o}$ is the smallest of the two, we have

$$E_{T_o} = E^*_{T_o} , \qquad E_{T_o} = C_{T_o} - E^*_{T_o} .$$

We give several values for the subcycles of satellites discussed earlier.

Important Note. Care must be taken not to confuse the two quantities D_{T_o} and $E^*_{T_o}$. For the SPOT satellites, often described in the literature, it happens that $D_{T_o} = 5$ and $E^*_{T_o} = 5$. One cannot conclude, however, that D_{T_o} represents a subcycle, contrary to what is often claimed. It is quite clear in Example 7.12 that, for ADEOS-1, these two quantities are different: $D_{T_o} = 11$ and $E^*_{T_o} = 15$.

Example 7.12. *Recurrence subcycles for various Earth-observation satellites.*

- For SPOT-4, we have $E^*_{\mathrm{T_o}} = 5$. Indeed, $E^*_{\mathrm{T_o}} D_{\mathrm{T_o}} = 5 \times 5 = 25$ and $C_{\mathrm{T_o}} - 1 = 26 - 1 = 25$. This means that, 5 days before ($J = -5$ [26] $= 21$) or 5 days after ($J = 5$) the day origin, the ground track passes at the grid interval from the ground track origin (see Fig. 7.9).
- For Landsat-3, ADEOS-2, IRS-1A and all satellites for which $D_{\mathrm{T_o}} = \pm 1$, we have $E^*_{\mathrm{T_o}} = 1$, whence $E_{\mathrm{T_o}} = 1$ and $E_{\mathrm{T_o}} = C_{\mathrm{T_o}} - 1$ (see Fig. 7.11 for ADEOS-2, Fig. 7.13, upper, for IRS-1A).
- For ADEOS-1, we have $E^*_{\mathrm{T_o}} = 15$. Indeed, $E^*_{\mathrm{T_o}} D_{\mathrm{T_o}} = 15 \times 11 = 165$ [41] $= 1$. We thus have $E_{\mathrm{T_o}} = 15$ and $E_{\mathrm{T_o}} = 41 - 15 = 26$, which can be confirmed from Example 7.7, with $u^*(15) = 1$ and $u^*(26) = 1$.
- For TOPEX/Poseidon, $E^*_{\mathrm{T_o}} = 3$, whence $E_{\mathrm{T_o}} = 3$ and 7 (see Fig. 7.10).
- For ICESat, the cycle is very long in order to obtain a very short grid interval: $\delta = 14.7$ km. We have $E_{\mathrm{T_o}} = 25$, because $E^*_{\mathrm{T_o}} D_{\mathrm{T_o}} = 25 \times (-22) = -550$ [183] $= -1$ (Fig. 7.12).

Note Concerning Cartography. Figures 7.9 and 7.10 show equal area projections in interrupted form. For a land-observing satellite, the map is cut in the region of the oceans in such a way as to minimise angular distortion on the continents. For an ocean-observing satellite, it is cut in the region of the land masses to minimise angular deformations across the oceans.

Time Required to Cover the Base Interval

The main reason for introducing the subcycle is to show how long it takes to cover the base interval. For SPOT-5, the value of the subcycle $E^*_{\mathrm{T_o}} = 5$ indicates that almost the whole of the base interval has been scanned in five days. However, for a satellite with $E^*_{\mathrm{T_o}} = 1$, we see that the ground track remains almost at the same place in the base interval after one day and that we need the whole cycle $C_{\mathrm{T_o}}$ (which can be of the order of a month) to cover the whole interval. This situation is generally considered as a drawback, as can be seen from the evolution of certain programmes.

The first Landsat satellites (Landsat-1, -2 and -3) had the subcycle $E^*_{\mathrm{T_o}} = 1$. Then, from Landsat-4, the orbit was changed to obtain $E^*_{\mathrm{T_o}} = 5$. The same happened with the Indian satellites in the IRS programme: the first IRS (IRS-1A and -1B) were replaced, from IRS-P2, with a change of subcycle (but keeping almost the same cycle). This change is very clearly shown in Fig. 7.13, where we have plotted the ground tracks of two IRS satellites IRS over 5 days. For IRS-1A, over this time span, a fraction 5/22 or less than one quarter of the interval has been covered. For IRS-P2, the whole interval has been covered. On the ground track of IRS-P2, in the vicinity of the origin ($\lambda = 0$, $\phi = 0$), we see the end of the ground track for the fifth day, since in 5 days, this satellite has accomplished $5 \times 14.208 = 71.04$ round trips. The distance between this ground track and the one going through the origin is equal to the interval δ, since $E^*_{\mathrm{T_o}} = 5$ for this satellite.

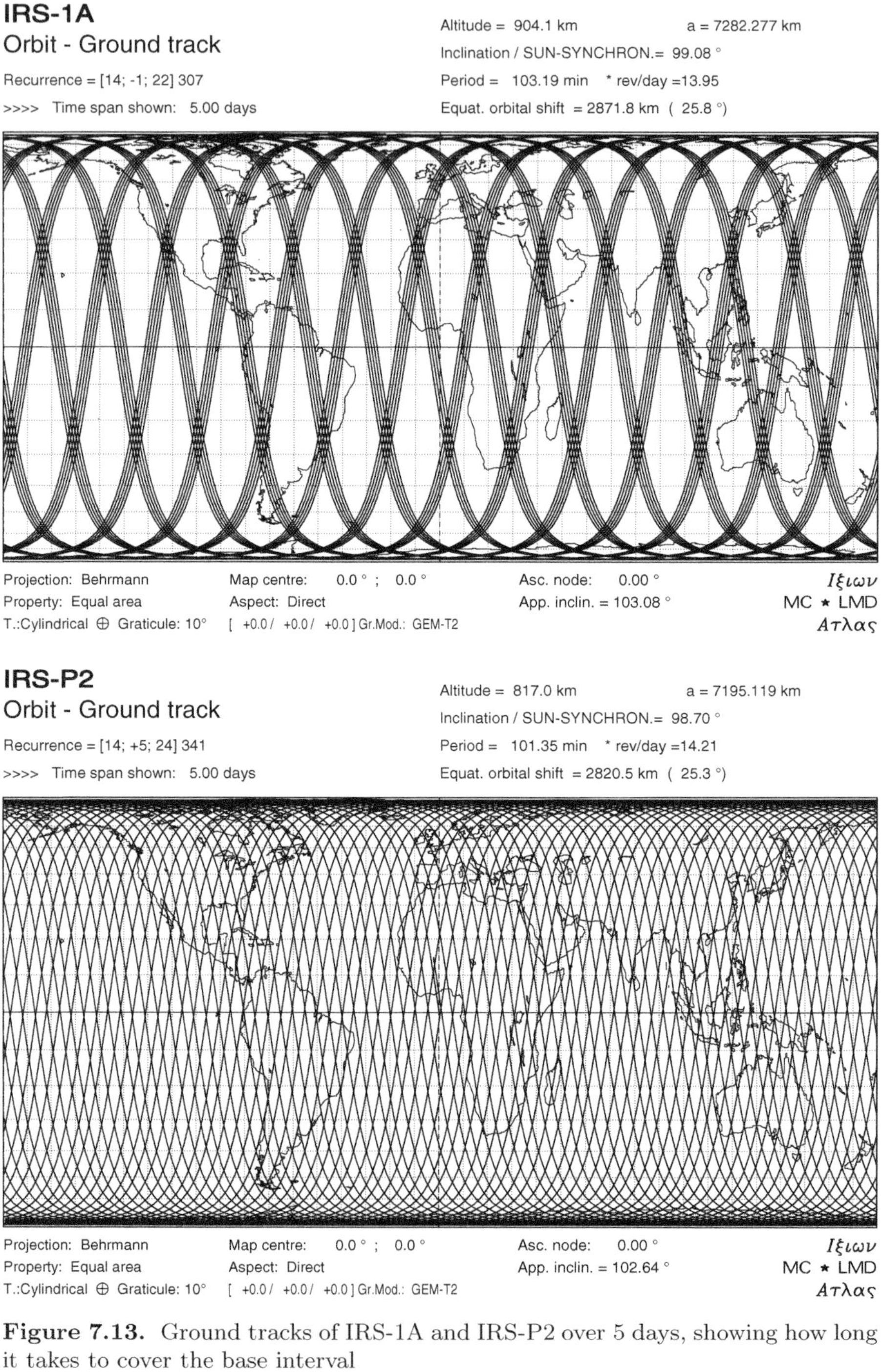

Figure 7.13. Ground tracks of IRS-1A and IRS-P2 over 5 days, showing how long it takes to cover the base interval

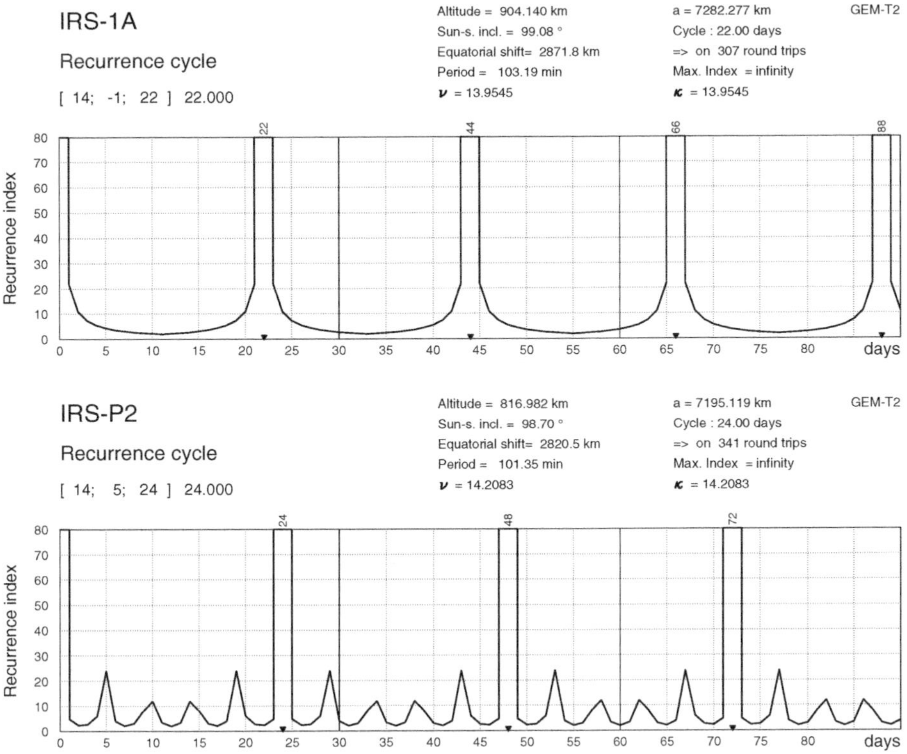

Figure 7.14. Recurrence index for two IRS satellites, showing the change of subcycle

The notion of subcycle is often used for recurrent satellites, since it is rather informative. It tells us that, after a certain number of days representing the subcycle, the ground track passes through almost the same place, where 'almost' means at the grid interval δ. However, it is just a special case of the notion of recurrence index to be discussed next.

7.6 Recurrence Index

7.6.1 Definition of Recurrence Index

Consider a day origin $J = 0$ and its base interval defined by λ_1^0 and λ_0^0, separated by a distance δ_{R}, the equatorial shift. The ascending node crossing in this interval on a given day J determines an ascending node of longitude λ_1^J, as discussed above. The distance between λ_1^J and the bounds λ_1^0 and λ_0^0 will be representative of the recurrence conditions. If the distance to one of the two bounds is zero, this means that the ground track passes once again through a certain point and we have recurrence with a cycle of J days, or else J is a multiple of the cycle. If the distance is small compared with the length of the interval, we may be in a situation with a subcycle.

We set

$$v(J) = \frac{\lambda_1^J - \lambda_0^0}{\delta_\mathrm{R}} . \tag{7.26}$$

From (7.20), we then have

$$v(J) = \frac{u(J)}{C_{\mathrm{T}_\circ}} . \tag{7.27}$$

These two equivalent definitions show that $v(J)$ is a real number between 0 and 1, representing a relative distance.

Just as we defined $u^*(J)$ by (7.22), we now define $v^*(J)$ as the smallest of the two relative distances $v(J)$ from the ground track to one or other of the bounds, viz.,

$$v^*(J) = \min \left\{ \frac{\lambda_1^0 - \lambda_1^J}{\lambda_1^0 - \lambda_0^0}, \frac{\lambda_1^J - \lambda_0^0}{\lambda_1^0 - \lambda_0^0} \right\} , \tag{7.28}$$

or

$$v^*(J) = \min \Big\{ v(J), 1 - v(J) \Big\} = \frac{u_J^*}{C_{\mathrm{T}_\circ}} , \tag{7.29}$$

whence,

$$v^*(J) = J \frac{\delta}{\delta_\mathrm{R}} = J \frac{|D_{\mathrm{T}_\circ}|}{C_{\mathrm{T}_\circ}} . \tag{7.30}$$

We shall call this the relative recurrence distance. It lies between 0 and 0.5. It follows that:

- $v^*(J) = 0$, for a recurrence cycle of J days,
- $v^*(J) = 1$, for a recurrence subcycle for J,
- $v^*(J) = 2$, for a ground track passing at 2δ from a bound,

and so on.

To obtain a function of J which increases as we approach recurrence conditions, we define the function $\Phi(J)$, which is simply the reciprocal of $v^*(J)$, i.e.,

$$\Phi(J) = \frac{1}{v^*(J)} . \tag{7.31}$$

We shall call this function the recurrence index. It is a dimensionless quantity. We now have:

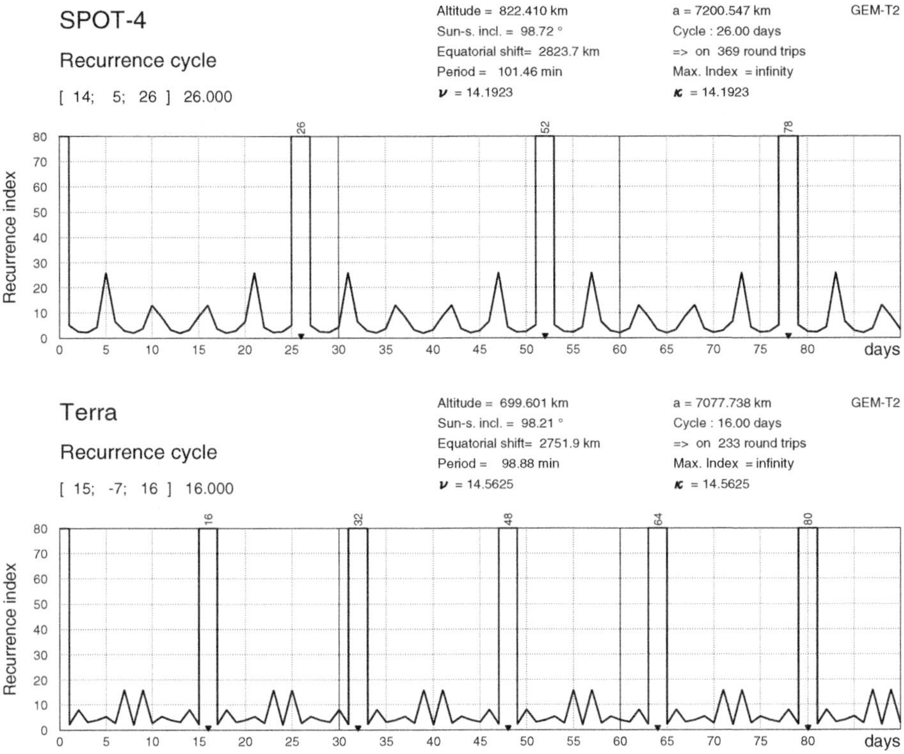

Figure 7.15. Recurrence index for satellites in the SPOT and Terra families

- $\Phi(J) = \infty \implies$ recurrence cycle of J days ,
- $\Phi(J) = C_{\mathrm{T_o}} \implies$ recurrence subcycle for J ,
- $\Phi(J) = C_{\mathrm{T_o}}/2 \implies$ ground track passing 2δ from a bound ,
- $\Phi(J) = C_{\mathrm{T_o}}/3 \implies$ ground track passing 3δ from a bound ,
- other cases,
- $\Phi(J) > 2$ in every case.

This index provides a useful way of specifying cycles, subcycles, and other quantities related to recurrence for any satellite, whether it is intentionally recurrent or not.

7.6.2 Perfect or Imperfect Recurrence

The methods discussed above concern satellites with known recurrence elements. They allow one to find the orbital characteristics from these elements. However, we may encounter another type of problem: given the orbit of a satellite, we may wish to find its recurrence cycle. For this satellite, h and i are known, so that P and ν are determined, and hence also κ.

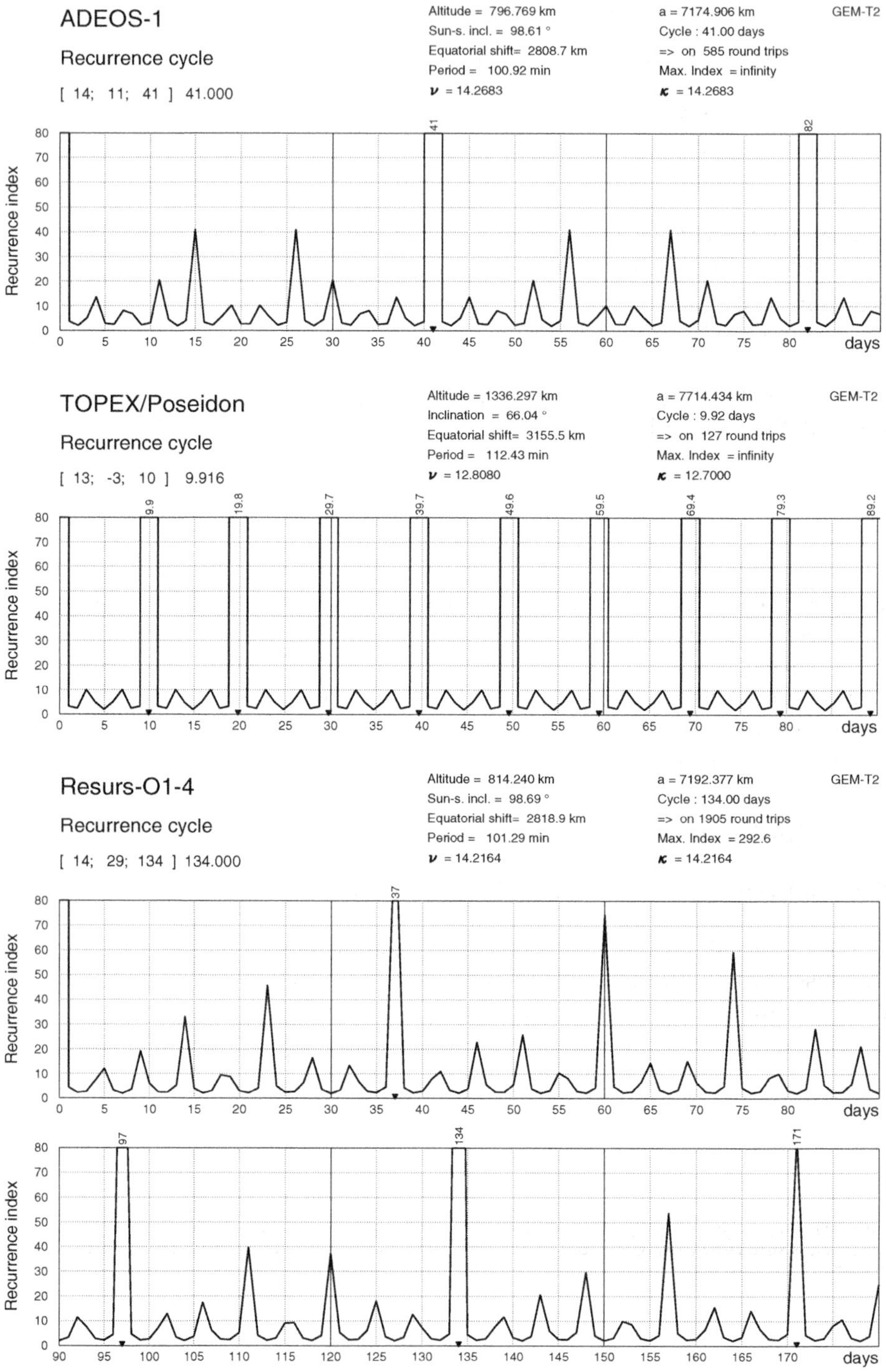

Figure 7.16. Recurrence index for ADEOS-1, TOPEX/Poseidon, Resurs-O1-4

The day J corresponding to the recurrence cycle $C_{\mathrm{T_o}}$ will be such that the product κJ is closest to a whole number. It will therefore be the day giving the highest value of the recurrence index $\Phi(J)$, which we shall write Φ_{m}.

Indeed, considering the expression (7.11) for κ as a function of the recurrence triple, the product κJ is

$$\kappa J = J\nu_{\mathrm{o}} + J\frac{D_{\mathrm{T_o}}}{C_{\mathrm{T_o}}} ,$$

and since $J\nu_{\mathrm{o}}$ is an integer, we have

$$\text{fractional part of } \kappa J = v(J) ,$$

which implies that

$$\text{distance between } \kappa J \text{ and the nearest integer} = v^*(J) .$$

If Φ_{m} is infinite, the satellite is recurrent, i.e., perfectly recurrent (and hence probably deliberately recurrent). If Φ_{m} is not infinite, the recurrence is said to be imperfect.

When we seek the recurrence characteristics from the orbital elements, the recurrence may turn out to be imperfect, and in this case, the quantities $u(J)$ and $u^*(J)$ are not whole numbers.

7.6.3 Applications of the Recurrence Index

We shall now give several example applications to show how the recurrence index is used, where $\Phi(J)$ is recorded over a time span of several months. The keys to the graphs carry the orbital characteristics of the satellite, the values of the cycle, with the maximal recurrence index, which shows whether recurrence is perfect or not, and the two basic quantities in this analysis, the daily orbital frequency ν and the daily recurrence frequency κ.

The graphs clearly show the cycles and the subcycles which stand out to varying degrees. It is also very easy to distinguish those satellites that are deliberately recurrent from those with a certain level of recurrence but which have not been intentionally attributed any recurrent behaviour.

Example 7.13. *Recurrence index for Sun-synchronous and non-Sun-synchronous satellites, whether or not recurrence is intended.*

Unless specifically stated, the satellites are Sun-synchronous and recurrence is intended.

- Change of subcycle leading to more rapid coverage of the base interval. We indicated earlier how, for the Landsat and IRS satellites, a modification of the subcycle $D_{\mathrm{T_o}}$ radically changed the way in which the base interval was covered. For IRS-1A and IRS-P2, shown in Fig. 7.13, the graph of the recurrence index in Fig. 7.14 shows this change. For IRS-1A, the ground track moves steadily across the base interval in 22 days. For IRS-P2, the ground track approaches the initial ground track on four occasions during the 24 day cycle, on days $J = 5, 10, 14$ and 19, or roughly every 5 days.

- Recurrence index for satellites in the SPOT and Terra families. These two families of remote-sensing satellites include a great many satellites, whose recurrence index is shown in Fig. 7.15. For SPOT, the index has 4 peaks in the 26 day cycle, on days $J = 5, 10, 16$ and 21, indicating a passage very close to the initial ground track roughly every 5 days. For Terra, there are two main peaks in the 16 day cycle, on days $J = 7$ and 9.
- Recurrence index for a satellite with very long cycle. The satellite ADEOS-1 has a relatively long recurrence cycle, with $C_{\mathrm{T}_o} = 41$ days. The recurrence index shown in Fig. 7.16 (upper) has peaks about every 4 days. Referring to Example 7.7, we see that the two main intermediate peaks, for $J = 15$ and 26, correspond to $u^* = 1$, the next two, for $J = 11$ and 30, correspond to $u^* = 2$, and so on.
- Recurrence index for a non-Sun-synchronous satellite. The non-Sun-synchronous satellite TOPEX/Poseidon has short recurrence cycle $C_{\mathrm{T}_o} = 10$ days, with two peaks at $J = 3$ and 7 days (see Example 7.9 and Fig. 7.16, centre). We note that κ and ν have different values: the cycle takes $C_{\mathrm{T}} = 9.916$ days. For ICESat, below the subcycle $J = 25$, there are two peaks at $J = 8$ and $J = 15$. The base interval is almost swept out in 8 days, as can be seen from Fig. 7.12.
- Recurrence index for a satellite which has not been designed to be recurrent. Resurs-O1-4 was launched without concern for recurrence conditions. The recurrence index shown in Fig. 7.16 (bottom) shows that there is no clear cycle. There is a main peak for $J = 134$ (with Φ_{m} which is not infinite) and two secondary peaks for $J = 37$ and 97 in this pseudocycle.

7.6.4 Recurrence Index and Orbital Characteristics

The recurrence index and recurrence itself are obviously sensitive to changes in inclination, and very sensitive to changes in altitude, even very small ones. The following example shows how a change in altitude of a few hundred metres can completely change the recurrence characteristics after a few weeks. Precisely recurrent satellites are moved back onto the nominal orbit as soon as the altitude varies by a fraction of a kilometre. This maneuver is required between one and four times a month.

Example 7.14. *Recurrence index and altitude variations for SPOT-5.*

Consider a Sun-synchronous satellite of the SPOT-5 type, with recurrence triple $[14, 5, 26]$. The recurrence index is plotted for various altitudes in Fig. 7.17, which graphs the function $\Phi(J, \Delta h)$. For $\Delta h = 0$, i.e., for the value of the altitude giving the required recurrence, we see that the main peaks occur for values of J that are multiples of $C_{\mathrm{T}_o} = 26$ days. The secondary peaks are clearly visible, 5 days before and after the main peak, corresponding to the recurrence subcycle. They become the main peaks for neighbouring altitudes. The recurrence triple $[14, 5, 26]$, for $\Delta h = 0$,

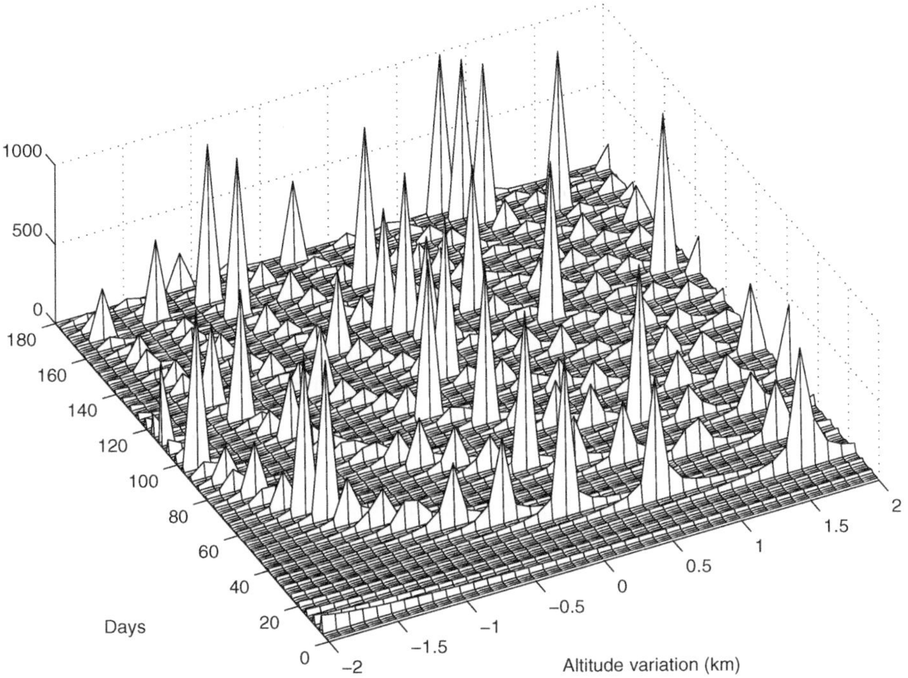

Figure 7.17. Recurrence index $\Phi(J, \Delta h)$ as a function of the day J and the altitude variation Δh for a satellite with the characteristics of SPOT-5. The central value $\Delta h = 0$ corresponds to the altitude giving the recurrence $C_{\mathrm{T_o}} = 26$, with the triple $[14, 5, 26]$. The value $\Phi = \infty$ is denoted by 1000

becomes $[14, 4, 21]$ for $\Delta h = +0.6$ km and $[14, 3, 16]$ for $\Delta h = +1.6$ km. Reducing the altitude, this initial recurrence behaviour becomes $[14, 6, 31]$ for $\Delta h = -0.4$ km, $[14, 7, 36]$ for $\Delta h = -0.7$ km, $[14, 8, 41]$ for $\Delta h = -1.0$ km, and so on.

7.7 Altitude Variations

The following analysis of altitude and frozen orbits is valid for any type of orbit, but only proves useful for near-circular LEO orbits. If the orbit is not close to circular, the altitude variations of the satellite during its revolutions are due to the eccentricity of the orbit, compared with which the flattening of the Earth is negligible. For MEO satellites, the altitude is not the relevant quantity. The same is true for GEO satellites, where the altitude is constant in time, since the satellite is stationary.

7.7.1 Altitude and Orbital Parameters

As we have seen, the altitude does not constitute a precise way of defining the position of the satellite, and indeed, it is not one of the six orbital parameters. The so-called near-circular, or even circular, orbit is never strictly circular and the Earth is not exactly spherical.

In previous chapters, devoted to the satellite ground track, the altitude has not been the main subject of discussion. However, it will be important in the following chapters, when we study the way in which the instruments aboard the satellite observe the Earth, i.e., how they 'see' it from a certain height.

The altitude of the satellite is found from the difference between the radius vector $r(a, e, v)$ defining the position of the satellite, as given by (1.39) and (1.53), and the Earth radius $R_\mathrm{T}(\phi)$ for the relevant latitude ϕ, treating the Earth as an ellipsoid of revolution with flattening[5] f. This value $R_\mathrm{T}(\phi)$ is given by (1.93), in which $\rho(\phi)$ represents $R_\mathrm{T}(\phi)$ and where the semi-major axis of the ellipse is taken to be $R = R_\mathrm{T}(0)$. We see that, with this definition, the altitude does not take into account the relief of the Earth's surface.

We write the altitude h of the satellite in the form

$$h = r(a, e, v) - R_\mathrm{T}(\phi) \ . \tag{7.32}$$

The latitude ϕ is related to i, ω, v by (3.47), and we obtain h in the form

$$h = h(a, e, i, \omega, v) \ . \tag{7.33}$$

The altitude is thus written as a function of five orbital parameters. The parameter Ω is not involved, since the terrestrial longitude is irrelevant here: we consider the ellipsoid $R_\mathrm{T}(\phi)$ rather than the geoid $R_\mathrm{T}(\lambda, \phi)$. We find the same case as was considered in (3.48).

The variation of the altitude is shown schematically in Fig. 7.18. The axis Ox lies in the Earth's equatorial plane, and the axis Oz is the polar axis. The difference between the two semi-axes of the ellipse representing the Earth is 21.4 km (see Example 1.2). The trajectory represented is that of a satellite in low, strictly polar orbit, with perigee over the North Pole ($\omega = 90°$). For an eccentricity of the order of 10^{-3}, the distance FC, between the focus F of the ellipse (centre of attraction, centre of the Earth) and the centre C of the ellipse, equal to ae, is of the order of 8 km. The orbit is near-circular. In Fig. 7.18, the eccentricities of the ellipses have been greatly exaggerated for illustrative purposes.

[5] The flattening f of the Earth is small enough to identify, in this application, the geocentric altitude, measured along a radius of the Earth, with the geodetic altitude, measured along the normal to the ellipsoid (see Sect. 1.10 and Fig. 1.6). Moreover, we neglect the dependence of $R_\mathrm{T}(\phi)$ on the longitude (related to the sectorial and tesseral terms in the expression for the geopotential). We saw in Chap. 3 that the difference in level of the ellipsoid and the geoid can reach several tens of metres, with a maximum of 100 m.

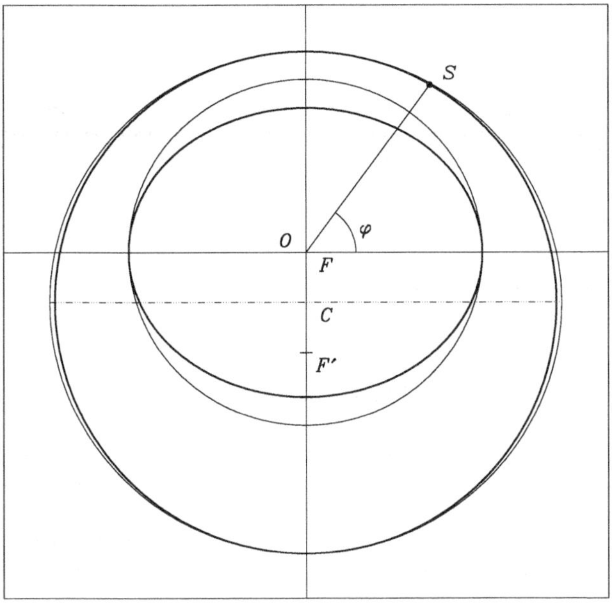

Figure 7.18. Schematic representation of the Earth ellipsoid, with centre O, and the elliptical trajectory of the satellite S, with centre C and focus F identified with O. The principal circles of the ellipses are indicated. The eccentricities used in the figure have been greatly exaggerated compared with the true eccentricities

For a given revolution, we consider in (7.33) the mean values of the orbital parameters a, e, i and ω. Instead of v, we have chosen α to determine the position of the satellite on its orbit. We have already seen that this angle, $\alpha = \omega + v$, specifies the position of the satellite as measured from the ascending node.

The altitude h is thus expressed in terms of the position on orbit α by

$$h(\alpha) = r(\alpha) - R_{\mathrm{T}}(\alpha) \,, \tag{7.34}$$

$$r(\alpha) = r\big[a, e, v(\omega, \alpha)\big] = \frac{a(1-e^2)}{1 + e\cos v} \,, \tag{7.35}$$

$$R_{\mathrm{T}}(\alpha) = R_{\mathrm{T}}\big[R, f, \phi(i, \alpha)\big] = \frac{R}{\sqrt{\cos^2\phi + \dfrac{\sin^2\phi}{(1-f)^2}}} \,, \tag{7.36}$$

where

$$v = v(\omega, \alpha) = \alpha - \omega \,, \qquad \phi = \phi(i, \alpha) = \arcsin(\sin i \, \sin\alpha) \,.$$

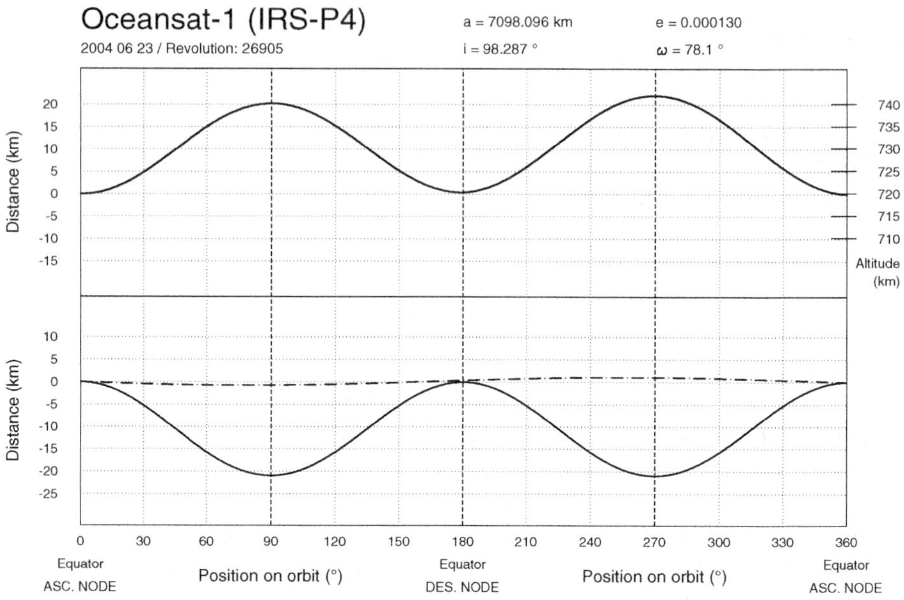

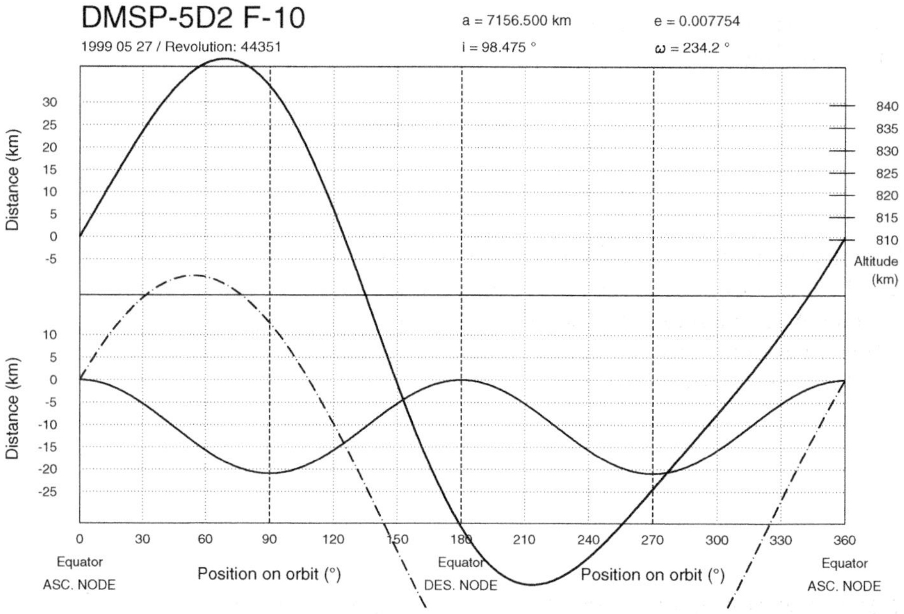

Figure 7.19. Altitude of the satellite as a function of its position on orbit during one revolution. For more detail, see Fig. 7.22. *Upper*: Oceansat-1 (IRS-P4). *Lower*: DMSP-5D2 F-10

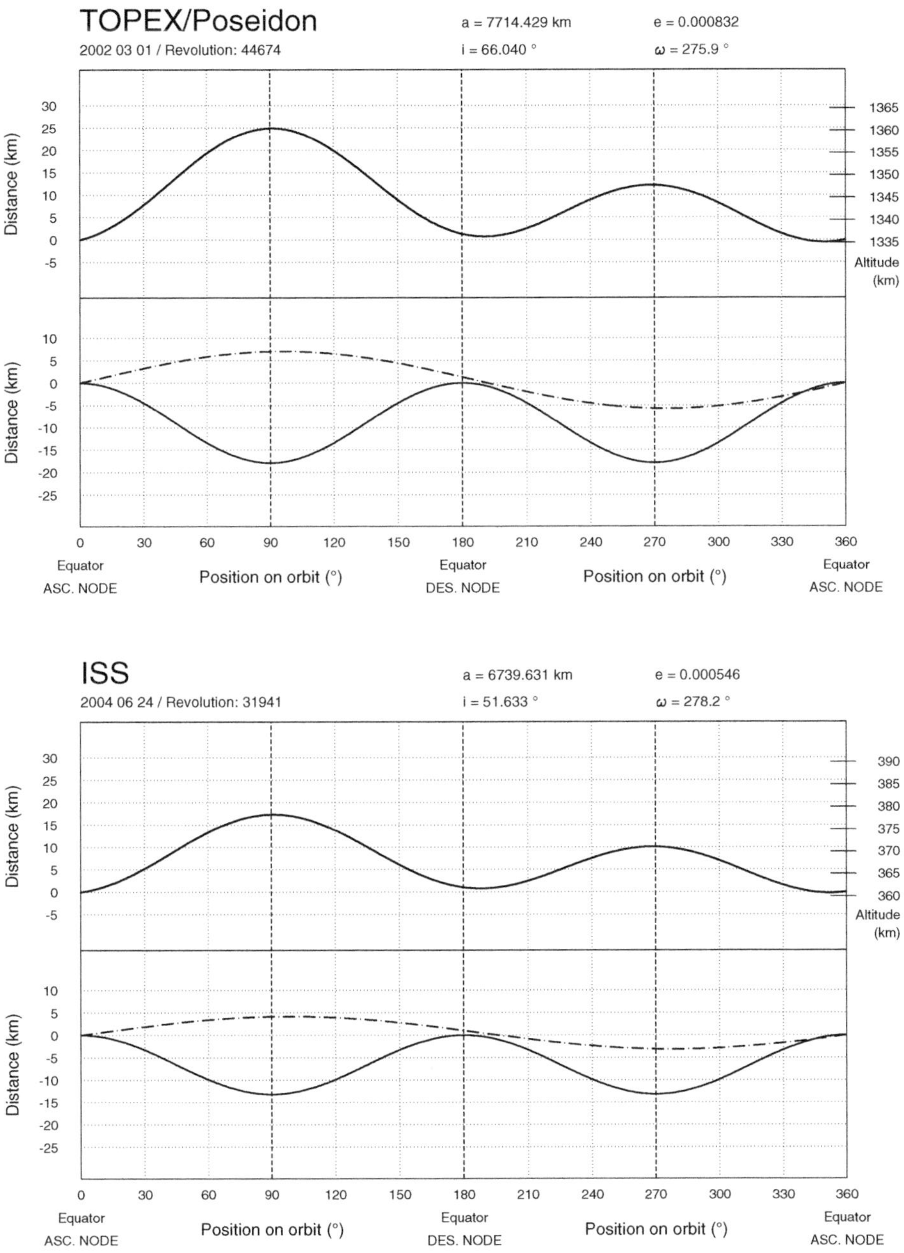

Figure 7.20. Altitude of the satellite as a function of its position on orbit during one revolution. For more detail, see Fig. 7.22. *Upper*: TOPEX/Poseidon. *Lower*: ISS, International Space Station

7.7.2 Altitude During One Revolution

Defining the function $h(\alpha)$ in this way, we note certain specific values of the altitude: $h(0)$ at the equator (ascending node), $h(\omega)$ at the perigee, $h(\pi)$ at the equator (descending node), $h(\omega + \pi)$ at the apogee. The function $r(\alpha)$ has period 2π and amplitude ae. The function $R_T(\alpha)$ has period π and its amplitude varies between 21.4 km for polar satellites (value of the product Rf) and 0 for equatorial satellites, because in this case, $R_T(\alpha) = R$ for all α.

When we give the height of a satellite as a function of its position on orbit, we must specify the revolution, or at least the day, because of the displacement of the perigee.

For the satellites in near-circular orbit that we are concerned with here, the difference between the anomalies v and M is very small (see Figs. 1.3, 1.4, and Example 2.1) and we will be able to replace α by the time t, using the relation $\alpha = 2\pi(t/T)$.

Example 7.15. *Altitude $h(\alpha)$ of the LEO satellites Oceansat-1, DMSP-5D2 F-10, TOPEX/Poseidon and ISS, during one revolution, as a function of the position on orbit.*

The metric orbit elements are practically constant over a time span of several days. However, the argument of the perigee ω can change quickly (apsidal precession). This is why the day and number of the revolution must be specified. The values of the orbital parameters are given in the corresponding figure. These data are provided by NORAD (North American Aerospace Defence Command).
For each satellite, the graphs are divided into two parts:

- on the lower part, the dashed curve gives

$$r(\alpha) - r(0) ,$$

  and the continuous curve gives

$$R_T(\alpha) - R_T(0) .$$

- on the upper part, we have plotted the altitude relative to the altitude at the equator, i.e.,

$$h(\alpha) - h(0) ,$$

  which is the difference between the two previous curves.

The altitude at the equator is obtained from

$$h(0) = \frac{a(1-e^2)}{1 + e\cos\omega} - R .$$

For easier comparison of these variations, all graphs are plotted on the same scale. Figure 7.19 shows the altitude of two Sun-synchronous satellites. As they have the same inclination, the contribution from the Earth ellipsoid in the variation of the

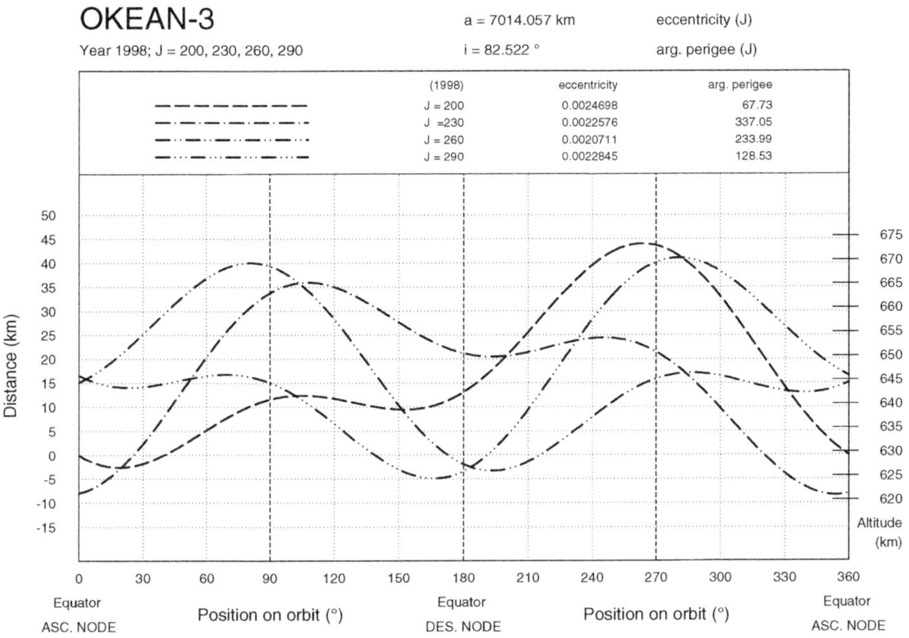

Figure 7.21. Representation of the altitude for the satellite Okean-3 as a function of its position on orbit α, expressed in degrees. Four revolutions are shown, with an interval of 30 days. *On the left*, the origin of the distance scale is the equatorial crossing for the day origin ($J = 200$). *On the right*, the scale indicates altitudes measured from the reference ellipsoid

altitude over one revolution is the same. It is the eccentricities of the orbits that are very different in these two cases. For Oceansat-1 (Fig. 7.19, upper), the eccentricity is very low. For DMSP-5D2 F-10 (Fig. 7.19, lower), it is relatively high for a near-circular orbit, with a value of $e \approx 8 \times 10^{-3}$, giving a fairly large amplitude between the apogee and the perigee. The altitude varies in this case between 730 and 850 km. Figure 7.20 shows the altitude of two prograde satellites. The oceanographic and geodetic satellite TOPEX/Poseidon has low eccentricity. The variation in altitude (between 1335 and 1360 km) is mainly due to the shape of the Earth (Fig. 7.20, upper). We shall see shortly that, for this satellite, the argument of the perigee remains practically fixed in time. For a satellite with medium inclination, such as the International Space Station ISS, the contribution from the Earth ellipsoid to the altitude variation is only 13 km (Fig. 7.20, lower). For satellites with still lower inclination, such as Rocsat-1 or TRMM (the maximal latitude attained is 35°), the difference between $R_{\mathrm{T}}(\alpha)$ and R is at most 7 km. It is then the value of the eccentricity e which is the determining factor in altitude variations.

7.7.3 Variation of the Altitude over a Long Period

If we consider the altitude over a period of several months, the elements a and i remain practically constant. The eccentricity e fluctuates around a central value and the argument of the perigee ω undergoes a secular variation to which one must add secondary periodic variations. The form of the altitude variation $h(\alpha)$ over one revolution depends mainly on the value of ω and to a lesser extent on the value of e, as we shall show in the following example.

Example 7.16. *Altitude as a function of the position on orbit $h(\alpha)$ for the satellite Okean-3, on various days of the year 1998.*

The satellite Okean-O1-3, generally called Okean-3, was launched on 4 June 1991. Figure 7.21 shows the altitude $h(\alpha)$ for four days during the year 1998, 30 days apart: $J = 200$ (19 July), $J = 230$ (18 August), $J = 260$ (17 September), $J = 290$ (17 October). During this 90 day period, a did not change significantly, whilst i varied between 82.521° and 82.523°. The eccentricity fluctuated between 2.07×10^{-3} and 2.47×10^{-3}. For its part, the argument of the perigee varied by −90.68°, −103.06° and −105.46° per 30 day interval. A small, long-period variation (of the order of a few months) is superposed on the secular variation (proportional to the time) in ω.
Calculating the mean apsidal precession rate, we find

$$\dot{\omega} = -\frac{360 + 67.73 - 128.53}{90} = -\frac{299.20}{90} = -3.32^\circ\text{day}^{-1} ,$$

which agrees with the value calculated in Example 4.3, where we found $\dot{\omega} = -3.31^\circ\text{day}^{-1}$. We see how, for ω, the periodic variations are superposed on the secular variation. The values of $e(J)$ and $\omega(J)$ are indicated in Fig. 7.21.
The graphs of $h(\alpha)$ have been plotted for the four days chosen. The right-hand scale takes its origin as the altitude of the equatorial crossing for day $J = 200$. The left-hand scale indicates the altitude relative to the reference ellipsoid.

7.8 Frozen Orbit

7.8.1 Definition of a Frozen Orbit

According to the expression for h given by (7.33), we see that the altitude of the satellite relative to its given subsatellite point varies in time, very slightly from one revolution to the next, but quite significantly over a time span of several days, as we saw in Example 7.16. When we require a satellite to be recurrent (cycle C_T), the aim is to obtain identical viewing conditions every C_T days. However, when we arrange for recurrence, it is the ground track which is fixed, not necessarily the altitude.

This may be a drawback for Earth-observation satellites, for which one generally requires the altitude to be constant for a given subsatellite point, from one crossing to the next, in order to compared the images obtained at different dates.

We would thus like to arrange for the altitude to depend only on the latitude of the subsatellite point, without variation in time. If these conditions are satisfied, we say that the orbit is frozen.[6] Note that the freezing of an orbit is independent of its recurrence characteristics, but that, in practice, only recurrent satellites (whether Sun-synchronous or otherwise) have frozen orbits.

7.8.2 Determining the Frozen Parameters

Consider now the relation (7.33) giving the altitude of the satellite above an arbitrary given point of latitude ϕ, as a function of the osculating, i.e., instantaneous, orbital parameters. Expressing the time directly in each of the orbital elements rather than via the true anomaly v, (7.33) becomes

$$h(t) = h\big[a(t), e(t), i(t), \omega(t)\big] \ .$$

We only take into account long period or secular variations here. Short period variations are averaged over one orbital revolution. In such conditions, as seen in Chap. 3 and shown schematically in Fig. 3.3, the semi-major axis a does not change. Equation (3.64) shows that, as long as the inclination is nonzero, the variation in i is negligible compared with the variation in e, which is illustrated in Example 7.16.

Equation (7.33) thus simplifies to

$$h(t) = h\big[e(t), \omega(t)\big] \ .$$

Consequently, the two parameters that concern us here, e and ω, undergo a long period variation due to the odd zonal terms, mainly J_3, whilst ω also undergoes a secular variation due to the even zonal terms, mainly J_2. Writing down the equations representing e and ω, whose values result from a complete treatment of the Lagrange equations, we must solve

$$\begin{cases} \dot{e} = 0 \ , \\ \dot{\omega} = 0 \ , \end{cases}$$

where $\dot{e} = \mathrm{d}e/\mathrm{d}t$ and $\dot{\omega} = \mathrm{d}\omega/\mathrm{d}t$ are functions of the unknowns e and ω, and to a lesser extent of the other orbital parameters.

It is a very complex matter to establish the conditions for the frozen orbit in the general case. We shall find them for an expansion up to degree 3.

[6] The first publications treating the subject of frozen orbits date back to 1965. They concerned satellites in low orbit around the Moon. The term 'frozen orbit' was first used to describe Seasat in 1976.

Using the perturbing potential $\mathcal{R} = \mathcal{R}_2 + \mathcal{R}_3$ calculated in Example 3.1 in the Lagrange equations, we obtain

$$\frac{\dot{e}}{n} = \frac{3}{2(1-e^2)^2} J_3 \left(\frac{R}{a}\right)^3 \left(1 - \frac{5}{4}\sin^2 i\right) \sin i \cos\omega \,, \tag{7.37}$$

$$\frac{\dot{\omega}}{n} = \frac{3}{(1-e^2)^2} J_2 \left(\frac{R}{a}\right)^2 \left(1 - \frac{5}{4}\sin^2 i\right) \left[1 + \frac{1}{2e(1-e^2)} \frac{J_3}{J_2} \left(\frac{R}{a}\right) \sin i \sin\omega\right] . \tag{7.38}$$

If we neglect J_3 in (7.37) and (7.38), we find $\dot{e} = 0$ and the relation (3.55), respectively.

Inclination Close to the Critical Inclination

By the very definition of the critical inclination given in (3.57), the term $[1 - (5/4)\sin^2 i]$ is zero. We thus have $\dot{\omega} = 0$ and $\dot{e} = 0$. The orbit is frozen, whether the eccentricity is low or high.

For inclinations close to the critical inclination, large oscillations in ω are due to the J_4 term (and following) since the contribution from J_2 and J_3 is very small. The expressions (7.37) and (7.38) must be expanded to higher degrees. It can then be shown that, for i between 53° and 74° (or between 106° and 127°), the eccentricity for a frozen orbit varies between 0 and 30×10^{-3}. For example, for $i = 66.4°$, we need $e_{\mathrm{F}} = 0.8 \times 10^{-3}$ and $\omega_{\mathrm{F}} = -90°$ (values chosen for TOPEX/Poseidon), where the frozen values have been given the subscript F.

Inclination Far from the Critical Inclination

For inclinations far from the critical inclination ($i < 53°$, $74° < i < 106°$, or $i > 127°$), which happens for typical Sun-synchronous satellites, referring to (7.37), we obtain $\dot{e} = 0$ by taking $\omega = \pm 90°$. Substituting this value, written ω_{F}, into (7.38), we obtain $\dot{\omega} = 0$ by setting the expression between square brackets equal to zero, i.e., setting

$$e_{\mathrm{F}} = e(\omega_{\mathrm{F}}) = -\frac{1}{2}\frac{J_3}{J_2}\frac{R}{a} \sin i \sin\omega_{\mathrm{F}} \,. \tag{7.39}$$

Since this calculation refers to near-circular orbits, we have neglected e^2 compared to 1 in (7.38). We could also use (3.67).

The sign of $\sin\omega_{\mathrm{F}}$ depends on the sign of J_3 in such a way that the expression for e comes out positive. For the Earth, with J_3 negative, we take

$$\omega_{\mathrm{F}} = +90° \,, \tag{7.40}$$

$$e_F = -\frac{J_3}{2J_2}\frac{R}{a}\sin i \ . \tag{7.41}$$

When the orbit is frozen, the perigee of a Sun-synchronous satellite thus lies practically above the North Pole.[7]

An example application of (7.41) for the SPOT satellites yields $e_F = 1.03 \times 10^{-3}$. The exact value of the frozen eccentricity is $e_F = 1.14 \times 10^{-3}$.

Frozen Eccentricity

For LEO satellites, the value e_F is always small (of the order of 10^{-3}), and the orbit is thus invariably near-circular. Many recurrent satellites, whether Sun-synchronous or otherwise, have a frozen orbit. For certain satellites, it is crucial to maintain[8] the parameters e and ω at the reference values[9] e_F and ω_F, respectivement.

We should also note that some satellites require a non-frozen orbit. To ensure that the perigee does not remain permanently in the vicinity of the North Pole and that the gravitational field is sampled at different altitudes, the geodetic satellite GOCE has variable eccentricity, ranging between 0 and 4.5×10^{-3}. Hence, the altitude varies between 236 and 282 km.

7.8.3 Altitude of a Satellite on a Frozen Orbit

In the case of a frozen orbit, where the ellipse representing the satellite trajectory is fixed in the orbital plane, we may calculate the altitude of the satellite

[7] However, we should not forget the special case of Sun-synchronous HEO satellites with critical inclination (orbits of the Ellipso Borealis type, with $e = 0.3463$). In this case, the orbit can be frozen with $\omega_F = -90°$.

[8] For example, for the satellite TOPEX/Poseidon, between 1992 and 2002, we note the following very narrow intervals of variation for the orbital elements: e_F from 0.73 to 0.83×10^{-3}, ω_F from 264 to 270°, i from 66.037 to 66.046°, a from 7714.422 to 7714.436 km.

[9] Frozen eccentricity for various satellites, announced theoretical value. The value of e_F given in brackets after each satellite should be multiplied by 10^{-3}. Satellites are listed in order of increasing altitude for each category:

- Non-Sun-synchronous satellites with inclination close to the critical inclination: Geosat (0.80), Seasat (0.80), TOPEX/Poseidon (0.75).
- Sun-synchronous satellites: Big Bird-1 (1.36), ERM (1.17), Aeolus-ADM (1.17), JERS-1 (1.23), LSPIM (1.17), ALOS (1.050), Landsat-4, -5, -7, Terra, Aqua, Calipso (1.1722), Oceansat-1 (1.13), CBERS-1, -2 (1.10), ERS-1, -2, Envisat (1.165), Radarsat-1 (1.15), ADEOS-1 (1.15), ADEOS-2 (1.055), MetOp-1 (1.1655), SPOT-1, -2, -3, -4, -5 (1.144), NOAA-6, -7, TIROS-N (1.13), Landsat-1, -2, -3 (1.11), MOS-1, -1B (0.9726), Nimbus-7 (1.00), NOAA-5 (0.97).
- Near-polar satellite: ICESat (1.30).

The true values (given by the NORAD elements) often differ somewhat from the values announced.

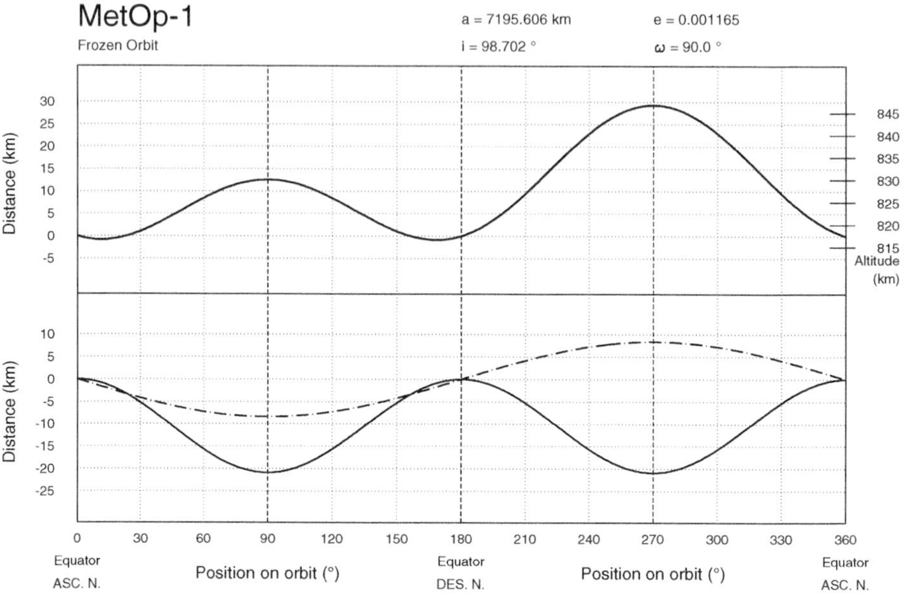

Figure 7.22. Representation of the altitude for the satellite MetOp-1, with frozen orbit. The altitude is given as a function of the position on orbit α, expressed in degrees. *Lower*: The difference between the distance $r(\alpha)$ of the satellite from the centre of the Earth and the distance $r(0)$ at its ascending node crossing is shown by a *dashed curve*. The difference between the Earth radius $R_T(\alpha)$ and the equatorial radius $R = R_T(0)$ is shown by a *continuous curve*. *Upper*: Difference between the first and second of the two curves in the lower part. *Left*: Scale for the differences. *Right*: Scale for the altitudes relative to the reference ellipsoid

as a function of a single variable, e.g., α, over a period T. The altitude variation will then repeat itself identically with period T.

We give here an example calculation of an altitude and altitude variation as a function of position on orbit.

Example 7.17. *Calculate the altitude as a function of the position on orbit for the satellite MetOp-1, which has a recurrent and frozen Sun-synchronous orbit.*

The orbital characteristics of MetOp-1 provided by ESA are as follows: recurrence of 412 revolutions in 29 days, Sun-synchronous inclination, frozen orbit: eccentricity $e = e_F = 0.001\,165\,5$, argument of the perigee $\omega = \omega_F = 90.0°$. The satellites MetOp-2 and MetOp-3 should have the same orbit. We calculate the orbital elements from the recurrence triple $[14, 6, 29]$ using the method described in Example 7.1. This yields

- semi-major axis $a = 7\,195.606$ km,
- Sun-synchronous inclination $i = i_{HS} = 98.702°$.

Knowing a and e, we calculate b and c:

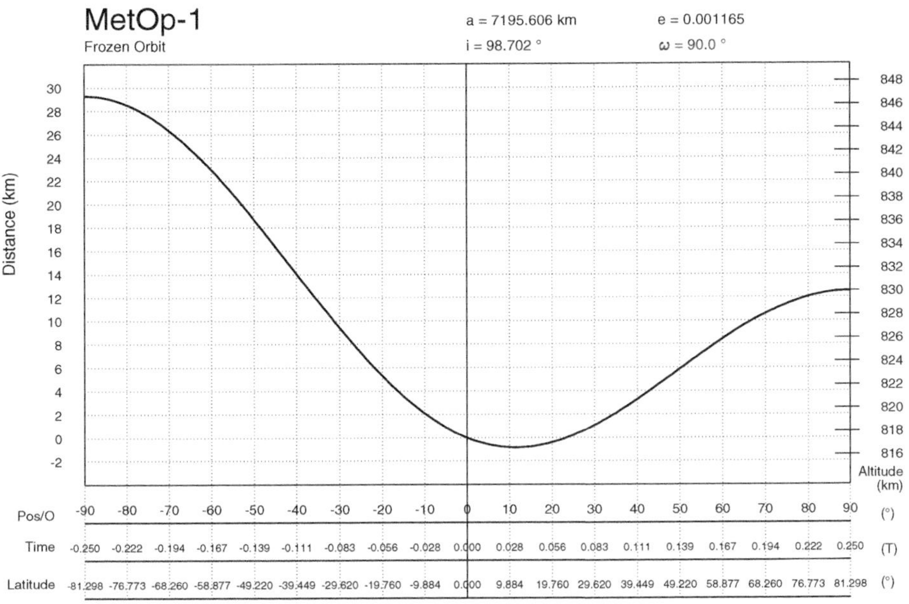

Figure 7.23. Altitude of the satellite MetOp-1, with frozen orbit. The altitude is given as a function of the position on orbit α, expressed in degrees. The geographic latitude ϕ is given for each value of α marked. The time is given as a fraction of the period, with origin at the equatorial crossing (ascending node). Detail from Fig. 7.22

$$c = ae = 8.386 \text{ km} , \quad b = a\sqrt{1-e^2} = 7195.596 \text{ km} , \quad a - b = 0.010 \text{ km} .$$

We have already noted that the orbit of a satellite of this kind is a circle shifted by 8 km relative to the centre of the Earth. The distance c is proportional to e, whereas the difference between the semi-axes is a function of e^2: a and b are thus equal to within 10 m. In Fig. 7.22, distances are expressed as a function of the position on orbit α. The graph of $r(\alpha) - r(0)$ shows a difference of $2ae = 16.772$ km between the radius at apogee $r_{\mathrm{a}} = r(\omega_{\mathrm{F}} + \pi)$ and the radius at perigee $r_{\mathrm{p}} = r(\omega_{\mathrm{F}})$. Concerning the Earth ellipsoid, the maximum latitude attained, for $\alpha = \omega_{\mathrm{F}}$, is $\phi = \pm\phi_{\mathrm{m}}$, with $\phi_{\mathrm{m}} = 180 - i = 81.298°$, which gives at the perigee

$$R_{\mathrm{T}}[\phi(\omega_{\mathrm{F}})] = R_{\mathrm{T}}(\phi_{\mathrm{m}}) = R_{\mathrm{m}} = 6357.240 \text{ km} , \quad R - R_{\mathrm{m}} = 20.897 \text{ km} .$$

We now calculate the values of the altitude for particular points on the orbit (equator, ascending and descending nodes, perigee, and apogee):

Asc. node	$h(0) = a(1-e^2) - R = 7195.596 - 6378.137 = 817.459$,
Perigee	$h(\pi/2) = a(1-e) - R_{\mathrm{m}} = 7187.223 - 6357.240 = 829.983$,
Desc. node	$h(\pi) = a(1-e^2) - R = 7195.596 - 6378.137 = 817.459$,
Apogee	$h(3\pi/2) = a(1+e) - R_{\mathrm{m}} = 7203.989 - 6357.240 = 846.749$,

$$h(\pi) = h(0) \, , \qquad h(3\pi/2) = h(\pi/2) + 2ae \, .$$

The minimum altitude is 816.623 km, reached when $\alpha = 11.512°$ and $\alpha = 168.488°$, corresponding to a latitude $\phi = 11.378°$. The maximum altitude is attained at the apogee.
Since the position of the perigee is symmetric with respect to the two nodes for this satellite, the altitude depends only on the latitude ϕ, and we do not have to specify the crossing direction (ascending or descending) of the orbit. In Fig. 7.23, we have indicated the latitude on the abscissa, between its two extreme values $-\phi_{\mathrm{m}}$ and $+\phi_{\mathrm{m}}$.

7.9 Appendix: Grid Points for Recurrent Satellites

When a satellite is recurrent, its ground track forms a grid fixed on the Earth (the reference grid discussed above). A grid point is any place where two ground tracks intersect, one in the ascending part and the other in the descending part. It can be very useful to know the latitudes of these points. We have studied this problem theoretically and summarise the results here.

During a recurrence cycle ($N_{\mathrm{T_o}}$ revolutions in $C_{\mathrm{T_o}}$ days), the satellite crosses a given meridian $M_{\mathrm{T_o}}$ times:

$$M_{\mathrm{T_o}} = N_{\mathrm{T_o}} - \sigma C_{\mathrm{T_o}} \, , \tag{7.42}$$

where[10] $\sigma = \mathrm{sgn}(\cos i)$, i.e., $+1$ if $i < 90°$ and -1 if $i > 90°$.

We define the ordinate of the grid point, y in degrees, by

$$y = j \frac{90}{N_{\mathrm{T_o}}} \, , \tag{7.43}$$

where

$$j = 0, 2, 4, 6, \ldots, M_{\mathrm{T_o}} \, , \qquad \text{if } M_{\mathrm{T_o}} \text{ is even} \, ,$$

$$j = 1, 3, 5, 7, \ldots, M_{\mathrm{T_o}} \, , \qquad \text{if } M_{\mathrm{T_o}} \text{ is odd} \, .$$

The ordinates y obtained in this way are related to the latitudes ϕ of the grid points by

$$y = g_{\mathrm{S}}(\phi) \, , \tag{7.44}$$

where g_{S} is the function defined by

---

[10] The method does not work if $\cos i$ and $\cos i'$ have opposite signs, where i' is the apparent inclination. This situation never arises with existing recurrent LEO satellites.

Table 7.9. Latitude ϕ for the first values of the grid points of four recurrent satellites (latitudes north and south). The variable j is defined by (7.43). The last value in the table, for $j = M_{\mathrm{T_o}}$, defined by (7.42), corresponds to the maximal latitude ϕ_{m} attained by the satellite

Rocsat-2		Oceansat-1		T/P & Jason-1		ICESat [cal]	
j	ϕ	j	ϕ	j	ϕ	j	ϕ
1	26.415	1	14.198	1	1.977	1	5.4990
3	58.545	3	38.380	3	5.907	3	16.2909
5	70.714	5	54.317	5	9.763	5	26.4828
7	75.961	7	63.804	7	13.504	7	35.7413
9	78.615	9	69.553	9	17.096	9	43.8404
11	80.046	11	73.234	11	20.513	11	50.6936
13	80.770	13	75.720	13	23.740	13	56.3484
		15	77.471	15	26.770	15	60.9440
		⋮	⋮	⋮	⋮	⋮	⋮
15	80.992	31	81.706	117	66.040	127	86.0000

$$g_{\mathrm{S}}(\phi) = \sigma_\phi \left| \arcsin \frac{\tan\phi}{\tan i} - \frac{1}{\kappa} \arcsin \frac{\sin\phi}{\sin i} \right| , \tag{7.45}$$

and $\sigma_\phi = \mathrm{sgn}(\phi)$.

The relevant latitudes ϕ can thus be obtained from the inverse function by numerical methods:

$$\phi = g_{\mathrm{S}}^{-1}(y) . \tag{7.46}$$

Table 7.9 shows the results for several satellites. One can check these values on ground track maps such as Fig. 5.25 for ICESat [cal].

The software Ixion on the attached CDROM can be used to calculate all the grid points of a recurrent satellite.

8 View from the Satellite

In the preceding chapters, we have discussed the satellite's orbit, position, and ground track. All this can be deduced from the satellite's position S as seen from the centre of attraction O, which is the centre of the Earth. The time has come to look at things from a different standpoint: we shall now be concerned with the view from an instrument carried aboard the satellite. The main difference here is that we are now looking at things from the point of view of the satellite S. As a consequence, this chapter is principally concerned with observation satellites.

8.1 Swath of an Instrument

8.1.1 Local Orbital Frame

Up to now, the satellite has been treated as a point, or at least, we have considered only the motion of its centre of gravity. But as a vehicle, the satellite can also move about its centre of inertia. Although this kind of motion is largely irrelevant for the purposes of calculating its trajectory, it is of course crucial when we come to ask what the instruments aboard will be able to view. If we want to produce an image of the Earth, we must not aim at the sky, and conversely!

Manipulation of the angular orientation of the satellite is referred to as attitude control. The attitude of the satellite tends to vary under the action of couples, which may be external, due to radiation pressure or atmospheric drag on solar panels, or internal, due to mechanical motion of the instrument motors. A stabilisation system is thus required to maintain the satellite in the right position relative to the local orbital frame.

For any point S on the orbit, this frame is defined by the following three axes, illustrated in Fig. 8.1:

- yaw axis $\boldsymbol{SZ}_{\mathrm{c}}$, directed towards the centre of the Earth, also called the nadir axis,
- pitch axis $\boldsymbol{SY}_{\mathrm{c}}$, directed normally to the orbital plane,
- roll axis $\boldsymbol{SX}_{\mathrm{c}}$, lying in the orbital plane and completing an orthogonal right-handed system of axes. This axis lies along the velocity vector of the satellite when the eccentricity is zero.

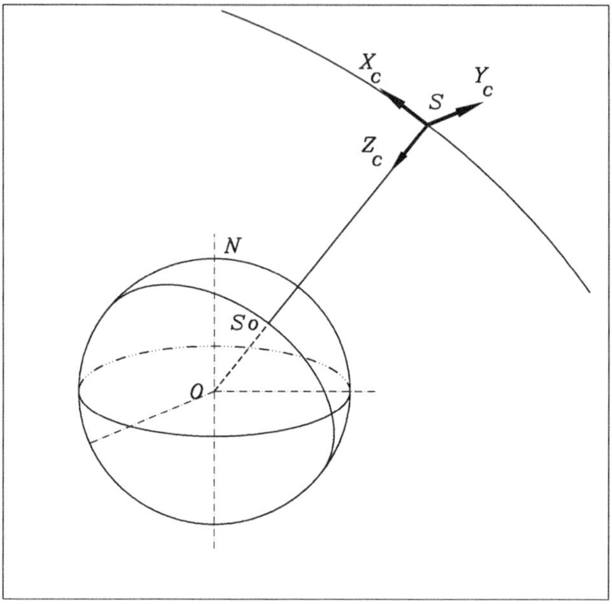

Figure 8.1. The Cardan frame centred on the satellite S. The ground track goes through the subsatellite point S_0. The axis $\boldsymbol{SZ}_{\mathrm{c}}$ points towards the centre of the Earth and the axis $\boldsymbol{SY}_{\mathrm{c}}$ is perpendicular to the orbit. If the orbit is circular, the axis $\boldsymbol{SX}_{\mathrm{c}}$ lies along the velocity vector

We shall refer to the axes of the local orbital frame as the Cardan axes, with the appropriate subscript. The angles obtained by rotation relative to these axes are the Cardan angles.[1]

8.1.2 Scanning Modes

There are various ways for an instrument to look at the Earth. The sensor can be equipped with a fixed objective relative to the satellite, but in most cases, the sensor is mobile (either itself or through the action of a mirror) along some axis of rotation.

To begin with, we may define three basic scanning modes, when the instrument rotates relative to one of the three Cardan axes. In the first two cases, the instrument axis and the axis of rotation are the same, whereas

[1] Gerolamo Cardano (1501–1576), sometimes known as Jerome Cardan in English, was an Italian mathematician. He provided a method for solving third order equations and invented a joint, now known as the Cardan or universal joint, which made the ship's compass insensitive to the yaw, pitch and roll of the vessel on the high seas. Note that the three terms all originate from the nautical context.

in the third, the instrument axis makes a constant angle with the axis of rotation. These three scanning modes are as follows:

- With a rotation about $\boldsymbol{SX}_{\mathrm{c}}$, the instrument scans perpendicularly to its displacement. This is orthogonal or across-track scanning.
- With a rotation about $\boldsymbol{SY}_{\mathrm{c}}$, the instrument scans along the ground track. This is along-track scanning.
- With a rotation about $\boldsymbol{SZ}_{\mathrm{c}}$, the instrument scans in conical mode, defined by the half-angle at the apex of the cone, which is the angle between the instrument axis and the axis of rotation.

During an observation, the smallest detected element is called a pixel, which is short for 'picture element'. The set of all such elements viewed on the ground in a single scan is called the swath.

The technical features of the various optical instruments and sensors are not the concern of this book. We shall consider only the geometrical aspects of scanning.

Scanning Mode for LEO Satellites

An instrument aboard an LEO satellite can use one of the three elementary scanning modes listed above. It can also alternate between the first two, or scan obliquely by a rotation of the instrument about an axis in the plane $SX_{\mathrm{c}}Y_{\mathrm{c}}$.

In orthogonal scanning, some instruments sweep from side to side across the swath, pixel by pixel as it were. Other instruments simultaneously record all the pixels in one row, and some can even record over several rows at once.[2]

Scanning Mode for GEO Satellites

In this case one cannot really say that there is a swath. Concerning the way images are taken, geostationary satellites fall into two main categories.

For satellites with three-axis stabilisation, such as GOES (from GOES-8) or GOMS, one axis is parallel to the polar axis, one axis points to the centre of the Earth, and one axis lies along the satellite's velocity vector. The sensor scans the disk presented by the Earth.

For rotating satellites, such as the METEOSAT, GMS or FY-2 series, the axis of rotation is parallel to the polar axis and the satellite rotates exactly

[2] Charge-coupled devices can detect a row of pixels (1D-CCD, one-dimensional) or several rows (2D-CCD, two-dimensional). Aboard SPOT-4, the HRVIR instrument uses the so-called push-broom mode with a 1D-CCD. The optical instrument is based on a telescope whose field of view is covered instantaneously by a row of 1728 detectors, each corresponding to one pixel. In the case of the POLDER instrument carried aboard ADEOS-1 and -2, and also planned for PARASOL, the use of 2D-CCDs will make it possible to obtain a set of lines simultaneously, instead of just one.

100 times per minute. A mirror is used to sweep across the Earth's disk. For the satellites METEOSAT-1 to -8, this (east–west) scanning is carried out from south to north in 2500 lines every 25 min. We note that, with this method, although it gives excellent results, the Earth is only viewed over 17.4° per revolution, i.e., the sensor views the Earth for 4.8% of the time, spending the other 95.2% of the time looking into the darkness of space (see Colour Plate III)!

Whatever method is used, the swath of a geostationary satellite will be treated like the across-track swath of a low-orbiting satellite: for a point P viewed on Earth, we consider the plane SS_0P (see Fig. 8.2) and define the angles of sight as for an LEO across-track swath. We thus define the angle S_0SP which plays the role of the half-swath angle f discussed below.

8.2 Swath Viewing Geometry

8.2.1 Definition of Angles

The region of the Earth covered by the swath of the instrument (we shall just say: viewed by the satellite) generally needs to be known with an accuracy of the order of a few kilometres. As far as our study of swath geometry is concerned, we shall thus treat the Earth as spherical.

Figure 8.2 shows all angles relevant to the satellite's view and swath. The satellite S is in orbit around the Earth at a distance $OS = d$ from the centre of the Earth O. The subsatellite point is denoted by S_0. Thus $OS_0 = R$ is the Earth radius and $SS_0 = h$ is the altitude of the satellite. We use the reduced distance

$$\eta = \frac{d}{R} = 1 + \frac{h}{R} , \tag{8.1}$$

i.e., the distance SO expressed in Earth radii and denoted by η, which we have already defined in (2.18). For a circular orbit of radius a, we have $d = a$.

At a given instant of time, the angle between the line of sight from the satellite and the nadir[3] is

$$f = (\boldsymbol{SS}_0, \boldsymbol{SP}) , \tag{8.2}$$

where the point P is the point the instrument is viewing, or target point. This angle is called the swath angle or scan angle.

[3] The nadir is the direction given by the vertical, looking downwards, i.e., towards the centre of the Earth. The opposite direction is the zenith. The word 'nadir' comes from the Arabic word *nāḍir*, from the root of the verb 'to look straight at'. اَلنَّاظِر

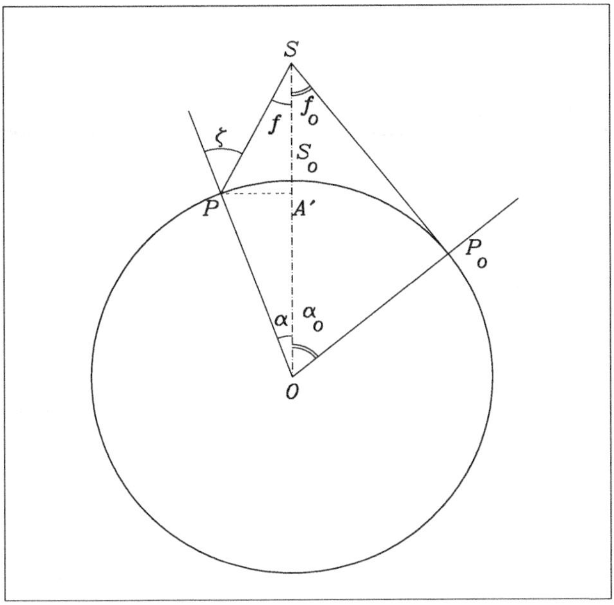

Figure 8.2. Definition of angles relevant to the swath of an instrument aboard a satellite. The satellite is S and the subsatellite point is S_0. The instrument looks at points P, such as the point P_0 on the Earth's limb. The Earth is spherical with centre O

For the point P, we define the viewing zenith angle[4] by

$$\zeta = (\boldsymbol{OP}, \boldsymbol{PS}) , \tag{8.3}$$

which is the angle at which the satellite is seen from the surface, measured from the local vertical. The elevation or site angle is the complementary angle of ζ, i.e., $90° - \zeta$.

We also use the angle α, which is the angle at the centre of the Earth defined by

$$\alpha = (\boldsymbol{OS}, \boldsymbol{OP}) . \tag{8.4}$$

These three angles are related by

$$f + \alpha = \zeta , \tag{8.5}$$

by considering the triangle OSP. The maximum value of f is obtained when the target point P is on the Earth's limb. We denote this point by P_0 and the

[4] In Arabic, *semt er-rās* means 'the path of the head'. This gives the word 'zenith', the point on the sky just above the head. The word 'azimuth' comes from *as-semt*, 'the path', with assimilation of the article. سمت الرّاس ، السمت

corresponding angles are given the subscript zero. Considering the triangle OSP_0, we obtain the relations

$$\sin f_0 = \frac{R}{d} = \frac{R}{R+h} , \qquad \alpha_0 = \frac{\pi}{2} - f_0 , \qquad \zeta_0 = \frac{\pi}{2} ,$$

or, using the reduced distance η,

$$\sin f_0 = \cos \alpha_0 = \frac{1}{\eta} . \tag{8.6}$$

Note on Terminology. The angle f defined above was called the half-swath angle. When we speak of the swath of an instrument, we generally mean the angle moved through by the instrument at the apex, i.e., $2f_{\mathrm{M}}$, where the angle f_{M} is the maximum value reached by f when the instrument arrives at the limit of its orthogonal scan. The angle at the apex is called the field of view. If f_{M} is greater than f_0, we must obviously take $f_{\mathrm{M}} = f_0$. To avoid confusion, we will therefore speak of the maximum possible half-swath to describe f_0 and the maximum instrument half-swath to speak of f_{M}.

8.2.2 Relations Between Angles

Let us now establish relations giving one of the angles f, ζ, α as a function of one of the other two and the altitude via η. We thus obtain six relations.

Relations between f and ζ. In the triangle OSP, we have the relation

$$\frac{\sin f}{R} = \frac{\sin \zeta}{d} ,$$

which yields

$$f = f(\zeta, \eta) , \qquad \sin f = \frac{1}{\eta} \sin \zeta = \sin f_0 \sin \zeta , \tag{8.7}$$

$$\zeta = \zeta(f, \eta) , \qquad \sin \zeta = \eta \sin f = \frac{\sin f}{\sin f_0} . \tag{8.8}$$

f and ζ as functions of α. To obtain f as a function of α, consider the triangle OSP, and express the segment PA' in two different ways (where A' is the projection of P on OS) to deduce that

$$(d - R\cos\alpha) \tan f = R \sin \alpha .$$

To obtain ζ as a function of α, consider the triangle OPA, where A is the intersection of OS with the line through P perpendicular to OP. This yields

$$f = f(\alpha, \eta) , \qquad \tan f = \frac{\sin \alpha}{\eta - \cos \alpha} = \frac{\cos \alpha_0 \sin \alpha}{1 - \cos \alpha_0 \cos \alpha} , \tag{8.9}$$

$$\zeta = \zeta(\alpha, \eta) \,, \qquad \tan\zeta = \frac{\sin\alpha}{\cos\alpha - 1/\eta} = \frac{\sin\alpha}{\cos\alpha - \cos\alpha_0} \,. \tag{8.10}$$

Expressions for α. We immediately obtain the values of α with (8.5) and the above relations:

$$\alpha = \alpha(f, \eta) \,, \qquad \alpha = -f + \arcsin(\eta \sin f) \,, \tag{8.11}$$

$$\alpha = \alpha(\zeta, \eta) \,, \qquad \alpha = \zeta - \arcsin\left(\frac{1}{\eta}\sin\zeta\right) \,. \tag{8.12}$$

8.2.3 Ground Swath

The ground half-swath is the distance F on the Earth's surface between the subsatellite point and the target point at angle f. The ground swath is then $2F$. The maximum ground half-swath is denoted by F_0. These lengths are given by

$$F = R\alpha \,, \tag{8.13}$$

$$F_0 = R\alpha_0 = R \arccos\frac{1}{\eta} \,. \tag{8.14}$$

In Fig. 8.2, F corresponds to the arc S_0P and F_0 to the arc S_0P_0.

Example 8.1. *Calculate the ground swath for an instrument viewing with angle $f = 45°$, aboard satellites at different altitudes: $h = 350, 700, 1050$ km, etc.*

With $350/R = 5.487 \times 10^{-2}$, we calculate η and the angles by the above formulas. The results are given in Table 8.1. We can then compare the swaths of satellites like TRMM ($h = 350$ km) or Terra ($h = 700$ km). For satellites with altitude less than 1000 km, the roundness of the Earth does not account for more than 10% of the value of the ground swath.
The limb is viewed with $f = 45°$ when $h = 2642$ km. Indeed, according to (8.6), we have $\sqrt{2} = \eta = (R + h)/R$, and hence $h = R(\sqrt{2} - 1) = 2642$ km. We can check from Table 8.1 that, for $h = 2800$ km, the swath $f = f_0$ is less than 45°.

8.2.4 Latitudes Viewed and Latitude Overlap

Viewed Latitude Range

For a satellite of inclination i, we defined the maximum latitude attained ϕ_{m} by (5.18). The ground track of the orbit lies within the latitude range

$$[-\phi_{\mathrm{m}}, +\phi_{\mathrm{m}}] \,.$$

Table 8.1. Sight angles for a half-swath $f = 45°$ for an instrument aboard a satellite at various altitudes h. Across-track swath. The angles f, ζ, α and f_0 defined in this chapter are in degrees. The altitude h and distances F and F_0 are in kilometres

h	0	350	700	1 050	1 400	2 800
f_0	90.0	71.4	64.3	59.2	55.1	44.0
f	45.0	45.0	45.0	45.0	45.0	44.0
ζ	45.0	48.2	51.7	55.4	59.6	90.0
α	0.0	3.2	6.7	10.4	14.6	46.0
F_0	0	2 066	2 861	3 433	3 887	5 118
F	0	360	745	1 162	1 623	5 118
$2F$	0	720	1 491	2 324	3 245	10 236

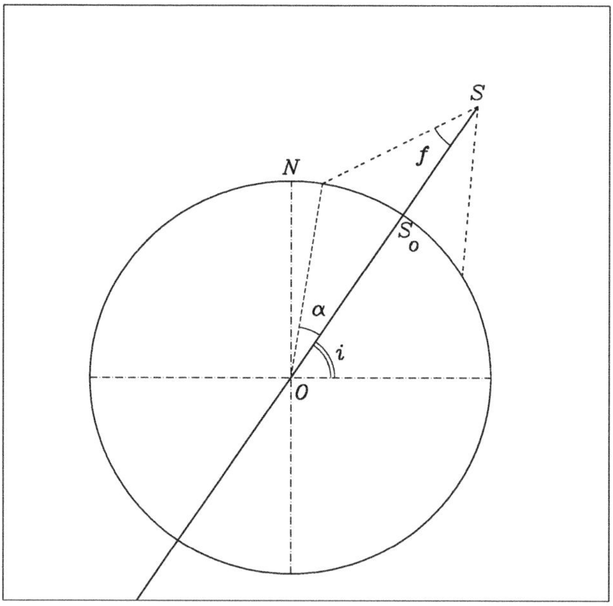

Figure 8.3. Swath orthogonal to the ground track of satellite S, with half-swath angle f. The plane of the diagram is the plane perpendicular to the orbit (of inclination i) passing through the polar axis ON, i.e., the plane of the meridian of S_0. The satellite S is at its maximum latitude

Consider the plane perpendicular to the orbit passing through the polar axis, as shown in Fig. 8.3. This is a meridian plane. With across-track scanning, it is when the satellite crosses this plane that it sees points on the Earth at the extreme latitudes. For an instrument with maximum half-swath f_M, the swath track lies in the interval

$$[-\phi_v, +\phi_v] ,$$

where the angle ϕ_v is the maximum latitude viewed as defined by

$$\begin{cases} \phi_\mathrm{v} = \phi_\mathrm{m} + \alpha_\mathrm{M} \,, & \text{if } \phi_\mathrm{m} + \alpha_\mathrm{M} < 90^\circ \,, \\ \phi_\mathrm{v} = 90^\circ \,, & \text{if } \phi_\mathrm{m} + \alpha_\mathrm{M} \geqslant 90^\circ \,, \end{cases} \tag{8.15}$$

where $\alpha_\mathrm{M} = \alpha_\mathrm{M}(f_\mathrm{M}, \eta)$ calculated from (8.11).

Latitude Overlap

When $\phi_\mathrm{m} + \alpha_\mathrm{M}$ is greater than 90°, we say that there is latitude overlap. This overlap concerns latitudes in the ranges:

$$[+90^\circ, 180^\circ - (\phi_\mathrm{m} + \alpha_\mathrm{M})] \quad \text{in the northern hemisphere} \,,$$

$$[-90^\circ, (\phi_\mathrm{m} + \alpha_\mathrm{M}) - 180^\circ] \quad \text{in the southern hemisphere} \,.$$

For a satellite in near-circular orbit, if a pole is viewed during an across-track scan, the two poles are viewed during each revolution.

Example 8.2. *Calculate the maximum latitude viewed and, in the relevant situations, the range of latitudes covered by the ScaRaB instrument ($f_\mathrm{M} = 48.91^\circ$) aboard Meteor-3-07, Resurs-O1-4 and Megha-Tropiques.*

For Meteor-3-07, $i = 82.6^\circ$, $\phi_\mathrm{m} = i$, $\eta = 1.187$, we obtain $\phi_\mathrm{m}+\alpha = 97.1$. All latitudes are viewed. Moreover, there is overlap between the pole and latitude $180 - 97.1 = 82.9^\circ$, for each hemisphere.
For Resurs-O1-4, $i = 98.7^\circ$, $\phi_\mathrm{m} = 180 - i = 81.3^\circ$, $\eta = 1.128$, we obtain $\phi_\mathrm{m} + \alpha = 90.6$. All latitudes are viewed. The overlap is very slight, between the pole and latitude 89.4°, for each hemisphere.
For Megha-Tropiques, $i = 20.0^\circ$, $\phi_\mathrm{m} = i$, $\eta = 1.136$, we obtain $\phi_\mathrm{m} + \alpha = 30.0$. The latitudes viewed lie in the band [30.0°S, 30.0°N], which corresponds to the region between the tropics (whence the name of the satellite).

8.3 Pixel Distortion

8.3.1 Calculating the Distortion Index

Consider an instrument which observes the Earth with across-track scanning. The axis of rotation of the instrument is perpendicular to the plane defined in Fig. 8.2.

Each angular interval δf of the half-swath angle corresponds to a half-swath interval δF on the ground. It is clear that, for a given constant interval, say 1°, the value of δF is smaller at the nadir (for $f = 0$) than when viewing the limb (for $f = f_0$): the satellite–target distance increases and, furthermore, the roundness of the Earth is relevant here.

The pixel, which depends on the value of the elementary interval δf of the instrument, has size δF in the scanning direction whilst its width can be considered as constant in the perpendicular direction, parallel to the axis of rotation of the instrument. To find the pixel distortion, we calculate the variation of the ratio $\delta F/\delta f$ as a function of the target point, which amounts to finding the variation of the ratio $\delta\alpha/\delta f$ as a function of α.

Differentiating (8.9), we obtain

$$(1+\tan^2 f)\,\mathrm{d}f = \frac{\eta\cos\alpha - 1}{(\eta - \cos\alpha)^2}\,\mathrm{d}\alpha\ ,$$

which yields, replacing $\tan f$ by its value as a function of α,

$$\frac{\mathrm{d}\alpha}{\mathrm{d}f} = \frac{\eta^2 - 2\eta\cos\alpha + 1}{\eta\cos\alpha - 1}\ . \tag{8.16}$$

Considering increments δf and $\delta\alpha$ small enough to identify them with $\mathrm{d}f$ and $\mathrm{d}\alpha$, we obtain

$$k(\alpha,\eta) = \frac{\delta\alpha}{\delta f}\ .$$

To measure the pixel distortion, denoted by $K(\alpha,\eta)$, we set

$$K(\alpha,\eta) = \frac{k(\alpha,\eta)}{k(0,\eta)}\ ,$$

thereby expressing $k(\alpha,\eta)$ relative to its value at the nadir. This value is $k(0,\eta) = \eta - 1$, which can be checked by calculating δF in two different ways at the nadir (using small angles), from the standpoint of S or O : $\delta F = h\delta f = R\delta\alpha$.

This pixel distortion index is thus

$$K(\alpha,\eta) = \frac{\eta^2 - 2\eta\cos\alpha + 1}{(\eta - 1)(\eta\cos\alpha - 1)}\ , \tag{8.17}$$

which can also be written using the bounding angle α_0 as

$$K = \frac{d}{h}\,\frac{\sin^2\alpha + (\cos\alpha - \cos\alpha_0)^2}{\cos\alpha - \cos\alpha_0}\ .$$

The pixel distortion index thus measures a one-dimensional distortion, the width being constant, for fixed increment δf. (We are not concerned here with scanning in which δf varies with f in such a way that δF remains roughly constant.) The function K is plotted in the following examples.

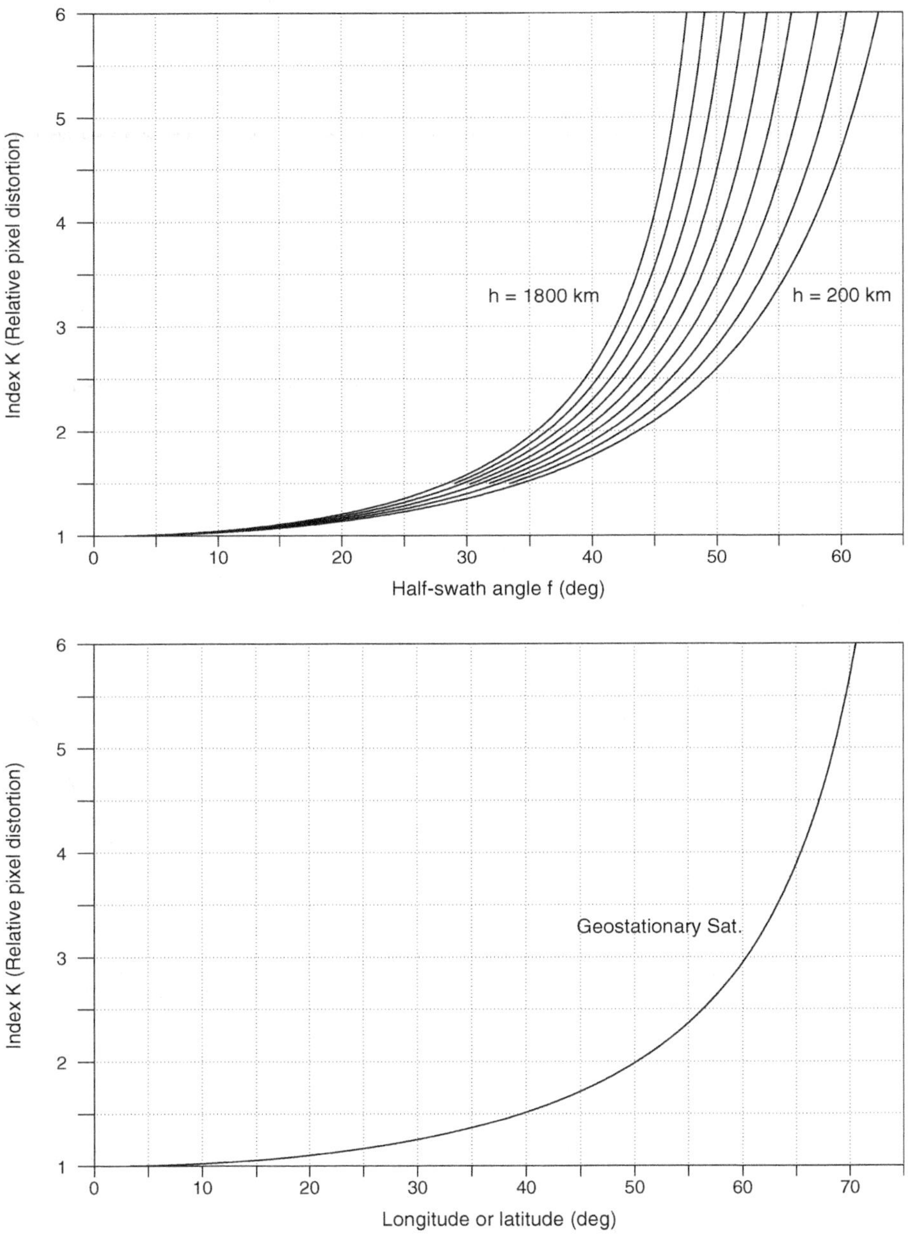

Figure 8.4. Relative pixel distortion as represented by the index K. *Upper*: For LEO satellites with altitude $h = 200$ km to $h = 1800$ km, in steps of 200 km, as a function of the half-swath angle f. *Lower*: For any geostationary satellite as a function of the angle α, representing the latitude or longitude from the subsatellite point

8.3.2 Pixel Distortion for LEO Satellites

The index K is calculated as a function of α, but the results are generally expressed in terms of the variables ζ or f. The distortion index $K(f)$ is plotted in Fig. 8.4 (upper) for LEO satellites in near-circular orbit and for altitudes between 200 and 1800 km, in steps of 200 km. The distortion index becomes large ($K > 2$) when f reaches roughly two thirds of its maximum value f_0.

Example 8.3. *Calculate the pixel size and distortion index for the ScaRaB instrument aboard Meteor-3-07 and Resurs-O1-4.*

Two identical ScaRaB instruments are carried aboard the Russian satellites Meteor-3-07 and Resurs-O1-4. Scanning is across the ground track. The maximum instrument half-scan angle, beyond which the instrument cannot view, is $f_{\mathrm{M}} = 48.91°$, giving a field of view of 97.82°. The complete scan is divided into 51 increments, which gives

$$\delta f = \frac{2f_{\mathrm{M}}}{51} = 1.92° = 33.5 \text{ milliradians} ,$$

for the pixel size. (This may be described as the effective pixel, whilst the true pixel is larger to give overlap.) This corresponds at the nadir to

$$\delta F = 40 \text{ km} \quad \text{for ScaRaB on Meteor-3-07} ,$$

$$\delta F = 27 \text{ km} \quad \text{for ScaRaB on Resurs-O1-4} .$$

At the limiting value of the scan angle, the pixel length is

$$K(f_{\mathrm{M}}) = 4.0 \quad \Longrightarrow \quad \delta F = 161 \text{ km} \quad \text{for ScaRaB on Meteor-3-07} ,$$

$$K(f_{\mathrm{M}}) = 3.2 \quad \Longrightarrow \quad \delta F = 86 \text{ km} \quad \text{for ScaRaB on Resurs-O1-4} .$$

The ground swath is $2F_{\mathrm{M}} = 3254$ km for Meteor-3-07, which is greater than the equatorial shift $\Delta\lambda_{\mathrm{E}} = 3059$ km. However, for Resurs-O1-4, the ground swath $2F_{\mathrm{M}} = 2078$ km is well below the equatorial shift $\Delta\lambda_{\mathrm{E}} = 2819$ km. The ScaRaB instrument was originally designed for satellites of type Meteor-3, at 1200 km altitude, the aim being to scan the whole planet in one day. To obtain this result aboard satellites of type Resurs-O1, at only 800 km altitude, the instrument would have been able to scan up to angles $f_{\mathrm{M}} = 55°$, and this would have given a pixel distortion of $K = 5.3$.

8.3.3 Pixel Distortion for GEO Satellites

Although a geostationary satellite sees almost half the Earth's surface, around the edge of the observed 'disk', the pixel distortion is large.

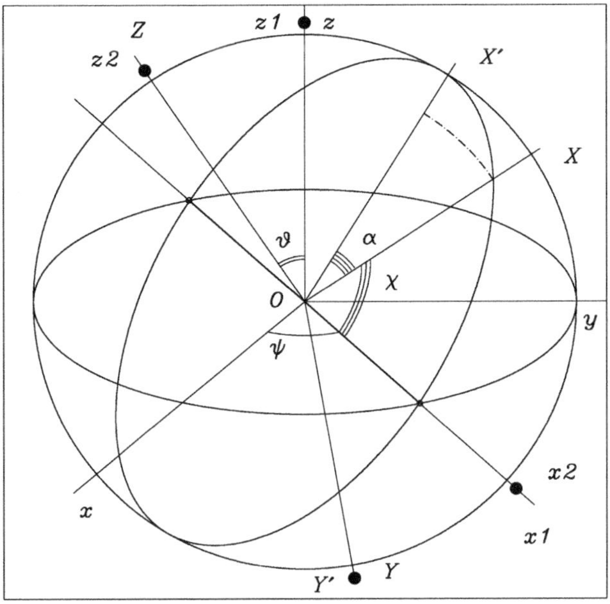

Figure 8.5. Rotation through α corresponding to the swath f of an instrument aboard the satellite. This figure complements Fig. 5.2. *Black dots* denote the four axes of rotation. The axis of rotation of the instrument is $\boldsymbol{OY}$ in the orbital plane Ox_2X, perpendicular to $\boldsymbol{OX}$. The ground track of a half-swath is shown by a *dashed curve*

The graph of $K(\alpha)$ in Fig. 8.4 (lower) shows that, for a geostationary satellite, the distortion index becomes greater than 2 beyond 50°. α can be replaced by $|\lambda - \lambda_S|$, the longitude along the equator measured from the subsatellite point, or by $|\phi|$, the latitude measured along the meridian λ_S. These graphs are clearly going to be the same for all geostationary satellites.

8.4 Swath Track for an LEO Satellite

8.4.1 Across-Track Swath

In order to calculate the coordinates of the points viewed and thus plot the ground track of the across-track scan, we return to the Euler angles first discussed when determining the subsatellite point on the ground track.

We consider that the scan with instantaneous angle f from the satellite S is equivalent to a scan with instantaneous angle α from the centre O of the Earth. The scan is in the plane perpendicular to the orbit, passing through S and O. Viewed from O, it is thus a scan with angle α, and axis $\boldsymbol{OY}$ (parallel to $\boldsymbol{SX}_c$ in the orbital plane), where the orthogonal axes $OXYZ$ were defined previously, in Fig. 5.2.

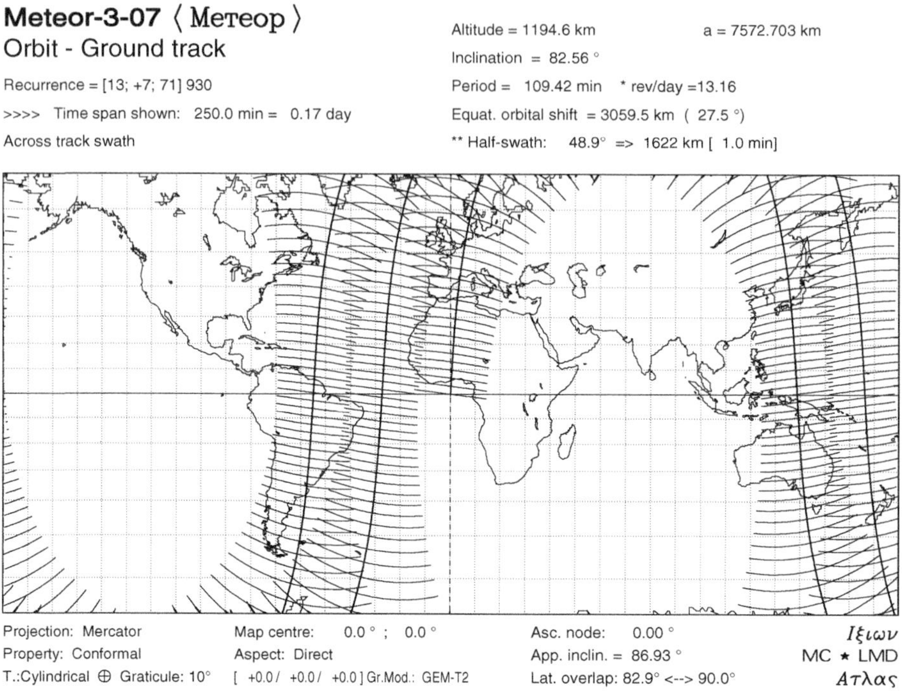

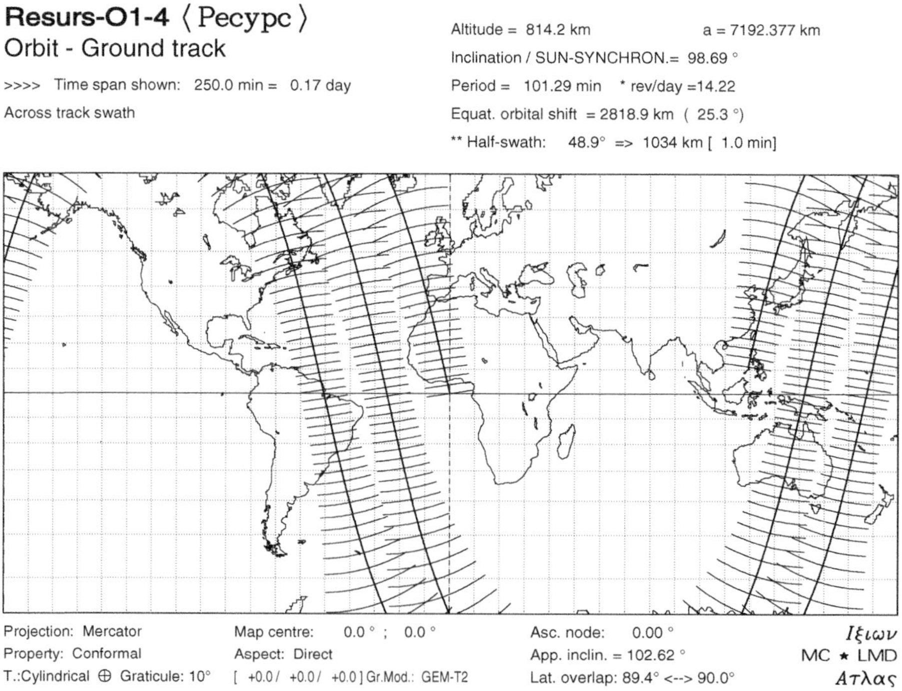

Figure 8.6. Across-track scan – LEO satellites equipped with the same instrument

This rotation corresponds to the rotation matrix

$$P_4 = \begin{pmatrix} \cos\alpha & 0 & -\sin\alpha \\ 0 & 1 & 0 \\ \sin\alpha & 0 & \cos\alpha \end{pmatrix} . \tag{8.18}$$

If (X', Y', Z') are the Cartesian coordinates of the target point on the Earth relative to the frame $\Re(Oxyz)$, we obtain these new coordinates from

$$\begin{pmatrix} X' \\ Y' \\ Z' \end{pmatrix} = P_4 \begin{pmatrix} X \\ Y \\ Z \end{pmatrix} ,$$

where the coordinates (X, Y, Z) of the subsatellite point were obtained using the product of the three rotations

$$P = P_1 P_2 P_3$$

defined by (5.8). Figure 8.5 completes Fig. 5.2 with the fourth rotation. Using (5.12) and (5.13), the Cartesian coordinates (X', Y', Z') can be used to calculate the polar coordinates λ' and ϕ' of the target point.

Example 8.4. *Swath track of the ScaRaB instrument aboard Meteor-3-07 and Resurs-O1-4.*

The value of the half-swath angle is $f_{\mathrm{M}} = 48.91°$ in both cases. For Meteor-3-07, the swaths overlap at the equator during two consecutive nodal crossings. This does not happen for Resurs-O1-4 (see Fig. 8.6). This effect is expressed (see below) by the fraction of equatorial overlap, which is greater than 1 in the first case, and less than 1 in the second. For clarity, swaths have been plotted at 60 s intervals, whereas ScaRaB actually scans every 6 s.

Scanning and Ground Track of the Across-Track Swath

When an instrument scans, the scan moves extremely quickly across the ground. For example, the instrument ScaRaB aboard Meteor-3-07 completes one scan in 6 s, so that the average speed of the scan on the ground is $3254/6 = 542\ \mathrm{km\,s^{-1}}$. Compared with the displacement of the subsatellite point along the satellite ground track, which is $6\ \mathrm{km\,s^{-1}}$, each swath track can be treated as instantaneous. For HRVIR aboard SPOT-4, the scan is effectively instantaneous.

The ground track of the orthogonal swath, perpendicular to the orbital plane, thus makes an angle of $90° - i$ with the equator. However, as we saw previously, the ground track makes an angle i' with the equator. This is the apparent inclination. In diagrams showing the ground tracks, the normal to the ground track of the swath thus makes an angle $i - i'$ with the satellite ground track at the equator.

Note on Cartography. This angular difference only shows up true to scale on maps plotted with a conformal projection.

Equatorial Overlap

Consider a full swath of an instrument to its viewing limit. Its width on the ground is $2F_{\mathrm{M}}$. Let L_{E} be the portion of the equator covered by the swath during one crossing by the satellite. To a first approximation,

$$L_{\mathrm{E}} \approx \frac{2F_{\mathrm{M}}}{\sin i} \,. \tag{8.19}$$

In fact, the exact relation for the orbit and swath ground tracks at the equator is

$$L_{\mathrm{E}} = \frac{2F_{\mathrm{M}}}{\sin i + \cos i \, \tan(i - i')} \,. \tag{8.20}$$

It is interesting to compare this distance L_{E} with the equatorial shift: both lengths are measured along the equator and their ratio Q_{E} thus measures the fraction of the equator seen by the satellite in one day, during the ascending node crossing:

$$Q_{\mathrm{E}} = \frac{L_{\mathrm{E}}}{D_{\mathrm{E}}} \,, \tag{8.21}$$

where the distance D_{E} is the equatorial shift, $\Delta\lambda_{\mathrm{E}}$ calculated from (5.21) and expressed in the same units as L_{E} (usually in km), without regard for sign. If Q_{E} is greater than unity, certain points on the equator are viewed more than once a day during the ascending node crossing (and likewise, of course, for the descending node crossing).

We may thus calculate the half-swath on the ground, F_1 (in km), or α_1 (in degrees), representing the necessary threshold to obtain full equatorial overlap, i.e., $Q_{\mathrm{E}} = 1$:

$$\alpha_1 = \frac{\Delta\lambda_{\mathrm{E}}}{2} \big[\sin i + \cos i \, \tan(i - i') \big] \,. \tag{8.22}$$

With (8.9), this gives the threshold value of the half-swath f_1.

Example 8.5. *Fraction of equatorial overlap for the satellites Megha-Tropiques and Oceansat-1.*

On the graphs of Fig. 8.7, we have shown the overlap fraction as a function of the half-swath f. We have also indicated the value of the half-swath F on the ground for

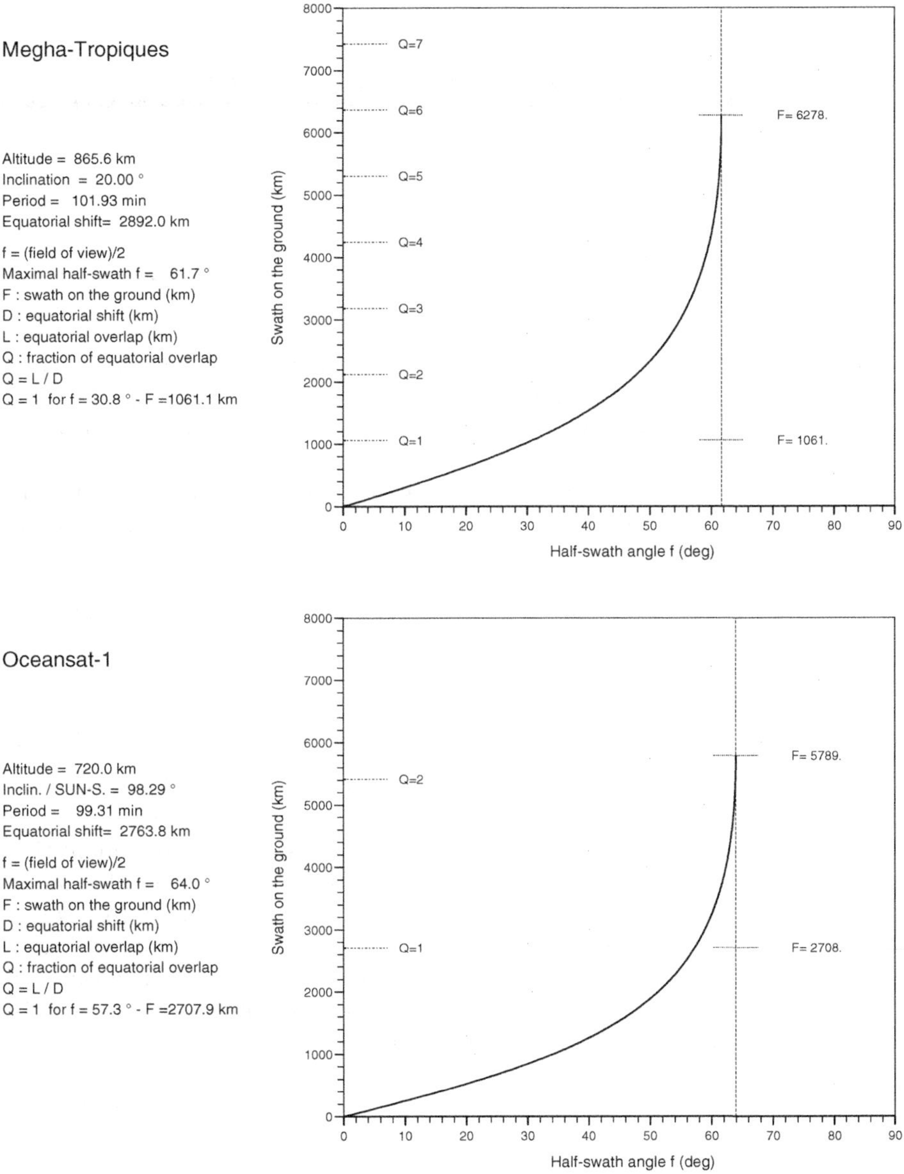

Figure 8.7. Fraction of equatorial overlap for two LEO satellites

$Q_E = 1$, denoted above by F_1, and the maximum value F_0 for limb viewing. For an LEO satellite, the value of Q_E increases as the satellite altitude increases (for F_M) and as the angle between the orbital plane and the equator decreases (for $1/\sin i$). We give here an example for a near-polar satellite (Oceansat-1) and another for a satellite with a small inclination at the equator (Megha-Tropiques).

Swath and Mission Constraints

The swath of the main instrument on a satellite and the orbital characteristics of that satellite are related. This constraint is particularly important if the satellite is recurrent. We shall give here several examples for the very different cases of wide, narrow, and very narrow swaths.

Example 8.6. *Fulfilling mission requirements with regard to swath and recurrence for Oceansat-1, SPOT-1 and ICESat.*

Oceansat-1. The Indian satellite Oceansat-1 (IRS-P4) is Sun-synchronous with equatorial crossing at noon and midnight, and recurrent, with triple $[14, 1, 2]$, corresponding to a cycle of 29 revolutions over 2 days. The equatorial shift is

$$\Delta\lambda_E = -\frac{360}{14.5} = -24.83^\circ \,, \qquad D_E = 2763.8 \text{ km} \,.$$

The aim of the mission is to view the equator during daytime every two days. To a first approximation (the orthogonal swath is practically parallel to the equator), the ground swath $2F_M$ must be at least half the equatorial shift, i.e.,

$$\alpha_M = \frac{180}{29} = 6.21^\circ \,, \qquad F_M = \frac{D_E}{4} = 691 \text{ km} \,.$$

Equation (8.9) gives $f_M = 42.3^\circ$. The OCM instrument aboard this satellite has a full swath of 1420 km, or $F_M = 710$ km, a few kilometres more than the strict minimum swath. Calculation gives

$$F_M = 710 \text{ km} \quad \Longrightarrow \quad f_M = 43.0^\circ \,, \qquad Q_E = 0.52 \,.$$

The fraction of equatorial overlap is thus slightly more than 1/2. Figure 8.7 (lower) can be used to estimate f_M directly for $Q = 0.5$.

SPOT-1. When the SPOT project was under development, the satellite was planned to fly between 800 and 850 km altitude, low enough for good resolution, but high enough to avoid too much atmospheric drag. The HRV instrument aboard SPOT-1 was designed with a field of view of 8.4°, or $f_M = 4.2^\circ$. The aim was then to ensure that the grid interval δ defined by (7.19) was slightly less than the ground swath.
We calculate the interval $\delta \approx 2h \tan f_M \approx 1.06^\circ$, which gives the number of round trips in a recurrence cycle as $N_{T_o} = 360/1.06 \simeq 340$. The value of ν then lies between 14.26 for $h = 800$ km and 14.11 for $h = 850$ km. This implies, for the cycle $C_{T_o} = N_{T_o}/\nu$, the bound $C_{T_o} = 24$. Moreover, when the mission was set up, it was hoped to have a cycle shorter than one month, and this implies

$$24 \leqslant C_{T_o} \leqslant 30 .$$

The recurrence triple thus had to have the form $[14, D_{T_o}, C_{T_o}]$ with $340 \leqslant N_{T_o} \leqslant 427$. Further considerations regarding subcycles then led to the choice of $[14, +5, 26]$, with 369 revolutions.

ICESat. This satellite is equipped with a laser designed for altimetry. The instrument, known as GLAS, points at the nadir with an almost pointlike field of view, since the pixel measures only 66 m on the ground. Recurrence with a very long cycle, 2723 round trips in 183 days, giving a grid interval $\delta = 15$ km, guarantees that, throughout the cycle, the satellite never goes over its ground track.

8.4.2 Variable-Yaw Swath

The across-track swath corresponds to a yaw angle of 0. The yaw angle is measured in a plane perpendicular to the yaw axis, or nadir axis, which points from the satellite to the centre of the Earth. There are scanning modes along the swath, called along-track scanning, and these correspond to a yaw angle of 90°. In this case, the swath does not exactly cover the ground track, for the same reasons that the across-track swath is not exactly perpendicular to the ground track – see Chap. 5 for the question of inclination and apparent inclination (angles i and i', or angles j and j'). By adjusting the yaw angle as a function of the latitude overflown, the ground track can be covered by the swath – see (5.32).

Yet another scanning mode varies the yaw angle continuously, as described in the next example.

Example 8.7. *Ground swath with variable yaw (PAP and RAP mode) of the CERES radiometer aboard the satellite Terra.*

NASA uses the following abbreviations for scanning modes:

- XT: across-track scanning, perpendicular to the orbit plane,
- AT: along-track scanning,
- PAP: programmable azimuth plane scanning,
- RAP: rotating azimuth plane scanning.

The satellites TRMM, Terra, and Aqua carry two CERES instruments: one is fixed (FM-1), with across-track swath (XT), while the other (FM-2), has variable yaw.

AT and PAP Mode. In order to carry out radiometric assessments at the Anchor Station Area, Valencia (Spain), we asked NASA's CERES team to program the line of sight of the instrument (PAP mode) according to (5.32), and this was done on 19 August 2004. The results are shown in Fig. 8.8 (upper). In AT mode, the line of sight does not exactly retrace the track [see Fig. 8.8 (lower)].

RAP Mode. Figure 8.9 (upper) shows the instrument track of FM-2 aboard Terra, when the satellite overflew California, at 18:13 UT on 9 February 2003, and Fig. 8.9

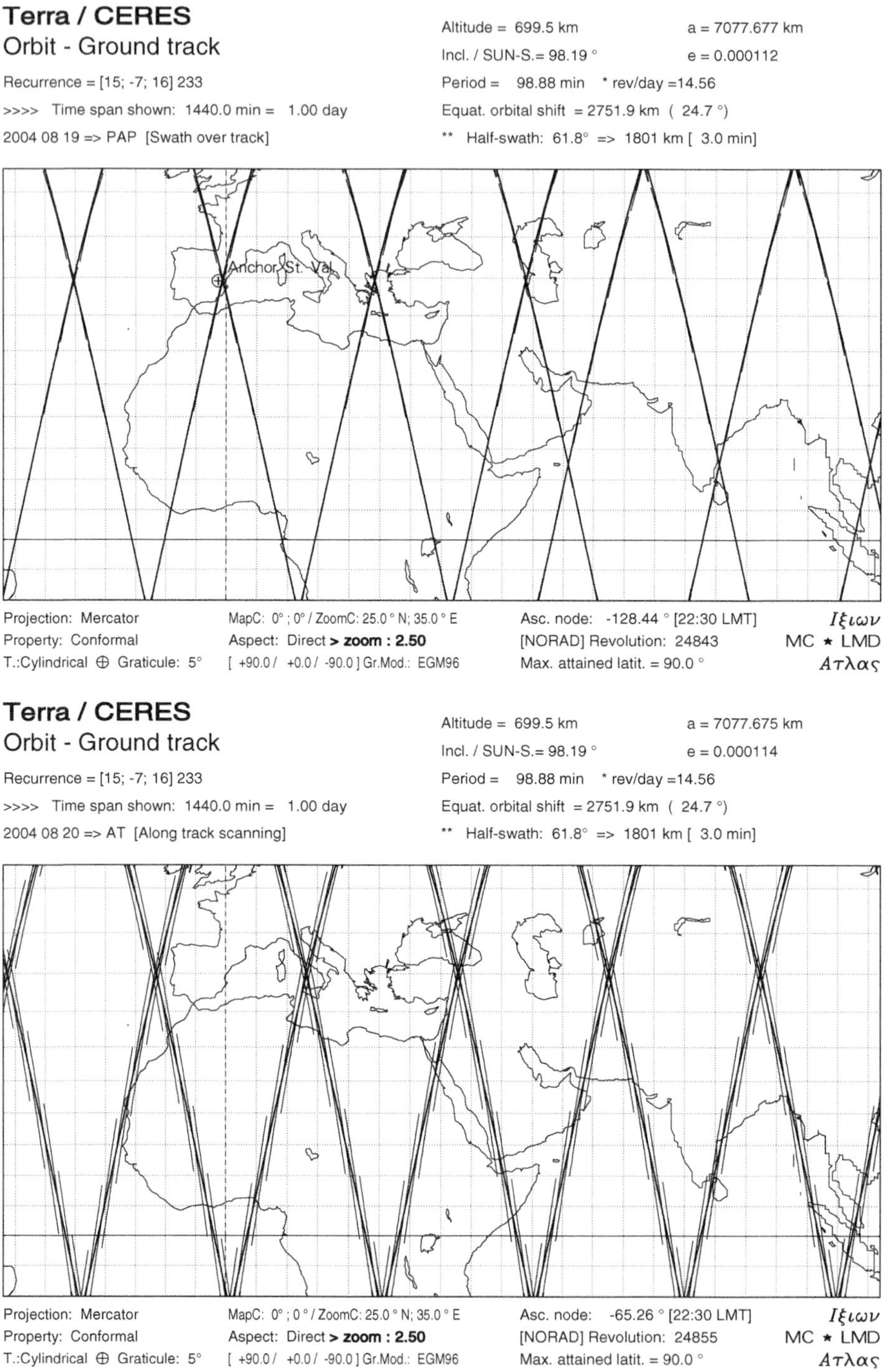

Figure 8.8. Variable-yaw swath (PAP mode) for an LEO satellite. Difference between along-track (AT), *lower*, and adjusted along-track (PAP), *upper*

Terra / CERES
Orbit - Ground track

Recurrence = [15; -7; 16] 233
2003 02 09 => RAP [Rotating Azimuth Plane Scanning]
Alternated variable-yaw swath [+12.0 min]

Altitude = 699.6 km a = 7077.704 km
Incl. / SUN-S.= 98.20 ° e = 0.000092
Period = 98.88 min * rev/day =14.56
Equat. orbital shift = 2751.9 km (24.7 °)
** Half-swath: 56.0° => 1215 km [0.1 min]

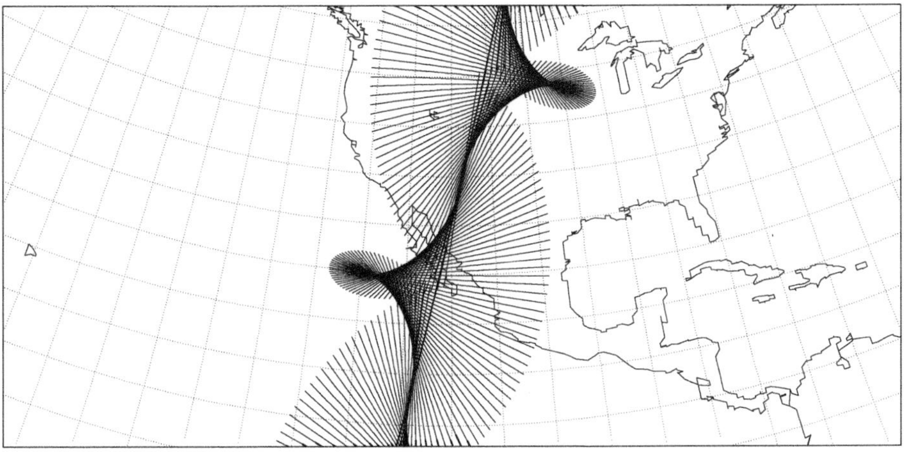

Projection: Mercator
Property: Conformal
T.:Cylindrical ⊕ Graticule: 5°

Map centre: 30.0 ° N; 110.0 ° W
Aspect: Oblique **> zoom : 4.00**
[+90.0 / +30.0 / +20.0] Gr.Mod.: EGM96

Asc. node: -111.51 ° [22:32 LMT]
[NORAD] Revolution: 16731
Max. attained latit. = 90.0 °

Iξιων
MC ⋆ LMD
Aτλας

Terra / CERES
Orbit - Ground track

Recurrence = [15; -7; 16] 233
>>>> Time span shown: 720.0 min = 0.50 day
Alternated variable-yaw swath [+12.0 min]

Altitude = 699.6 km a = 7077.738 km
Inclination / SUN-SYNCHRON.= 98.21 °
Period = 98.88 min * rev/day =14.56
Equat. orbital shift = 2751.9 km (24.7 °)
** Half-swath: 61.8° => 1801 km [0.5 min]

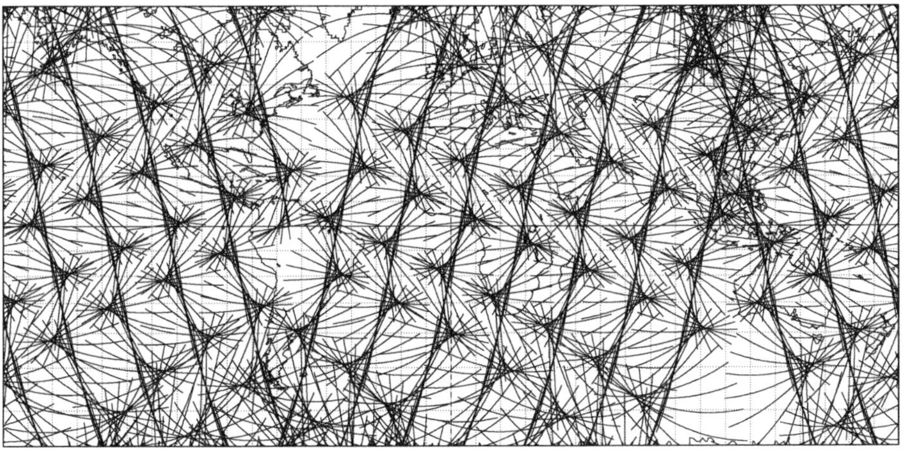

Projection: Mercator
Property: Conformal
T.:Cylindrical ⊕ Graticule: 10°

Map centre: 0.0 ° ; 0.0 °
Aspect: Direct
[+0.0 / +0.0 / +0.0] Gr.Mod.: EGM96

Asc. node: -64.60 ° [22:30 LMT]
App. inclin. = 102.06 °
Max. attained latit. = 90.0 °

Iξιων
MC ⋆ LMD
Aτλας

Figure 8.9. Variable-yaw swath (RAP mode) for an LEO satellite

DMSP-5D2 F-11 / SSM/I
Orbit - Ground track

>>>> Time span shown: 1440.0 min = 1.00 day
Ground track - Conical swath / VZA=53.1°

Altitude = 849.8 km a = 7227.887 km
Inclination / SUN-SYNCHRON.= 98.84 °
Period = 102.04 min * rev/day =14.11
Equat. orbital shift = 2839.8 km (25.5 °)
** Effect. h-ap.: 38.9 ° - Radius/grnd 915 km [1.0 min]
** Half-aperture: 52.1 ° = 721 km - Effect. swath: 1442 km

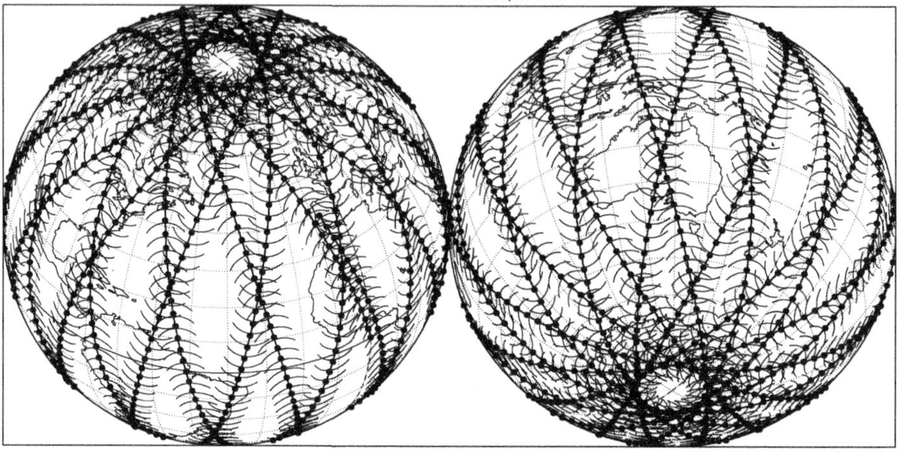

Projection: Orthographic
Property: none
T.:Azimuthal ⊕ Graticule: 10°

Map centre (r.): 42.0 ° S; 139.0 °E
Aspect: Oblique
[-90.0 / +132.0 / -49.0] Gr.Mod.: JGM-3

Asc. node: 0.00 °
Max. attained latit. = 87.6 °

Ιξιων
MC ★ LMD
Ατλας

Megha-Tropiques / MADRAS
Orbit - Ground track

Recurrence = [14; -1; 7] 97
>>>> Time span shown: 203.9 min = 0.14 day
Ground track - Conical swath / VZA=53.1°

Altitude = 865.6 km a = 7243.700 km
Inclination = 20.00 °
Period = 101.93 min * rev/day =14.13
Equat. orbital shift = 2892.0 km (26.0 °)
** Effect. h-ap.: 42.3 ° - Radius/grnd 928 km [1.0 min]
** Half-aperture: 65.0 ° = 841 km - Effect. swath: 1682 km

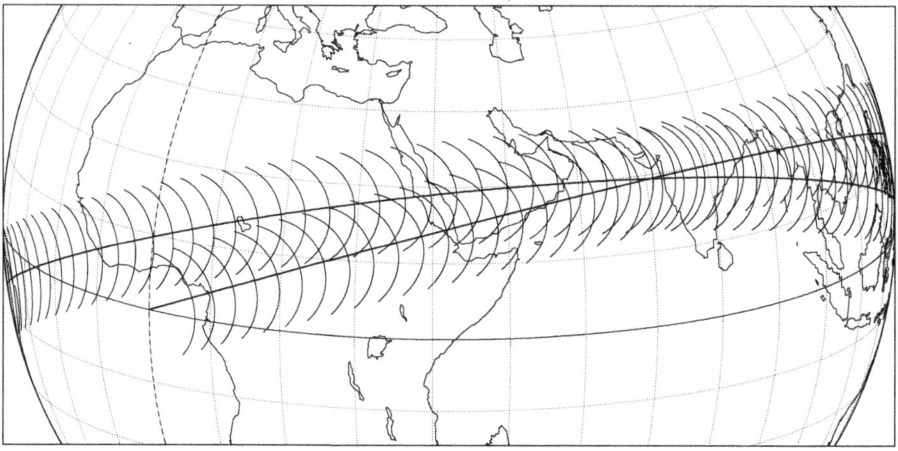

Projection: Orthographic
Property: none
T.:Azimuthal ⊕ Graticule: 10°

Map centre: 15.0 ° N; 42.0 °E
Aspect: Oblique **> zoom : 2.00**
[-90.0 / +75.0 / +48.0] Gr.Mod.: GRIM5-C1

Asc. node: 0.00 °
Max. attained latit. = 27.6 °

Ιξιων
MC ★ LMD
Ατλας

Figure 8.10. Conical swath – LEO satellites equipped with similar instruments (lower) shows the track over half a day. This RAP scanning mode consists in making a half-turn in 6 min, followed by a half-turn in the opposite direction. The swath represented here [Fig. 8.9 (upper)], every 6 s, corresponds to $f = 56°$.

8.4.3 Conical Swath

Conical swaths are used by microwave radiometers, in particular. In this case, for physical reasons connected with the phenomenon, the target points must be viewed at a constant angle. The maximum half-swath f_M must therefore be adapted to the altitude of the satellite and also the angle ζ_M. We now give several illustrations of the conical swath.

Example 8.8. *Conical swath tracks of the radiometer SSM/I aboard the satellite DMSP-5D2 F-11 and the radiometer MADRAS aboard the satellite Megha-Tropiques.*

SSM/I. The instrument SSM/I is a passive radiometer. The angle f_M between its axis and the axis of rotation, the nadir axis $\boldsymbol{SZ}_c$, is constant, in such a way that the viewing zenith angle is also constant, with $\zeta_M = 53.1°$. For this satellite, the calculation gives $f_M = 44.9°$ and $F = R(\pi/180)\alpha_M = 914.6$ km. Scanning is not through a complete circle of radius F, but over an arc of angle $\beta_M = 102.4°$, on either side of the axis $\boldsymbol{SX}_c$ along the velocity vector. For this satellite, scanning is in the forward direction. The chord of the scanned arc is called the effective or useful swath $2F_e$. It is given by

$$2F_e = R(\pi/180) \arcsin\left(\sin\alpha_M \sin\beta_M\right) = 1442 \text{ km} ,$$

where angles are in degrees. When β_M is greater than 180°, we find $2F_e = 2F$. Figure 8.10 (upper) shows the ground track in steps of one minute for greater clarity, although the radiometer turns at 31.6 rev/min. During one revolution of the instrument, i.e., 1.9 s, the subsatellite point moves through 12.5 km. Looking at the ground track over one day, we see that a large fraction of the Earth's surface is viewed every day.

MADRAS. The sighting geometry of this radiometer, planned for the satellite Megha-Tropiques, is practically the same. The ground track is shown over two revolutions in Fig. 8.10 (lower).

8.5 View from a GEO Satellite

When a geostationary satellite views the Earth, the maximum swath in the sense that we have defined f_0 is

$$f_0 = \arcsin\frac{1}{\eta_{GS}} , \tag{8.23}$$

where the reduced distance here is η_{GS} defined by (4.58). This implies, with $\eta_{GS} = 6.611$,

$$f_0 = 8.700^\circ = 0.1518 \text{ rad} . \tag{8.24}$$

The corresponding angle at the centre of the Earth is

$$\alpha_0 = \arccos \frac{1}{\eta_{GS}} = 90^\circ - 8.7^\circ = 81.3^\circ \quad \Longrightarrow \quad 2F_0 = 18\,100 \text{ km} . \tag{8.25}$$

The part of the Earth viewed by a geostationary satellite is called the Earth's disk, although it is called the slot in the context of remote sensing.

Let λ_S be the longitude of the satellite S (the parking longitude or longitude of the subsatellite point). Then the longitudes viewed on the equator by S lie in the interval

$$[\lambda_S - 81.3^\circ, \lambda_S + 81.3^\circ] .$$

Along the meridian λ_S, the latitudes viewed occupy the same interval of 81.3° on either side of the equator.

For an arbitrary point P on the Earth, with geographic coordinates λ and ϕ, we write the distance D to the subsatellite point S_0 (we mean, of course, the distance on the sphere, measured along a great circle, viz., $D = R\alpha$) using the spherical triangle S_0PP', where P' is the intersection of the meridian of P with the equator:

$$\cos \overset{\frown}{S_0P} = \cos \overset{\frown}{S_0P'} \cos \overset{\frown}{PP'} ,$$

$$\cos\alpha = \cos(\lambda - \lambda_S) \cos\phi . \tag{8.26}$$

This corresponds to (ST I).

The locus of points P viewed at distance D from the subsatellite point is thus defined by

$$D = R\frac{\pi}{180} \arccos\left[\cos(\lambda - \lambda_S) \cos\phi\right] , \tag{8.27}$$

where the angles are in degrees. This is the locus of points viewed at the same angle from the satellite, and hence viewed with the same pixel distortion.

The condition for the point P to be viewed at all is given by (8.6) and (8.26) as

$$\eta_{GS} \cos(\lambda - \lambda_S) \cos\phi \geqslant 1 . \tag{8.28}$$

The area s of the Earth which is viewed, for a given value of the angle α, is

$$s(\alpha) = 2\pi R^2(1 - \cos\alpha) .$$

The maximal area viewed is thus $s(\alpha_0)$, which represents the fraction

$$\frac{s(\alpha_0)}{4\pi R^2} = \frac{1}{2}\left(1 - \frac{1}{\eta_{\text{GS}}}\right) = 0.424\,, \tag{8.29}$$

or about 42% of the total area.

Example 8.9. *Represent the locus of points on the Earth that are equidistant from the subsatellite point of a GEO satellite.*

The distance D between a point on the Earth seen by the geostationary satellite and the subsatellite point of the same satellite is defined by (8.27). The locus of points on the Earth at the same distance D has been represented with steps of 500 km in the value of D, and 200 km in the enlarged maps. We denote these loci by $\mathcal{L}(D)$.

METEOSAT. The European satellites METEOSAT, in their operation phase, are stationed at longitude $\lambda_S = 0°$. Figure 8.11 (upper) shows the Earth as it is viewed by the satellite, i.e., the Earth's disk. The curves $\mathcal{L}(D)$ for given D are circles, represented by circles on this map, which has a non-conformal but axisymmetric projection. The Guyou projection, based on elliptic functions, presents the globe in a rectangle, whilst preserving angles, i.e., it is a conformal projection. The curves $\mathcal{L}(D)$ are represented in the direct aspect in Fig. 8.11 (lower).

FY-2A, FY-2B. The Chinese satellites FY-2A then FY-2B are stationed at longitude $\lambda_S = 105°$E. Figure 8.12 (upper) shows the locus of points $\mathcal{L}(D)$ viewed at the same angle in an orthographic representation centered on Peking (right-hand image).

GOES. The US satellite GOES-East is stationed at longitude $\lambda_S = 75°$W. This position was previously occupied by the succession of satellites SMS-1, SMS-2, GOES-5, GOES-7, GOES-8 (partial or total occupation during their operating lifetimes). The Argentinian meteorological office (*Servicio Meteorológico Nacional*) represents data in a stereographic projection centered on the point (34.8°S, 68.6°W), situated in the centre of the country. We have used this projection in Fig. 8.12 (lower) to represent the locus of points viewed at the same angle. This locus $\mathcal{L}(D)$ is thus represented by circles, since the stereographic projection preserves angles. Note that the GOES-East and Feng Yun-2 satellites are diametrically opposite one another with respect to the centre of the Earth.

Elektro-1 (GOMS-1). The Russian satellite Elektro-1 (GOMS-1) is stationed at longitude $\lambda_S = 76°$E. We have represented $\mathcal{L}(D)$ in a Guyou transverse projection centered on Moscow in Fig. 8.13. It is clear that geostationary satellites are not of much interest to Russia.

METEOSAT

Locus of points
equidistant
from the subsatellite point

Altitude =35787.6 km a_GS = 42165.785 km
Inclination = 0.00 ° Parking Longit.= 0.0 °
Period = 1435.91 min * rev/day = 1.00
Equat. orbital shift =40072.1 km
** Half-swath: 8.7° - On ground 9050.2 km [500.0 km]

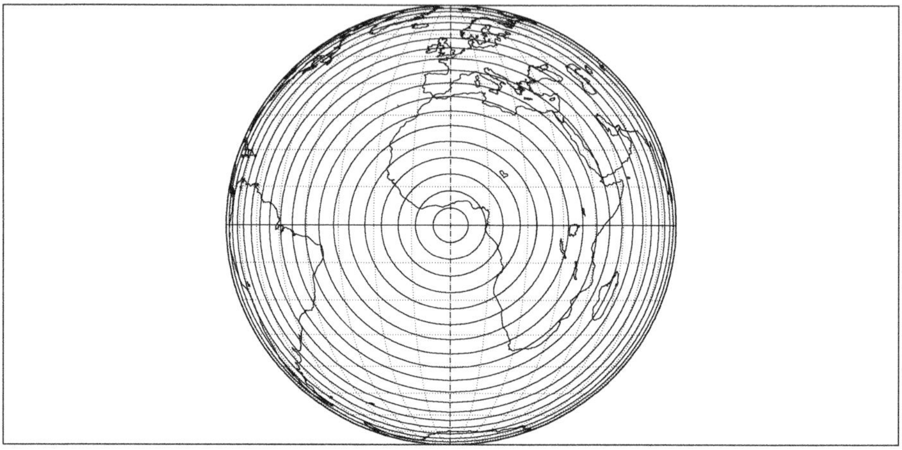

Projection: Orthographic Map centre: 0.0 ° ; 0.0 ° Geostationary $I\xi\iota\omega\nu$
Property: none Aspect: Equatorial Max. attained latit. = 81.3 ° MC ★ LMD
T.:Azimuthal ⊕ Graticule: 10° [-90.0 / +90.0 / +90.0] Gr.Mod.: GEM-T2 $A\tau\lambda\alpha\varsigma$

METEOSAT

Locus of points
equidistant
from the subsatellite point

Altitude =35787.6 km a_GS = 42165.785 km
Inclination = 0.00 ° Parking Longit.= 0.0 °
Period = 1435.91 min * rev/day = 1.00
Equat. orbital shift =40072.1 km
** Half-swath: 8.7° - On ground 9050.2 km [500.0 km]

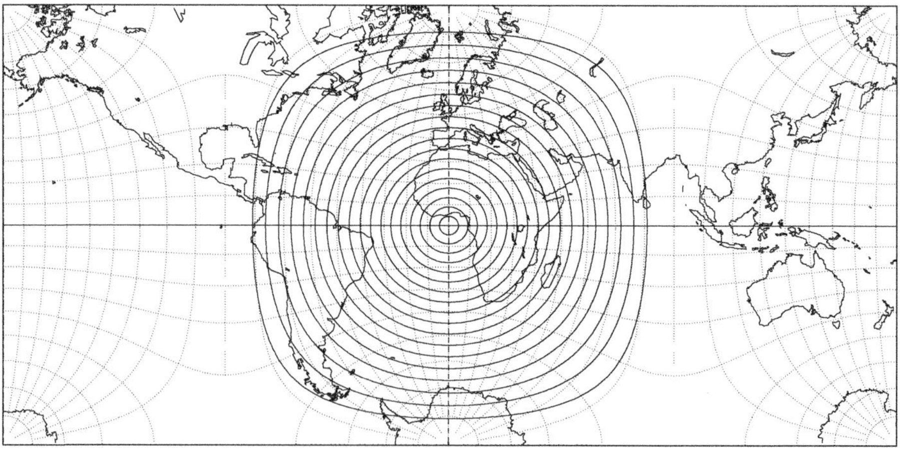

Projection: Guyou Map centre: 0.0 ° ; 0.0 ° Geostationary $I\xi\iota\omega\nu$
Property: Conformal Aspect: Direct App. inclin. = 0.00 ° MC ★ LMD
T.:(various) ⊕ Graticule: 10° [+90.0 / +0.0 / -90.0] Gr.Mod.: GEM-T2 Max. attained latit. = 81.3 ° $A\tau\lambda\alpha\varsigma$

Figure 8.11. Locus of points equidistant from the subsatellite point of a GEO satellite

Feng Yun-2 〈風雲二〉

Locus of points
equidistant
from the subsatellite point

Altitude =35787.6 km a_GS = 42165.785 km
Inclination = 0.00 ° Parking Longit.=105.0 ° E
Period = 1435.91 min * rev/day = 1.00
Equat. orbital shift =40072.1 km
** Half-swath: 8.7° - On ground 9050.2 km [500.0 km]

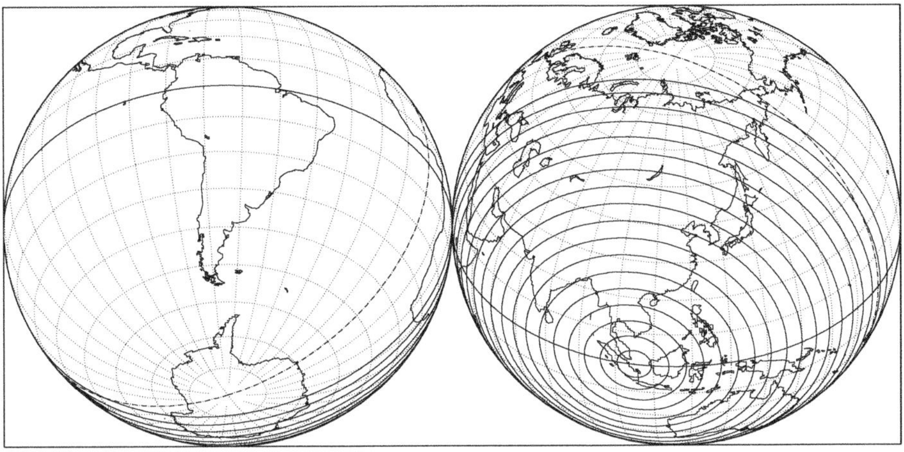

Projection: Orthographic | Map centre (r.): 40.0 ° N; 116.0 °E | Geostationary | Ιξιων
Property: none | Aspect: Oblique | Max. attained latit. = 81.3 ° | MC ★ LMD
T.:Azimuthal ⊕ Graticule: 10° | [-90.0 / +50.0 / -26.0] Gr.Mod.: GEM-T2 | Ατλας

GOES-East

Locus of points
equidistant
from the subsatellite point

Altitude =35787.6 km a_GS = 42165.785 km
Inclination = 0.00 ° Parking Longit.= 75.0 ° W
Period = 1435.91 min * rev/day = 1.00
Equat. orbital shift =40072.1 km
** Half-swath: 8.7° - On ground 9050.2 km [500.0 km]

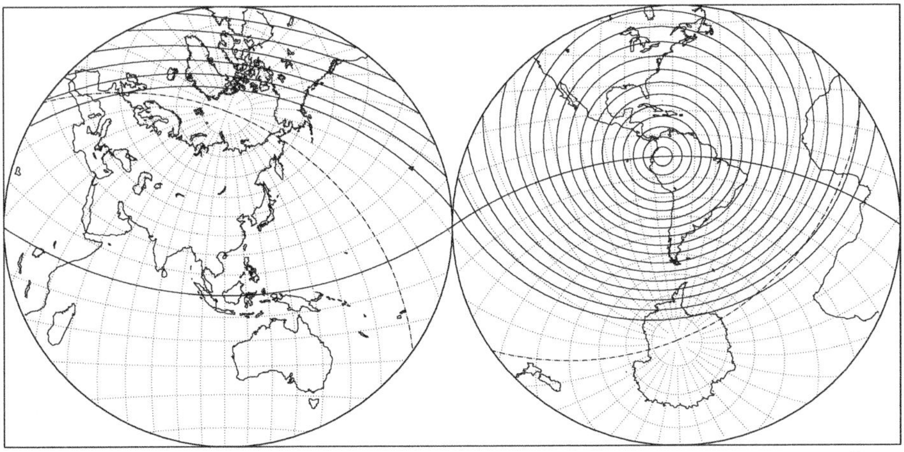

Projection: Stereographic | Map centre (r.): 34.8 ° S; 68.6 ° W | Geostationary | Ιξιων
Property: Conformal | Aspect: Oblique | Max. attained latit. = 81.3 ° | MC ★ LMD
T.:Azimuthal ⊕ Graticule: 10° | [-90.0 / +124.8 / +158.6] Gr.Mod.: GEM-T2 | Ατλας

Figure 8.12. Locus of points equidistant from the subsatellite point of a GEO satellite

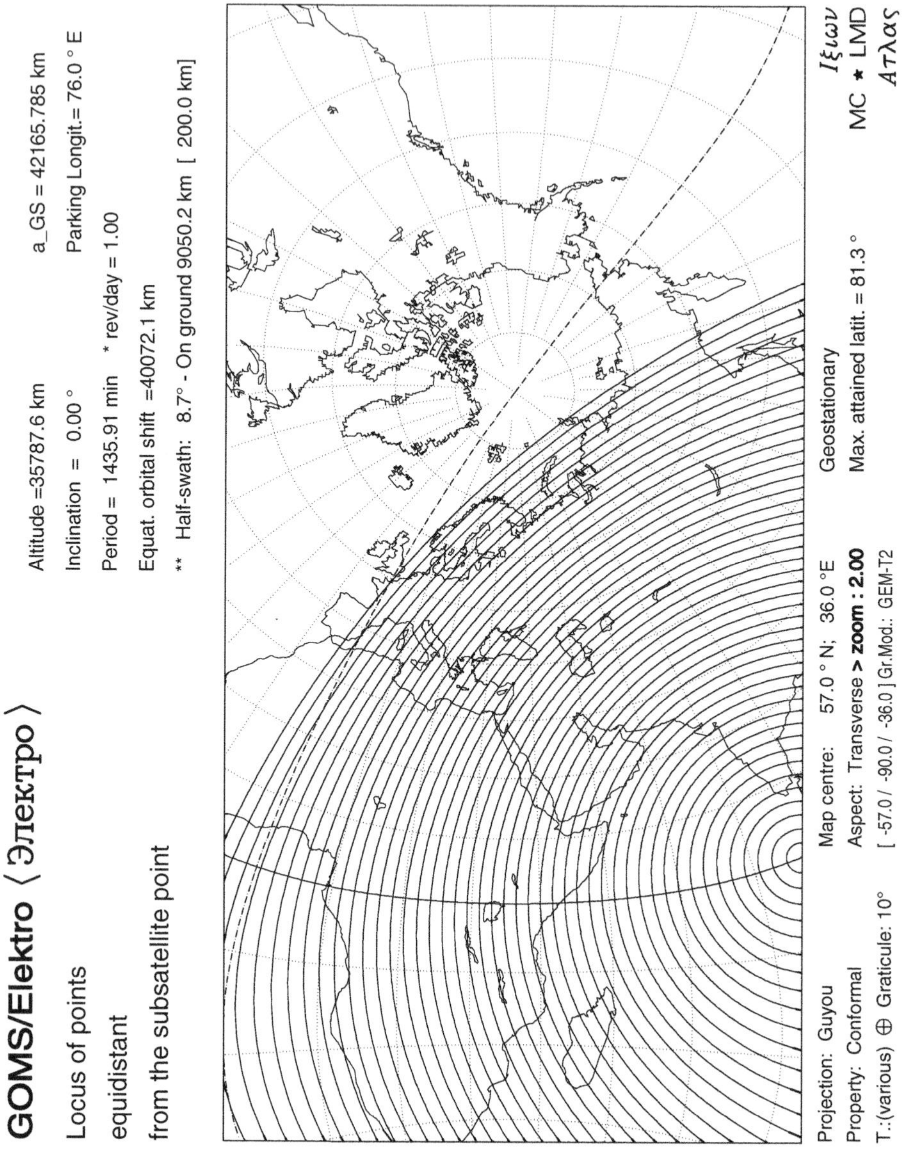

Figure 8.13. Locus of points equidistant from the subsatellite point of the Russian GEO satellite Elektro-1. The map is centered on Moscow (Russia)

9 Temporal and Angular Sampling

Once again, we shall change our point of view! From an arbitrary point P on the Earth, we now note the time and angular conditions of our view of the satellite S. This is the opposite problem to determining the ground track of the swath: we must now establish the satellite sampling for a given instrument. We shall only consider across-track swaths. We shall also determine, for this point P, the direction of the Sun at the instant of time when P is viewed by the satellite.

9.1 Basic Principles of Sampling

To obtain the sampling, our method consists in noting all the intersections of the swath track with a given meridian, called the reference meridian. This method, which underlies the 'sampling' function of the Ixion software, can then be used to make various comparisons and produce statistics depending on the latitude of the target point.

For each point viewed, called the target point and denoted by P, we determine, in terms of the satellite position S, the time at which it is seen and the direction PS, which is called the line-of-sight direction. This straight line is defined by the angles of the spherical coordinate system: the zenith angle and the azimuth angle. We call this the overpass of the satellite S at the point P. We can then refer to the overpass time, and the zenith and azimuth angles of overpass.

Temporal sampling is achieved when we know the overpass times in the sense mentioned above for any point on the Earth, for a given satellite and instrument, over a certain period of time, e.g., one month. When we speak of angular sampling, we mean a record of the sight angles for each overpass. When the word 'sampling' appears without further specification, we are generally referring to both sets of data. Sampling data is complemented by mentioning the conditions of solar illumination, i.e., the angles determining the position of the Sun for each target point P.

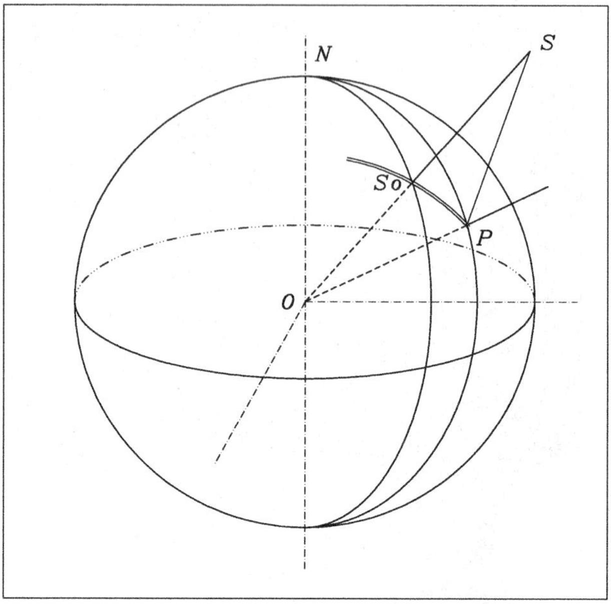

Figure 9.1. Directions relating to the swath of an instrument aboard a satellite. The Earth is shown with centre O, as is the North Pole N and the equator. The satellite S, with subsatellite point S_0, views the point P in a swath S_0P (*doubled curve*). This swath lies in the plane OS_0SP, called the swath plane $\mathcal{F}$, orthogonal to the direction of displacement of the satellite S. Figure 8.2 shows the angles in the plane $\mathcal{F}$

9.2 Satellite–Target Direction

9.2.1 Line-of-Sight Direction of the Satellite

Calculating Angles

Figure 9.1 shows the subsatellite point S_0, which is the intersection of OS with the Earth (centre O). The swath plane $\mathcal{F}$ is the plane OS_0SP, orthogonal to the direction of displacement of the satellite S. The velocity vector defines a right-handed orientation in $\mathcal{F}$.

At each time, we know the position of the satellite, and hence the geographic coordinates of S or S_0, denoted by λ_S and ϕ_S, longitude and latitude, and the swath angle f, through the angle α at the centre of the Earth. The two angles defining the direction $\boldsymbol{PS}$ are the zenith and azimuth angles, which belong to the spherical coordinate system centered on P, with the local horizontal plane $\mathcal{H}$ as reference plane. This plane, perpendicular to $\boldsymbol{OP}$, is the tangent plane to the sphere at P.

Zenith angle ζ. This is defined by

$$\zeta = (\boldsymbol{OP}, \boldsymbol{PS}) \,, \tag{9.1}$$

measured in the positive sense for the orientation defined by the satellite velocity vector, with the straight line $\boldsymbol{OP}$ (the local vertical) as origin, as discussed in relation to Fig. 8.2. The zenith angle ζ can be calculated[1] immediately from the angle α using (8.10).

Azimuth angle β. This is defined in the local horizontal plane $\mathcal{H}$. The local vertical confers an orientation on $\mathcal{H}$. The angle β is measured in the positive sense for this orientation, taking the north as origin. It is the dihedral angle between the swath plane $\mathcal{F}$ and the meridian plane OPN of P, denoted by $\mathcal{M}$:

$$\beta = \text{dihedral angle } \{\mathcal{M}, \mathcal{F}\} \,, \tag{9.2}$$

that is, the angle in $\mathcal{H}$ between the swath track and the meridian at P. The angle β is called the line-of-sight azimuth.

To calculate the angle β, we consider the spherical triangle NPS_0, shown in Fig. 9.1. If λ and ϕ are the geographic coordinates of the target point P, the three known elements of the triangle are two sides (two arcs) and an angle, the dihedral angle between the two meridian planes through P and S_0:

$$\overset{\frown}{NS_0} = \frac{\pi}{2} - \phi_S \,, \quad \overset{\frown}{NP} = \frac{\pi}{2} - \phi \,, \quad \widehat{N} = \lambda - \lambda_S \,.$$

We seek the angle β which is the angle of the spherical triangle at P.

Using (ST X), we have

$$\cot \overset{\frown}{NS_0} \, \sin \overset{\frown}{NP} = \cos \overset{\frown}{NP} \, \cos N + \sin N \, \cot P \,,$$

or in the present case,

$$\tan \phi_S \, \cos \phi = \sin \phi \, \cos(\lambda - \lambda_S) + \frac{\sin(\lambda - \lambda_S)}{\tan \beta} \,,$$

and finally,

$$\tan \beta = \frac{\sin(\lambda - \lambda_S) \, \cos \phi_S}{\cos \phi \, \sin \phi_S - \sin \phi \, \cos \phi_S \, \cos(\lambda - \lambda_S)} \,. \tag{9.3}$$

[1] As the swath plane $\mathcal{F}$ is oriented, ζ can take positive or negative values. Indeed, ζ varies in the interval $[-\pi/2, +\pi/2]$. Insofar as the azimuth angle β is defined (below) throughout the whole plane, it would suffice to define ζ in the interval $[0, +\pi/2]$. However, this redundancy allows one to say whether the satellite is in the ascending or descending stretch of the orbit. One then assigns the sign of ζ to f and α.

From the tangent, the angle β' between the straight line PS_0 and the north is obtained in the interval $[-\pi/2, +\pi/2]$. To obtain the angle β between $\boldsymbol{PS}_0$ and the north in the interval $(-\pi, +\pi]$, we observe that β has the same sign as $\lambda - \lambda_S$. Hence, if $\beta' = \arctan \beta$ has the same sign as $\lambda - \lambda_S$, we have $\beta = \beta'$. In the other case, $\beta = \beta' + \pi \; [2\pi]$.

Note. The altitude of the satellite does not appear in the relation giving the azimuth angle β.

Example 9.1. *Calculate the azimuth angle when the target point and the satellite lie on the same geographic parallel.*

When $\phi_S = \phi$, (9.3) becomes

$$\cot \beta = \sin \phi \tan \frac{\lambda - \lambda_S}{2} \,.$$

We take the following points: P (45°N,10°W) and S_0 (45°N,20°E). With the above relation, we obtain $\tan \beta = -5.278$, whence $\beta' = -79.3°$. With $\lambda - \lambda_S < 0$, we have $\beta = -79.3°$.

We observe that the direction $\boldsymbol{PS}_0$ is not at right-angles to the north. This direction is taken on the arc of the great circle passing through P and S_0. As the chosen parallel is situated in the northern hemisphere, this arc is north of the parallel, making an angle of 10.7° with the parallel at P. The angle is negative (S_0 is east of P). Swapping P and S_0, the angle β is positive.

If we choose points in the southern hemisphere, P (45°S,10°W) and S_0 (45°S,20°E), we obtain $\beta' = 79.3°$, hence $\beta = 79.3° - 180° = -100.7°$. The north is still taken as the origin.

Polar Viewing Conditions

For a circular orbit, if the pole is attained by the swath, it always sees the satellite with the same zenith angle. A prograde satellite, oriented in the direction of its displacement, always sees the North Pole to its left and the South Pole to its right. Conversely, a retrograde satellite (hence any Sun-synchronous satellite) always sees the South Pole to its left and the North Pole to its right. With the sign conventions mentioned above, ζ is positive in the case when a prograde satellite is viewed from the North Pole.

When viewing from the pole, the angles α and i are complementary (the sum of the angles is a right-angle), as can be seen from Fig. 8.3. Equations (8.9) and (8.10) thus become

$$\tan f_{(\mathrm{NP})} = \frac{\cos i}{\eta - \sin i} \,, \qquad \tan \zeta_{(\mathrm{NP})} = \frac{\cos i}{\sin i - 1/\eta} \,, \tag{9.4}$$

where the subscript (NP) indicates that the point under consideration is the North Pole. The signs of $\zeta_{(\mathrm{NP})}$ and $f_{(\mathrm{NP})}$ are given by the sign of $\cos i$.

For the South Pole, with subscript (PS), we find the negatives of these values, viz.,

$$f_{(\mathrm{SP})} = -f_{(\mathrm{NP})} \,, \quad \zeta_{(\mathrm{SP})} = -\zeta_{(\mathrm{NP})} \,. \tag{9.5}$$

If a satellite in circular orbit is to see the pole during a revolution, and hence both poles in each revolution, the half-swath f of its instrument must be greater than the threshold value f_{P} given by

$$\text{pole viewed} \quad \Longleftrightarrow \quad |f| \geqslant |f_{\mathrm{P}}| \quad \text{with } f_{\mathrm{P}} = \arctan\left(\frac{\cos i}{\eta - \sin i}\right) \,. \tag{9.6}$$

9.2.2 Geostationary Satellites

Although the scanning mode is different, geostationary satellites can be treated using the general relations with the substitutions $\eta = \eta_{\mathrm{GS}} = 6.611$ and $\phi_{\mathrm{S}} = 0$, taking λ_{S} to be the parking longitude.

For the zenith angle, we first calculate α from (8.26). We then obtain ζ from (8.10), which becomes here

$$\tan\zeta = \frac{\sin\alpha}{\cos\alpha - 1/\eta_{\mathrm{GS}}} \,. \tag{9.7}$$

For the angle β, (9.3) yields

$$\tan\beta = -\frac{\tan(\lambda - \lambda_{\mathrm{S}})}{\sin\phi} \,. \tag{9.8}$$

The two graphs in Fig. 9.2 give the sight angles ζ and $|\beta|$ as a function of the latitude $|\phi|$ and the longitude difference $|\lambda - \lambda_{\mathrm{S}}|$. Figure 9.2 (upper) representing the values of ζ should be compared with Fig. 8.11 (lower) representing the values of α.

Example 9.2. *Calculate the line-of-sight directions of METEOSAT-7 from Paris and of FY-2A from Sydney.*

METEOSAT-7 ($\lambda_{\mathrm{S}} = 0°$). Viewing from Paris (48°52′N, 2°20′E), with $\lambda_{\mathrm{S}} = 0°$, $\lambda = +2.33°$, $\phi = +48.87°$, we obtain

$$\alpha = 48.91° \,, \quad \zeta = 56.1° \,, \quad \beta' = -3.1° \,, \quad \beta = 176.9 \,.$$

The line of sight thus has zenith angle 56° (or elevation angle 34°) and azimuth angle 177° with the north (or 3° with the south, directed slightly westward).

FY-2A ($\lambda_{\mathrm{S}} = 105°$E). Viewing from Sydney (33°55′S, 151°10′E), with $\lambda_{\mathrm{S}} = +105°$, $\lambda = +151.17°$, $\phi = -33.92°$, we obtain

$$\alpha = 54.92° \,, \quad \zeta = 62.6° \,, \quad \beta' = 61.8° \,, \quad \beta = 61.8 \,.$$

The line of sight thus has zenith angle 63° (or elevation angle 27°) and azimuth angle 62° with the north. With $|\phi| = 34°$ and $|\lambda - \lambda_{\mathrm{S}}| = 46°$, we find the calculated values of ζ and β in Fig. 9.2.

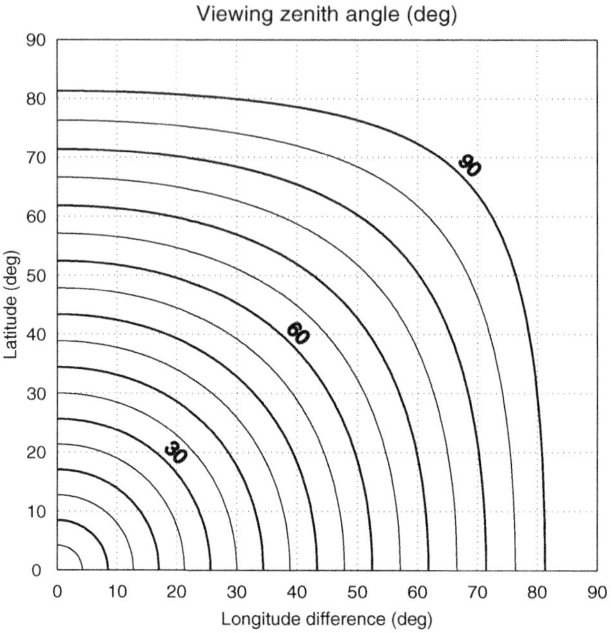

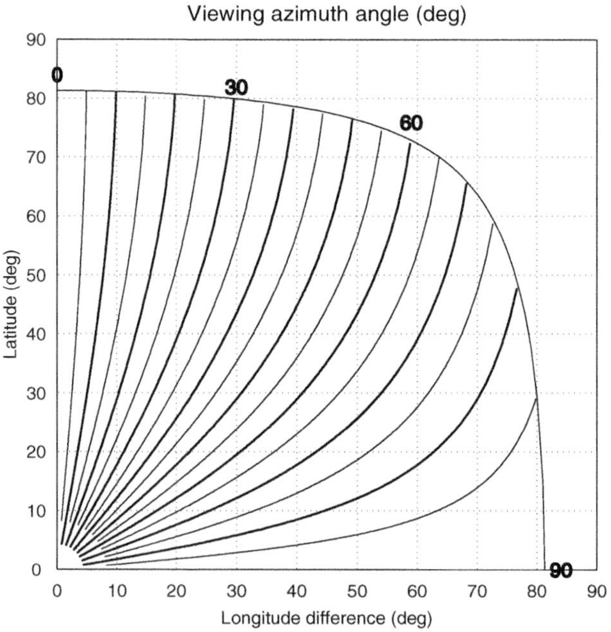

Figure 9.2. Sight angles for the target–satellite direction when the satellite is geostationary, as a function of the latitude $|\phi|$ and the longitude difference $|\lambda - \lambda_S|$. All angles are in degrees. *Upper*: $|\zeta|$. *Lower*: $|\beta|$

9.3 Target–Sun Direction

9.3.1 Solar Line-of-Sight Direction

For an arbitrary point P on the Earth's surface, with geographic coordinates λ and ϕ, we have already defined the local horizontal plane $\mathcal{H}$. In this section, quantities carrying the subscript s refer to the direction of the Sun. We now calculate the spherical coordinates β_s and ζ_s of the direction PS_s, where S_s represents the position of the Sun.

To this end, we consider the celestial sphere with centre O, relative to the point considered, as shown in Fig. 9.3. The direction of the zenith is OZ, normal to the horizontal plane $\mathcal{H}$ represented by the horizon circle at that place. The direction of the celestial north pole is ON, normal to the equatorial plane $\mathcal{E}$ represented by the celestial equator. The half-great circle passing through N and Z is the geographical meridian $\mathcal{M}$ at the place in question. This is the plane of Fig. 9.3.

The angle between the two straight lines OZ and ON, or the dihedral angle $(\mathcal{H}, \mathcal{E})$, is equal to the colatitude at the point (the complementary angle of the latitude ϕ):

$$(\boldsymbol{OZ}, \boldsymbol{ON}) = \frac{\pi}{2} - \phi \,. \tag{9.9}$$

Consider now an arbitrary direction OM. The half-great circle through Z and M is the vertical of M. The half-great circle through N and M is the celestial meridian of M.

When describing the direction of the Sun, we consider M at the intersection of OS_s with the celestial sphere. This direction OM can be specified relative to $\mathcal{E}$ using celestial equatorial coordinates, i.e., the right ascension α (or the hour angle H) and the declination δ. Relative to $\mathcal{H}$, in local horizontal coordinates, it is specified by the azimuth β_s and the zenith angle ζ_s. We shall express the horizontal coordinates as a function of the time (via H), the date (via δ), and the geographical position of the point P (via ϕ).

In the spherical triangle NZM, the sides are given by

$$\overset{\frown}{NZ} = \frac{\pi}{2} - \phi \,, \quad \overset{\frown}{ZM} = \zeta_s \,, \quad \overset{\frown}{NM} = \frac{\pi}{2} - \delta \,,$$

and the angles are

$$\widehat{N} = H \,, \quad \widehat{Z} = \pi - \beta_s \,, \quad \widehat{M} \,,$$

but note that the angle at M is not used here. Concerning the azimuth angles, the origin is taken when the point M lies in the meridian plane $\mathcal{M}$. The azimuth β_s, like β, is measured relative to the northerly direction.

The horizontal coordinates are expressed in terms of the equatorial coordinates and the latitude in the form $(\beta_s, \zeta_s) = f(H, \delta; \phi)$ by the relations:

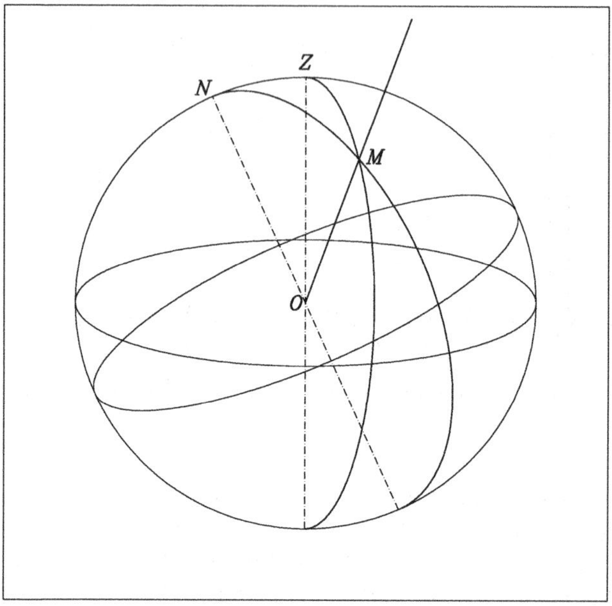

Figure 9.3. Celestial sphere relative to a given point on Earth. The axis OZ is the local vertical, normal to the local horizontal plane $\mathcal{H}$. The axis ON is the celestial polar axis, normal to the celestial equatorial plane $\mathcal{E}$. The straight line OM represents an arbitrary direction from the point in question

$$\cos\zeta_{\mathrm{s}} = \sin\phi\,\sin\delta + \cos\phi\,\cos\delta\,\cos H\ , \tag{9.10}$$

$$\sin\zeta_{\mathrm{s}}\,\sin\beta_{\mathrm{s}} = \cos\delta\,\sin H\ , \tag{9.11}$$

$$\sin\zeta_{\mathrm{s}}\,\cos\beta_{\mathrm{s}} = -\cos\phi\,\sin\delta + \sin\phi\,\cos\delta\,\cos H\ . \tag{9.12}$$

The reader is referred to Sect. 3.16, where the triangle ABC corresponds here to the triangle NZM. The above three relations are the fundamental relations of spherical trigonometry, sometimes known as Gauss' relations.

Equation (9.10) gives ζ_{s}, an angle in the interval $[0, \pi/2]$. Substituting its value into (9.12) yields the azimuth β'_{s} in $[0, \pi]$. The value of β_{s}, which lies in $(-\pi, \pi]$, is then given by comparing with the sign of H : $\beta_{\mathrm{s}} = \beta'_{\mathrm{s}}$ if $H \geqslant 0$, and $\beta_{\mathrm{s}} = -\beta'_{\mathrm{s}}$ if $H \leqslant 0$.

Equation (9.11) yields ζ_{s} in the interval $[-\pi/2, +\pi/2]$, whenever we need to know this angle, whether the Sun is above ($\zeta_{\mathrm{s}} \geqslant 0$) or below ($\zeta_{\mathrm{s}} \leqslant 0$) the local horizon. In certain cases, it is preferable to use the solar elevation h_{s} rather than ζ_{s}. These two angles are complementary. See Colour Plate VI.

Example 9.3. *Calculate the position of the Sun on 10 July 1998, at 06:30 UT, at the Baikonur launch site in Kazakhstan.*

In Example 4.4, we calculated that the instant of time 06:30 UT corresponded on this date with 10:58 LAT. Naturally, one must use the local apparent time in this situation. We thus have for the hour angle

$$H = 10 \text{ h } 58 \text{ m} - 12 \text{ h } 00 \text{ m} = -1 \text{ h } 02 \text{ m} ,$$

or $H = -62/4 = -15.5°$. Regarding the other quantities:

$$\delta(J = 10 \text{ July}) = +22.3° \quad \text{and} \quad \phi = 45°38'\text{N} = +45.6° .$$

Equation (9.10) gives $\cos\zeta_s = 0.8947$, whence $\zeta_s = 26.5°$, implying a solar elevation $h_s = 63.5°$. Regarding the azimuth, (9.12) implies that $\cos\beta_s = 0.8329$, whereupon $\beta'_s = 33.6°$, and since H is negative, $\beta_s = -33.6°$. The straight line PS_s thus points eastward (we are before the apparent noon, i.e., 12:00 LAT).

9.3.2 Sunrise, Sunset and Apparent Noon

Sunrise and Sunset

The obliquity ε is used to define circles on the Earth's surface (small circles called parallels) at certain significant latitudes: the polar circles, which are the arctic circle at $\phi = 90° - \varepsilon = 66°34'$N and the antarctic circle at $\phi = 66°34'$S, and the tropics, which are the Tropic of Cancer at $\phi = \varepsilon = 23°26'$N and the Tropic of Capricorn at $\phi = 23°26'$S.

Between the tropics, the Sun passes through the zenith at noon on those two days of the year when the declination is equal to the latitude. With $\delta = \phi$ and $H = 0$, (9.10) then gives $\cos\zeta_s = 1$, whence $\zeta_s = 0$ or $h_s = 90°$. Beyond the polar circles, there are days when the Sun never rises, and others when it never sets.

To study the sunrise and sunset in the general case, we write (9.10) in the form

$$\cos\zeta_s = \sin h_s = \cos\delta \, \cos\phi \, (\cos H - T) , \tag{9.13}$$

with

$$T = -\tan\delta \, \tan\phi .$$

The case of the poles can be treated immediately. Equation (9.10) shows that, for any H, we have $h_s = \delta$. Having eliminated this case, we see that solution of $\sin h_s = 0$ is equivalent to solution of $\cos H = T$.

We consider two cases, depending on whether $|T|$ is greater than or less than 1.

The Case $|T| > 1$. If $|T| > 1$, i.e., $|\phi| + |\delta| > 90°$, the quantity $\sin h_s$ cannot be zero for any H. In this case there is no sunset or sunrise.

- If T is negative, i.e., $T < -1$, or $|\phi + \delta| > 90°$, (ϕ and δ: same sign),

$$\sin h_{\mathrm{s}} > 0 \quad \Longleftrightarrow \quad h_{\mathrm{s}} > 0 \ .$$

  The Sun is above the horizon all day, a phenomenon known as the polar day.
- If T is positive, i.e., $T > +1$, or $|\phi - \delta| > 90°$, (ϕ and δ: opposite sign),

$$\sin h_{\mathrm{s}} < 0 \quad \Longleftrightarrow \quad h_{\mathrm{s}} < 0 \ .$$

  The Sun is below the horizon all day, a phenomenon known as the polar night.

The Case $|T| \leqslant 1$. If $|T| \leqslant 1$, i.e., $|\phi| + |\delta| < 90°$, (9.13) has two roots:

$$\sin h_{\mathrm{s}} = 0 \quad \Longleftrightarrow \quad \cos H = T \ ,$$

determining two opposite values of H, denoted H_{rise} and H_{set} :

$$\begin{cases} H_{\mathrm{set}} = \arccos(-\tan\delta\,\tan\phi) \ , \\ H_{\mathrm{rise}} = -H_{\mathrm{set}} \ . \end{cases}$$

With $\zeta_{\mathrm{s}} = 90°$ in (9.12), we obtain the corresponding azimuth values:

$$\begin{cases} \beta_{\mathrm{S_{set}}} = \arccos\left(-\dfrac{\sin\delta}{\cos\phi}\right) \ , \\ \beta_{\mathrm{S_{rise}}} = -\beta_{\mathrm{S_{set}}} \ . \end{cases}$$

We have the following special cases:

- At the equator ($\phi = 0$), for the whole year,

$$H_{\mathrm{set}} = 90° \ , \quad \beta_{\mathrm{S_{set}}} = 90° - \delta \ .$$

- At the equinoxes ($\delta = 0$), for the whole Earth,

$$H_{\mathrm{set}} = 90° \ , \quad \beta_{\mathrm{S_{set}}} = 90° \ .$$

The value $H_{\mathrm{set}} = 90°$ corresponds to sunrise at 06:00 LAT and sunset at 18:00 LAT.

Note. In these calculations, we have not taken into account atmospheric refraction. For medium latitudes ($|\phi| < 55°$), and on average over the year, refraction brings the sunrise forward by about 3 min and delays the sunset by the same amount.

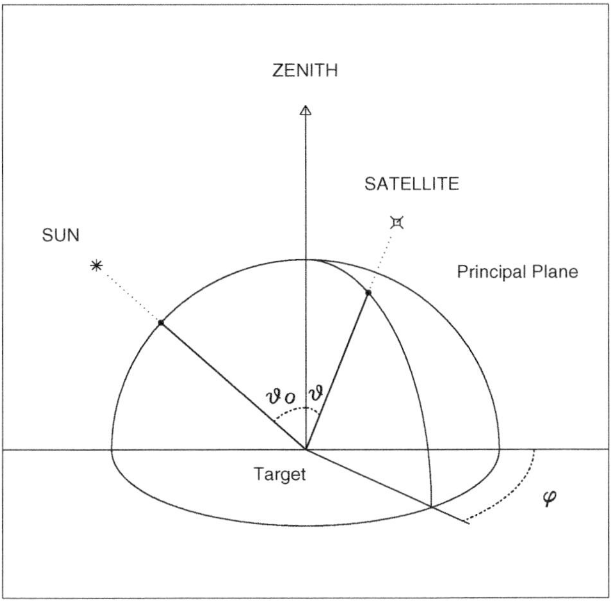

Figure 9.4. Angles used to describe the geometry of the Sun–target–satellite configuration

Apparent Noon

Apparent noon, or 12:00 LAT, corresponds to $H = 0$. In all the monthly tables showing overpass time at the end of this chapter, where the time (LMT) is shown on the abscissa, we thus have

$$\text{noon LMT} = \text{noon LAT} + E_{\text{T}} \, ,$$

where E_{T} is the equation of time discussed in Chap. 4 [see (4.47) and (4.48)].

In these tables, the dashed curve shows the times (LMT) of the sunrise, sunset, and apparent solar noon for the relevant place during the month.

9.4 Sun–Target–Satellite Configuration

Angles Describing the Sun–Target–Satellite Configuration

For any point P, the directions of the Sun and satellite at a given time are defined by the four angles β, ζ, β_{S}, ζ_{S}. When studying certain physical phenomena, such as radiation phenomena or those related to questions of remote sensing, it is useful to know that three angles are in fact sufficient to characterise the geometry of the Sun–target–satellite configuration. These are the two zenith angles and the relative azimuth. Using the standard relations in this area of study, we set

$$\begin{cases} \theta_0 = \zeta_s \;, \\ \theta = |\zeta| \;, \\ \varphi_A = \beta_s - \beta + \pi \quad [2\pi] \;. \end{cases}$$

Defined in this way, the relative azimuth is zero when the directions are opposite and it is equal to π when the Sun and the satellite are on the same side with respect to the target point (see Fig. 9.4). Most of the phenomena studied are symmetrical with respect to the principal plane, which is the plane spanned by the direction of the Sun and the vertical at the target point. We then consider the relative azimuth in the interval $[0, \pi]$. For reasons of clarity, we use φ_A to denote the relative azimuth defined in $[0, 2\pi]$ and φ_B the one defined in $[0, \pi]$ by

$$\varphi_B = \varphi_A \ \text{ if } \ \varphi_A \leqslant \pi \;, \qquad \varphi_B = 2\pi - \varphi_A \ \text{ if } \ \varphi_A > \pi \;.$$

In radiation-related studies, when we speak of the relative azimuth without further specification, we are referring to φ_B, written φ.

Scattering Angle

In certain studies, there is reference to the scattering angle. This is the angle between the two directions denoted here by PS and PS_s. This angle γ takes values in the interval $[0, \pi]$. Its value is found by taking the scalar product of the two vectors $\boldsymbol{PS}$ and $\boldsymbol{PS}_s$:

$$\cos\gamma = \sin\zeta \, \sin\zeta_s \, \cos(\beta - \beta_s) + \cos\zeta \, \cos\zeta_s \;, \tag{9.14}$$

or using the above notation,

$$\cos\gamma = \cos\theta_0 \, \cos\theta - \sin\theta_0 \, \sin\theta \, \cos\varphi \;. \tag{9.15}$$

9.5 Monthly Sampling Tables

For an arbitrary point on the Earth, the monthly sampling tables allow one to visualise all the overpasses of a given satellite, carrying an instrument with well-specified swath. These tables are extremely useful, as much for the preparation of missions as for the exploitation of data transmitted by the satellite. We now give a series of examples.

Table 9.1. Angles used to describe the geometry of the satellite–target–Sun configuration in the case of an across-track swath. The satellite is Terra, recurrent with a 16 day cycle, initialised at the date 2000 02 29 22:30 TSM. Instrument CERES: $f_M = 61.8°$. Table for April 2000: $J = 1$ corresponds to the date 2000 04 01, for P ($\lambda = 64.6°$W, $\phi = 45.0°$N). The number of the overpass in the month is n, the day of the month is J, the overpass time is in decimal hours LMT, and the angles f, ζ, β, ζ_s, β_s, φ_A and γ defined in the text are in degrees. Missing values, denoted by a dash, indicate nighttime overpass. The values in the table can also be found in Fig. 9.5

n	J	LMT	f	ζ	β	ζ_s	β_s	φ_A	γ
1	1	10.859	−18.3	−20	−100	44	−153	126	35
2	1	12.482	+60.3	+75	+62	41	+171	288	90
3	1	20.366	+61.5	+77	−61	–	–	–	–
4	1	21.987	−3.9	−4	+101	–	–	–	–
5	2	9.936	−57.4	−69	−90	49	−136	134	43
6	2	11.572	+39.2	+45	+73	41	−169	299	72
7	2	21.072	+51.8	+61	−69	–	–	–	–
8	2	22.703	−49.4	−57	+94	–	–	–	–
9	3	10.655	−33.7	−38	−97	44	−149	128	34
63	17	10.859	−18.3	−20	−100	38	−152	128	30
64	17	12.482	+60.3	+75	+62	36	+168	285	86
65	17	20.366	+61.5	+77	−61	–	–	–	–
66	17	21.987	−3.9	−4	+101	–	–	–	–
67	18	9.936	−57.4	−69	−90	44	−133	137	43

Example 9.4. *Monthly sampling tables for a wide swath instrument aboard the Sun-synchronous satellites Terra and Aqua, for a point with latitude 45° N.*

The CERES instrument scans across track. It has half-swath $f_M = 61.8°$ so that $\zeta_M = 78.0°$. The overlap fraction is $Q_E = 1.34$ when the instrument is aboard a satellite on a Terra-type orbit.

Terra. For the satellite Terra, launched in December 1999, we consider the following initial conditions (ascending node crossing, doc. LaRC/NASA): $\lambda_0 = 29.06°$W, date 2000 02 29 22:30 LMT. The monthly table shows the month of April, with $J = 1$ corresponding to the date 2000 04 01. The point P has coordinates $\lambda = 64.6°$W, $\phi = 45.0°$N. For each overpass, we calculate the overpass time (UT then LMT) and the angles describing the satellite–target–Sun geometry. See Fig. 9.5. The results for the first few overpasses are shown in Table 9.1. The values for the whole month (31 consecutive days, whatever the month) are shown in the monthly sampling table in Fig. 9.5. In this table, the LMT times are given from 0 to 24 on the abscissa axis and the days from 1 to 31 on the ordinate axis. Each point (triangle with apex at the top or the bottom) corresponds to an overpass, dashes (long or short) refer to the target–satellite direction, and small circles (white or black) refer to the

target–Sun direction. Dot-dashed lines indicate the times of sunrise and sunset, as well as the local apparent noon.
The recurrence cycle $C_{T_o} = 16$ days is clearly visible. On days J and $J+16$, we find the same values of ζ and β. Moreover, since the satellite is Sun-synchronous, we also find the same values of the LMT time.
Note that, for this latitude, on almost every day, there are two overpasses around 11 a.m. and two around 10 p.m. (120 overpasses in 31 days, or a daily average of 3.9 overpasses).

Aqua. For the satellite Aqua, on the same orbit as Terra, we consider the same ascending node, with crossing at 13:30 LMT. We calculate the sampling with an identical CERES instrument to the one aboard Terra (across-track scanning). The monthly table shown in Fig. 9.6 shows a sampling that can be described as 'symmetrical' with respect to the one for Terra.

Average Number of Overpasses per Day. On a given meridian, Fig. 9.7 shows the average number N of overpasses per day as a function of the latitude ϕ, from the North Pole to the South Pole, for the CERES instrument aboard Terra (or Aqua). For the maximum half-swath $f_M = 61.8°$ (continuous curve), the graph of $N(\phi)$ has an almost flat minimum around the equator, then increases towards the poles. Beyond $\phi = 82°$, each point is viewed on each revolution of the satellite. In the same figure, we have plotted the graph of $N(\phi)$ for $f = 3f_M/4$, $f = f_M/2$, and $f = f_M/4$.

Example 9.5. *Asymmetry between the northern and southern hemispheres regarding the overpass time of a Sun-synchronous satellite.*

Let us calculate the sampling for the VMI instrument, across-track swath, $f_M = 50.5°$, aboard the Sun-synchronous ($\tau_{AN} = 22{:}30$) and recurrent ($C_{T_o} = 26$ day) satellite SPOT-4. Fig. 9.8 shows a monthly table with the days of the month on the horizontal axis and the latitudes on the vertical axis. We consider a given time slot and note overpass times with triangles and angles ζ with line segments. The chosen time slot was of 2.5 hr on either side of noon. This is the most favorable period for image taking. It is easy to see why the northern hemisphere has the advantage over the southern, through the choice of τ_{AN}. The recurrence cycle of 26 days is clearly visible, as is the 5 day subcycle.

Example 9.6. *Monthly sampling tables for an instrument with average swath aboard a near-polar satellite (MetOp-1), for locations at various latitudes.*

MetOp-1 is Sun-synchronous ($\tau_{AN} = 21{:}30$) and recurrent ($C_{T_o} = 29$ day). We have chosen $\lambda_0 = 0°$ as ascending node. The instrument MHS has a swath of rather typical amplitude, intermediate between what one might call wide or narrow, with $f_M = 49.4°$, whence $\zeta_M = 59.0°$ and $Q_E = 0.77$.

- On the equator (Fig. 9.9), there is roughly one overpass per day (during the day or the night). The recurrence subcycle of 5 days is clearly visible.
- For high latitudes, such as $\phi = 70°$ (Fig. 9.10), there are 6 daily overpasses in two batches of consecutive overpasses.

Example 9.7. *Monthly sampling tables for an instrument with average (then wide) swath aboard a low-inclination satellite (Megha-Tropiques) over two consecutive months.*

Megha-Tropiques has inclination 20°. We have chosen $\lambda_0 = 0°$ as ascending node, with τ_{AN} = 12:00 at the date 2009 06 01.

Megha-Tropiques/ScaRaB. The instrument ScaRaB has swath of average amplitude, with $f_M = 48.9°$, whence $\zeta_M = 58.9°$ and $Q_E = 2.09$. We examine the overpasses for the point with geographic coordinates $\lambda = 0°$, $\phi = 15°$N. At such a latitude, with this satellite/instrument configuration, there are 5 or 6 overpasses per day, occurring in consecutive revolutions.
In the monthly tables for June, then July, shown in Figs. 9.11 and 9.12, we may observe the influence of nodal precession on overpass times [June (J=31) = July (J=1)]. The value of $\dot{\Omega}$ calculated above induces a value of $C_S = -51.3$ days for the cycle relative to the Sun. This means that the overpass time moves forward from one day to the next, until, after 51 days, we return to the same overpass times.

Megha-Tropiques/FOV max. Figure 9.13 shows the mean number N of overpass times as a function of the latitude ϕ. The ScaRaB instrument aboard Megha-Tropiques has been replaced by an instrument that would scan from limb to limb, called here FOV max (maximal field of view). For the maximal half-swath, $f_M = f_0 = 61.7°$ (continuous curve), the graph of $N(\phi)$ exhibits a flat maximum around the equator. This puts us in an interesting and novel situation: between 8°N and 8°S, every point is viewed during each revolution, i.e., 13.1 times per day. In the same figure, we have plotted the graphs of $N(\phi)$ for $f = 3f_M/4$ (this value corresponds roughly to that for ScaRaB), $f = f_M/2$ and $f = f_M/4$. The 'wobbly' appearance of the graphs is due to the fact that the recurrence cycle is short (7 days).

Note. Since the nodal period T_d is 101.93 min, the number of round trips per day is $\nu = 14.1$. As the satellite moves in the prograde direction, in the same direction as the Earth, it will only overfly a meridian $\nu - 1$ times per day. The mean daily frequency of intersection of a meridian is thus 13.1. (If the satellite had inclination $i = 160°$, it would cross the meridian 15.1 times per day.) The period T' obtained from this frequency is the synodic period of the satellite and the Earth.

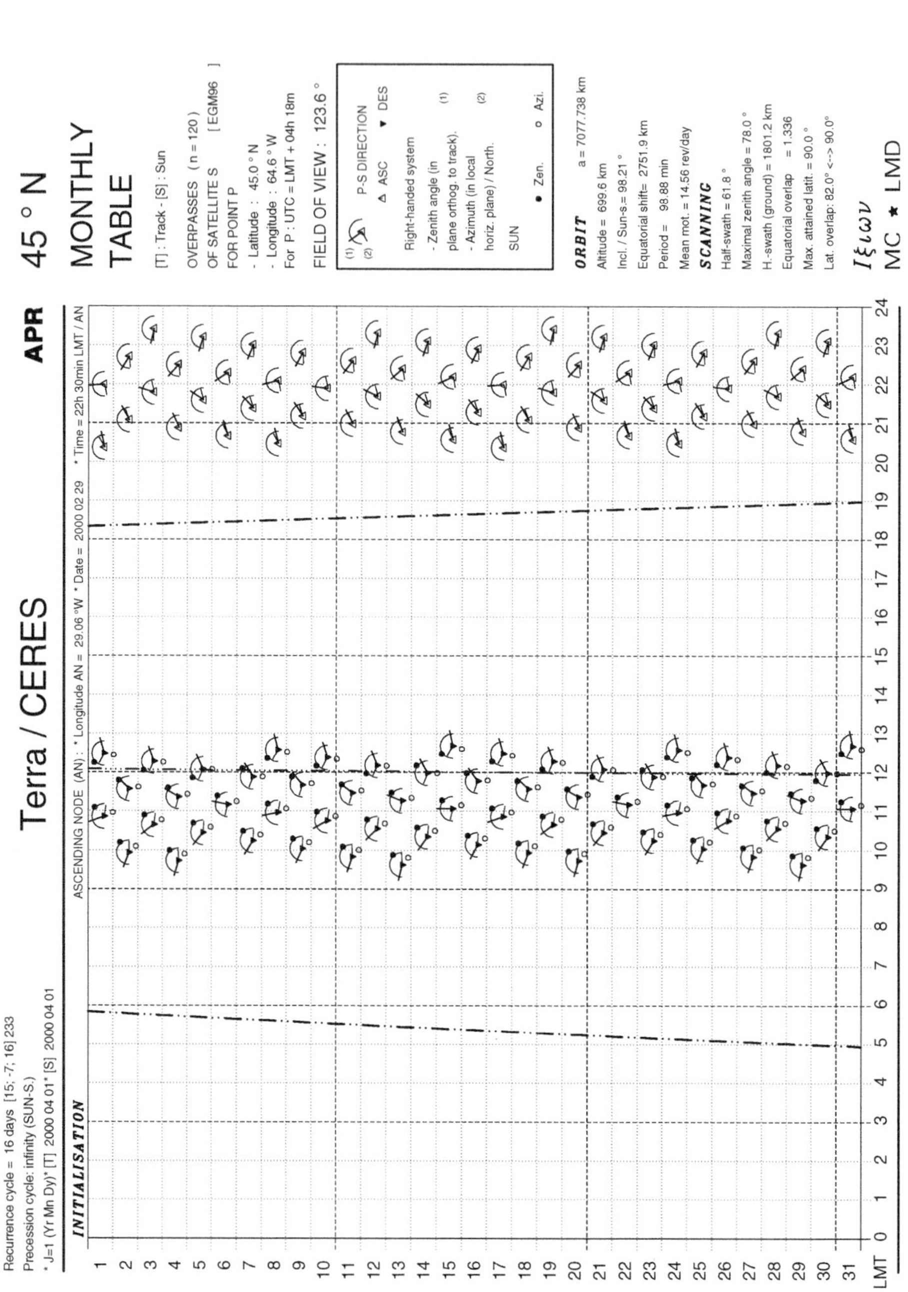

Figure 9.5. Monthly table. Overpass times

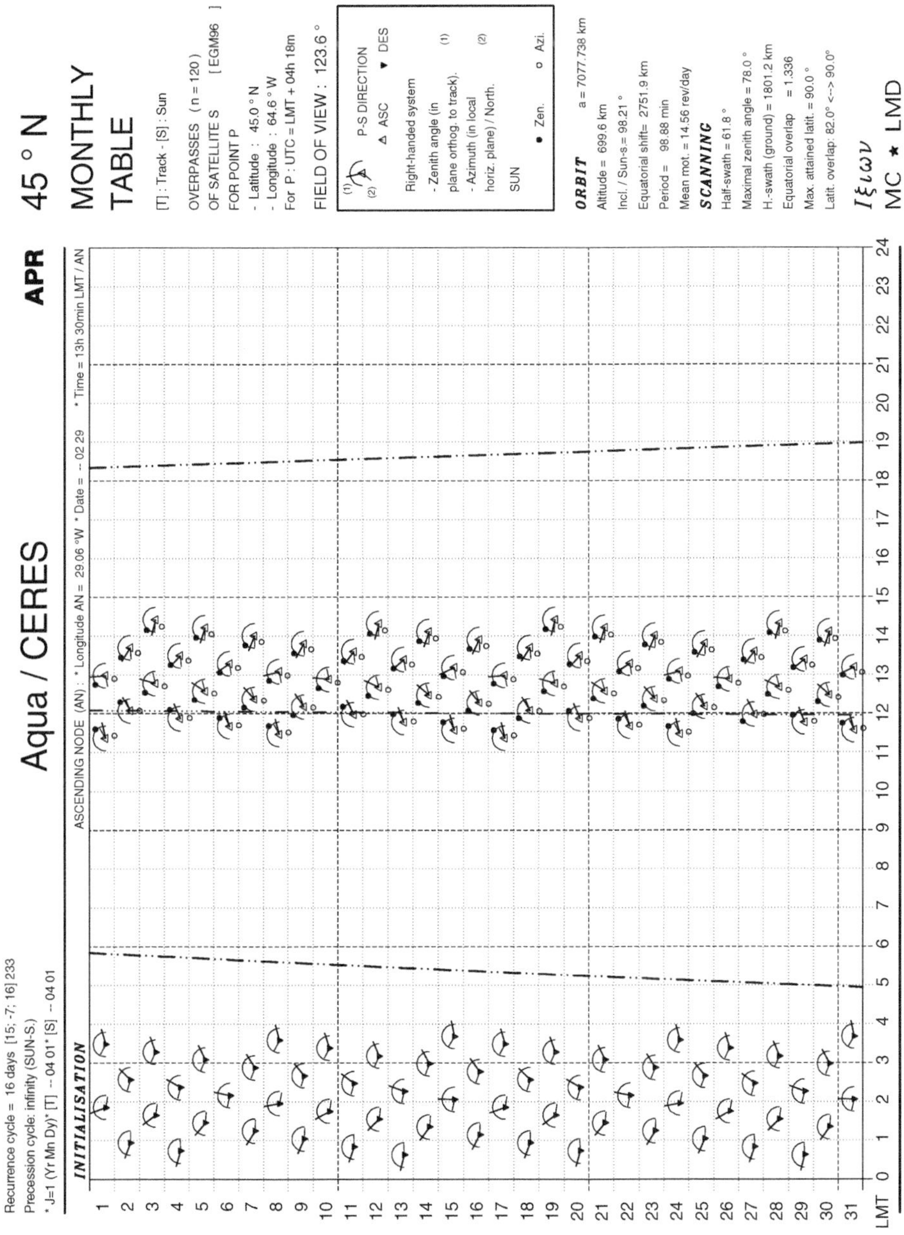

Figure 9.6. Monthly table. Overpass times

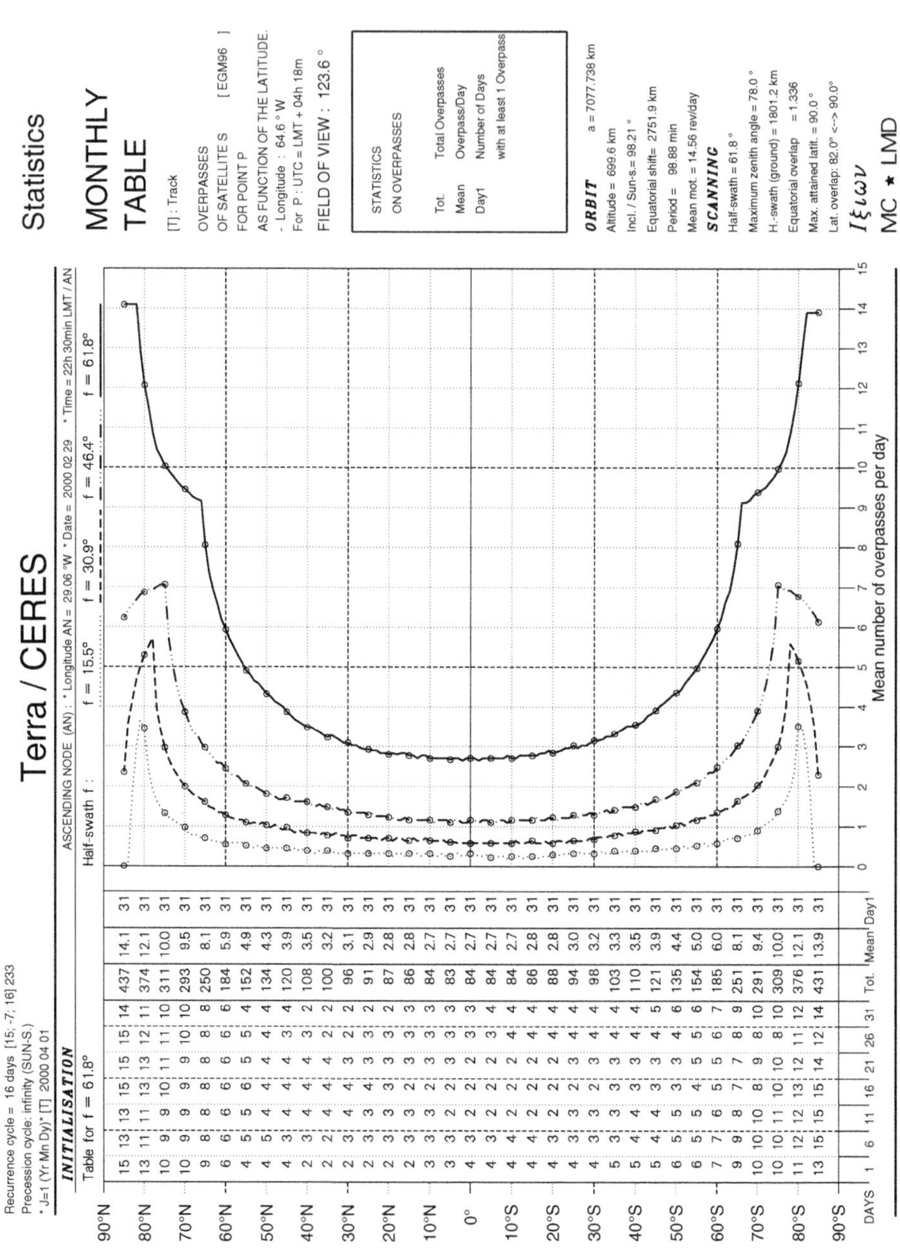

Figure 9.7. Monthly table. Overpass statistics

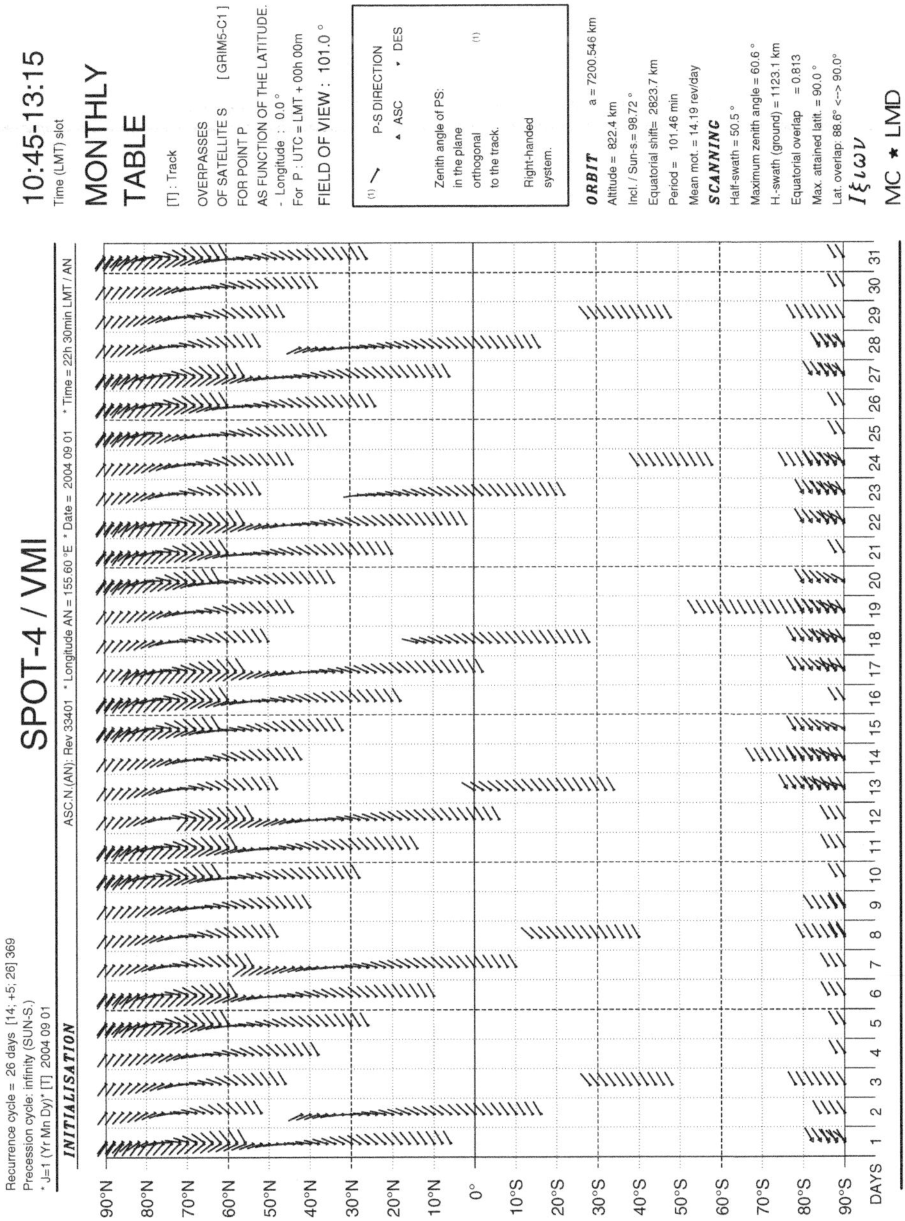

Figure 9.8. Monthly table. Overpasses in a time slot

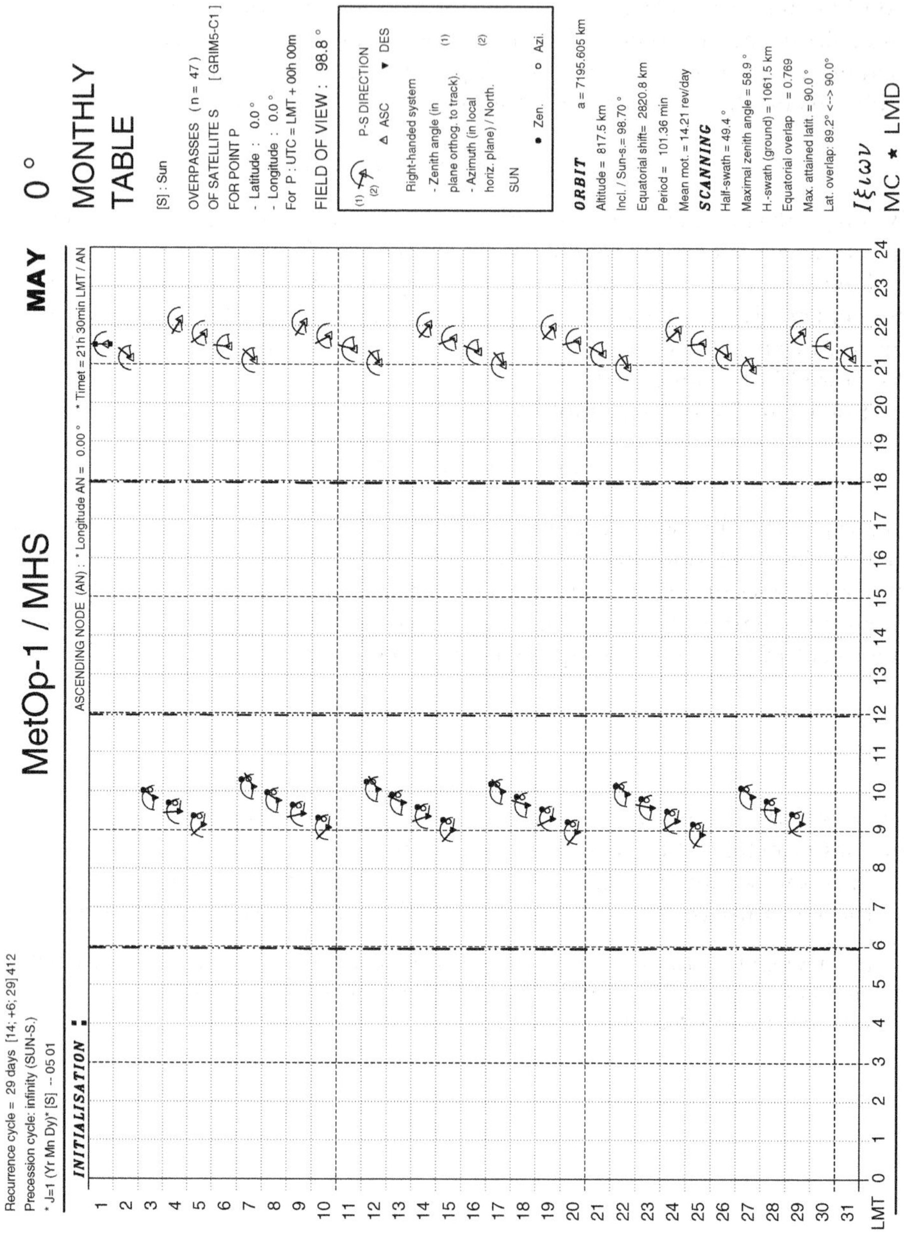

Figure 9.9. Monthly table. Overpass times

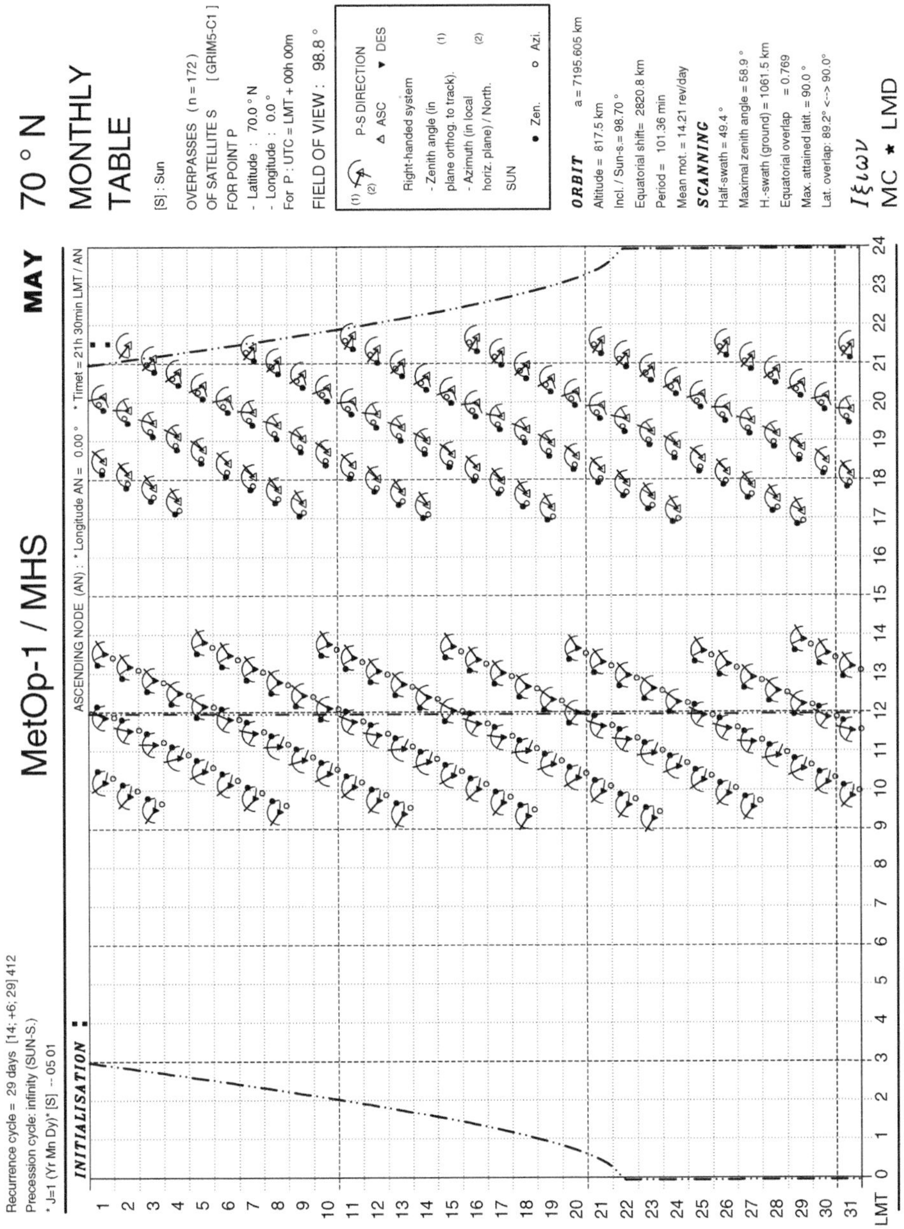

Figure 9.10. Monthly table. Overpass times

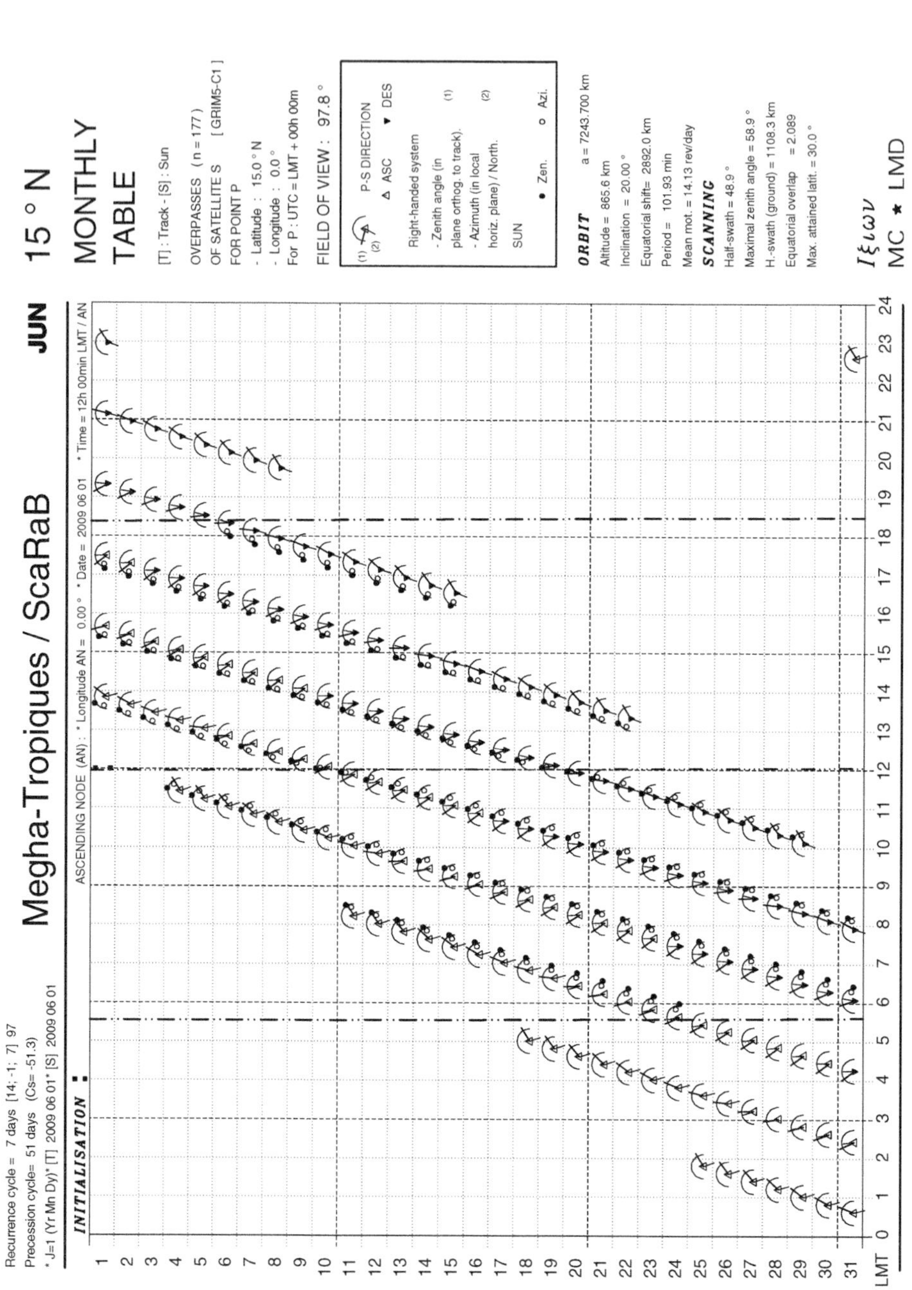

Figure 9.11. Monthly table. Overpass times

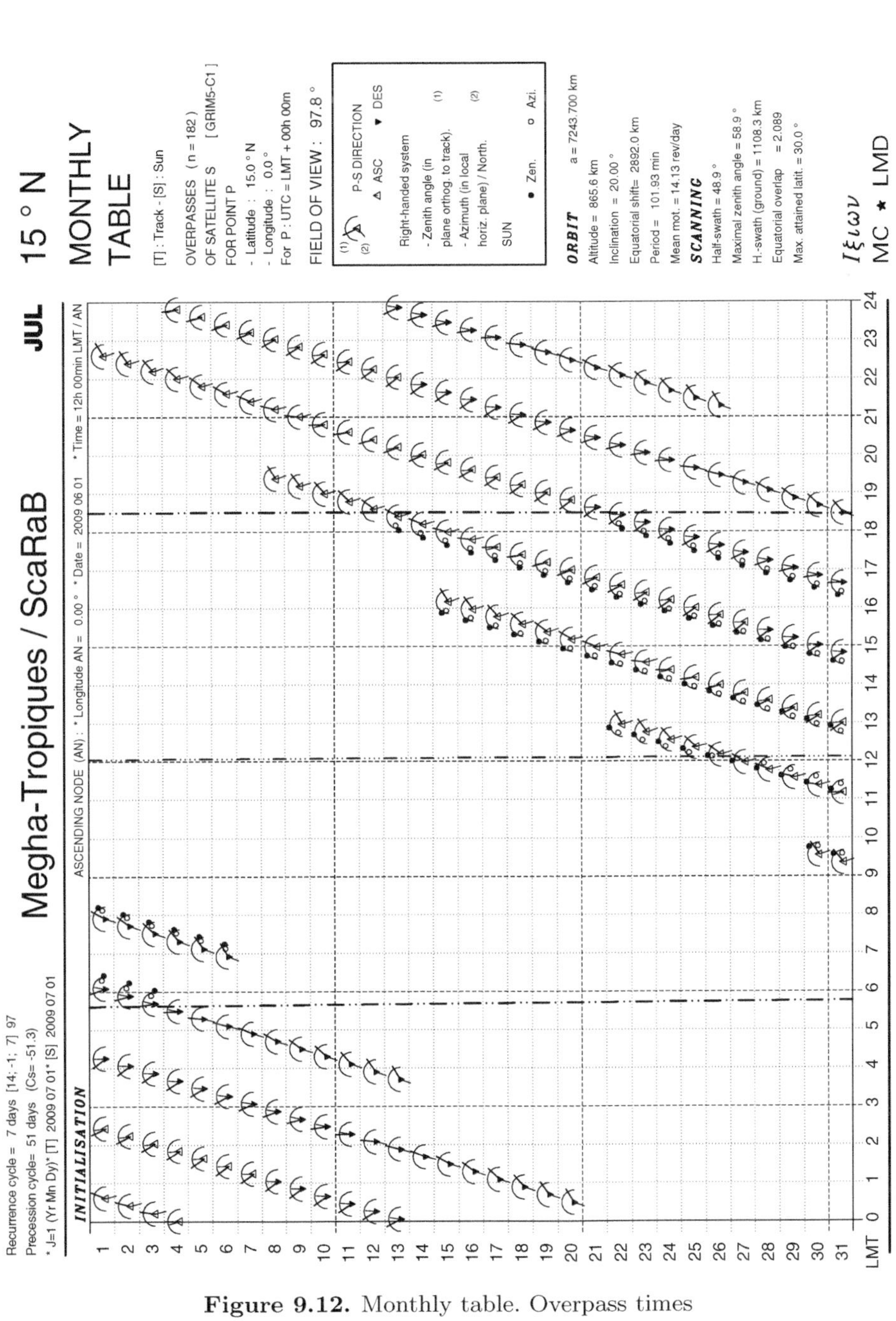

Figure 9.12. Monthly table. Overpass times

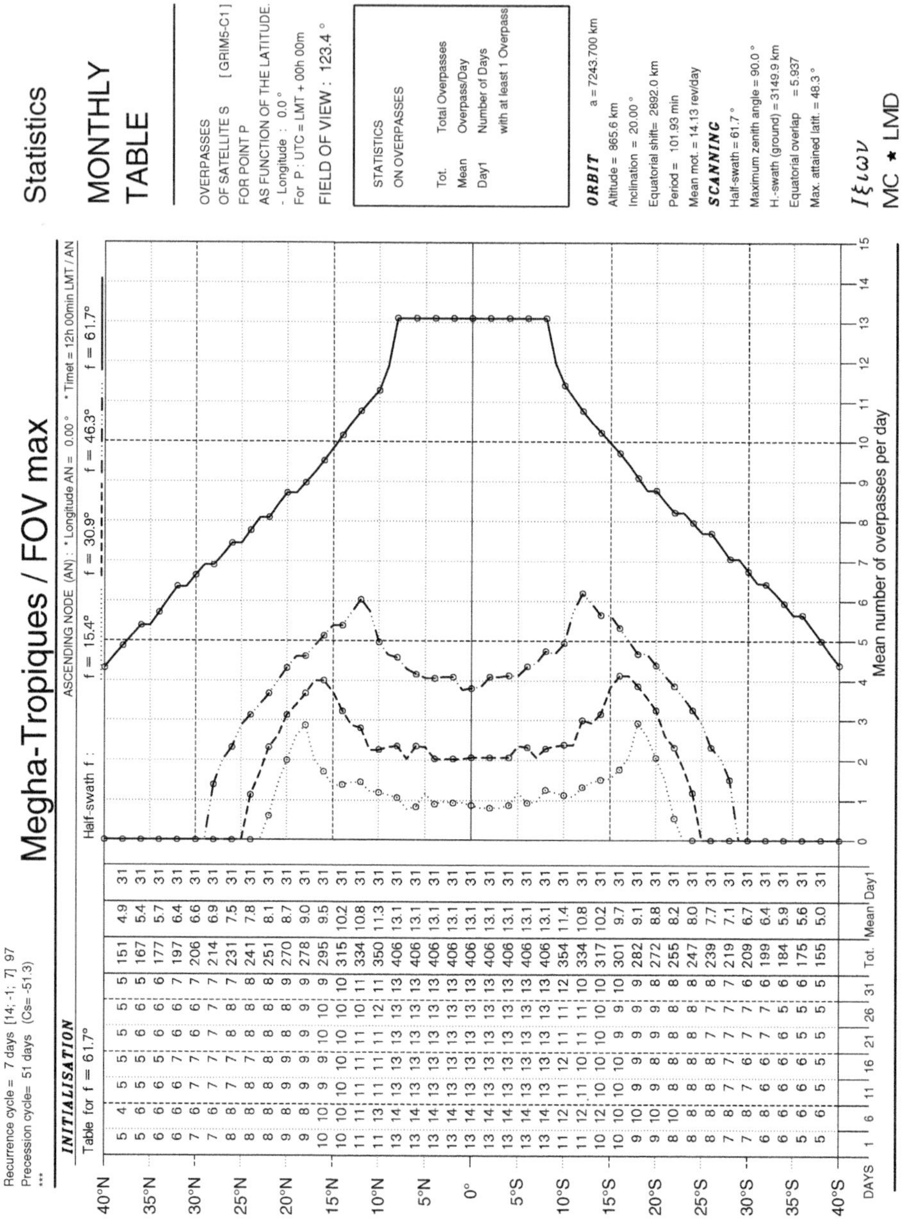

Figure 9.13. Monthly table. Overpass statistics

10 Satellites of Mars

10.1 Presenting the Planet Mars

10.1.1 Mars and Space Exploration

The planet Mars, situated after the Earth when we move away from the Sun, is often described as Earth's sister planet. Little brother would perhaps be more appropriate. The diameter of this telluric planet is only half the diameter of Earth, and its mean density is lower. The length of the Martian day and the obliquity are very close to the values for the Earth. Climatic phenomena related to the seasons can be observed on both planets, since Mars does indeed possess an atmosphere, although much less dense and of a different composition to the Earth's atmosphere.[1] The remoteness of Mars, about 1.5 astronomical units further out than the Earth, means that it is much colder than the Earth, the greenhouse effect due to the atmosphere also being lesser. In addition, the year is almost twice as long.

The planet has been observed since ancient times. It doubtless owes its association with the god of war to its red colour: Ares for the Greeks (ὁ Ἄρης, εως, whence the prefix 'areo' for attributes pertaining to this planet) and Mars for the Romans. Later, telescopic observation revealed some detail on its surface and it was suggested that Martians had been digging canals. In the twentieth century, more precise telescopic observations were improved still further by data from probes at the beginning of the space age.

In October 1960, the URSS attempted to send two probes (sometimes called Marsnik-1 and -2), just three years after the first Sputnik, to overfly Mars. However, they exploded at launch. (In fact it is not clear whether there were one or two probes.) All fourteen subsequent probes failed, either at launch or later by loss of contact. Sputnik-29, Mars-1 (which was the first probe to get near to Mars, but silent), and Sputnik-31 in 1962, Zond-2 and -3 in 1965, two probes without clearly attributed names (Mars-1969-A and

[1] The main components of the Martian atmosphere are, in molar fractions: carbon dioxide CO_2 (0.95), nitrogen N_2 (0.03), argon Ar (0.02), water H_2O (< 0.0005). The mean pressure at the surface is 6 hPa. Due to the condensation of carbon dioxide in the polar ice caps, the atmospheric pressure varies by as much as 30 % with the seasons.

-B) in 1969, Kosmos-419, Mars-2 and -3 in 1971, and Mars-4, -5, -6 and -7 in 1973.

Further attempts were made fifteen years later with Phobos-1 and -2, launched in July 1988. The mission was to observe the moon Phobos from an orbit around Mars. The probe Phobos-2 fulfilled part of its mission. Mars was a forbidden planet for the Soviets and it remained so for the Russians. The very ambitious mission Mars-96 ended on 16 November 1996, the day of its launch, in the Pacific. In all, they clocked up 18.5 failures for 19 attempts.

The United States launched Mariner-3 and -4 in November 1964. The first probe was lost, but Mariner-4 overflew Mars and sent the first photographs on 14 July 1965. The probes Mariner-6 and -7, launched in February and March 1969, overflew the Red Planet and provided a great many photographs. Launched in May 1971 (like Mariner-8, lost at launch), Mariner-9 was the first to go into orbit around the planet ($h_{\mathrm{p}} = 1\,650$ km, $h_{\mathrm{a}} = 17\,100$ km, $T \approx 12$ hr), on 14 November 1971. From then until 27 October 1972, it sent back 7329 photographs which totally changed our understanding of Mars.

The two probes Viking-1 and -2, launched in August and September 1975, also successfully accomplished their missions. For each probe, there was an orbiter and a lander. The landers transmitted data concerning the Martian atmosphere and surface over several Martian years.

The following probe, seventeen years after Viking, was the first not to be sent in tandem. This was Mars Observer, launched on 25 September 1992, lost as it was being placed in orbit.

The probe known as Mars Global Surveyor, launched on 7 November 1996, took with it many of the instruments designed for the previous mission. It went into orbit around Mars after a ten month journey to reach its destination at 14 light-minutes from the Earth. We shall discuss this satellite further in the present chapter (referring to it as MGS). Its orbit was made circular by air braking.[2] Such an orbit allows an instrument like MOLA to obtain a very accurate topographical survey of Mars, and we shall use its results here. The MOC camera produced high-precision photography, with a resolution of a few metres on the ground.

Launched on 4 December 1996, the probe Mars Pathfinder made an opportune softlanding on 4 July 1997 and released the small robot Sojourner which subsequently investigated the immediate vicinity of the landing site.

---

[2] The probe, whose motion is mainly governed by the planetary attraction, goes into a highly eccentric orbit with one focus at the centre of the planet. To obtain a circular orbit, one uses the drag of the planetary atmosphere on the spacecraft. The satellite loses energy, mainly at the periastron, since this is where it moves most quickly and there is the most atmosphere. The apoastron is thus reduced upon each revolution. This maneuver is known as air braking. It has the disadvantage of being very slow. In fact it takes several months. However, it is very economical in terms of energy. Since it does not require use of retro rockets, there is no need to carry propellant. And it is well known that it costs a great deal of propellant to take on extra propellant.

The orbiter and lander of the Mars Surveyor-98 programme both failed. Concerning the orbiter, Mars Climate Observer launched on 11 December 1998 was lost due to an incorrect estimate of the altitude on 23 September 1999, as it approached Mars. As for the lander, Mars Polar Lander (MPL) launched on 3 January 1999 went silent on 3 December 1999, just as it arrived near Mars. Contact was lost with the probe and the two Deep Space-2 penetrators.

The probe Mars Odyssey, which took over the orbiter part of the Mars Surveyor-2001 (MSP'01) project, was launched on 7 April 2001 and reached Mars on 24 October. The orbit was circularised by air braking.

The Japanese failed to get their probe Nozomi (hope) into orbit.[3]

The three missions which left in 2003 were highly successful. The European probe Mars Express, launched on 2 June, went into Martian orbit on 25 December.[4] The two US Mars Exploration Rovers, MER-A and -B, launched on 10 June and 8 July, arrived on 4 and 25 January 2004, releasing two robot vehicles of about 130 kg each, called Spirit and Opportunity, which immediately began to carry out their missions.

A brief glance at future missions reveal a considerable interest in Mars over the coming years. Unless otherwise stated, all projects are managed by NASA.

In 2005, Mars Reconnaissance Orbiter (MRO) will use air braking to circularise its orbit. In 2007, the Phoenix mission will raise something of the MPL and MSP'01 missions from the ashes, sending a lander to the higher latitudes of the planet.

In 2009, the mission Mars Science Laboratory (MSL) will send a rover as large as a car, which will run on nuclear fuel for one Martian year. Shortly afterwards, the satellite Mars Telecom Orbiter (MTO) will be placed in orbit as a relay for MSL. This mission replaces the US–Italian project for a telecommunications satellite, the G. Marconi Mission, cancelled in 2003.

The French mission Premier (*Programme de Retour d'Echantillons Martiens et Installation d'Expériences en Réseau*) was also cancelled in 2003. This was to prepare for the very ambitious US–French mission Mars Sample Return to bring back Martian rock samples. The latter is still programmed, for 2013 at the earliest.

In 2009, the European mission ExoMars should softland a rover (payload Pasteur), as part of the Aurora programme.

---

[3] Launched on 3 July 1998, the probe Planet-B (Nozomi) was to reach Mars on 11 October 1999, with a Moon–Earth–Moon gravity-assist maneuver. As this maneuver was not perfectly successful, the probe was placed in a heliocentric orbit so that it would, in principle, reach Mars four years behind schedule, in January 2004. However, this attempt also failed.

[4] Unfortunately, the lander Beagle-2 remained silent. The name Beagle given to this ground-based laboratory is explained in the note on Darwin.

These missions occur at a rate of about one every 26 months. Naturally, the trip from Earth to Mars is undertaken only when the necessary conditions contrive to make the journey as short as possible. This happens when the two planets are in opposition, i.e., the Sun, Earth and Mars lie in that order along a straight line. The interval of time between two consecutive oppositions is the synodic period. With the values for the periods given in Table 10.1, $T = 1$ for the Earth and $T_1 = 1.88$ for Mars (unit: Earth sidereal year), we obtain the synodic period T' from (2.23) as $T' = 1.88/0.88 = 2.135$, which gives $T' = 780$ day ≈ 2 yr 2 m.

An approximate value is sufficient here since the orbits of Earth and Mars are not exactly concentric. We calculate more precisely that the interval between two oppositions varies from 764 days (oppositions close to the aphelion of Mars) to 810 days (oppositions close to the perihelion of Mars) owing to the eccentricity of the orbit.

10.1.2 Geography of Mars

We have presented maps of our own beautiful planet Earth without specific indications. However, the features of Mars are distinctly less familiar to most Earthlings. How may we delineate these features? Although there is no separation into land and sea, different zones may appear brighter or darker through the telescope, depending on the albedo of the various terrains. The maps shown here are topographical charts, bearing no relation to the albedo. The curves are therefore contours, plotted from data obtained by the MOLA instrument and processed by the MOLA/NASA science team.

The zero meridian is arbitrarily chosen, as it is on Earth. It passes through the small crater[5] called Airy-0 in homage to the 'creator' of the Greenwich meridian.[6] The zero altitude was chosen in an even more arbitrary manner than on Earth. Today it is defined as the gravitational equipotential surface

[5] In his 1877 map of Mars, the Italian astronomer Schiaparelli measured longitudes from a meridian passing through a region which he considered characteristic and easily identifiable, called *Sinus Meridiani* by Camille Flammarion. When Mariner-9 mapped Mars in 1972, with a resolution of 1 km, a more precisely defined point had to be chosen in this region. The choice of this small impact crater was made by the Mariner-9 Team. It is 500 m across and located in the crater Airy. The coordinates of Airy-0 are 5.2°S, 0.0°E.

[6] George Biddell Airy (1801–1892) was an English astronomer who studied diffraction rings from the standpoint of astronomy, physics and mathematics. As the seventh director of the Royal Observatory at Greenwich, from 1835 to 1881, he considerably increased the importance of the institution when he established the meridian by means of a transit telescope in 1850 and persuaded the whole country to adopt the local mean solar time at Greenwich. In 1884, the international conference on the meridian chose the Greenwich meridian as the zero longitude. This spelt the end for other proposed prime meridians in Paris, Moscow and the Island of Ferro.

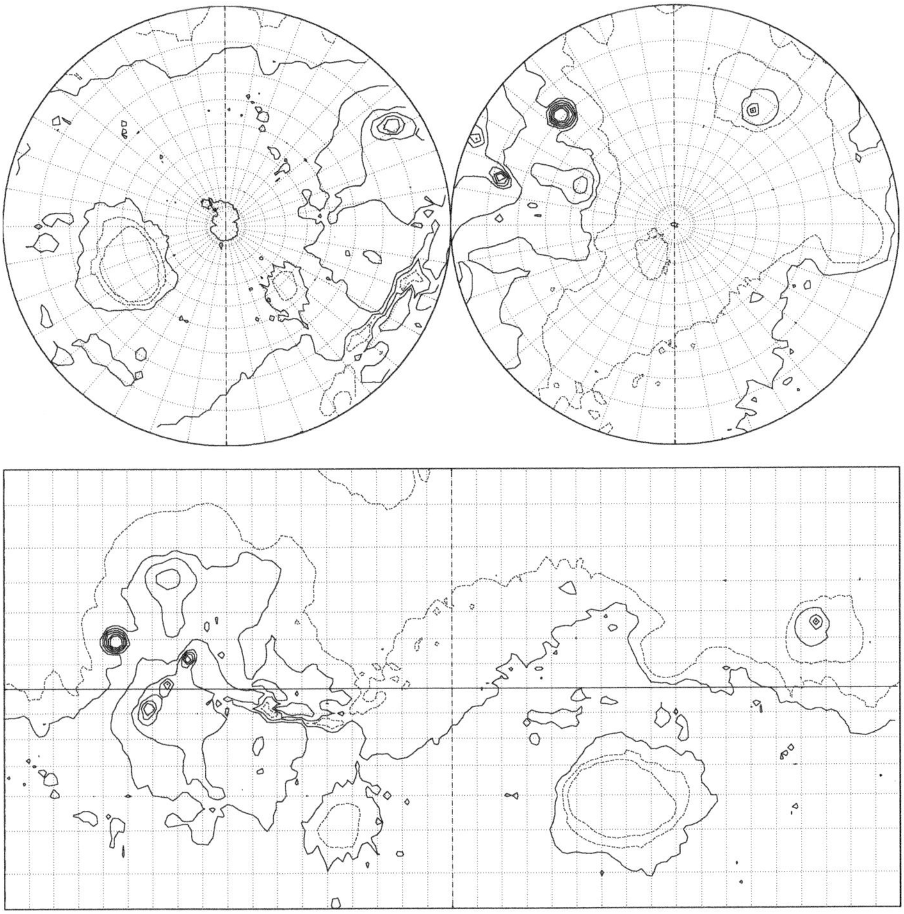

Figure 10.1. Topographical chart of the planet Mars compiled from data gathered by the MOLA instrument aboard MGS (processed by MOLA/NASA Team). Contours: steps of 2.5 km. Altitude 0 in *bold face*, negative altitudes *dashed*. MOLA data is used here in a downgraded mode with grid 2° for greater clarity. Geodetic grid: graticule 10° in latitude and longitude. *Upper*: polar stereographic projection (North Pole on the right and South Pole on the left). *Lower*: Equatorial Mercator projection. The highest point is Olympus Mons (18°N, 225°E). To the south-east is the Tharsis Bulge, a high region encompassing a line of three volcanoes: Ascraeus Mons (12°N, 254°E), Pavonis Mons (0°, 247°E), Arsia Mons (9°S, 239°E). It is bordered by Valles Marineris, stretching from 5°S, 265°E to 15°S, 310°E. North of this region is Alba Patera (42°N, 252°E). The main impact basins are Hellas (45°S, 70°E), Argyre (50°S, 320°E), Isidis (12°N, 88°E), Utopia (45°N, 110°E).
Location of successful missions: Viking-1 (22.48°N, 312.06°E), Viking-2 (47.97°N, 134.29°E), Pathfinder (19.17°N, 326.79°E), Spirit (14.57°S, 175.47°E), Opportunity (1.95°S, 354.47°E)

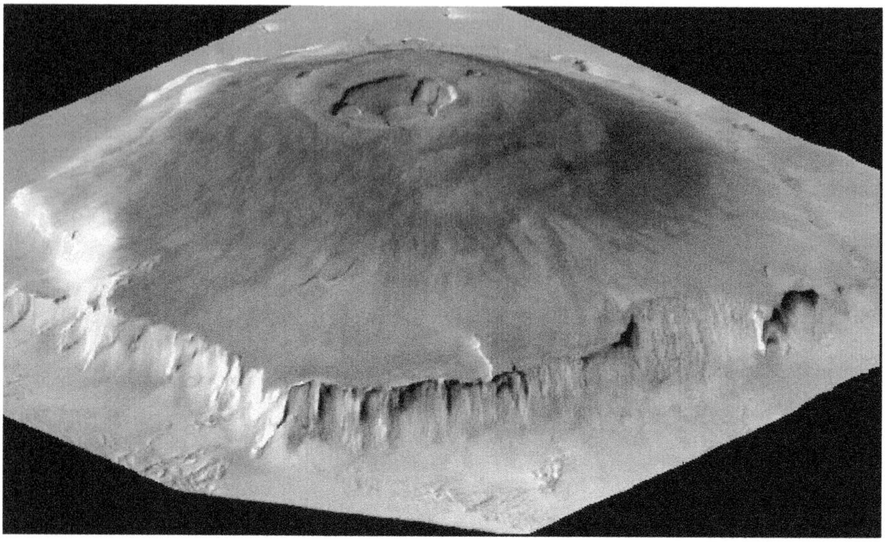

Figure 10.2. The volcano Olympus Mons. Image reconstructed from images obtained by the Viking orbiter. Altitudes are multiplied by a factor of 10. Credit: NASA/JPL

whose mean value at the equator is equal to the mean radius determined by MOLA, which implies an elevation of 2 km above the old zero level.

The topographical map in Fig. 10.1 and Colour Plate XIII show a very clear difference of altitude between the northern and southern hemispheres. The huge impact basis Hellas,[7] in the southern hemisphere is the site of the lowest point on Mars, some 7 825 m below the zero level. Near the equator, Olympus Mons (21 183 m) is the highest mountain in the Solar System[8] [see

[7] At the end of the nineteenth century, names were attributed on the basis of an Earthly design. Certain dark areas seemed to evoke the shape of the Mediterranean, whereupon Greece or the Gulf of Syrtis were placed there. Naturally, north and south were swapped, an artefact of telescopic vision! The astronomer Giovanni Schiaparelli may be considered as the father of the current Martian nomenclature, thanks to intensive observations he carried out from 1877. He borrowed names from ancient history and classical mythology. The International Astronomical Union unified the various appellations. They comprise two Latin nouns, a generic noun such as *mons* for 'mount', and a proper noun such as *Olympus*. This gives names like Olympus Mons, Mare Tyrrhenum, and so on. Valles Marineris is the valley discovered by Mariner-9. The impact basins (Latin noun: *planitia*), however, are generally designated by a single noun, e.g., Hellas, Argyre, Utopia.

[8] This volcano has a circular base with diameter 650 km. It has a very distinct caldera, 40 km wide and 4 km deep. It is now inactive, like all Martian volcanoes, but the small number of impacts on its surface indicate that it was active in the geologically recent past. The volume of Olympus Mons is about a hundred times

Fig. 10.2 and Colour Plate XVI (lower image)]. To the south-east of Olympus Mons is the Tharsis region, known as the Tharsis Bulge, with its alignment of three volcanoes (14 to 18 km high), and a little further east is the great scar of Valles Marineris.[9] The poles are covered by the ice caps.[10]

Geologically, the southern hemisphere, above the mean land level, is covered with large craters and composed of ancient landscapes, whilst the northern hemisphere, several kilometers below the mean land level, features wide plains under a layer of lava. In the equatorial zone, the Tharsis Bulge is a vast plateau at high altitude, incorporating the great volcanoes.

The caption to Fig. 10.1 indicates the locations of successful Mars missions. They are situated in low-altitude regions which provide sufficient atmospheric thickness to allow for air braking with parachutes.

10.2 Geodetic and Astronomical Quantities

10.2.1 Satellite in Keplerian Orbit

It is no easy task to put a satellite into orbit around Mars. However, if the probe, launched from Earth, is captured by the Martian gravitational attraction without actually crashing into the planet, so that Mars becomes the attractive centre of the satellite orbit, there is no fundamental difference in the way the motion is determined compared with a satellite around the Earth, as discussed at length in the first part of this book. To study the Keplerian motion of a given satellite (semi-major axis a), we may simply replace the value of the geocentric gravitational constant $\mu = \mathcal{G}M_{\mathrm{Earth}}$ by the areocentric gravitational constant $\mu = \mathcal{G}M_{\mathrm{Mars}}$, as given in Table 10.1.

If we also know the radius of the planet, we use (2.16) and (2.17) to define the periods T_0 and $T_{0(h=0)}$. Hence,

$$T_{0(h=0)} = 100.15 \text{ min} , \tag{10.1}$$

greater than the biggest terrestrial volcano. Eruptions of fluid lava have created an enormous volcanic shield over very long geological periods. This volcano, like the three others of the Tharsis region, has remained in the same position with respect to the source of the magma. This great stability tends to prove that tectonic activity is low or non-existent on Mars.

9 The Mariner valleys constitute a system of several parallel canyons, about 5000 km long. The greatest of them is 6 km deep over a width of about 200 km.

10 The two poles are covered by a polar deposit of radius several hundred kilometers. These are undoubtedly made up of sediments and water ice. The whole thing is then covered over with a permanent cap of water ice in the case of the North Pole and frozen carbon dioxide (CO_2) at the South Pole. On top of this, a carbon dioxide ice layer condenses seasonally, in the Martian winter, to sublime in summer. This alternation of summer and winter in the different hemispheres has long been observed from Earth, as attested by drawings due to Huygens in 1672.

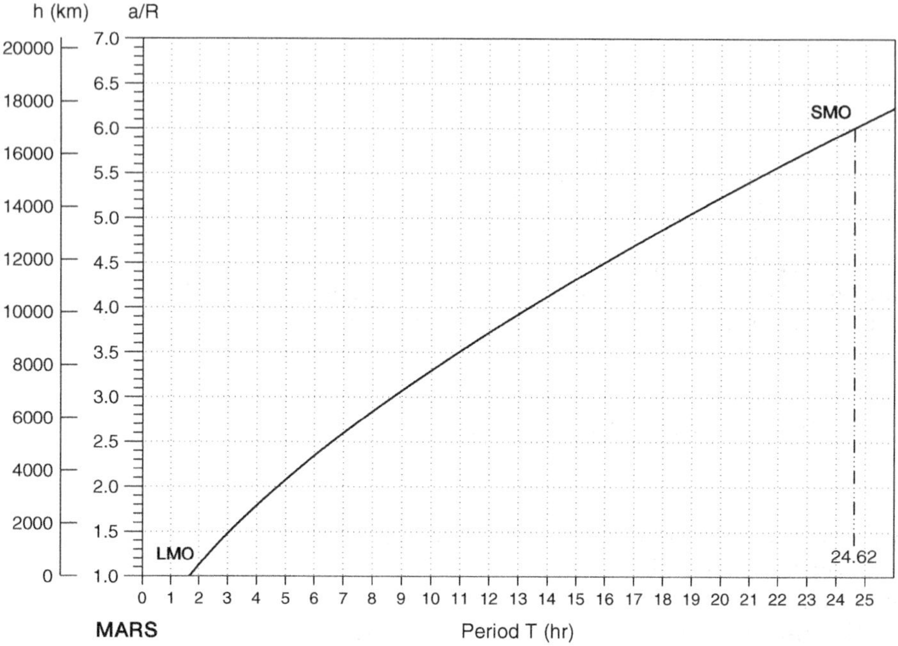

Figure 10.3. Relation between period and altitude

and we obtain T_0 as a function of $\eta = a/R$ or $h = a - R$ using (2.20). On the graph of the function, shown in Fig. 10.3, we have noted specific orbits LMO and SMO which will be discussed further below. This figure should be compared with Fig. 2.2.

10.2.2 Geodetic and Astronomical Data

To study the true (perturbed) motion, geodetic characteristics of the planet must be brought into consideration, in particular, higher terms in the expansion of the attractive potential. To characterise specific orbits, e.g., stationary or Sun-synchronous, and study instrument sampling, astronomical features such as the periods of planetary motions, eccentricity and obliquity are required.

Table 10.1 provides geodetic[11] and astronomical data for Mars, comparing them with the same data for Earth. Table 10.2 shows how the evaluation of the gravitational constant μ has evolved, from the first estimated values obtained after the discovery of Phobos and Deimos to the Goddard Martian Model (GMM) which makes use of MGS observations.

[11] When Mars replaces the Earth as attractive centre, we sometimes replace the prefix 'geo' by 'areo'. However, certain terms such as 'geophysics', 'geodesy' and 'geography' are generally retained for all planets.

Table 10.1. Geodetic and astronomical data for Mars compared with the same for Earth. For the units, n.d. means dimensionless

Quantity	Symbol	Unit	Mars	Earth
Gravitational constant	$\mu = \mathcal{G}M$	m^3s^{-2}	4.282 837 2 10^{13}	3.986 004 4 10^{14}
Equatorial radius	R	km	3 396.200	6 378.137
Flattening	$1/f$	n.d.	169.8	298.3
Acceleration at ground	g	$m\,s^{-2}$	3.7	9.8
Gravitational potential	J_2	n.d.	1.960×10^{-3}	1.083×10^{-3}
Gravitational potential	J_3	n.d.	$+36.0 \times 10^{-6}$	-2.5×10^{-6}
Gravitational potential	J_4	n.d.	-32.0×10^{-6}	-1.6×10^{-6}
Period of revolution				
sidereal	N_{sid}	day	686.980 0	365.256 4
tropical	N_{tro}	day	686.972 5	365.242 2
anomalistic	N_{ano}	day	686.995 1	365.249 6
Period of rotation				
sidereal	J_{sid}	hr	24.622 962	23.934 471
		s	88 642.663	86 164.090
mean solar day	J_{M}	hr	24.659 8	24.000 0
		s	88 775.245	86 400.000
Obliquity	ε	deg	25.19	23.44
Eccentricity	e	n.d.	0.093 40	0.016 71

Table 10.2. Measured areocentric gravitational constant $\mu = \mathcal{G}M$ and estimated error. Historical evolution, with method used and year

Method	Year	μ ($km^3\,s^{-2}$)	Error
Phobos, Deimos (Hall)	1878	42 900	±70
Mariner-4	1969	42 828.32	±0.13
Mariner-6	1970	42 828.22	±1.83
Mariner-9	1973	42 828.35	±0.55
MGS/GMM-1	1993	42 828.3580	±0.0512
MGS/GMM-2B	2001	42 828.371 901	±0.000 074

To carry out this study for Mars, we repeat the calculations made for Earth, without change of notation. For example, the angular speed of Mars in its orbit around the Sun and in its rotation about the polar axis will be denoted by $\dot{\Omega}_{\text{S}}$ and $\dot{\Omega}_{\text{T}}$, respectively. From (4.18) and (4.22), we obtain

$$\dot{\Omega}_{\text{S}} = 0.52404°\text{day}^{-1} , \tag{10.2}$$

$$\dot{\Omega}_{\text{T}} = 350.891\,98°\text{day}^{-1} . \tag{10.3}$$

The word 'day' refers to the unit of time equal to 86 400 s. As the mean day on Earth lasts one day, it is better to find another name for the Martian day. The mean solar day on Mars is tradiational called 'sol'. We then have

$$J_{\mathrm{M}} = 1 \text{ sol} = 1.027\,491\,27 \text{ day} . \tag{10.4}$$

It is useful to express the tropical year in sols, since it concerns the recurrence of the seasons:

$$N_{\mathrm{tro}} = 668.592\,1 \text{ sol} . \tag{10.5}$$

10.2.3 Areocentric Longitude and Martian Day

Relation Between Longitude and Day

To determine a given day on Mars, i.e., specify the position of the planet on its heliocentric orbit, we do not use the day of the month, as on Earth. We consider the ecliptic longitude l, defined in ecliptic coordinates in Chap. 4. It only differs from the true anomaly v by choice of origin. The origin for l is at the vernal equinox[12] and for Mars it is traditional to denote this angle l by L_{S}. It is called the areocentric solar longitude, or areocentric longitude for short.

Hence, on Mars, the date is specified by the true anomaly with the spring equinox as origin, whereas on Earth, it is specified by the mean anomaly with 1 January as origin. The true anomaly v has its origin at the periastron (perihelion). The solar longitude of the periastron L_{Sp} is

$$L_{\mathrm{Sp}} = 250.98° .$$

The mean anomaly M also has its origin at the periastron and we shall denote it by M_{S} when the origin is at the vernal equinox. Its value at the periastron is then denoted by M_{Sp}.

We have the relations

$$v = L_{\mathrm{S}} - L_{\mathrm{Sp}} , \tag{10.6}$$

$$M = M_{\mathrm{S}} - M_{\mathrm{Sp}} . \tag{10.7}$$

Recall that the mean anomaly M is obtained from the true anomaly v analytically, but that the converse problem of obtaining the true anomaly v from the mean anomaly M requires us to solve the Kepler problem:

[12] The orbit of Mars is itself subject to a precession of the equinoxes. This motion is slower than it is for Earth: $7''.51$ per year, or one round trip in 173 000 yr. For Mars, only the Sun contributes to this precession, whereas on Earth, not only is the Sun's effect greater, but one must add the even more significant contribution due to the Moon, as we have already seen.

$$\begin{aligned} v \longmapsto M : &\quad M = M(v) \quad \text{by (1.54)} , \\ M \longmapsto v : &\quad v = v(E) , \quad E = E(M) \quad \text{by iteration (1.72)} . \end{aligned}$$

With these relations, we obtain

$$M_{\mathrm{Sp}} = 261.32° .$$

Likewise, at the vernal equinox, we have $v = v_\gamma = 360 - L_{\mathrm{Sp}} = 109.02°$, whence $M = M_\gamma = M(109.02) = 98.68°$. Conversely, with $M = 98.68°$, we obtain $v = 109.02°$ as shown in Example 1.1.

Day Obtained from the Areocentric Longitude. The day J (zero at the vernal equinox) is obtained from the mean anomaly by

$$J = \frac{N_{\mathrm{tro}}}{360} M_{\mathrm{S}} . \tag{10.8}$$

We obtain the day in days (J_{d}) or in sols (J_{s}) depending on whether N_{tro} is expressed in days or in sols. We then express the Martian days in sols. Using the relations between angles discussed above and the relation (1.54), we obtain J_{s} from L_{S}:

$$\begin{aligned} J_{\mathrm{s}} = J_{\mathrm{sp}} + \frac{N_{\mathrm{tro}}}{360} \Bigg[& 2 \arctan \left(\sqrt{\frac{1-e}{1+e}} \tan \frac{L_{\mathrm{S}} - L_{\mathrm{Sp}}}{2} \right) \\ & - \frac{180}{\pi} \frac{e\sqrt{1-e^2} \sin(L_{\mathrm{S}} - L_{\mathrm{Sp}})}{1 + e \cos(L_{\mathrm{S}} - L_{\mathrm{Sp}})} \Bigg] \quad [\mathrm{mod}\ N_{\mathrm{tro}}] , \end{aligned} \tag{10.9}$$

where

$$\frac{N_{\mathrm{tro}}}{360} = 1.85720 , \quad J_{\mathrm{sp}} = \frac{N_{\mathrm{tro}}}{360} M_{\mathrm{sp}} = 485.32 . \tag{10.10}$$

The values of the angles and days[13] are given in Table 10.3.

The table also shows the reduced distance r/a, the distance to the Sun divided by the semi-major axis of the orbit, and the eccentric anomaly E, which are related by (1.57), and the quantity $E_{\mathrm{CS}} = L_{\mathrm{S}} - M_{\mathrm{S}}$, comparable with the equation of centre, $E_{\mathrm{C}} = v - M$, defined in Chap. 1. The two angular differences E_{CS} and E_{C} only differ by a constant; E_{CS} is zero at the vernal equinox and E_{C} is zero at the periastron.

The extreme values of E_{CS} are:

- maximum $E_{\mathrm{CS}} = +0.35°$ for $L_{\mathrm{S}} = 94.01 + L_{\mathrm{Sp}} = 344.99°$,

[13] Concerning the calculation of J_{s} using (10.9), the object in square brackets must be expressed in degrees, since it multiplies $N_{\mathrm{tro}}/360$. Note also that, since $L_{\mathrm{S}} = 0$ implies $J_{\mathrm{s}} = 0$, we can calculate the value of J_{sp} from (10.9) without using (10.10).

Table 10.3. Correspondence between the areocentric longitude L_S and the day (expressed in days or in sols). The passage at periastron occurs for $L_S = L_{Sp} = 250.98°$, and at apoastron for $L_S = L_{Sa} = L_{Sp} - 180 = 70.98°$ (minimum and maximum of the reduced distance r/a, respectively). The areocentric longitude L_S, the mean anomaly M_S, and hence the days J have their origin at the vernal equinox. The anomalies v, E and M have their origin at the periastron passage. The difference $L_S - M_S$ is the equation of centre E_{CS}

L_S [deg]	M_S [deg]	E_{CS} [deg]	J_d [day]	J_s [sol]	v [deg]	E [deg]	M [deg]	r/a –
0.0	0.0	0.0	0.0	0.0	109.0	103.9	98.7	1.0224
10.0	10.7	−0.7	20.4	19.8	119.0	114.2	109.3	1.0383
20.0	21.7	−1.7	41.3	40.2	129.0	124.7	120.3	1.0532
30.0	32.9	−2.9	62.9	61.2	139.0	135.4	131.6	1.0665
40.0	44.5	−4.5	84.9	82.6	149.0	146.1	143.2	1.0776
50.0	56.2	−6.2	107.3	104.5	159.0	157.0	154.9	1.0860
60.0	68.2	−8.2	130.1	126.6	169.0	167.9	166.8	1.0913
70.0	80.1	−10.1	152.9	148.8	179.0	178.9	178.8	1.0934
80.0	92.1	−12.1	175.8	171.1	189.0	189.9	190.8	1.0920
90.0	104.1	−14.1	198.6	193.3	199.0	200.8	202.8	1.0873
100.0	115.9	−15.9	221.1	215.2	209.0	211.7	214.5	1.0794
110.0	127.5	−17.5	243.2	236.7	219.0	222.5	226.1	1.0688
120.0	138.8	−18.8	264.9	257.8	229.0	233.2	237.5	1.0560
130.0	149.8	−19.8	285.9	278.3	239.0	243.7	248.5	1.0413
140.0	160.6	−20.6	306.4	298.2	249.0	254.1	259.3	1.0256
150.0	171.0	−21.0	326.3	317.5	259.0	264.3	269.7	1.0092
160.0	181.0	−21.0	345.5	336.2	269.0	274.4	279.7	0.9929
170.0	190.8	−20.8	364.1	354.3	279.0	284.3	289.5	0.9770
180.0	200.2	−20.2	382.1	371.8	289.0	294.0	298.9	0.9620
190.0	209.4	−19.4	399.6	388.9	299.0	303.6	308.1	0.9483
200.0	218.3	−18.3	416.6	405.4	309.0	313.1	317.0	0.9362
210.0	227.0	−17.0	433.2	421.6	319.0	322.4	325.7	0.9260
220.0	235.5	−15.5	449.5	437.4	329.0	331.7	334.2	0.9178
230.0	243.9	−13.9	465.5	453.0	339.0	340.9	342.6	0.9118
240.0	252.2	−12.2	481.4	468.5	349.0	350.0	350.9	0.9080
250.0	260.5	−10.5	497.1	483.8	359.0	359.1	359.2	0.9066
260.0	268.8	−8.8	512.9	499.2	9.0	8.2	7.5	0.9076
270.0	277.1	−7.1	528.7	514.6	19.0	17.3	15.8	0.9108
280.0	285.5	−5.5	544.7	530.1	29.0	26.5	24.1	0.9164
290.0	294.0	−4.0	560.9	545.9	39.0	35.8	32.6	0.9242
300.0	302.6	−2.6	577.5	562.0	49.0	45.1	41.3	0.9341
310.0	311.5	−1.5	594.4	578.5	59.0	54.5	50.2	0.9458
320.0	320.6	−0.6	611.8	595.4	69.0	64.1	59.3	0.9592
330.0	330.0	0.0	629.7	612.9	79.0	73.8	68.7	0.9740
340.0	339.7	0.3	648.2	630.8	89.0	83.7	78.3	0.9897
350.0	349.7	0.3	667.3	649.4	99.0	93.7	88.3	1.0060
360.0	360.0	0.0	687.0	668.6	109.0	103.9	98.7	1.0224

- minimum $E_{\rm CS} = -21.05°$ for $L_{\rm S} = 259.99 + L_{\rm Sp} - 360 = 156.97°$.

The amplitude of the variation is $(21.05 + 0.35)/2 = 10.70°$. According to (2.13), this value, expressed in radians as $(\pi/180)10.70 = 0.1868$, corresponds to twice the eccentricity $e = 0.0934$ of the orbit of Mars.

Areocentric Longitude Obtained from the Day. Knowing the day $J_{\rm s}$, we can calculate the difference with the day of passage at perigee. This difference, $J_{\rm s} - J_{\rm sp}$, gives the difference in mean anomaly, $M_{\rm S} - M_{\rm Sp}$. By iteration (Kepler's problem), we obtain the true anomaly v and this in turn gives the areocentric longitude $L_{\rm S} = v + L_{\rm Sp}$.

Relation with the Date on Earth. To find the longitude $L_{\rm S}$ from the date expressed in the form D = year month day hour, it is convenient to begin by transforming the date D into a Julian date denoted by JD. We calculate the difference with a date $\rm JD_0$, known as the time of passage of Mars at the vernal equinox. We can take

$$\mathrm{D} = 2002 \text{ April } 18.7 \quad \longmapsto \quad \mathrm{JD_0} = 2\,452\,383.2 \quad \Longrightarrow \quad L_{\rm S} = 0\,.$$

This difference gives

$$J_{\rm d} = \mathrm{JD} - \mathrm{JD_0} \quad [N_{\rm tro}]\,.$$

All these quantities are in Earth days.

We can go from $J_{\rm d}$ to $J_{\rm s}$ (in sol) using the coefficient given by (10.4), obtaining $L_{\rm S}$ as indicated above.

Example 10.1. *Calculate the solar areocentric longitude for the date 1 October 2008.*

The time calculated is for the relevant day at 0 h. The date D = 2008 10 01 00:00 gives JD = 2 454 740.5. Taking the zero value $\rm JD_0$ mentioned above, we obtain

$$J_{\rm d} = \mathrm{JD} - \mathrm{JD_0} = 2\,357.3\ [686.97] = 296.3\,.$$

From the sol/day coefficient, we have $J_{\rm s} = 288.4$. Solving the Kepler problem, we eventually obtain $L_{\rm S} = 135.0°$.

Definition of the Seasons

A season is the length of time corresponding to an interval of solar longitude equal to 90°, starting from the vernal equinox. The seasons are named as on Earth, e.g., the northern spring or southern autumn corresponds to $L_{\rm S}$ between 0° and 90°, etc. On Mars, the lengths of the seasons are not the same, as on Earth, but the differences in their lengths are greater than on

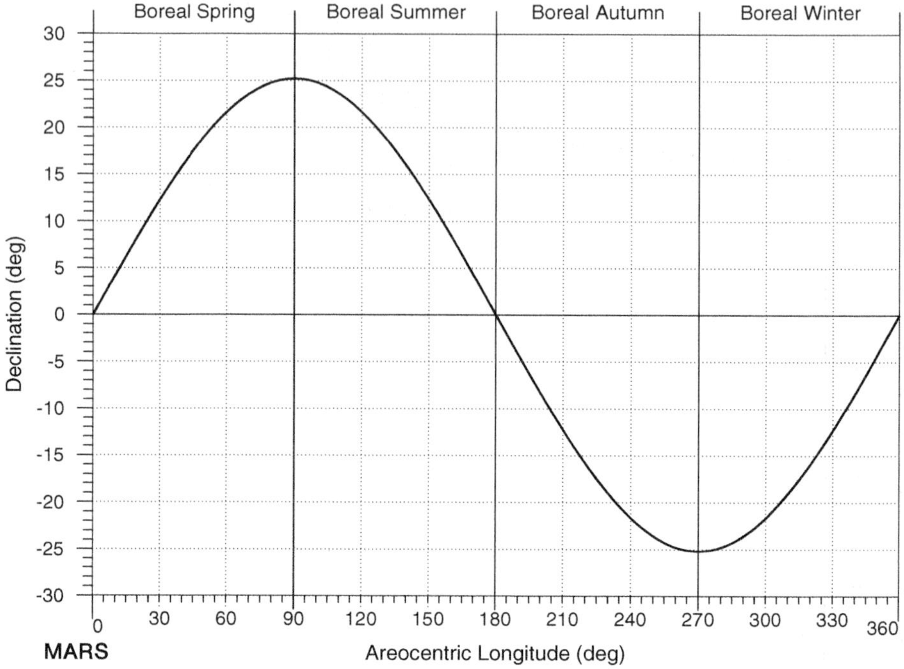

L_S [deg]	J [sol]	δ [deg]	δ	Beginning of season	Length of season
0.0	0.0	0.00	$\delta = 0$	Spring equinox	
24.1	48.7	10.00			
53.5	112.1	20.00			193.3
90.0	193.3	25.19	$\delta = \varepsilon$	Summer solstice	
126.5	271.2	20.00			
155.9	328.7	10.00			178.6
180.0	371.8	0.00	$\delta = 0$	Autumn equinox	
204.1	412.1	−10.00			
233.5	458.4	−20.00			142.7
270.0	514.6	−25.19	$\delta = -\varepsilon$	Winter solstice	
306.5	572.7	−20.00			
335.9	623.4	−10.00			154.0
360.0	668.6	0.00	$\delta = 0$	Spring equinox	

Figure 10.4. Graph of the declination δ as a function of the areocentric solar longitude L_S. *Table*: significant values of δ, with corresponding values of L_S in degrees and the date J in sols. The obliquity of Mars is $\varepsilon = 25.19°$. Note the unequal lengths of the seasons (in sols). The seasons indicated are those in the northern hemisphere

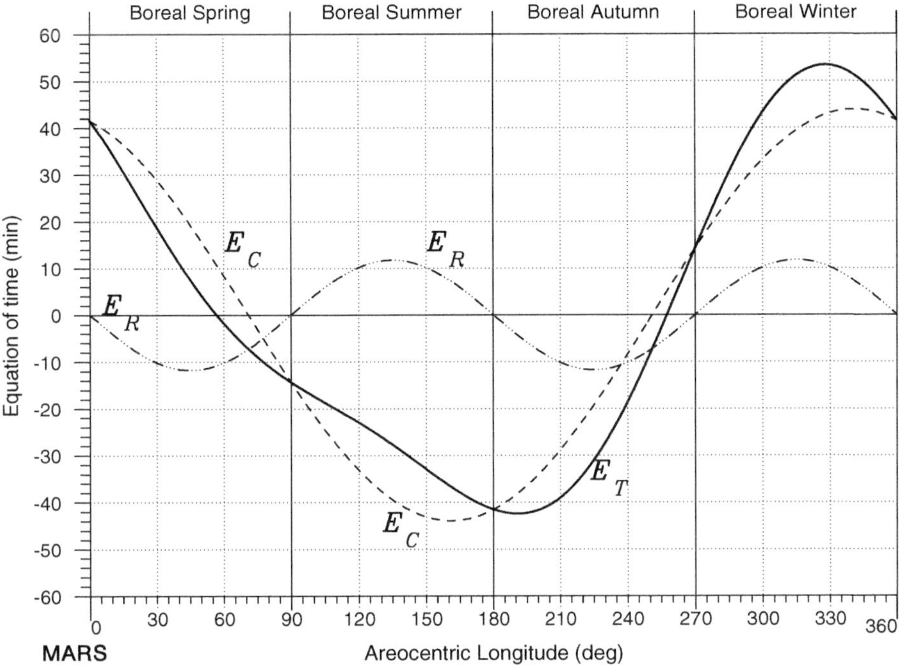

L_S [deg]	J [sol]	E_T [min]
0	0	+41.52
2	4	+40.00
28	57	+20.00
57	120	0.00
110	237	−20.00
173	360	−40.00
191	391	−42.43
207	417	−40.00
238	465	−20.00
257	495	0.00
275	522	+20.00
296	556	+40.00
328	609	+53.36
360	669	+41.52

Figure 10.5. Graph of the equation of time E_T, the sum of the equation of centre E_C and the reduction to the equator E_R, as a function of the areocentric solar longitude L_S. *Table*: significant values of E_T, with corresponding values of L_S in degrees and the date J in sols. Minutes (min) used for E_T, E_C and E_R are minutes of 60 seconds (1 sol = 1479.6 min)

Earth: 193 sols for the spring and 143 sols for the autumn. The precise length of the seasons is given in the table associated with Fig. 10.4.

It is sometimes useful to define the month in the same way, by an interval of 30° in the solar longitude, starting from the vernal equinox. (No name is attributed to these months, apart from the bounding values of L_S.) Table 10.3 can be used to calculate the length of these months. The shortest (L_S : 240°–270°) lasts 46.1 sols and the longest (L_S : 60°–90°) lasts 66.7 sols, i.e., 45% longer. These correspond to the passages at periastron and apoastron, respectively, illustrating Kepler's second law. The passage at periastron occurs at the end of the northern autumn.

The exact equivalent on Earth of this division into twelve months is the partitioning of the year into twelve signs of the zodiac. The shortest of these signs (l: 270°–300°) is the one containing the passage at perigee ($l = 282°$), which lasts 29.45 days. The longest (l: 90°–120°) lasts 31.45 days, i.e., 7% longer. This note should in no way be interpreted as an advertisement for astrology!

10.2.4 Declination

When considering the Sun and the Earth in Chap. 4 [see Fig. 4.7 and (4.51)], we have already calculated the declination, expressing it as a function of the solar longitude. With the notation used there, we can write

$$\sin\delta = \sin L_S \sin\varepsilon \ . \tag{10.11}$$

The declination is thus very simply obtained as a function of L_S :

$$\delta = \arcsin\left(0.42562 \sin L_S\right) \ . \tag{10.12}$$

The graph of this function is plotted in Fig. 10.4, which also shows key values of the declination.

Calculations to find the sunrise and sunset are strictly identical to those for the Earth, provided that we continue to ignore atmospheric refraction. We also define specific parallels on Mars: polar circles (64°49′N and S) and the tropics (25°11′N and S), at slightly different values to their terrestrial counterparts.

10.2.5 Equation of Time

In Chap. 4, we discussed the definition of the equation of time E_T, the sum of the equation of centre E_C and the reduction to the equator E_R. To express the equation of centre $E_C = v - M$, we return to (2.10). Stopping at the first order, we can replace E by v in the argument of the sine to obtain

$$\begin{cases} v - E \approx e \sin v \ , \\ E - M \approx e \sin v \ , \end{cases} \tag{10.13}$$

whence

$$v - M \approx 2e \sin v \ , \tag{10.14}$$

and with (10.6),

$$E_{\mathrm{C}} \approx 2e \sin(L_{\mathrm{S}} - L_{\mathrm{Sp}}) \ . \tag{10.15}$$

The expression for the reduction to the equator $E_{\mathrm{R}} = \alpha - l$, defined by (4.41), is obtained directly with (4.40). Using the areocentric solar longitude, this gives

$$E_{\mathrm{R}} \approx -\tan^2 \frac{\varepsilon}{2} \sin(2L_{\mathrm{S}}) \ . \tag{10.16}$$

The equation of time is then

$$E_{\mathrm{T}} = 2e \sin(L_{\mathrm{S}} - L_{\mathrm{Sp}}) - \tan^2 \frac{\varepsilon}{2} \sin(2L_{\mathrm{S}}) \ . \tag{10.17}$$

To express E_{T} in minutes, we convert radians into minutes of time: 2π rad is equivalent to one sol, or 1 479.6 min. Expressing L_{S} in degrees, we end up with

$$E_{\mathrm{T}}(L_{\mathrm{S}}) \ [\mathrm{min}] = 43.92 \sin(L_{\mathrm{S}} - 251) - 11.74 \sin(2L_{\mathrm{S}}) \ . \tag{10.18}$$

The graph of the function E_{T} is plotted in Fig. 10.5, with the graphs for E_{C} and E_{R}. Significant values of the equation of time are given in the associated tables.

Of course, as on the Earth, E_{C} has an annual period and E_{R} a period of half the length. However, on Mars, the amplitude of E_{C} is four times the amplitude of E_{R}, and E_{T} reaches large values (with a maximum of 53 min). It is worth remembering that $E_{\mathrm{T}} = \mathrm{LMT} - \mathrm{LAT}$.

Note on Expressions for the Declination and the Equation of Time. To end this section, note that the expressions for the declination and the equation of time are simpler for Mars than they are for the Earth. This happens because, for Mars, we use L_{S} as the variable, which amounts to using the true anomaly, specifying the position of the Sun directly. For the Earth, we use the day, related to the mean anomaly, which is only indirectly related to the position of the Sun.

10.3 Satellite in Real Orbit

10.3.1 Perturbative Accelerations

The only difference in the calculation of the Keplerian orbit for the Earth and for Mars is the value of μ, as noted earlier. For the true orbit, perturbing

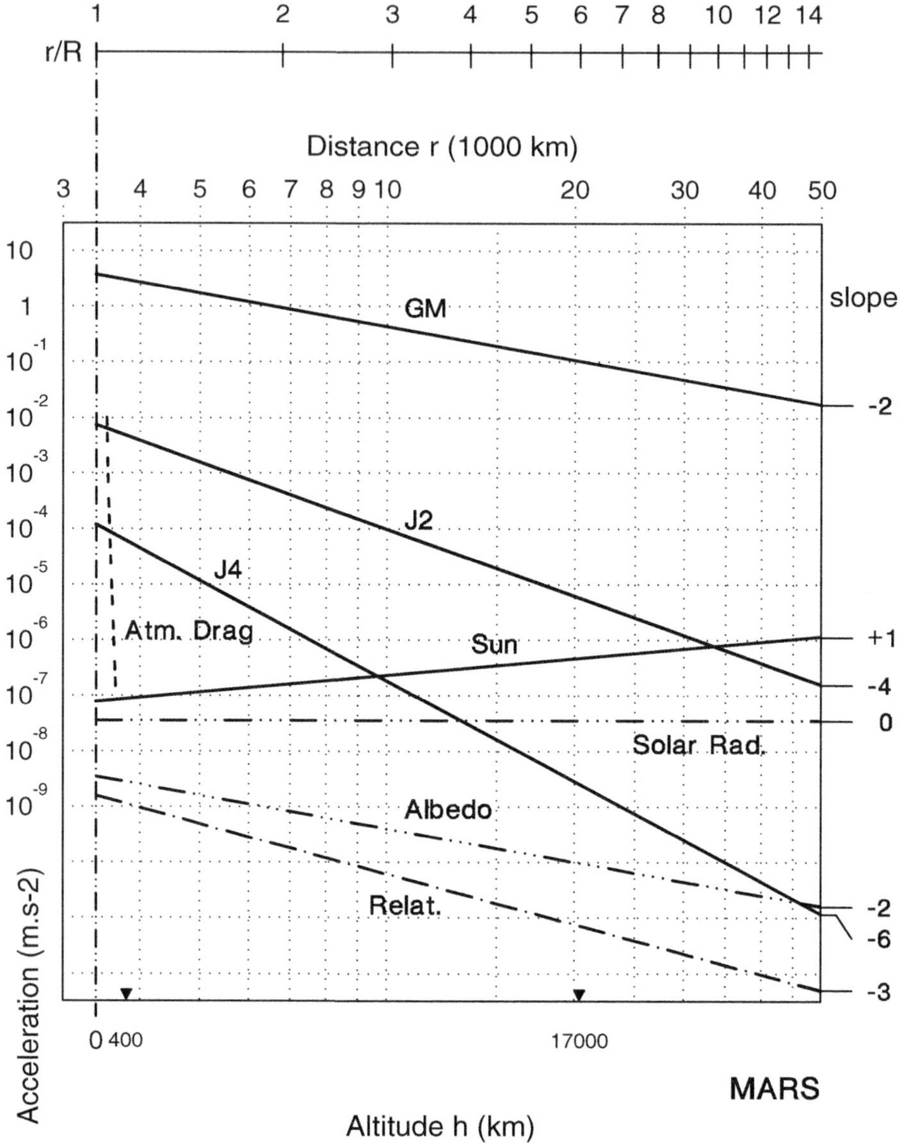

Figure 10.6. Central and perturbative accelerations as a function of the distance r of the satellite from the centre of Mars. Log–log scale. In the range of variation considered, the curves are approximately straight lines and their slope is indicated. The altitudes of two types of satellite have also been noted

terms must be taken into account, and these are of the same kind for Mars as for Earth. Figure 10.6 shows the various accelerations affecting the motion as a function of the distance r from the satellite to the centre of the planet. This corresponds to Fig. 3.1 for the Earth. The notation for the accelerations is the same as in Table 3.3.

Conservative Forces

The central acceleration γ_{CCC} has slope $p = -2$ for a log–log scale. Likewise, the acceleration due to the term in J_2 and those due to the following terms, i.e., J_4, J_6, etc., follow the same pattern as their counterparts for the Earth, i.e., $p = -4$ for $\gamma_{\mathrm{CCN}.J2}$, $p = -6$ for $\gamma_{\mathrm{CCN}.J4}$, etc. At the origin $h = 0$, numerical values are

$$\gamma_{\mathrm{CCC}}(R) = g(R) = g_0 = 3.73\ \mathrm{m\,s^{-2}}\ ,$$

$$\gamma_{\mathrm{CCN}.J2}(R) = 7.4 \times 10^{-3}\ \mathrm{m\,s^{-2}}\ , \qquad \gamma_{\mathrm{CCN}.J4}(R) = 1.2 \times 10^{-4}\ \mathrm{m\,s^{-2}}\ .$$

The lunisolar attraction on a terrestrial satellite is simply replaced here by the solar attraction, which is weaker than on Earth. As for the Earth, its value is obtained from (3.96). The slope of the curve is $p = +1$ and its value at the origin is

$$\gamma_{\mathrm{CS}}(R) = 2\frac{\mu_{\mathrm{S}}}{a_{\mathrm{S}}^3} R = 8 \times 10^{-8}\ \mathrm{m\,s^{-2}}\ ,$$

where a_{S} is the semi-major axis of the heliocentric orbit of Mars.

Tidal effects on the satellite are due to land tides caused by the Sun, which are much less marked than on Earth. The relativistic effect γ_{CR}, with slope $p = -3$, is calculated as for the Earth.

Dissipative Forces

The effect of drag due to the Martian atmosphere is less than on Earth for the same reduced altitude, because the atmosphere on Mars is less dense and the thermosphere is cooler. Solar radiation pressure is only half the value on Mars as compared with the Earth (term going as a_{S}^{-2}). The albedo effect depends on the region overflown. The albedo of Mars is rather low (mean value 0.22).

10.3.2 Secular Variation of Orbital Elements

We use the theory of perturbations to determine the evolution of the six orbital elements of the satellite. We showed in Chap. 3 that the three metric elements a, e and i remained constant (ignoring short- and long-period periodic variations). The secular variation of the angular elements is given by (4.1) or (4.4) for $\dot{\Omega}$, (4.2) or (4.10) for $\dot{\omega}$, and (4.3) for $\dot{M}$, as a function of the inclination i and the semi-major axis a of the orbit.

The periodic variations can be expressed with the help of the coefficient K_0 defined by (4.6). For Mars, this coefficient can be expressed as follows, depending on the units used:

$$K_0 = 3.074\,84 \times 10^{-6}\ \mathrm{rad\,s^{-1}}\,, \tag{10.19}$$

$$K_0 = 15.222°\mathrm{day}^{-1}\,, \tag{10.20}$$

$$K_0 = 15.640°\mathrm{sol}^{-1}\,, \tag{10.21}$$

$$K_0 = 29.047\ \mathrm{rev}\,(\mathrm{Martian\ yr})^{-1}\,. \tag{10.22}$$

For given values of i and a, the precession rates are higher on Mars than on Earth because of the value of the J_2 term, which is twice as great.

The unit for calculations is rad/s, but in graphs showing these quantities as a function of the inclination, we have used units of deg/sol. The nodal precession rate $\dot{\Omega}$ shown in Fig. 10.7 (upper) has a maximum of $15.6°\mathrm{sol}^{-1}$ for $a = R$ and $i = 0°$ or $180°$. With the same conditions, the apsidal precession rate $\dot{\omega}$ shown in Fig. 10.7 (lower) has a maximum of $31.2°\mathrm{sol}^{-1}$. The value of the critical inclination is independent of the attracting planet (if the geoid is close to ellipsoidal, i.e., if $J_n < 10^{-2} J_2$ for all $n > 2$).

Example 10.2. *Calculate the nodal precession rate of a satellite at altitude 340 km, with inclination 50°.*

This altitude corresponds to one tenth of the radius of Mars, i.e., $\eta = a/R = 1.10$. Applying (4.4) and converting units, we find $\dot{\Omega} = -7.2°\mathrm{sol}^{-1}$. We also obtain the result with Fig. 10.7 (upper), where the required value can be read off directly. A complete round trip is made in $360/7.2 = 50$ sol. The orbit of the satellite thus accomplishes $669/50 = 13.3$ round trips in the retrograde direction every Martian year.

10.3.3 Classification of Satellites

The classification criteria are the same as for terrestrial satellites.

Areosynchronous Satellite

We define an areosynchronous satellite by $n = \dot{\Omega}_\mathrm{T}$, and an areostationary satellite by adjoining the condition $i = 0$. Considering the mean Keplerian motion, we obtain

$$a_0^3 = \frac{\mu}{\dot{\Omega}_\mathrm{T}^2} = 8.524\,26 \times 10^{21}\,, \quad a_0 = 20\,428\ \mathrm{km}\,, \quad h_0 = 17\,031\ \mathrm{km}\,.$$

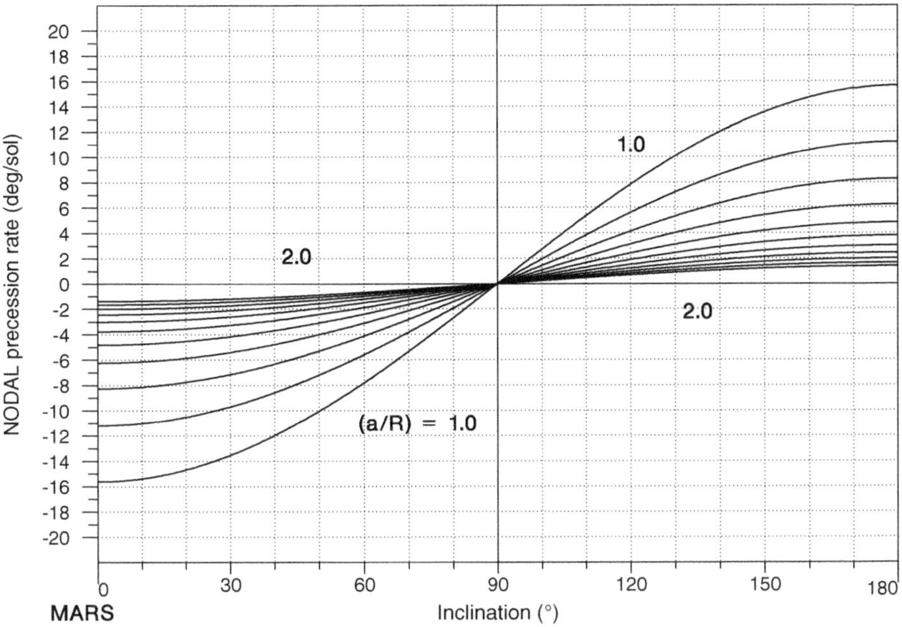

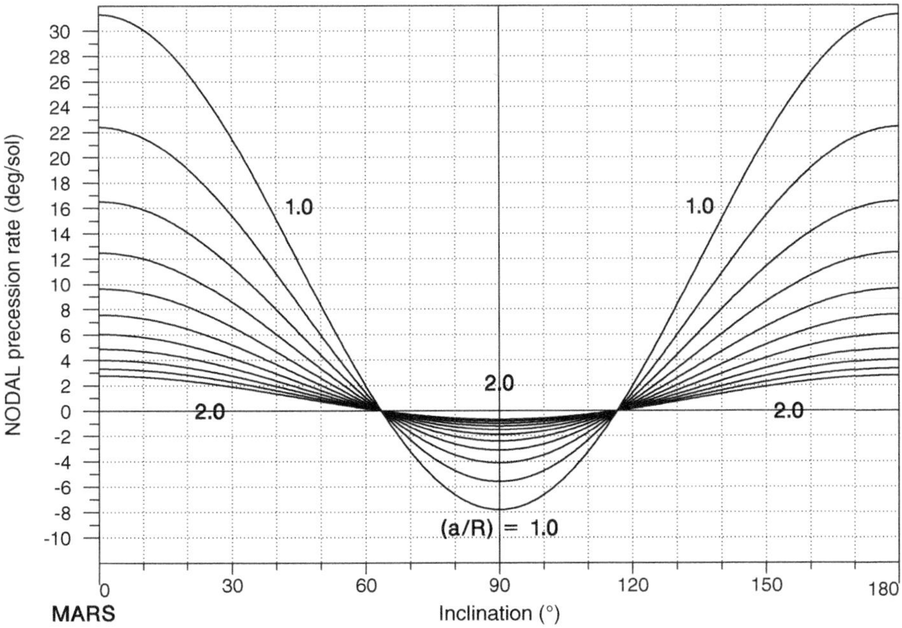

Figure 10.7. Precession rate in deg/sol as a function of the inclination i for various values of the reduced distance from $a/R = 1.0$ to $a/R = 2.0$, in steps of 0.1. *Upper*: nodal precession rate $\dot{\Omega}$. *Lower*: apsidal precession rate $\dot{\omega}$. Note that the value of the critical inclination is independent of the attractive source

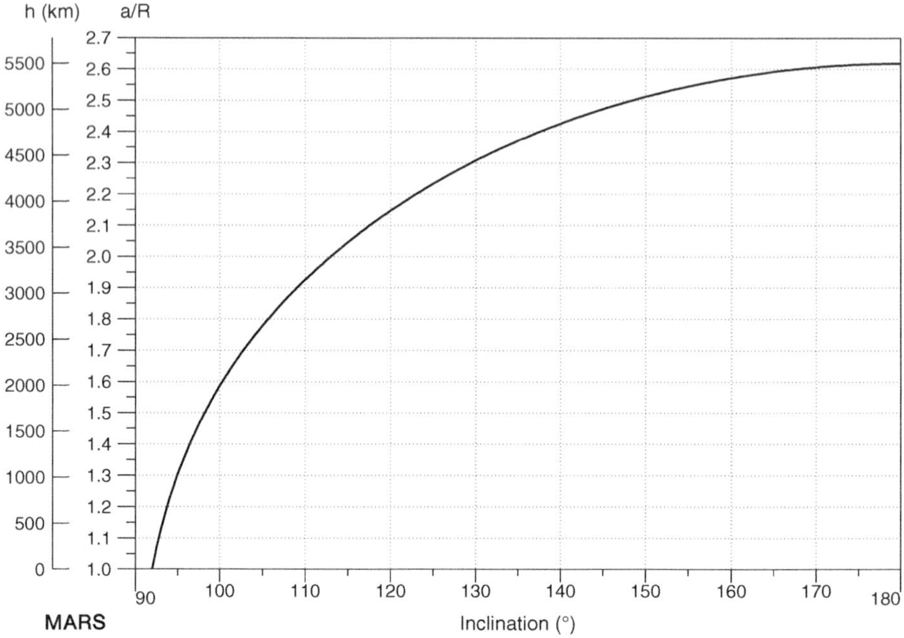

Figure 10.8. Altitude of a Sun-synchronous satellite as a function of the angle of inclination. The whole possible range is shown

Bringing the J_2 term into the calculation of the period, we obtain the value a_1, which is slightly bigger than a_0. In the case of a stationary satellite for Mars, the perturbing accelerating due to the term in J_2 is greater than that due to the Sun, as can be seen from Fig. 10.6. Recall that this is not so for a terrestrial stationary satellite, where the lunisolar perturbation is greater than that due to J_2. We shall take this value a_1 to be the one for an areostationary satellite, denoting it by a_{GS} :

$$a_{\mathrm{GS}} = 20\,430.99 \text{ km} , \quad h_{\mathrm{GS}} = 17\,034.79 \text{ km} , \tag{10.23}$$

$$\eta_{\mathrm{GS}} = \frac{a_{\mathrm{GS}}}{R} = 6.016 . \tag{10.24}$$

The value of η_{GS} for the Earth, given by (4.58), is close to the value found for Mars, because the same is true of the diurnal rotation periods and the mean densities of these two planets.

The possibility of placing satellites in areostationary orbits is currently under study, e.g., the satellite MARSat (Mars Areostationary Relay Satellite).

Sun-Synchronous Satellite

Concerning Sun-synchronous satellites, we have seen that the condition is $\dot{\Omega}(a,i) = \dot{\Omega}_\mathrm{S}$. We begin by calculating the constant of Sun-synchronicity, using (4.63), which gives for Mars

$$k_\mathrm{h} = 29.0403\ .$$

This is three times the value for the Earth, because J_2 is greater for Mars, and in addition, the planet moves more slowly around the Sun.

From (4.67) or (4.68), we thus obtain the relation between the inclination and the altitude.[14] Figure 10.8 shows the altitude as a function of the inclination for a Sun-synchronous satellite, which is necessarily retrograde.

The minimum value of i_HS, denoted by $i_{\mathrm{HS\,min}}$, is obtained for a (fictitious) satellite revolving at ground level ($\eta = 1$ or $h = 0$):

$$i_{\mathrm{HS\,min}} = \arccos\left(-\frac{1}{k_\mathrm{h}}\right) = \arccos(-0.0344) = 92.0°\ .$$

The maximum value of h is obtained when $i = 180°$:

$$\eta_{\mathrm{HS\,max}} = \frac{a}{R} = k_\mathrm{h}^{2/7} = 2.6182\ ,$$

$$a_{\mathrm{HS\,max}} = 8\,892\ \mathrm{km}\ , \quad h_{\mathrm{HS\,max}} = 5\,496\ \mathrm{km}\ .$$

It is not therefore possible to place a Sun-synchronous satellite (in circular orbit) at an altitude greater than 5 500 km (roughly the same bounding altitude as on Earth).

Terminology for Martian Satellites

The same terminology is used for satellites whether they are revolving around the Earth or Mars. The orbit can be prograde or retrograde. It can be Sun-synchronous, recurrent or frozen, or none of these.

Concerning altitude, a high orbit refers to areostationary satellites, also known as Stationary Mars Orbiting (SMO) satellites, and a low orbit means any satellite revolving below 1000 km, also known as Low Mars Orbiting (LMO) satellites.

[14] As for the Earth, there is a slight difference in the value of i_HS depending on the degree to which the planetary potential is expanded. For $h = 400$ km, $i_\mathrm{HS}(J_4) = 92.991°$ and $i_\mathrm{HS}(J_2) = 92.914°$, i.e., a difference of 0.077°.

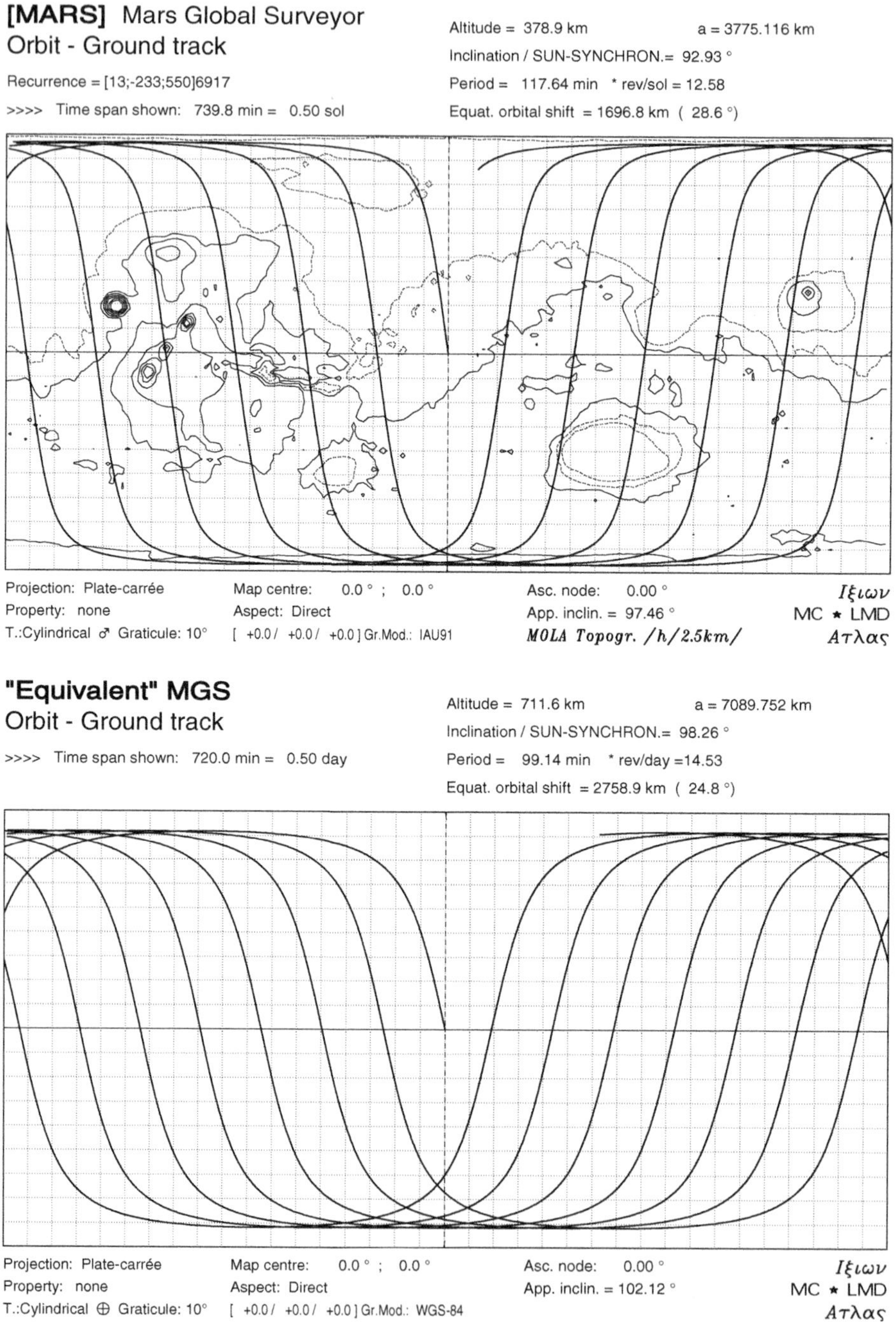

Figure 10.9. *Upper*: ground track of the Sun-synchronous satellite MGS over half a sol. *Lower*: ground track of the fictitious Sun-synchronous satellite equivalent to MGS but with Earth as the attracting body

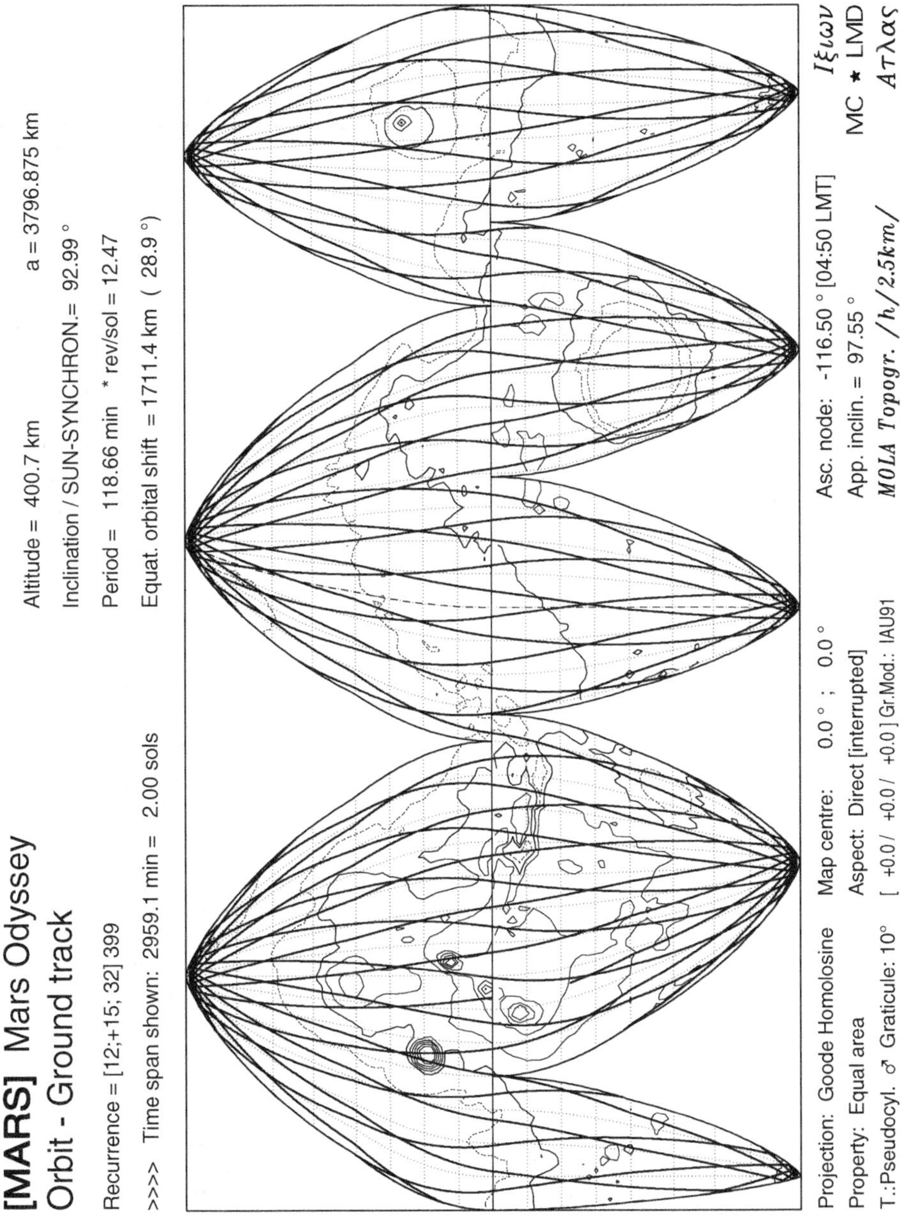

Figure 10.10. Ground track of the Mars Odyssey satellite over two sols, during the mapping phase

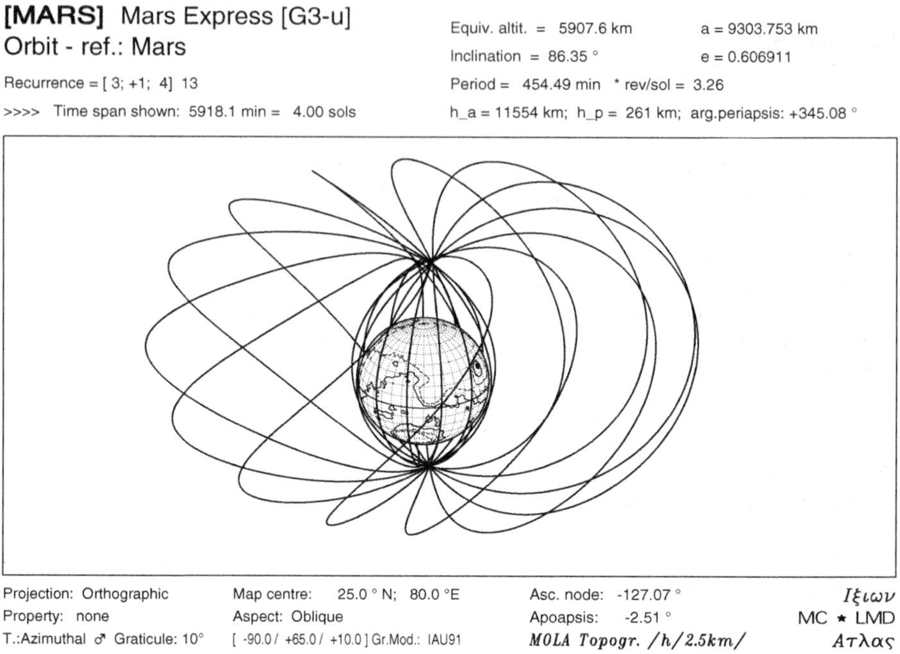

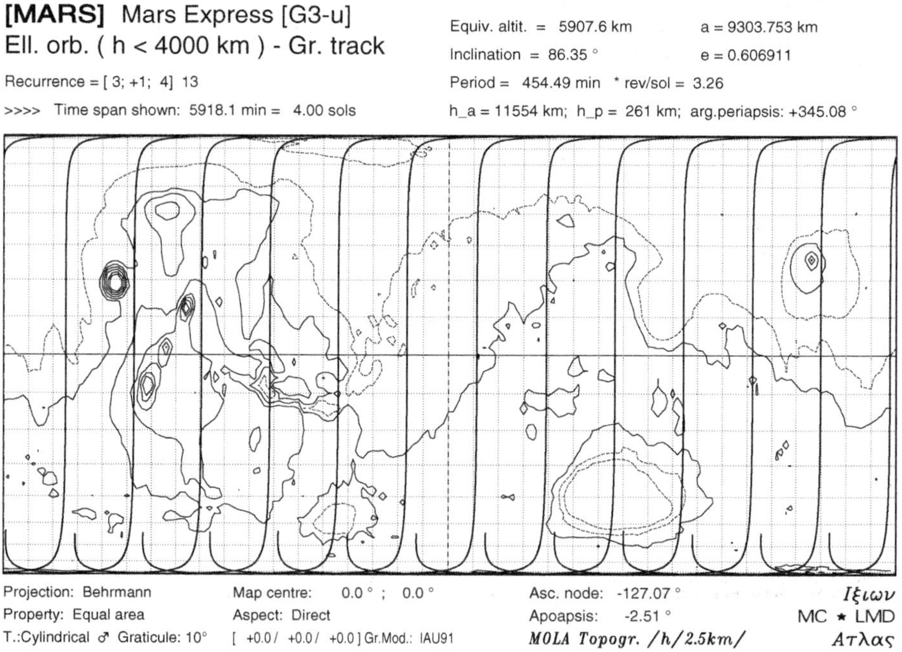

Figure 10.11. Orbit and ground track of the satellite Mars Express (orbit G3-u) over 4 sols, starting from 9 January 2004 ($L_S = 330°$). In the *lower image*, the ground track is only indicated if the altitude of the satellite is less than 4 000 km

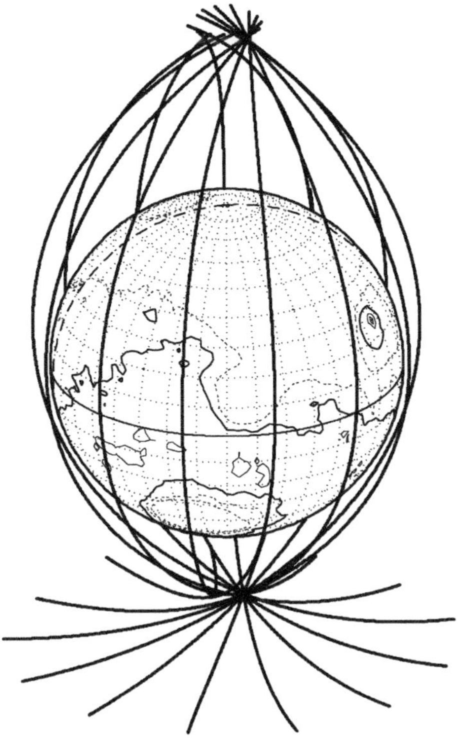

Figure 10.12. Orbit of Mars Express when its altitude is less than 4 000 km, over a cycle of 4 sols, from 9 January 2004 ($L_S = 330°$). Type of orbit: 3G-u. This figure synthesises the two figures in Fig. 10.11

10.4 Representing the Ground Track

For all considerations related to the ground track of a satellite in orbit around Mars, we use the same relations as for a terrestrial satellite. This goes for the equations for the ground track, the equatorial shift, apparent inclination, and so on.

The ground track of an LMO satellite has the same general appearance as the ground track of an LEO satellite. The following example brings out this similarity.

Example 10.3. *Comparison between the ground track of Mars Global Surveyor (MGS), in near-circular Sun-synchronous orbit around Mars at an altitude of 379 km, with the ground track of a fictitious terrestrial satellite at the same reduced altitude, also on a Sun-synchronous orbit.*

MGS. During its topographical phase, MGS followed a Sun-synchronous orbit with $h = 379$ km. The reduced distance is

$$\eta = a/R = 3775.1/3196.2 = 1.111547 \, .$$

We obtain the inclination of the Sun-synchronous satellite MGS using (4.68):

$$i_{HS} = \arccos(-1.111547^{2/7}/29.0403) = 92.86^\circ \ .$$

Expanding beyond the J_2 term for the nodal precession rate, we obtain

$$i_{HS} = 92.93^\circ \ .$$

The ground track of the satellite thus lies between 87.07°N and 87.07°S. It is shown in Fig. 10.9 (upper), over half a Martian day (half a sol). With 12.6 round trips per sol, the equatorial shift is 29°.

Equivalent MGS. To compare the orbits and ground tracks on Mars and on Earth, we calculate the characteristics of a terrestrial satellite at the same reduced altitude (i.e., the same reduced distance from the attractive centre of the planet, and hence the same value for η). We call this fictitious satellite the equivalent MGS (for the Earth). We obtain $a = 1.1116R = 7\,090$ km, or $h = 712$ km. We deduce the inclination $i_{HS} = 98.26^\circ$. The ground track of the satellite lies between 81.74°N and 81.74°S. It is shown in Fig. 10.9 (lower). With 14.5 round trips per day, the equatorial shift is 25°. We note here the main differences:

- the inclination of the Sun-synchronous Martian satellite is more polar (k_h is greater for Mars than for the Earth),
- the equatorial shift is a little greater for the Martian satellite, because its period is longer, since Mars has a lower mean density than Earth [see (11.3)].

Example 10.4. *Ground track of Mars Odyssey during the mapping phase.*

In circular orbit for its mapping phase, the instruments of the orbiter Mars Odyssey are carrying out a geological mapping mission. The ground track of the satellite is shown in Fig. 10.10 over two sols, which corresponds to a recurrence quasi-cycle (see Example 10.6 below). We have chosen an interrupted equivalent projection, dividing the surface in such a way as to bring out the major geological features.

Example 10.5. *Orbit and ground track of Mars Express during a cycle of 4 sols.*

Mars Express has a highly eccentric orbit. Its near-polar inclination is a long way from the critical inclination. This leads to an apsidal precession rate of $\dot{\omega} = -0.557^\circ$ per sol, corresponding to precisely one round trip of the pericentre in one Martian year. The representations given here concern the orbit type known as 3G-u (see Table 10.5). The period represented begins on 9 January 2004 ($L_S = 330^\circ$) and lasts for a cycle of 4 sols. The orbit shown in Fig. 10.11 (upper) is plotted in a frame moving with the planet. Its ground track, shown in Fig. 10.11 (lower), is only plotted if the altitude of the satellite is less than 4 000 km. Above this altitude, observation of the planet is not accurate enough. Figure 10.12 shows the representation limited to $h < 4\,000$ km for the orbit itself.

10.5 Orbit Relative to the Sun. Crossing Times

To study the ground track in relation to the Sun, we apply the methods of Chap. 6 to the Martian satellite. We need only note that the length of the mean day on Mars is the sol. The second is still the unit of time, wherever we are located, and one minute is equal to 60 seconds.

But a word of warning is in order for times. It is usual to note the overpass time LMT on Mars as on Earth, e.g., 22:30 LMT. In this case, the overpass occurred 22.5 hr after midnight, where the hour here is the fraction 1/24 of a sol. To avoid confusion, we shall not use the word 'hour' for Mars, but rather 'sol/24'.

The quantity P, in round trips per (Martian) year, is used to calculate the cycle C_{S} in sols.

Overpass Time for a Sun-Synchronous Satellite

For Sun-synchronous satellites, the overpass time is obtained as a function of the latitude using (6.9). The value of the coefficient to adjust for units is the same as on Earth: if times are in sol/24 and angles in degrees, then $K = 15$, since 1 sol corresponds to 360 degrees.

The graph of the curve giving the latitude ϕ as a function of $\Delta\tau = \tau - \tau_{\mathrm{AN}}$ is shown in Fig. 10.13. Comparing with Fig. 6.5 (lower), the equivalent for the Earth, we see that on Mars there is no significant change with altitude for a satellite in low orbit: this is due to the high value of the constant of Sun-synchronicity k_{h}.

Overpass times are chosen for Sun-synchronous satellites on Mars[15] on the basis of roughly the same considerations as on Earth. To obtain useful lighting conditions, one requires the satellite to cross the equator about two hours before or after midday; in order to benefit from a maximal illumination of the solar panels, a dawn–dusk orbit is favoured.

The ground track of the satellite MGS, noting LMT times, is shown in Colour Plate XII for its topographical phase (with $\tau_{\mathrm{AN}} = 14{:}00$).

10.6 Orbit Relative to Mars. Recurrence and Altitude

10.6.1 Recurrence

When we consider the question of recurrence on Mars, we encounter the same advantages and constraints as we have already seen on Earth. Everything

[15] At the beginning of a mission, a dawn–dusk orbit is generally used. Indeed, the transfer of a space probe from Earth to Mars is made along a trajectory which is tangential to the orbit of Mars, as viewed in the heliocentric frame, at the time of insertion when the probe becomes a satellite of Mars. The orbit of the satellite is then perpendicular to the Sun–Mars direction, and is therefore a dawn–dusk orbit. The same reasoning would apply to a satellite of Venus.

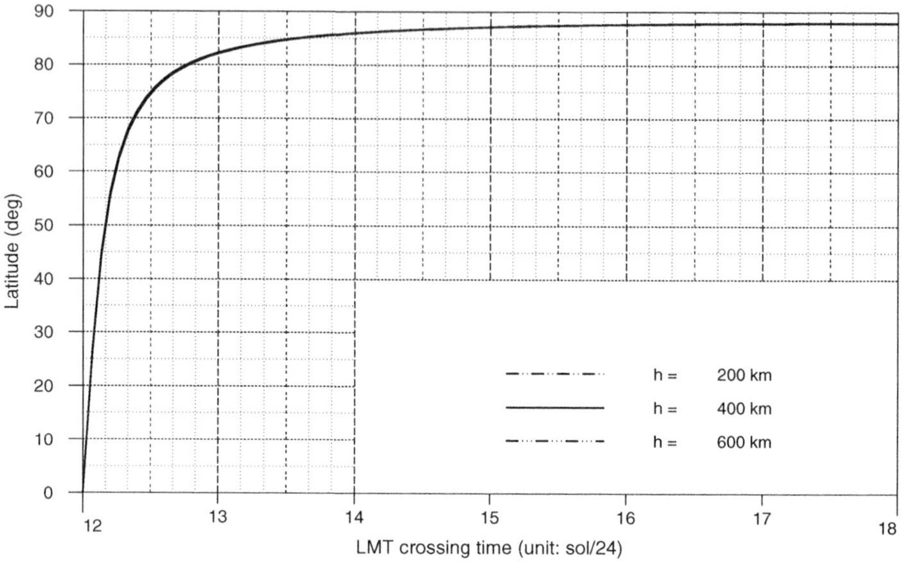

Figure 10.13. Graph of $\phi(\Delta\tau)$: relation between the latitude of the point under consideration and the LMT time difference between the ascending node crossing and overpass at this latitude for a Sun-synchronous satellite, for altitudes $h = 400$ km and $h = (400 \pm 200)$ km (curves practically superposed). Compare with Fig. 6.5

discussed in Chap. 7 applies equally to Mars, provided of course that we use sols instead of days.

Recurrence for a Sun-Synchronous Satellite

We have just seen that the altitude of a Martian Sun-synchronous satellite lies between the theoretical bounds $h = 0$ and $h = 5\,496$ km, which corresponds to the values $\nu = 14.73$ and $\nu = 3.49$ of the daily orbital frequency, respectively. The altitudes chosen for Sun-synchronous satellites in missions under development or already carried out are generally less than 600 km, with the exception of telecommunications projects, such as MTO ($h = 4\,450$ km, period 1/4 sol).

The recurrence diagram in Fig. 10.14 helps us to visualise the altitudes leading to different recurrence configurations on Mars. This diagram is the counterpart of Fig. 7.1, the recurrence diagram for the Earth.

Recurrent satellites on Mars are shown in Table 10.4 and in Fig. 10.15. Given the recurrence triple (which is in principle maintained throughout the mission), one can accurately determine the characteristics of the orbit.

Example 10.6. *Calculate the orbital characteristics of the satellites Mars Global Surveyor, Mars Odyssey and InterMarsNet.*

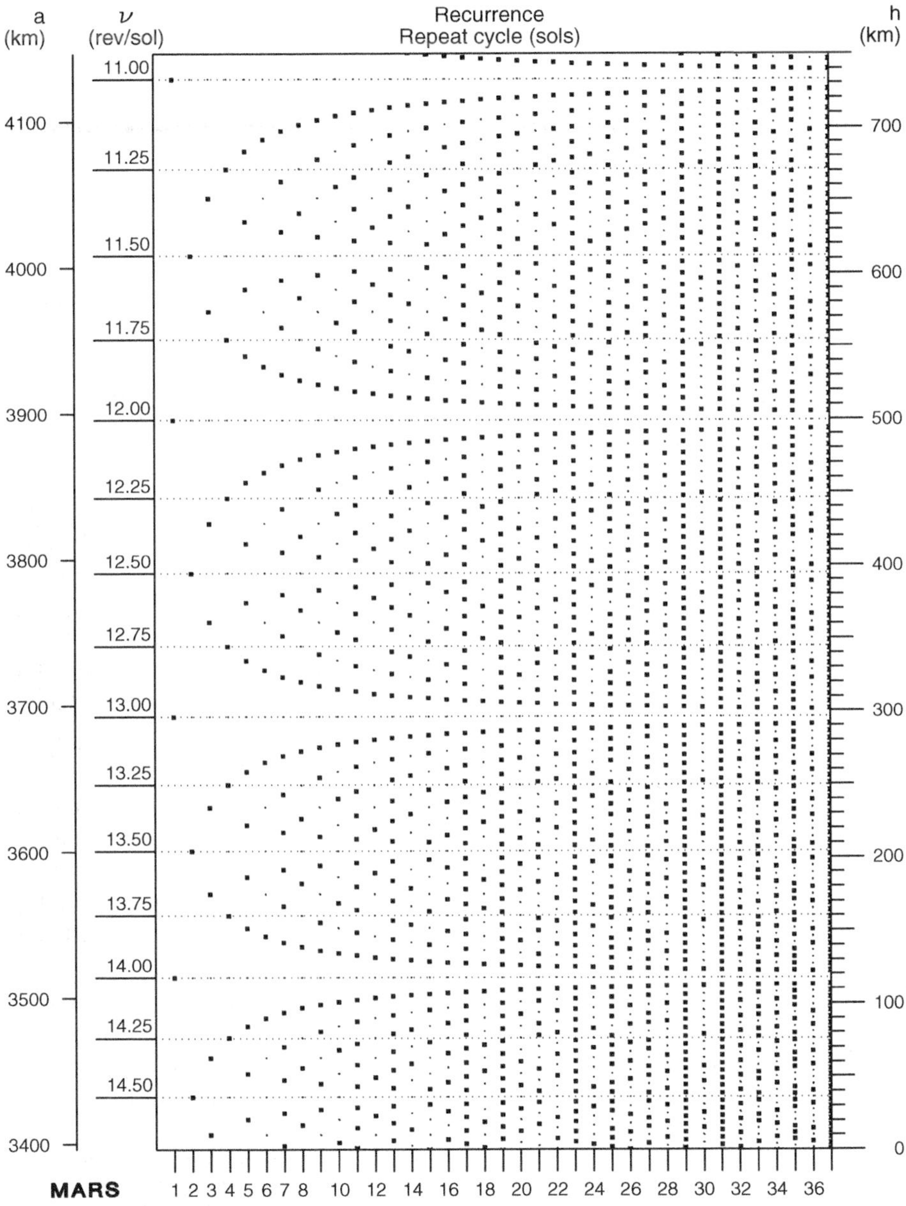

Figure 10.14. Recurrence diagram for Sun-synchronous satellites. For altitudes between 0 et 750 km, *small squares* indicate values of the altitude h (and the semi-major axis a) for which recurrence is possible. The *abscissa* records the value of the recurrence cycle in sols. The daily frequency ν is in revolutions per sol

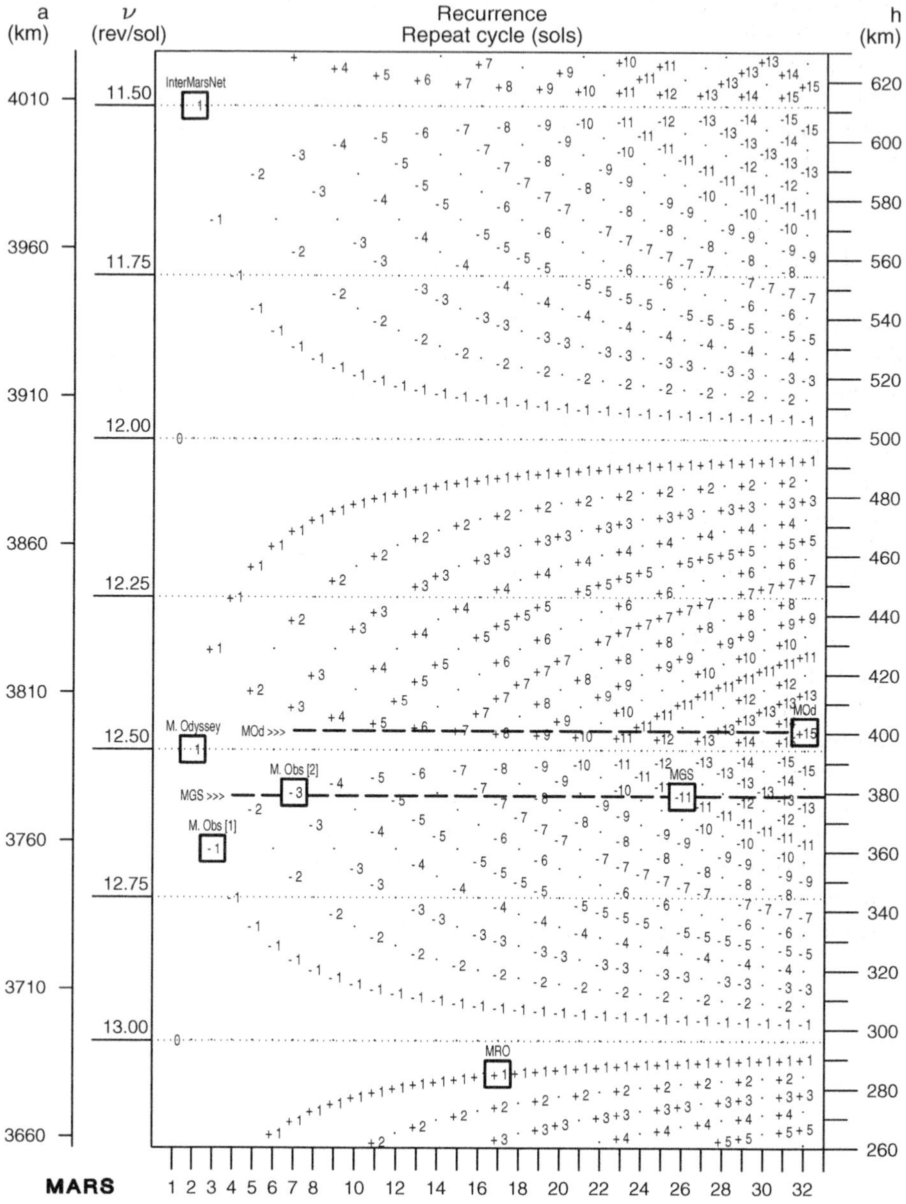

Figure 10.15. Recurrence diagram for Sun-synchronous satellites. For altitudes between 260 and 630 km, the possible recurrences are indicated by the value of D_{T_o}. *Boxed values* correspond to the satellites appearing in Table 10.4. For example, for MRO, we find the triple [13, +1, 17], i.e., $\nu_o = 13$ (the integer closest to ν, as ordinate), $D_{T_o} = +1$ (indicated on the diagram) and $C_{T_o} = 17$ (abscissa)

Table 10.4. Orbital characteristics for Sun-synchronous satellites obtained from the recurrence triple $[\nu_o, D_{T_o}, C_{T_o}]$. For a complete description, see the caption of Table 7.1. Orbit of satellites in orbit: MGS, Mars Odyssey; under development: MRO. Planned orbit for a satellite which failed insertion: Mars Observer (recurrence over 3 and 7 sols); for an abandoned project: InterMarsNet

Satellite	ν_o	D_{T_o}	C_{T_o}	N_{T_o}	T_d	a	h	i_{HS}
MGS	13	−233	550	6 917	117.64	3 775.116	379	92.93
Mars Odyssey	12	+1	2	25	118.36	3 790.524	394	92.98
Mars Odyssey	12	+15	32	399	118.66	3 796.875	401	92.99
MRO	13	+1	17	222	113.30	3 681.252	285	92.69
Mars Observer	13	−1	3	38	116.80	3 757.095	361	92.89
Mars Observer	13	−3	7	88	117.69	3 776.107	380	92.94
InterMarsNet	12	−1	2	23	128.65	4 007.867	612	93.61

MGS. The probe Mars Global Surveyor (MGS) left its heliocentric orbit for a highly eccentric ($h_p = 258$ km, $h_a = 54\,021$ km, $e = 0.88$, $T = 45$ hr) areocentric orbit known as MOI (Mars Orbit Insertion) on 12 September 1997. The air-braking maneuver lasted for 16 months, during which many scientific measurements were carried out, in particular, on the magnetic field. The solar panels did not deploy correctly, extending the time required for air braking. In February 1999, the final orbit was near-circular, Sun-synchronous, recurrent and frozen. The satellite MGS then entered its mapping phase. The instrument MOLA (Mars Orbiter Laser Altimeter) began to make very precise topographic measurements of the planet.
With recurrence maintained at 6 917 revolutions in 550 sols, the draconitic period is given by

$$T_d = 550/6917 \text{ sol} = 117.6436 \text{ min} .$$

To begin with, we set $T_0 = T_d$ and thereby deduce the inclination and radius $a_0 = 3\,781$ km of the Keplerian orbit. We calculate the rate of secular variation of the angle elements. We deduce the anomalistic period $T_a = 117.5090$ min, then the values of the semi-major axis and the inclination, viz., $a = 3\,775.116$ km and $i = 92.93°$, respectively.

Mars Odyssey. The probe Mars Odyssey reached Mars on 24 October 2001 and went into a high and highly eccentric Martian orbit ($T \approx 18.6$ hr). Air braking eventually led on 30 January 2002 to a Sun-synchronous ($\tau_{AN} = $ 04:30) circular orbit close to 400 km ($T \approx 2$ hr), with the satellite ground track repeating every 2 days. The recurrence diagram in Fig. 10.14 shows that, at this altitude, a recurrence cycle over 2 sols corresponds to the triple [12,1,2], $\nu = 12.5$, with 25 revolutions per cycle. The draconitic period is therefore

$$T_d = 2/25 \text{ sol} = 118.36 \text{ min} ,$$

and the various intervals discussed in Chap. 7 are $\delta = 360/25 = 14.4°$, $\delta_D = \delta$, and $\delta_R = 2\delta$. By iterative calculation, we obtain the semi-major axis $a = 3\,790.524$ km ($h = 394$ km) and the inclination $i_{HS} = 92.98°$. In fact, during the science part of the mission, for each 2-sol repeat cycle, the ground track shifts a small amount (53 km) at the equator. This shift by 0.9°, which we denote by δ', can be considered as the grid interval of a longer cycle. After $\delta'/\delta = 0.9/14.4 = 16$ cycles of 2 sols, the ground track has thus shifted by an interval δ, which means that, after 16 cycles of 25 revolutions, the satellite, with this new recurrence characteristic, has made one revolution less, i.e., 399 revolutions in 32 sols. The recurrence triple is then [12,15,32]. The draconitic period is

$$T_\mathrm{d} = 32/399 \text{ sol} = 118.66 \text{ min} ,$$

and we recover the grid interval $\delta = 360/399 = 0.9^\circ$. We then obtain $a = 3\,796.875$ km ($h = 401$ km) and the inclination $i_\mathrm{HS} = 92.99^\circ$.
It should be noted that the satellite Mars Odyssey is not strictly Sun-synchronous, since its LMT crossing time varies continuously from 03:23 to 05:20 during the 917 (Earth) days of the science mission, then from 05:20 to 06:50 during the 800 following days.

InterMarsNet. The InterMarsNet mission was a European project in collaboration with the United States. This project, planned for 2003, was abandoned in 1996 and more or less replaced by Mars Express, for launch in 2003. We have included it here as an example because of its interesting orbit, which is Sun-synchronous, dawn–dusk, with 2-sol recurrence, frozen and eclipse-free. With a recurrence of 23 revolutions in 2 sols, we obtain

$$T_\mathrm{d} = 2/23 \text{ sol} = 128.66 \text{ min} .$$

After some calculation, we obtain $a = 4\,007.867$ km and the inclination $i_\mathrm{HS} = 93.61^\circ$.

Example 10.7. *Characteristics of a Sun-synchronous satellite with one-sol recurrence and critical inclination.*

During preliminary orbital studies for the Russian satellite Mars-96, an orbit was examined that would both have been Sun-synchronous and have the critical inclination (like Ellipso-Borealis for the Earth). We re-examine this idea here, but adding the further constraint that the satellite should be one-day recurrent, an interesting configuration for links with ground bases. We have calculated the following orbital characteristics:

$$a = 6\,996.822 \text{ km} , \quad e = 0.471\,873 , \quad i = 116.72^\circ \quad \text{(critical and Sun-sync.)} .$$

We deduce the following: $h_\mathrm{p} = 299$ km, $h_\mathrm{a} = 3\,601$ km, period $T_\mathrm{d} = T_\mathrm{a} = 296.21$ min, recurrence triple $[5, 0, 1]$.

Recurrence for a Non-Sun-Synchronous Satellite

After its insertion, the satellite Mars Express was placed in a recurrent near-polar orbit with 13 revolutions in 4 sols, an orbit known as 3G-u, then 11 revolutions in 3 sols, the 3G-b orbit. The transition from 3G-u to 3G-b was

Table 10.5. Orbital and recurrence characteristics for the Mars Express baseline orbit. Semi-major axis a, altitudes at perigee and apogee, h_p and h_a in km, period T in minutes. For these two orbits, with eccentricity e, the inclination is $i = 86.35°$. N_{T_o} is the number of revolutions in the cycle

Satel. / Orbit	a	e	T	Phase triple, N_{T_o}	h_p	h_a
MEx G3-u	9 303.8	0.606 91	454.49	$[3, +1, 4]$, 13	261	11 554
MEx G3-b	8 584.4	0.573 93	402.83	$[4, -1, 3]$, 11	261	10 115

Table 10.6. Orbital and recurrence characteristics for satellites with one-sol recurrence cycles. Non-Sun-synchronous satellites with three different inclinations and comparison with Sun-synchronous satellites. Altitudes h in km, angles i in degrees. Daily orbital frequency ν in round trips per sol

ν_o	$i = 20$ h	$i = 65$ h	$i = 110$ h	$i = i_{HS}$ h	i_{HS}
14	48.6	81.2	146.0	117.6	92.3
13	237.0	263.8	321.0	296.2	92.7
12	449.0	470.6	520.6	499.2	93.3
11	690.5	707.4	750.7	732.6	94.0

made between 6 and 10 May 2004 (revolutions 372 to 386). The characteristics of these orbits are given in Table 10.5.

As for Mars Odyssey, the orbits are not strictly recurrent. For G3-u, with $\delta = 360/13 = 27.70°$, $\delta_R = 4\delta = 110.77°$, the ground track slips by $\delta' = 1.09°$ after each cycle of 4 sols.

Satellites with One-Day Recurrence

As on Earth, low-orbiting (LMO) satellites around Mars should a priori avoid one-day recurrence cycles. For example, a Sun-synchronous satellite at altitude $h = 500$ km would find itself in this situation, as is clear from the recurrence diagram. Every day it would pass over the same ground track, unable to observe other regions. Table 10.6 lists altitudes leading to such recurrence, for Sun-synchronous and non-Sun-synchronous satellites.

Grid Points for Recurrent Satellites

The latitudes of grid points are calculated from the function g_S defined by (7.45). There is no difference with a terrestrial satellite, except that, for Sun-synchronous satellites, the inclination i_{HS} depends on the planet.[16]

[16] For the satellite Mars Odyssey, with a 2-sol recurrence cycle and using the notation of Chap. 7, we obtain the following results: for $j = 1, 3, 5, 7, \ldots, M_{T_o} = 27$, latitudes are $\phi = 26.443, 62.674, 75.858, 80.789, \ldots, \phi_m = 87.024$.

Recurrence Index

Figure 10.16 shows the recurrence index for MGS. The recurrence cycle is very long, $C_{T_o} = 550$ sol (over 6 917 revolutions). The graph has a secondary peak for 7 sols (over 88 revolutions), and other significant peaks for 26 sols (over 327 revolutions), ..., for 144 sols (over 1 811 revolutions). The MGS/JPL terminology picks out three cycles: Repeat Cycle (7 sols), Mapping Cycle (26 sols), and Super Cycle (550 sols).

The instrument MOLA aboard MGS is a laser with an extremely narrow beam. The very long cycle thus leads to a very small ground track interval,[17] ensuring a complete coverage of the planet, apart from a small disk centered on the poles (see Example 10.3), which has been dealt with by changing the orientation of the satellite. On Earth, too, where cycles C_{T_o} rarely exceed 40 days, altimetry satellites like ICESat have very long cycles.

10.6.2 Altitude

Recall that we define the altitude of a satellite as a function of the angle α giving the position on orbit (see Chap. 7). This altitude $h(\alpha)$, expressed by (7.34), is obtained as the difference between the distance $r(\alpha)$ to the centre of the planet (the centre of attraction) and the radius $R_T(\alpha)$ of the reference ellipsoid (the distance of the subsatellite point on the ellipsoid from the centre of the planet).

Frozen orbits are less circular than on Earth for two reasons:

- the flattening of Mars is more pronounced,
- the value of the frozen eccentricity e_F [see (7.41)] is greater, because the ratio $|J_3/J_2|$ is equal to 18.363×10^{-3} for Mars and 2.339×10^{-3} for the Earth.

We also note that, since J_3 and $\sin \omega_F$ must have opposite signs, the position of the frozen perigee is given by $\omega_F = 270°$ for recurrent Sun-synchronous satellites. The periastron of the frozen Sun-synchronous orbit is practically over the South Pole.

Example 10.8. *Altitude of the satellite MGS during the mapping phase.*

The variation in altitude of the satellite MGS is shown in Fig. 10.17 over one period as a function of the position on orbit α. The values of the orbital elements provided by NASA correspond to a given revolution as specified in the figure, during its mapping phase. As the orbit is maintained and frozen, we may consider that the

[17] The beam angle of the MOLA laser is $2f = 0.85$ mrad (or 3 arcsec), which gives a spot on the ground of width $\Delta = 2fh = 320$ m. With $N_{T_o} = 6\,917$, the ground track interval δ (expressed as a length) is $\delta = 2\pi R/6\,917 = 3.085$ km. At the equator, we thus have $\Delta/\delta \approx 1/10$.

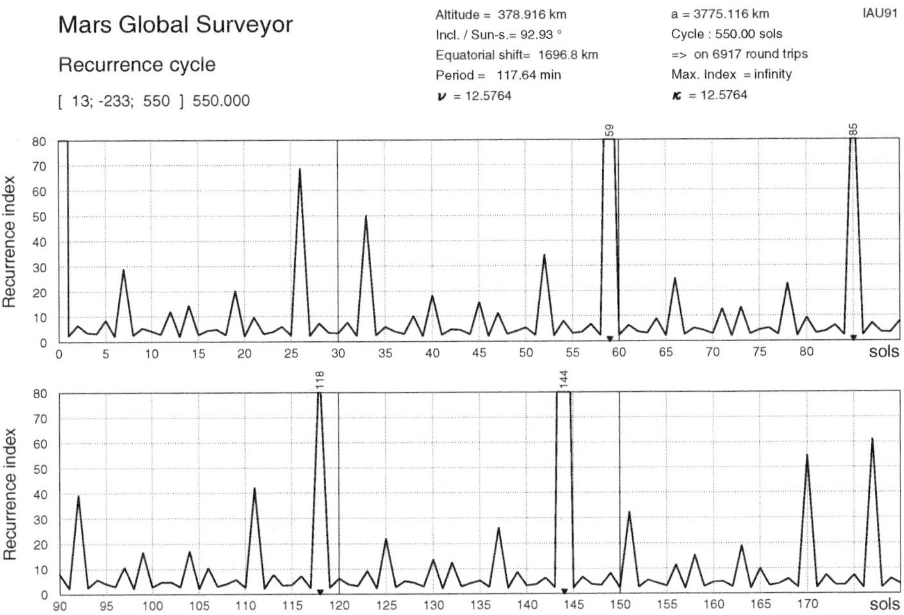

Figure 10.16. Recurrence index for the satellite MGS, with orbital parameters leading to the recurrence indicated

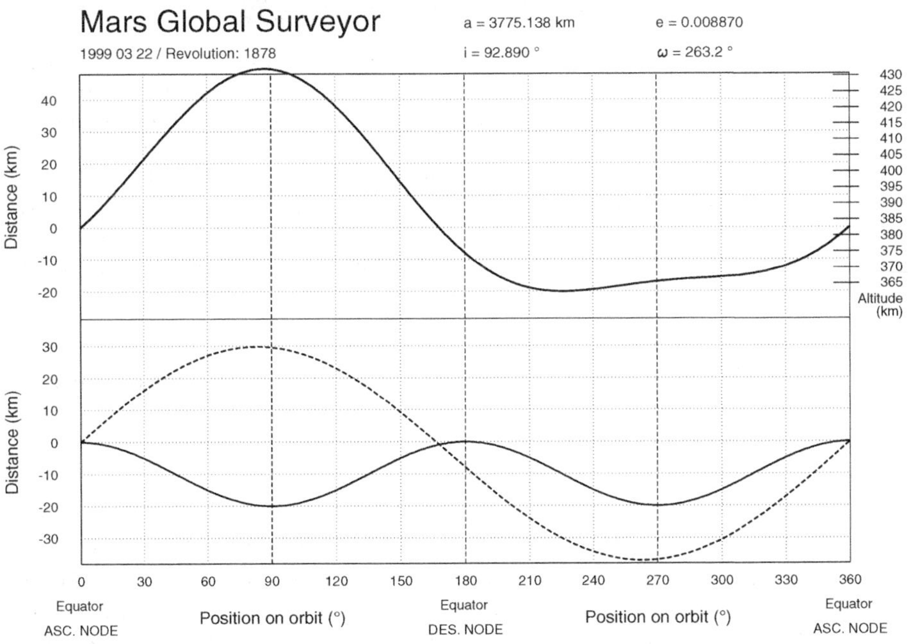

Figure 10.17. Altitude of the satellite MGS, with true orbital parameters (data NASA, revolution 1878, mapping phase). For details, see Fig. 7.22

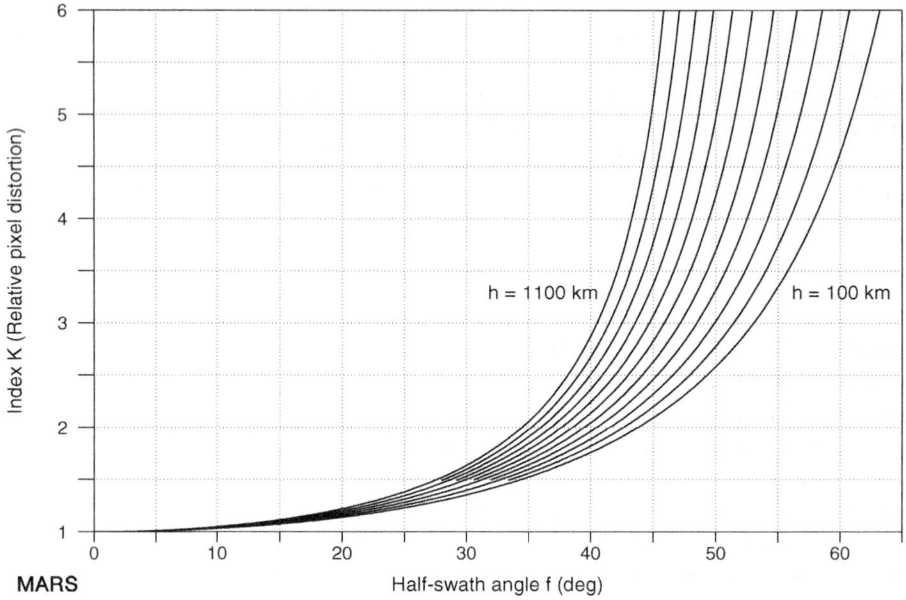

Figure 10.18. Pixel distortion index for LMO satellites with altitudes $h = 100$ km to $h = 1\,100$ km, in steps of 100 km, as a function of the half-swath angle f

variation $h(\alpha)$ does not change from one revolution to another. We note that the position ω_{F} of the frozen periastron is at 7° from the ideal position. If we calculate the frozen eccentricity using (7.41) with the orbital elements provided in Table 10.4, we find $e_{\mathrm{F}} = 8.25 \times 10^{-3}$. This result is close to the true value for the revolution under consideration, viz., $e_{\mathrm{F}} = 8.87 \times 10^{-3}$.

10.7 View from the Satellite

All the ideas discussed in Chap. 8 apply to Martian satellites. The adaptation is facilitated by the fact that we have used the variable η which represents the reduced distance a/R.

Pixel Distortion

For an LMO satellite in circular orbit, the pixel distortion index K is shown in Fig. 10.18 for a range of altitudes. For an SMO satellite, in high stationary orbit, the curve $K(\alpha)$ is almost the same as the one for a GEO satellite on Earth, shown in Fig. 8.4. We obtain $K = 2$ for $\alpha = 49°$, $K = 3$ for $\alpha = 59°$ (instead of $\alpha = 50°$, $\alpha = 61°$, respectively, for the terrestrial satellite). This similarity between the curves $K(\alpha)$ is due to the similar values of η_{GS}.

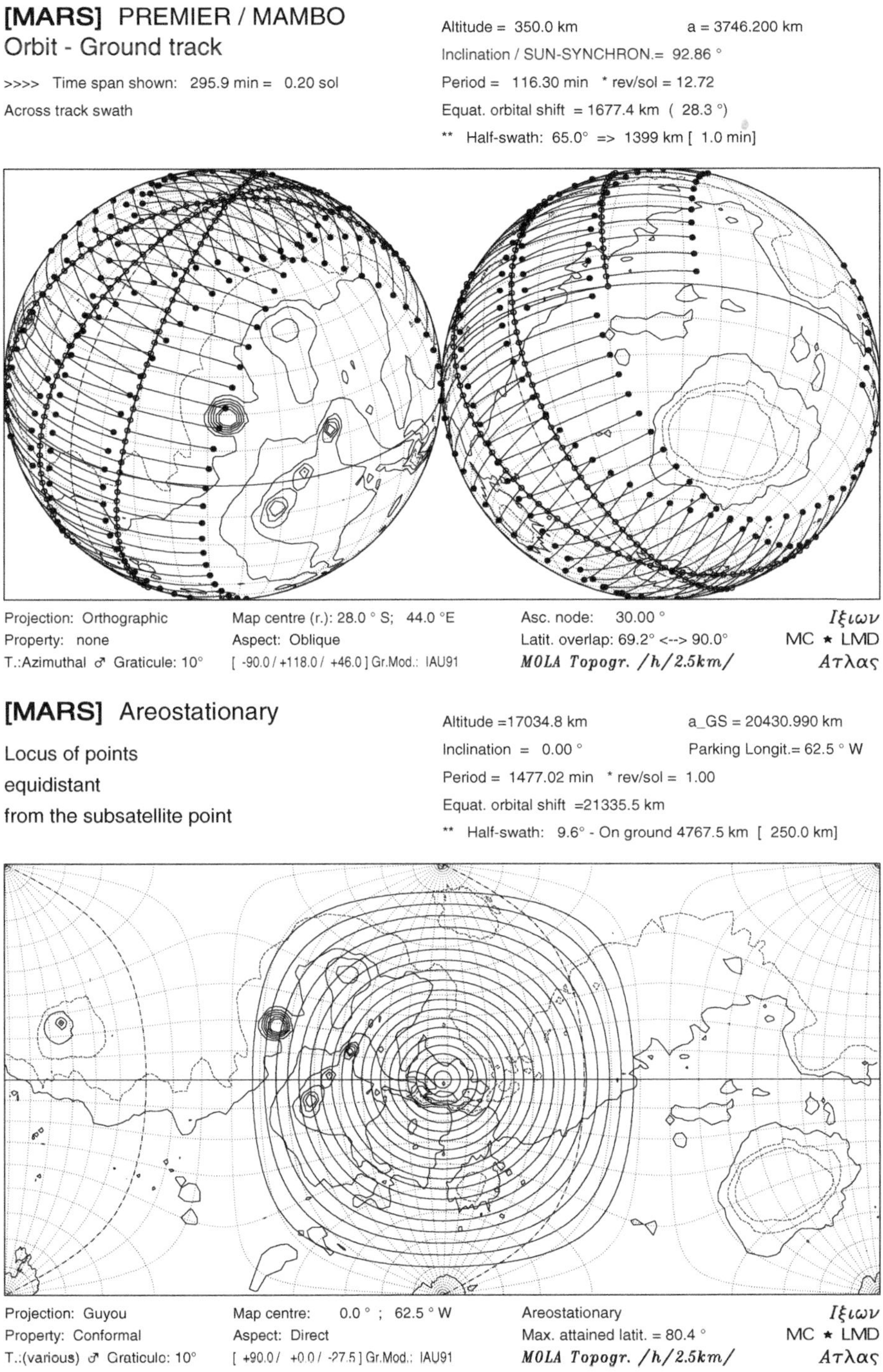

Figure 10.19. *Upper*: orbital track of the satellite Premier and the swath of the instrument Mambo. *Lower*: locus of equidistant points – SMO

Swath Track for an LMO Satellite

Possible scanning modes are the same for instruments aboard terrestrial and Martian satellites. Here we consider the ground track of an across-track swath.

Example 10.9. *Swath track of the instrument Mambo aboard Premier.*

The instrument Mambo (Mars Atmosphere Microwave Brightness Observer), proposed for the abandoned satellite Premier MO-07, views the limit of its across-track scan, alternately left and right, each minute. In the representation of the track shown in Fig. 10.19 (upper), black dots mark the place viewed at this limiting point while small circles indicate the nadir of the satellite (the subsatellite point).

View from an SMO Satellite

When an areostationary (SMO) satellite views Mars, the maximum swath in the sense that we have defined f_0 is given by (8.23). With the value of η_{GS} defined by (10.24), we obtain

$$f_0 = \arcsin \frac{1}{6.016} = 9.569° = 0.1670 \text{ rad} . \tag{10.25}$$

The corresponding angle at the centre of Mars is

$$\alpha_0 = 90° - 9.6° = 80.4° \quad \Longrightarrow \quad 2F_0 = 9\,535 \text{ km} . \tag{10.26}$$

If λ_{S} is the longitude of the satellite S (parking longitude or longitude of the subsatellite point), the longitudes viewed on the equator by S lie in the interval

$$[\lambda_{\mathrm{S}} - 80.4°, \lambda_{\mathrm{S}} + 80.4°] ,$$

and depending on the meridian λ_{S}, latitudes are viewed over the same interval of 80.4° on either side of the equator.

The fraction of the planetary surface viewed by the areostationary satellite, calculated using (8.29), is in this case 0.417 (or about 42%, as for the Earth).

Example 10.10. *Locus of points on Mars at an equal distance from the subsatellite point for an areostationary satellite.*

As in Example 8.8, we plot the locus $\mathcal{L}(D)$. The distance D defined by (8.27) represents here the (great circle) distance between a point on Mars viewed by the areostationary satellite and the subsatellite point of this same satellite [see Fig. 10.19 (lower)]. D varies in steps of 250 km. The subsatellite point chosen in this example corresponds to the position originally proposed for the equatorial NetLander of the Premier mission.

10.8 Temporal and Angular Sampling

All the ideas described in Chap. 9 can be adapted to Martian satellites. However, we should mention one point that will not be discussed here. When Martian missions are set up, the Sun–target–satellite–Earth geometry has to be taken into account. Indeed, whilst the eyes of the satellite are turned towards Mars, its ears must be oriented towards Earth, from whence it receives its instructions, and so must its mouth, for the transmission of data.

In the present section, we shall describe an example of temporal and angular sampling, in the context of our study of the Sun–target–satellite configuration.

Example 10.11. *Sampling tables for a target viewed by the THEMIS instrument aboard Mars Odyssey.*

The instrument THEMIS (Thermal Emission Imaging System) carried aboard Mars Odyssey uses limb-to-limb across-track scanning. We examine the sampling for the point with geographic coordinates $\phi = -26.938°$, $\lambda = 324.814°$, corresponding to Holden Crater in Uzboi Valles, photographed on 12 March 2002, during revolution 1063.

31-Sol Table. We begin by calculating the solar areocentric longitude L_S for the chosen date: D = 2002 03 12 gives $L_S = 341°$. In the table shown in Fig. 10.20, the abscissa gives the 'hour' LMT, which is in fact 1/24 sol, as mentioned above, while the ordinate gives the sol from $J = 1$ for sol 631 ($L_S = 341°$) to $J = 31$ for sol 661 ($L_S = 356°$). The sunrise, sunset and local apparent noon are marked by dot-dashed lines. Note the big difference between noon LMT and noon LAT, which is the equation of time.[18] Each overpass of the satellite (with the required viewing conditions) is indicated by a triangle (see the explanations in Example 9.4). We see that half of the overpasses occur during the day (the afternoon) and the other half at night.

Statistics over 31 Sols. We consider points of different latitudes on the meridian passing through the relevant point. This is in fact the meridian $\lambda = 324.814°$, or $\lambda = 35.2° W$. We set up another type of table, shown in Fig. 10.21. On the horizontal axis, we have the solar zenith angle from 0° (zenith) to 90° (sunrise or sunset), and on the vertical axis, latitudes from +90° (North Pole) to −90° (South Pole). We calculate the overpasses at these different latitudes, for the points on the chosen meridian. These overpasses are indicated by circles, squares, and so on, depending

[18] For L_S between 341° and 356°, we read the value of the equation of time from Fig. 10.5: $E_T \approx +50$ min, which can be written +0:50. Thus, applying the relation LMT = LAT + E_T, we have, for local apparent noon, LMT = 12:00 + 0:50 = 12:50. In the sampling table, the line indicating noon LAT does indeed go through 12:50 LMT (or more precisely, from 12:51 LMT for $J = 1$ to 12:44 LMT for $J = 31$).

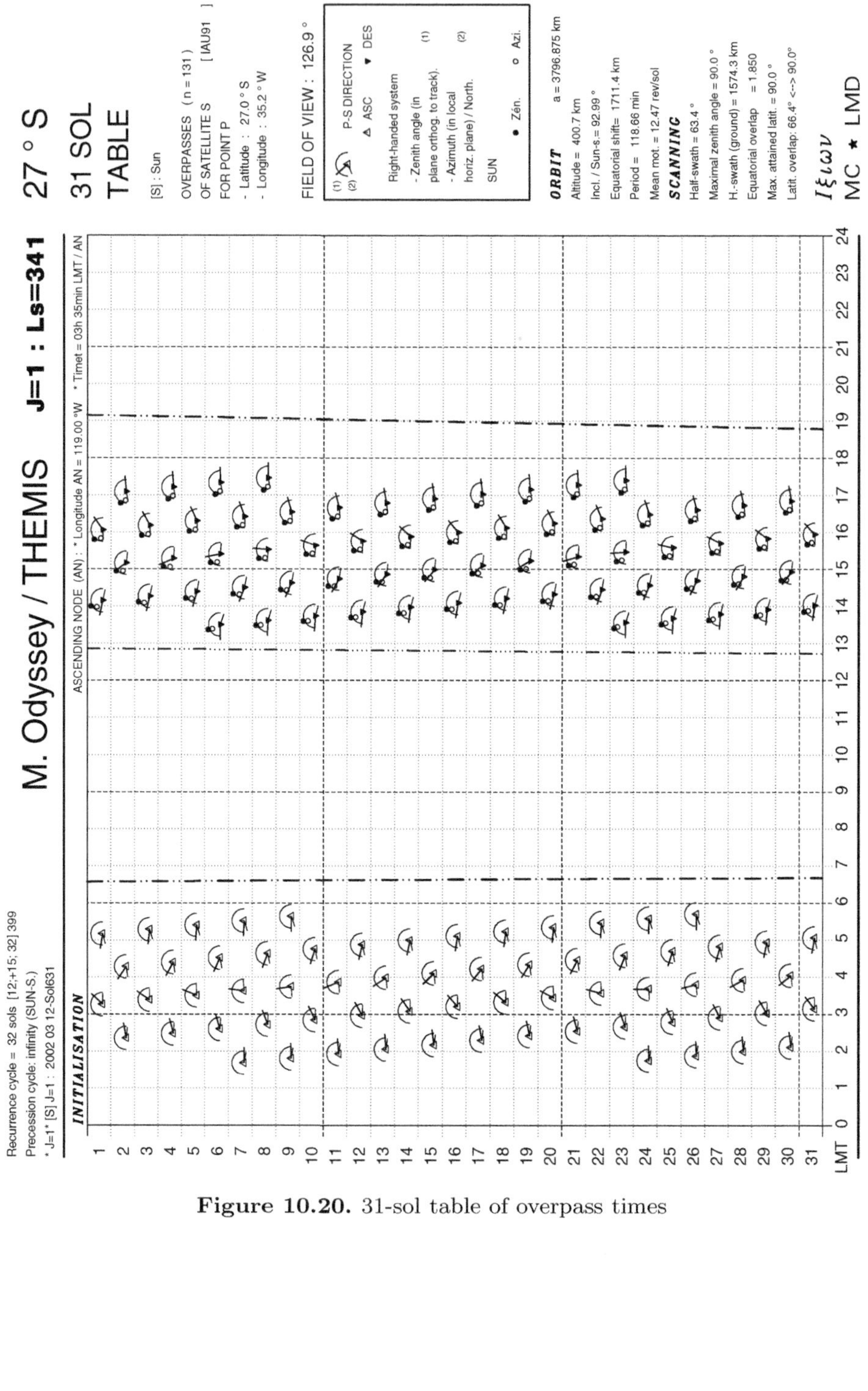

Figure 10.20. 31-sol table of overpass times

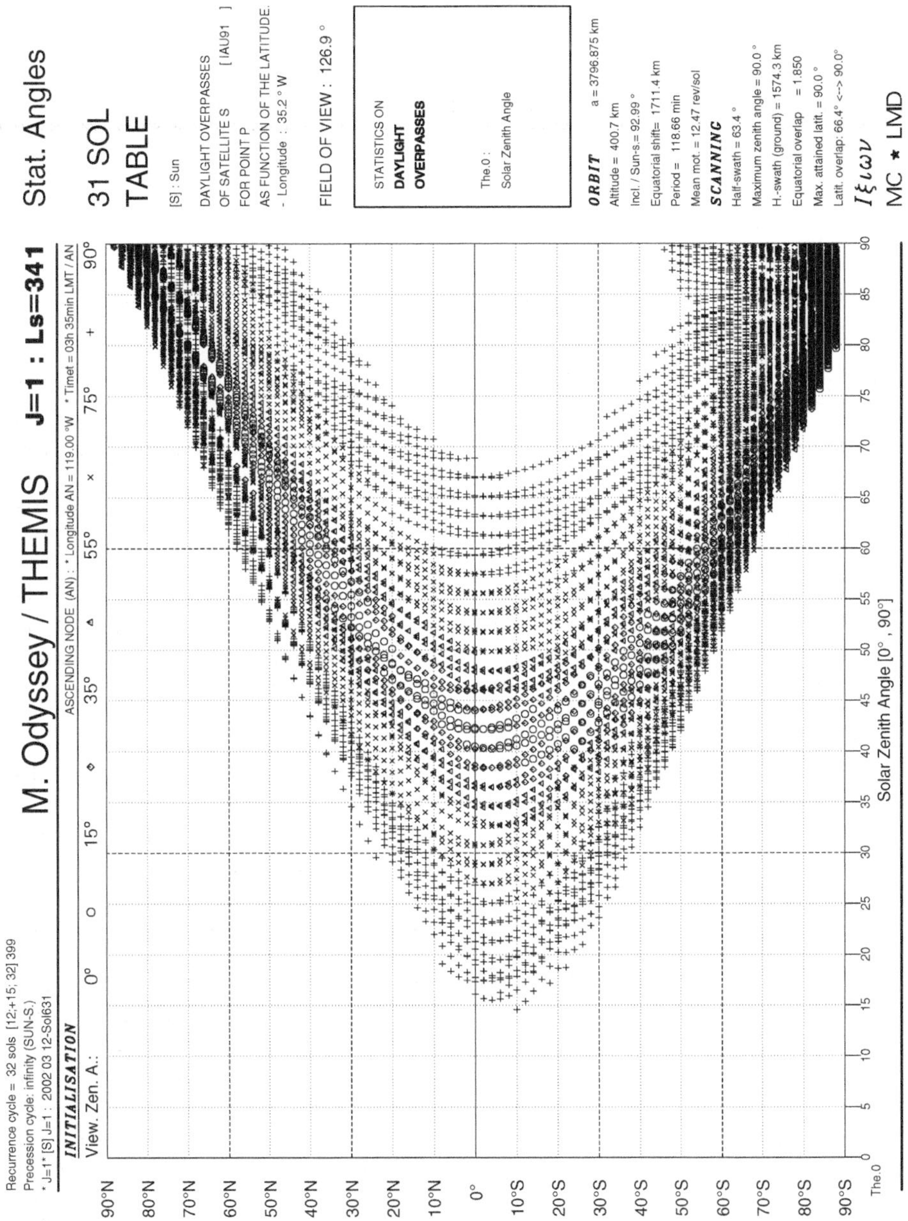

Figure 10.21. 31-sol table. Solar angles as a function of latitude, noting the viewing zenith angle, for all places on a given meridian

Table 10.7. Velocity of satellite and ground track, and relative velocity of the ground track for various satellites in circular (Keplerian) orbit. For each satellite, we give the altitude h (in km), the length of the semi-major axis a or distance to the centre of Mars (in km), the daily frequency ν (in round trips per sol), the Keplerian period T_0 (in min and in sol), the velocities V, V_0, w_E (for the two values 0° and 90° of the angle i), already defined. Type $\mathcal{T}$ of satellite: s (ground level), O (for observation), C (for communications), S (areostationary, i.e., SMO). Natural satellites: Φ (Phobos), Δ (Deimos)

h [km]	a [km]	ν [rev/sol]	T_0 [min]	T_0 [sol]	V	V_0	w_E 0	w_E 90	$\mathcal{T}$
0	3 396	14.77	100.15	0.068	3.55	3.55	3.31	3.56	s
100	3 496	14.14	104.61	0.071	3.50	3.40	3.16	3.41	O
200	3 596	13.56	109.13	0.074	3.45	3.26	3.02	3.27	O
300	3 696	13.01	113.71	0.077	3.40	3.13	2.89	3.14	O
400	3 796	12.50	118.36	0.080	3.36	3.00	2.76	3.01	O
500	3 896	12.02	123.06	0.083	3.32	2.89	2.65	2.90	O
600	3 996	11.57	127.83	0.086	3.27	2.78	2.54	2.79	O
700	4 096	11.15	132.66	0.090	3.23	2.68	2.44	2.69	O
4 450	7 846	4.21	351.68	0.238	2.34	1.01	0.77	1.04	C
5 983	9 379	3.22	459.63	0.311	2.14	0.77	0.53		Φ
17 031	20 427	1.00	1 477.38	0.999	1.45	0.24	0.00		S
20 063	23 459	0.81	1 818.16	1.229	1.35	0.20	−0.04		Δ

on the value of the viewing zenith angle.

Note. For the dates considered here, the declination varies from −8° to −2°. This is therefore the end of the austral summer. The high number of daytime overpasses shown on this graph for latitudes close to the South Pole are due to the particularly long polar day.

10.9 Appendix: Further Aspects of Martian Satellites

10.9.1 Velocity of Satellite and Ground Track

For a circular orbit, we calculate the velocity of the satellite and its ground track using (5.33) through (5.37). Recall that the velocities V (satellite) and V_0 (ground track) are defined in the Galilean frame $\mathfrak{R}$, and the velocity w_E in the Martian frame $\mathfrak{R}_T$. This velocity w_E is said to be relative because it represents the the velocity of the ground track relative to the planet. The results are shown in Table 10.7.

If we compare a terrestrial and a Martian satellite at the same reduced altitude, we see that the periods are roughly the same, but the velocities are halved for the Martian satellite (as is obvious from a glance at the equations).

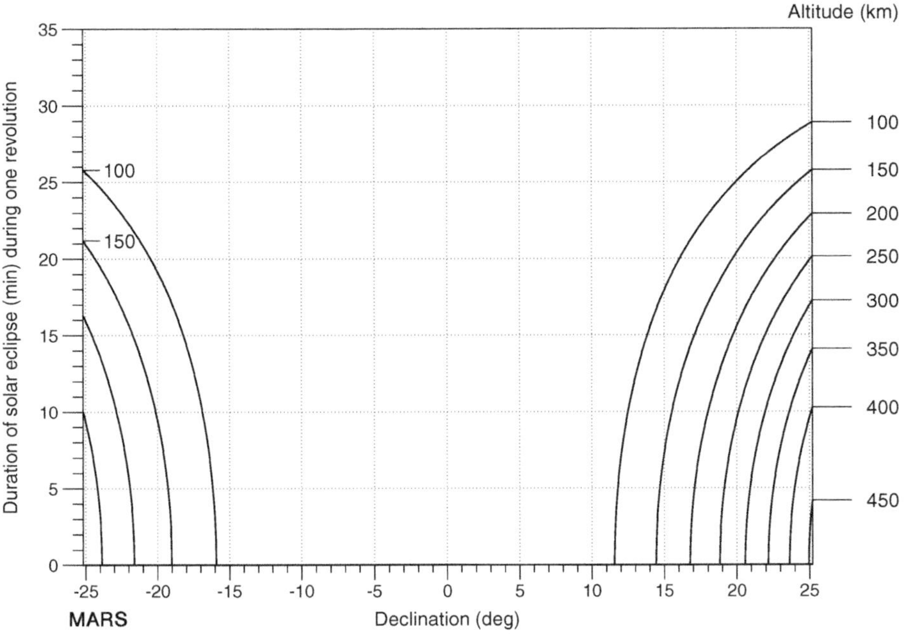

Figure 10.22. Sun-synchronous satellite with dawn–dusk orbit. Length of solar eclipse in minutes during one revolution for the given altitude as a function of the declination. Graphs for $\tau_{\mathrm{AN}} = 18{:}00$ (for $\tau_{\mathrm{AN}} = 06{:}00$, take the negative of the declination)

For an areostationary satellite, we do indeed find that the relative velocity $w_{\mathrm{E}}(0)$ is zero. We also note that $T_0 = J_{\mathrm{sid}}$, which is slightly different from $J_{\mathrm{M}} = 1$ sol.

The two moons of Mars are shown in Table 10.7, Φ for Phobos and Δ for Deimos. They are discussed further below.

10.9.2 Duration of Solar Eclipse

Solar eclipse is a critical problem for a Martian satellite, primarily because the solar constant is halved on Mars in comparison to the value on Earth, but also because the accumulators used to store electrical energy greatly increase the mass of the probe to be put in orbit.

LMO Dawn–Dusk Orbit

We consider a low-orbiting, Sun-synchronous dawn–dusk satellite ($\tau_{\mathrm{AN}} = 06{:}00$ or $18{:}00$). Equation (6.14) is satisfied when η lies between 1.1354 and 2.2930, which gives, using the altitude,

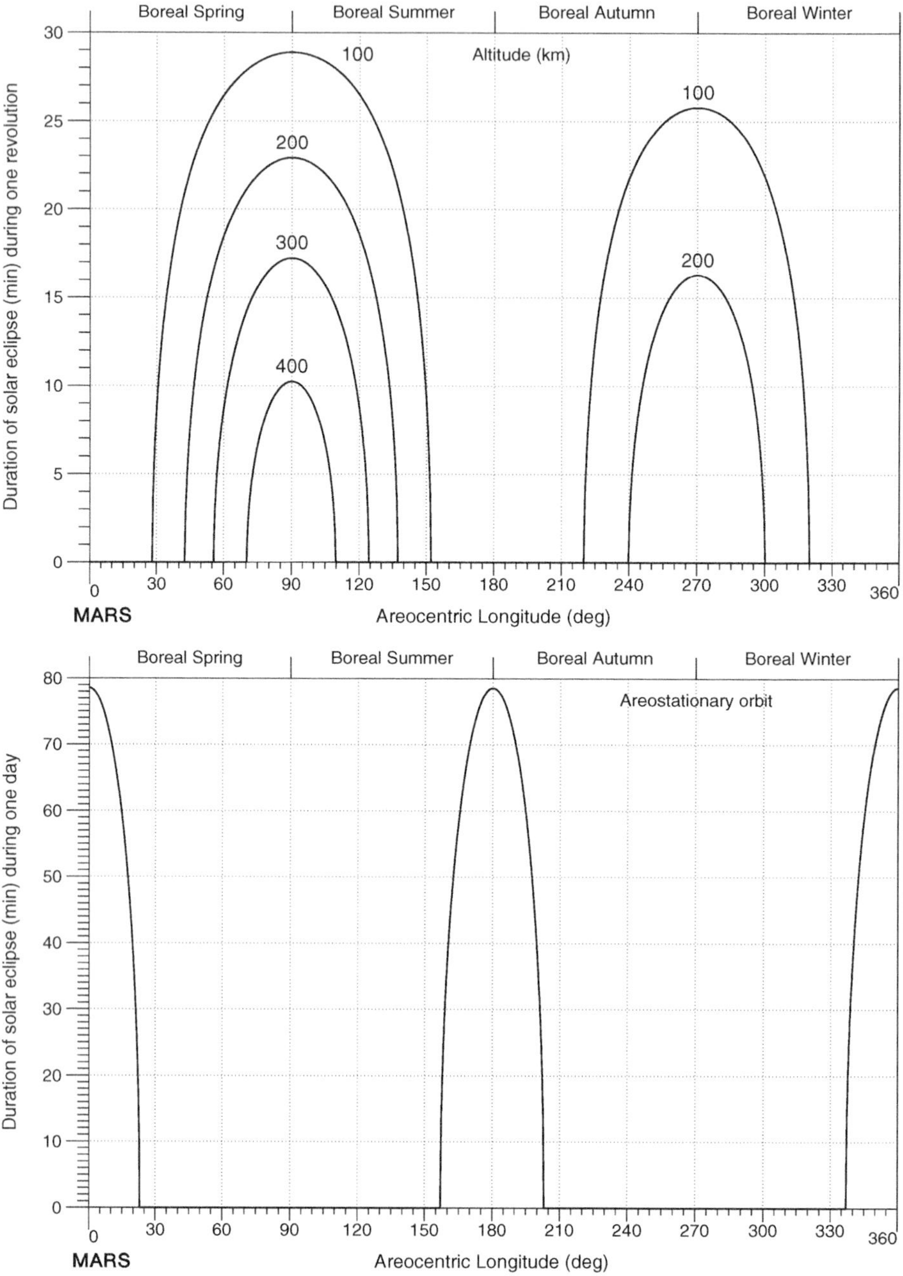

Figure 10.23. Length of solar eclipse in minutes as a function of areocentric solar longitude. *Upper*: Sun-synchronous satellite in dawn–dusk orbit. Length of eclipse during one revolution for the given altitude. Graphs for $\tau_{AN} = 18{:}00$ (for $\tau_{AN} = 06{:}00$, take the negative of the declination). *Lower*: areostationary satellite. Length of eclipse during one sol

$$\text{no eclipse} \iff 460 < h < 4\,391 \text{ km} .$$

The interval of altitudes where there is no eclipse is much wider than on Earth. This is explained by the fact that the Sun-synchronous Martian orbit is more polar than the equivalent terrestrial orbit, for the same reduced altitude (due to the higher value of k_h).

The satellite InterMarsNet, in dawn–dusk Sun-synchronous orbit at $h = 612$ km, would have been eclipse-free. The communications satellite MTO, planned for altitude $h = 4\,450$ km, Sun-synchronous with $i_{HS} = 130.2°$, would suffer a very short eclipse in 06:00 or 18:00 dawn–dusk orbit.[19] With $h = 4\,391$ km, $i_{HS} = 128.9°$, there is no eclipse for this type of orbit.

We have plotted representative graphs of the length of solar eclipse over one revolution for various low altitudes, as a function of the declination (see Fig. 10.22), or as a function of the solar areocentric longitude, which can be related to the date [see Fig. 10.23 (upper)].

SMO Orbit

For an areostationary satellite, with the value of f_0 obtained from (10.25), there is eclipse if $|\delta| < f_0$. This situation occurs twice a year, when the solar areocentric longitude is close to that of the equinoxes, with a longitude difference[20] less than $\Delta L_S = 23°$:

$$\text{eclipse for SMO} \iff [L_S : 337\text{–}23] , \quad [L_S : 157\text{–}203] .$$

The maximum eclipse is found from (6.18), whence

$$\Delta t_{e0} = \frac{9.6}{180} J_{sid} = 78.8 \text{ min} . \tag{10.27}$$

The value of Δt_e as a function of the declination is given by (6.19).

[19] With the proposed configuration, the eclipse would occur on 395 sols per Martian year, with a daily maximum of 47 min.

[20] The value ΔL_S of this bounding difference is easy enough to find. Equation (10.11) relates δ and L_S. The condition $\delta = f_0$ yields

$$\sin \Delta L_S = \frac{\sin f_0}{\sin \varepsilon} .$$

Expressing f_0 in terms of the reduced distance, we obtain

$$\Delta L_S = \arcsin \frac{1}{\eta_{GS} \sin \varepsilon} = \arcsin \frac{1}{2.560} = 22.99° ,$$

and hence the value $\Delta L_S = 23°$.

10.9.3 Natural Satellites

Phobos and Deimos

The planet Mars has two natural satellites, discovered in August 1877 by A. Hall.[21] They have circular, equatorial orbits (i is about 1°). Their small size (of the order of ten kilometres)[22] and lumpy shape makes them look like large rocks. Like the Moon or other natural satellites, they always turn the same face towards their planet.

The largest, Phobos (a = 9 377.2 km, i = 1.082°, e = 0.0151, T = 0.318910 day), with η = 2.76, is well below the areostationary orbit.[23] The second moon, Deimos (a = 23 463.2 km, i = 1.791°, e = 0.00033, T = 1.262441 day), with $\eta = 6.91$, is slightly higher.

One consequence of tidal effects is that the distance between the planet and its moon varies slightly in time. If the satellite is beyond the synchronous orbit, it moves away. If it is within, it moves closer until it breaks under tidal forces when it reaches the so-called Roche limit.[24]

---

[21] Hall named them after the two male offspring of Ares and Aphrodite. They had the same weak points as their father: Phobos, ὁ Φόβος, ου, 'fear', Deimos, ὁ Δεῖμος, ου, 'dread'.

> ὣς φάτο, καί ῥ ' ἵππους κέλετο Δεῖμόν τε Φόβον τε
> ζευγνύμεν, αὐτὸς δ ' ἔντε ' ἐδύσετο παμφανόωντα.
> *Ὁμήρος, Ἰλιάδα, Ο (119-120)*

According to the Iliad, Book XV, transl. Ian Johnston:

> Then he [Ares] told Terror and Flight to harness his horses,
> while he dressed himself in his glittering armour.

[22] Phobos – Mean diameter: 22.2 km (26.8 x 21.0 x 18.4), mean density d relative to water: $d = 1.9$, albedo: 0.07
Deimos – Mean diameter: 12.6 km (15.0 x 12.0 x 10.4), $d = 2.2$, albedo: 0.07.

[23] Phobos is one of the rare known examples of a natural satellite whose orbital angular speed is greater than the angular speed of the planet about its own axis: it is thus below the planetostationary orbit. Until certain discoveries made by Voyager-1 and 2, it was the only natural satellite known to have this property. For a Martian observer, Phobos rises in the west and sets in the east. Seen from Phobos, Mars covers roughly half of the sky. Indeed, the planet is seen from the apex of a cone with angle $2f_0 = 2\arcsin(1/\eta) = 42°$.

[24] Edouard Roche (1820–1883) was a French astronomer. He devised a cosmogonic theory of the Solar System and studied the internal structure of the Earth. This combination of astronomical and geophysical knowledge led him in 1849 to establish an expression for the tidal forces exerted by a central body on a natural satellite. He calculated that the satellite in orbit is destroyed if the radius of the orbit falls below a certain threshold, since called the Roche limit. The tidal forces are then greater than the cohesive gravitational forces. This limit depends on the relative densities of the satellite and the central body. In July 1994, the comet P/Shoemaker–Levy-9 flew very close to Jupiter, in fact, within the Roche

The Soviet probe Phobos-2 made a novel attempt to approach the moon Phobos. In an equatorial orbit for two months from 29 January 1989, the probe gradually moved towards Phobos. When it was only 50 m from the surface of Phobos, on 27 March, it ceased to transmit, just as it was about to launch its landing modules.

The Lagrange points of the Mars–Phobos system have been calculated in Example 3.2.

View and Sampling

Phobos can only be viewed from those points on Mars with latitude less than $|\phi| = \phi_{\mathrm{V}} = 69.8°$. Indeed, according to (8.15), we have

$$\phi_{\mathrm{V}} = i + \arccos(1/\eta) = 1.08 + 68.77 = 69.85° \ .$$

The 'sampling' of a natural satellite is represented by its synodic period. This period T' is obtained using (2.23):

$$\frac{1}{T'} = \frac{1}{1.027} - \frac{1}{0.319} = -2.161 \text{ day}^{-1} \ ,$$

which gives $T' = -0.463$ day or -0.451 sol. The negative sign of the synodic period indicates that the relative motion occurs in the retrograde direction.

Concerning Deimos, which is rather close to the areostationary orbit, only those points close to the poles cannot view it. Indeed, we now have

$$\phi_{\mathrm{V}} = i + \arccos(1/\eta) = 1.79 + 81.68 = 83.47° \ .$$

The synodic period is calculated from

$$\frac{1}{T'} = \frac{1}{1.027} - \frac{1}{1.262} = +0.1811 \text{ day}^{-1} \ ,$$

which gives $T' = +5.52$ day or $+5.37$ sol.

---

limit, whereupon it broke up into about twenty fragments which crashed into the planet. See also the note on Saturn and its rings.

11 Satellites of Other Celestial Bodies

In the present chapter, we shall examine the motion of a satellite around another celestial body, but more concisely than we have done for the Earth and Mars. The chapter is divided into two parts, each following practically the same plan. Part A concerns the satellites of planets, whilst Part B deals with satellites of natural satellites. In this second part, to avoid confusion, a natural satellite or moon of a planet will be referred to by the term 'natural satellite', and the term 'satellite' will be reserved for an artificial or technological satellite.

A: Satellite of a Planet

11.1 Planets of the Solar System

11.1.1 Presenting the Planets

The planets of the Solar System gravitate around its central star, the Sun. They fall into two categories: the telluric planets and the giant planets.[1]

The telluric planets[2] are the four planets closest to the Sun: Mercury, Venus, Earth, and Mars. They are rather similar to one another in size and composition (iron and siliceous rocks). The giant planets are more remote: Jupiter, Saturn, Uranus, and Neptune. They are much bigger than the telluric planets and have totally different compositions (mainly hydrogen and helium). The radius of the giant planets is determined as being the radius of

[1] One also finds the terms 'terrestrial' and 'Jovian', the latter being a reference to Jupiter, from the Latin *Jupiter, Jovis*. Up until the Middle Ages, in the time of the geocentric astronomy in which the roles of the Sun and Earth were swapped over, the planets between the Earth and the Sun (the Moon, Mercury, and Venus) were known as inferior planets, whilst those considered to lie beyond the Sun (Mars, Jupiter, and Saturn) were known as superior planets.

[2] In Latin, *Terra, æ* (f.) means 'the Earth', but *Tellus, uris* (f.) represents poetically the Earth goddess, or mother Earth, 'the fertility goddess of crop and cattle' (Horace).

an isobaric surface (at 1 bar). It is not possible to define a ground level or obtain a geographical representation of the planet.

The asteroid belt gravitates between these two groups.[3] It is composed of thousands of rocky bodies, sometimes with quite strange shapes. Beyond the giant planets lie a small planet, Pluto, and the asteroids of the Kuiper Belt.[4] Still further out lies the Oort cloud,[5] a mysterious world of comets. At the present time, astronomers tend to classify Pluto with the planetoids of Kuiper and Oort, in the category of trans-Neptunian objects.

The two telluric planets furthest from the Sun both have their own natural satellites: the Earth has one, the impressive Moon, whilst Mars has two minuscule followers. The giant planets each have a whole series of moons, some of which are larger than Mercury or Pluto.

Mercury, Venus, Mars, Jupiter and Saturn have been known[6] since ancient times (the Earth itself has not been considered a planet for such a long time), and have inherited in English and most European languages the names of the main divinities in the Roman pantheon.[7] The same custom applied to

---

[3] Most of these asteroids belong to the Main Belt, between Mars and Jupiter, but some actually lie on the orbit of Jupiter (the Trojan asteroids mentioned in Sect. 3.14), whilst others cross the trajectory of Mars and even the Earth orbit (near-Earth asteroids).

[4] Gerard Kuiper (1905–1973) was an American astronomer of Dutch extraction. He discovered the atmosphere of Titan (1945) and showed that it was made up of methane. He also showed that the atmosphere of Mars was mainly composed of carbon dioxide (1947). He hypothesised that the Solar System was encircled by an asteroid belt, of which Pluto was a representative. Since 1992, hundreds of trans-Neptunian objects have been discovered, some with a diameter of several hundred kilometres. They form what is now called the Kuiper Belt (or Kuiper–Edgeworth Belt).

[5] Jan Hendrik Oort (1900–1992) was a Dutch astronomer. From his studies of a great many very long period comets with orbits well outside the plane of the ecliptic, he deduced in 1950 that one should find, beyond the orbit of Neptune and the Kuiper Belt (which lie largely within the ecliptic), a ball-shaped rather than ring-shaped 'cloud' made up of small celestial bodies, which could be considered as a kind of cometary reservoir. The diameter of this so-called Oort cloud would be some 10^5 astronomical units, or one light-year. Gravitational perturbations, even very weak effects, would cause these comets to move either towards the centre of the Solar System, or out towards others stars.

[6] The word 'planet' comes from the Latin *planeta*, which comes in turn from the Greek πλάνητες, plural of ὁ πλάνης, ητος, which means 'traveller, wanderer'. The root means 'to wander, to leave the path'. Indeed, Greek astronomers distinguished those bodies in movement (the planets, πλάνητες ἀστέρες) from fixed objects (the stars, ἀστέρες). They recognised seven planets, the five mentioned above, plus the Moon and Sun. Replacing the name of the Sun by the name of God, one obtains the days of the week as they are found in most Latin languages, although not of course in English.

[7] The planet with the shortest period is associated with Mercury (Hermes in Greek): always in motion, appearing and disappearing in quick leaps and bounds

planets discovered later: Uranus by W. Herschel in 1781, Neptune by J.G. Galle in 1846 using the famous calculations of Le Verrier,[8] and Pluto by C.W. Tombaugh on 18 February 1930.

As soon as astronomers had switched to a heliocentric vision of the Solar System, they were struck by the regularity or harmony (to use the term employed by Kepler) of this system. The orbits[9] of the first eight planets all

during the seasons, it is indeed well represented by the messenger of the gods. The next, the brightest and most beautiful, is Venus (Aphrodite), goddess of love. Then, red as blood, her belligerent companion Mars (Ares), god of war, followed by Jupiter (Zeus), the master of Olympus, plenitude incarnated by a planet, benevolence in person. Beyond him, with his characteristic slow gait and pale countenance, Saturn, the father of Jupiter. The old man recalls the idea of time through a Greek play on words between his name (Cronus, Kronos in Greek) and the word 'chronos' for time.

[8] Urbain Le Verrier (1811–1877) was a French astronomer, who remains famous for having discovered a planet, not with any optical instrument, but with mathematical calculations: 'Le Verrier spied Neptune with the nib of his pen', in the words of Arago. The first tables for Uranus, discovered some sixty years earlier, contained too many disagreements with observation. Le Verrier suggested that these errors could be explained by the presence of an unknown planet, which he assumed to be in the plane of the ecliptic and at a distance given by the empirical Bode's law (or Titius–Bode law). He used the perturbation method to determine the position of this heavenly body, obtaining the orbital elements to great accuracy. Following his indications, the planet Neptune was discovered a few days later, on 31 August 1846, by the German astronomer J.G. Galle. The English astronomer J.C. Adams obtained very close results at the same time, but was not supported by his director, Airy. Le Verrier then devoted himself to the theory of the Solar System and setting up ephemerides. See also the notes on Delaunay and Airy.
On 14 November 1854, a hurricane destroyed part of the French and allied fleet in the Crimean war. Le Verrier understood that this storm had crossed Europe from west to east, and set up the first network of weather stations, beginning in 1855.

[9] As an example of the sought regularity, one must mention Bode's law, also called the Titius–Bode law. This was an empirical relation discovered in 1766 by the German astronomer J.D. Tietz (whose name is latinised to Titius), then formulated and established as a law by his colleague Bode in 1778. It can be formulated as follows. Let a_S be the semi-major axis of the planet's orbit, expressed in astronomical units. The relation gives a_S for the six planets known at the time:

$$a_S = 0.4 + 0.3 \times 2^n ,$$

with $n = -\infty$ for Mercury, $n = 0$ for Venus, $n = 1$ for the Earth, $n = 2$ for Mars, $n = 4$ for Jupiter, and $n = 5$ for Saturn. These values can be compared with the values of a_S in Table 11.2b. It appears that the result is not so bad. However, we do not have here the accuracy of measurement that forms the basis for astronomy! This law, which is not strictly a law in the scientific sense of the term, has raised much controversy. Is it purely fortuitous? Does it reflect lie, to within a few degrees, in the same plane, known as the ecliptic. The motion, along almost circular orbits, is always in the same direction, i.e., anticlockwise as viewed from above the north pole of the Sun. In most cases, the planets rotate in this direction about their own axes, and their natural satellites orbit them in this direction as well. Moreover, the fact that this direction is also the direction of rotation of the Sun itself has long suggested that there is a connection with the formation of the Solar System, as proposed in Laplace's theory in *Exposition du système du monde* of 1796.

11.1.2 Space Exploration of the Planets

One of the main motivations for space exploration today is the desire to find traces of life, something usually associated with the presence of water in the liquid state. This explains in part projects to investigate Mars, or indeed Europa, a Galilean moon of Jupiter.

A further motivation is sometimes the study of the atmosphere in the case of the telluric planets: Venus with its gases heated to tremendous temperatures by the greenhouse effect [mean pressure and temperature at ground level: 90 bar, 750 K; composition: carbon dioxide CO_2 (96%), nitrogen N_2 (3%)] and Mars with its rarified gases [mean pressure and temperature at ground level: less than 10^{-2} bar, about 250 K; composition: carbon dioxide CO_2 (95%), nitrogen N_2 (3%)]. Two natural satellites also carry an atmosphere: Titan, the largest moon of Saturn [mean pressure and temperature at ground level: 1.5 bar, 90 K; composition: nitrogen N_2 (98%), methane CH_4 (2%)], and Triton, the largest satellite of Neptune [mean pressure and temperature at ground level: about 10^{-5} bar (1.5 Pa), 37 K; composition: nitrogen N_2 (95%)]. Not much is known about the atmosphere of Pluto and its companion Charon, except that, if there is an atmosphere, it must be extremely tenuous. The same can be said for the atmosphere of Io, a Galilean satellite of Jupiter with high levels of volcanic activity.

At the beginning of space exploration beyond the confines of the Earth, the aims were many and varied: there was a thirst for knowledge about the

---

the action of physical forces during the formation of the Solar System? Does it reveal the action of some gravitational phenomenon after the formation of the planets? Today, astronomers have a preference for the first of these three hypotheses, but one cannot deny at least one point in favour of this 'law': it has played a considerable role in the historical development of the subject. It was while looking for the planet $n = 6$, shortly after the formulation by Bode, that Herschel found Uranus. Then, assuming that the body perturbing Uranus was in the orbit $n = 7$, Le Verrier calculated the position of Neptune. His own account shows this beyond doubt, and Adams had used the same arguments. The most surprising thing is that the least good agreement between the 'law' and measurement is precisely for the case of Neptune! Many astronomers had sought to fill the space at $n = 3$, until Piazzi discovered a first asteroid in the space between Mars and Jupiter.

various bodies making up the Solar System, without any specific programme, but driven by the quest for technological and ideological ascension by the two superpowers of the day, as will be seen from the brief chronology below.

During 1959, the probe Luna-1 achieved the first lunar flyby in January, Luna-2 landed on the Moon in September, and Luna-3 sent the first photos of the dark side of the Moon back to the USSR from its October flyby (see Fig. 11.9). In 1966, Luna-9 made the first softlanding in January, and Luna-10 was the first lunar satellite in March. The United States had their revenge with the astronauts of Apollo-11, on 20 July 1969.

After the Moon, the next target was Venus: for the United States, successful flybys were operated by Mariner-2 in December 1962, Mariner-5 in November 1967, and the two probes Pioneer Venus-1 (or Pioneer Venus Orbiter, or Pioneer 12) and Pioneer Venus-2 (or Pioneer Venus Probe Bus, or Pioneer-13) in December 1978; for the USSR, successful missions were carried out by Venera-4 in October 1967, Venera-5 and -6 in May 1969 with atmospheric capsules, whilst the first softlanding was made by Venera-7 on 15 December 1970, followed by other successful missions Venera-8 to -16, a record in perfect contrast to their Mars programme. Finally, the Soviets produced Vega-1 and -2 in June 85, releasing balloons and a landing module. As the atmosphere of Venus is very opaque, it was mapped by radar, from 1990 to 1994, by the US probe Magellan, launched on 4 May 1989 by the shuttle Atlantis (STS-30) and placed in orbit around the planet on 10 August 1990. Before it, three probes had carried out measurements in orbit: Pioneer Venus-1 from 1979 to 1992 and Venera-15 and -16 from 1983 to 1986.

The probe Mariner-10 accomplished the first mission to two planets. It was launched on 3 November 1973 and flew by Venus on 5 February 1974, whereafter it encountered Mercury three times, on 29 March 1974, 21 September 1974 and 16 March 1975.[10] The Venus flyby was the first use of a gravity-assist maneuver.[11]

---

[10] The overpasses were separated by 176 days. The probe had been placed in an eccentric heliocentric orbit with exactly twice the period T_1 of Mercury ($T =$ 88 day). In this case, the synodic period is $T' = 2T = T_1 = 176$ day.

[11] To model the trajectory of a probe from the Earth to Venus, for example, we use the idea of patched conics. As long as the probe is in the sphere of influence of the Earth or of Venus, its motion is described by a conic section with the relevant planet at the focus (in motion relative to the Sun). Between the two, the motion is heliocentric, described by a conic section whose focus (the Sun) is fixed. The three conics are then patched together. However, to go from the Earth to a non-neighbouring planet (other than Venus or Mars), one can fly close by some intermediate planet. To model the trajectory from the Earth to Mercury in these conditions, five conic sections are then patched together. The third corresponds to a flight close by Venus (generally, a branch of a hyperbola). The velocity relative to Venus has the same magnitude when it enters and when it leaves the sphere of influence, but the direction is significantly changed. This is called gravitational deflection. In this way the (vectorial) velocity relative to

After Mars, discussed in the last chapter, the next target was the more distant planets. Jupiter was overflown on 1 December 1973 by Pioneer-10, launched on 3 March 1972, and 1 December 1974 by Pioneer-11, launched on 6 April 1973. The latter continued out to Saturn, which it reached on 1 September 1979.

It was the Voyager probes, making full use of the gravity-assist technique, that really revolutionised our knowledge of the outer planets. NASA took advantage of an exceptional alignment of the planets from Jupiter to Neptune, at the beginning of the 1980s, to accomplish with these two probes what became known as the Grand Tour. Such a favourable configuration only occurs about once every 180 years.

Voyager-1, launched on 5 September 1977, flew past Jupiter on 5 March 1979 and Saturn on 12 November 1980, before making a closer investigation of Titan. Voyager-2 was the subject of a remarkable round of gravitational billiards: launched on 20 August 1977, it overflew Jupiter on 9 July 1979, Saturn on 26 August 1981, Uranus on 24 January 1986 and Neptune on 24 August 1989, all the while sending back photos of great quality. The four probes (Pioneer-10 and -11, Voyager-1 and 2) are currently on their way out of the System System.

The probe Galileo, launched[12] in 1989, was placed in orbit around Jupiter in December 1995 to study the giant planet and its four large Galilean moons.[13] Although originally programmed to last two years, the mission of the Galileo orbiter actually went on for eight years. When the fuel reserves (hydrazine) had almost run out, the probe was sent into Jupiter to remove all risk of collision with Europa, which might have contaminated it with elements of terrestrial life. On the way from the Earth to Jupiter, Galileo photographed Venus, the Earth – see Colour Plates I and II – and several asteroids.

---

the Sun can be greatly modified without energy expenditure. To reach Mercury, the speed must be reduced. To reach Jupiter or the more remote planets, the speed is increased. This kind of maneuver is called a gravitational sling-shot. It is used quite systematically for long-distance journeys. The probe Ulysses, launched on 6 October 1990 to study the Sun, thereby reach a speed of 125 km/s (or 450 000 km/hr). Using a Jupiter swing-by, it left the plane of the ecliptic to overfly the south pole of the Sun in 1994, then the north pole in 1995. Further examples are given below for the Galileo and Cassini probes.

[12] VEEGA trajectory: Venus–Earth–Earth gravity-assist sequence. We give the date (year month day) for each planet overflown. Launch: 1989 10 18, by Atlantis shuttle, STS-34. Venus flyby: 1990 02 10. Earth fl. 1: 1990 12 08. Earth fl. 2: 1992 12 08. Jupiter Orbit Insertion (JOI): 1995 12 07).

[13] These four natural satellites were discovered in 1610 by Galileo, an event which had important scientific and philosophical consequences. Other much smaller moons were discovered from 1892.

The US probe Cassini[14] was launched on 15 October 1997 to study the Saturn system. Using four gravity-assist maneuvers,[15] it went into orbit around Saturn on 1 July 2004. On 25 December 2004, the European module Huygens[16] should separate from the main spacecraft and, three weeks later, on 14 January 2005, descend towards the surface of Titan with the help of a parachute.

Cassini's mission consists in making 76 revolutions around Saturn over 4 years, which will bring it into 52 close encounters with seven of Saturn's 31 known moons, and also involve 45 Titan flybys.

---

[14] Giovanni Domenico Cassini (1625–1712) was a French astronomer of Italian origins (called Jean-Dominique Cassini by the French). As a famous planetary astronomer in Italy, he was invited to France by Colbert and appointed to the Academy of Science (1669). At the request of Louis XIV, he directed the Paris Observatory from its foundation. This post was then successively occupied by several generations of the Cassini family (see notes on Cassini II and Cassini III). As the founder of this dynasty, G.D. Cassini is often called Cassini I. He wrote a great many works on Venus, Mars, and Jupiter. His results concerning the moons of Jupiter were used by Roemer to calculate the speed of light in 1675. Cassini discovered four moons of Saturn between 1671 and 1684, and the gap between rings A and B, known as the Cassini division.
Concerning the rings, note that Galileo had observed in 1610 that Saturn appeared to have 'two ears', which he considered to be moons. However, a few years later, they were no longer visible to him. Huygens realised that they were looking at a ring, which was visible at various different inclinations. But it was Cassini who first saw two concentric rings, A and B. Other rings have since been observed and they are currently labelled with letters up to G. Cassini put forward the hypothesis, since confirmed, that the rings were made up of a multitude of small bodies gravitating on very close orbits. The current theories about Saturn's rings allow for two possibilities: either they consist of icy particles that have never been able to accrete because they orbit Saturn within the Roche limit, or they result from the break-up of a natural satellite whose orbital evolution has brought it within the Roche limit.

[15] VVEJGA trajectory: Venus–Venus–Earth–Jupiter gravity-assist sequence. For each planet overflown, we give the date (year month day), and speeds in km/s, in the form [before/after]. Speeds are expressed in a heliocentric frame. Launch: 1997 10 15. Venus/1: 1998 04 26 [37.2/40.9]. Venus/2: 1999 06 24 [39.2/42.3]. Earth: 1999 08 18 [35.0/39.1]. Jupiter: 2000 12 30 [11.6/13.7]. Saturn: 2004 07 01 (SOI, insertion in orbit).

[16] Christiaan Huygens (1629–1695) was a Dutch physicist, mathematician and astronomer. In addition to his great treatises on probability, dynamics (*Horologium oscillatorium* in 1673), and optics (*Treatise on Light* in 1690), he also wrote fundamental works on astronomy. By eliminating chromatic aberration, he improved the refracting telescope (Huygens' eyepiece), and with this enhancement, made fundamental discoveries in astronomy. For example, he discovered Saturn's moon Titan and the rings, as well as the rotation of Mars.

Table 11.1. Exploration of the Solar System. Dates (Gregorian year or $\mathcal{A}$, since ancient times) marking the beginning of the following events: observation from Earth, flyby of celestial body, landing on object, orbit around object, sample return. Dates in brackets refer to the year when projects should be achieved, with x representing a number between 0 and 9. The number n is the number missions, successful or otherwise, launched before 2005. For the Earth, N represents thousands of missions

Object	Observation from Earth	Flyby	Landing	Orbit around	Sample return	n
Moon	$\mathcal{A}$	1959	1959	1966	1969	64
Mercury	$\mathcal{A}$	1974		2011		2
Venus	$\mathcal{A}$	1962	1966	1975		24
Earth	–	1957	1961	1957	–	N
Mars	$\mathcal{A}$	1965	1971	1971	(201x)	32
Jupiter	$\mathcal{A}$	1973		1995		6
Saturn	$\mathcal{A}$	1973		2004		4
Uranus	1781	1986				1
Neptune	1842	1989				1
Pluto	1930	(201x)				0
Asteroids	1801	1991	2001	2000	(2007)	5
Comets	$\mathcal{A}$	1985	(2011)	(2011)	(2006)	8

Table 11.2. Following page. Planets of the Solar System. (**a**) Geodetic characteristics. Geodetic data: planetocentric gravitational attraction μ, equatorial radius R of planet. Deduced quantities: central acceleration at ground level g_0, escape velocity V_e, period of a satellite in Keplerian orbit at ground level $T_{0(h=0)}$, mean density d. (**b**) Astronomical characteristics. Data relating to planetary orbit: semi-major axis a_S, sidereal period of revolution N_{sid}, eccentricity e, inclination i with respect to the ecliptic. Data relating to the rotation of the planet: obliquity ε. Sphere of influence ρ_Σ. (**c**) Planetosynchronicity. Astronomical data: period of rotation J_{sid}. Deduced quantities: reduced distance η_{GS} for the orbit of a stationary satellite (whence a_{GS} and h_{GS}). The distance η_{GS} should be compared with ρ_Σ/R. (**d**) Sun-synchronicity. Astronomical data: period of revolution N_{sid}. Geodetic data: terms J_2, J_3 and J_4 in the expansion of the gravitational potential (values to be multiplied by 10^{-6}). Deduced quantities: constant of Sun-synchronicity k_h, maximum value of the reduced distance $\eta_{HS\,max}$, denoted here by η_m for a Sun-synchronous satellite. Dimensionless quantity n.d.

Planet	$\mu = \mathcal{G}M$ [$m^3\,s^{-2}$]	R [km]	g_0 [$m\,s^{-2}$]	V_e [$km\,s^{-1}$]	$T_{0(h=0)}$ [min]	d n.d.
Mercury	$2.203\,208 \times 10^{13}$	2 439.7	3.70	4.25	85.02	5.44
Venus	$3.248\,586 \times 10^{14}$	6 051.8	8.87	10.36	86.50	5.27
Earth	$3.986\,004 \times 10^{14}$	6 378.2	9.80	11.18	84.49	5.52
Mars	$4.282\,831 \times 10^{13}$	3 396.2	3.71	5.02	100.15	3.94
Jupiter	$1.266\,865 \times 10^{17}$	71 492.	24.79	59.53	177.85	1.34
Saturn	$3.794\,063 \times 10^{16}$	60 268.	10.45	35.48	251.54	0.69
Uranus	$5.794\,549 \times 10^{15}$	25 559.	8.87	21.29	177.76	1.29
Neptune	$6.836\,540 \times 10^{15}$	24 764.	11.15	23.50	156.08	1.64
Pluto	$8.261\,000 \times 10^{11}$	1 195.	0.58	1.18	150.51	1.73

Planet	a_S [a.u.]	N_{sid} [yr]	e n.d.	i [deg]	ε [deg]	ρ_Σ [km]	ρ_Σ/R n.d.
Mercury	0.387	0.241	0.205 63	7.00	2.0	9.79×10^4	40
Venus	0.723	0.615	0.006 77	3.39	177.4	5.37×10^5	89
Earth	1.000	1.000	0.016 71	0.00	23.4	8.05×10^5	126
Mars	1.524	1.881	0.093 41	1.85	25.1	5.03×10^5	148
Jupiter	5.201	11.862	0.048 39	1.31	3.1	4.20×10^7	587
Saturn	9.538	29.456	0.054 15	2.49	26.7	4.75×10^7	788
Uranus	19.183	84.019	0.047 17	0.77	97.9	4.50×10^7	1 763
Neptune	30.055	164.767	0.008 59	1.78	28.8	7.54×10^7	3 045
Pluto	39.440	247.689	0.248 81	17.14	122.0	2.68×10^6	2 244

Planet	J_{sid} [hr]	η_{GS} n.d.	a_{GS} [km]	h_{GS} [km]	Planeto-synchronous
Mercury	1 407.510	99.555	242 885	240 446	impossible
Venus	5 832.444	253.900	1 536 551	1 530 499	impossible
Earth	23.934	6.611	42 164	35 786	achieved
Mars	24.623	6.015	20 428	17 031	planned
Jupiter	9.925	2.238	160 009	88 517	possible
Saturn	10.659	1.863	112 271	52 003	possible
Uranus	17.240	3.235	82 689	57 130	possible
Neptune	16.110	3.372	83 514	58 750	possible
Pluto	153.294	15.515	18 540	17 345	–

Planet	N_{sid} [day]	J_2 [10^{-6}]	J_3 [10^{-6}]	J_4 [10^{-6}]	k_h n.d.	η_m n.d.	Sun-synchronous
Mercury	87.969	60	0	0	0.13	< 1	impossible
Venus	224.701	6	1	0	0.03	< 1	impossible
Earth	365.256	1083	−3	−2	10.11	1.94	achieved
Mars	686.980	1960	36	−32	29.04	2.62	achieved
Jupiter	4 332.59	14 736	1	−587	775.46	6.69	polar
Saturn	10 759.2	16 298	0	−915	1 505.78	8.09	polar
Uranus	30 688.5	3 339	0	−32	1 245.12	7.66	polar
Neptune	60 182.3	3 410	0	−35	2 840.11	9.70	polar
Pluto	90 469.7	0	–	–	0.00	–	–

As far as the outermost planet of the Solar System is concerned, US projects to explore Pluto[17] aim to include also an encounter with some of the planetoids in the Kuiper Belt, which is still more remote. If Pluto appears to be something of an intruder among the planets of the Solar System, thanks to the inclination and eccentricity of its orbit, it looks much less out of place as a large object belonging to the Kuiper Belt.

We shall mention space exploration of the asteroids a little later, when we come to consider the probe NEAR. Cometary exploration (with Vega, Rosetta, etc.) was discussed in Chap. 5, in the context of satellites classified by mission (space exploration).

The main dates of this exploration have been summarised in Table 11.1. The table gives the year marking the beginning of various types of mission (in brackets, if the mission is still under development).

11.2 Geodetic and Astronomical Quantities for Planets

Geodetic and Astronomical Data

Table 11.2a gives the two basic quantities associated with each planet, its constant of central attraction μ and its equatorial radius R, which allow one to calculate the following quantities:

- the acceleration due to gravity g_0 on the surface of the planet, from (3.22),
- the escape velocity V_e, from (1.32),
- the period of a satellite in circular Keplerian orbit at zero altitude, $T_{0(h=0)}$, from (2.17), to which we shall return shortly,
- the density d, that is, the mass per unit volume of the body relative to that of water ($\rho_{\text{water}} = 10^3\ \text{kg}\,\text{m}^{-3}$, density $d_{\text{water}} = 1$).

Table 11.2b gives basic astronomical quantities concerning the orbit of the planet around the Sun, including the semi-major axis a_S and the sidereal period N_{sid}.

Kepler's third law is expressed very simply by

$$[N_{\text{sid}}\ (\text{yr})]^2 = [a_S\ (\text{a.u.})]^3\ , \qquad (11.1)$$

with the chosen units.

We have also calculated the radius of the sphere of influence ρ_Σ and noted the characteristics of the planetary orbit, i.e., e, i and ε, quantities which are

[17] The mission Pluto–Kuiper Express was to send a probe towards Pluto in 2004, for approach in 2012, whereupon it would have extended its trip into the Kuiper Belt. This project, abandoned in 2000, has been replaced by New Horizons, also involving a JGA (Jovian Gravity Assist) orbit. With a departure in 2006, it is due to arrive in 2015. Following this, it should study at least one object beyond Pluto.

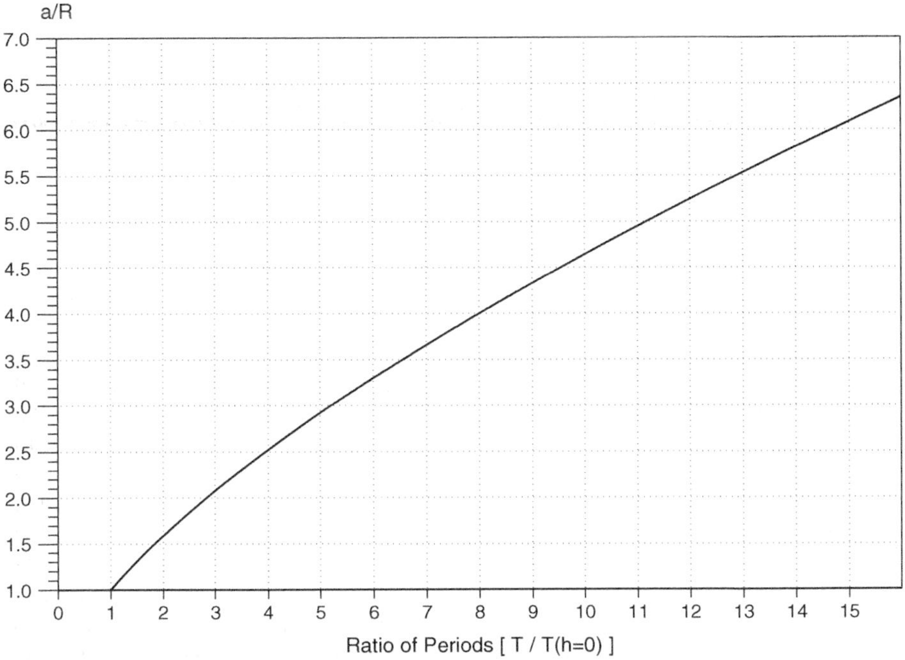

Figure 11.1. Relation between period and altitude. In this graph, the *abscissa* is the reduced period, i.e., the ratio of the period of the satellite to the period of a satellite at zero altitude, and the *ordinate* is the reduced distance $\eta = a/R$, i.e., the ratio of the semi-major axis of the orbit to the radius of the planet. Both quantities are dimensionless

not directly necessary in this study, but which allow one to compare the orbits of different planets. For example, we see that the orbit of Pluto is highly eccentric and that the rotation of Venus is retrograde ($\varepsilon > 90°$), etc.

To study the true motion and characterise special orbits, we use the geodetic and astronomical quantities given in Tables 11.2c and d.

Satellite in Keplerian Orbit

As already discussed, when a satellite is in orbit (semi-major axis a) around a planet, the period T_0 of its Keplerian motion is given by (2.16). We can also calculate the period $T_{0(h=0)}$ of a fictitious satellite at zero altitude. Considering the mean mass per unit volume ρ of the planet, we have $\mu = \rho V \mathcal{G}$, where V is the volume of the planet, assumed spherical. With (2.17), we obtain

$$T_{0(h=0)} = \sqrt{\frac{3\pi}{\mathcal{G}}}\frac{1}{\sqrt{\rho}} = 3.7584 \times 10^{-5} \rho^{-1/2} \, . \qquad (11.2)$$

This shows that the Keplerian period $T_{0(h=0)}$ can be expressed uniquely in terms of the mean density of the attractive body. As the Earth is the densest

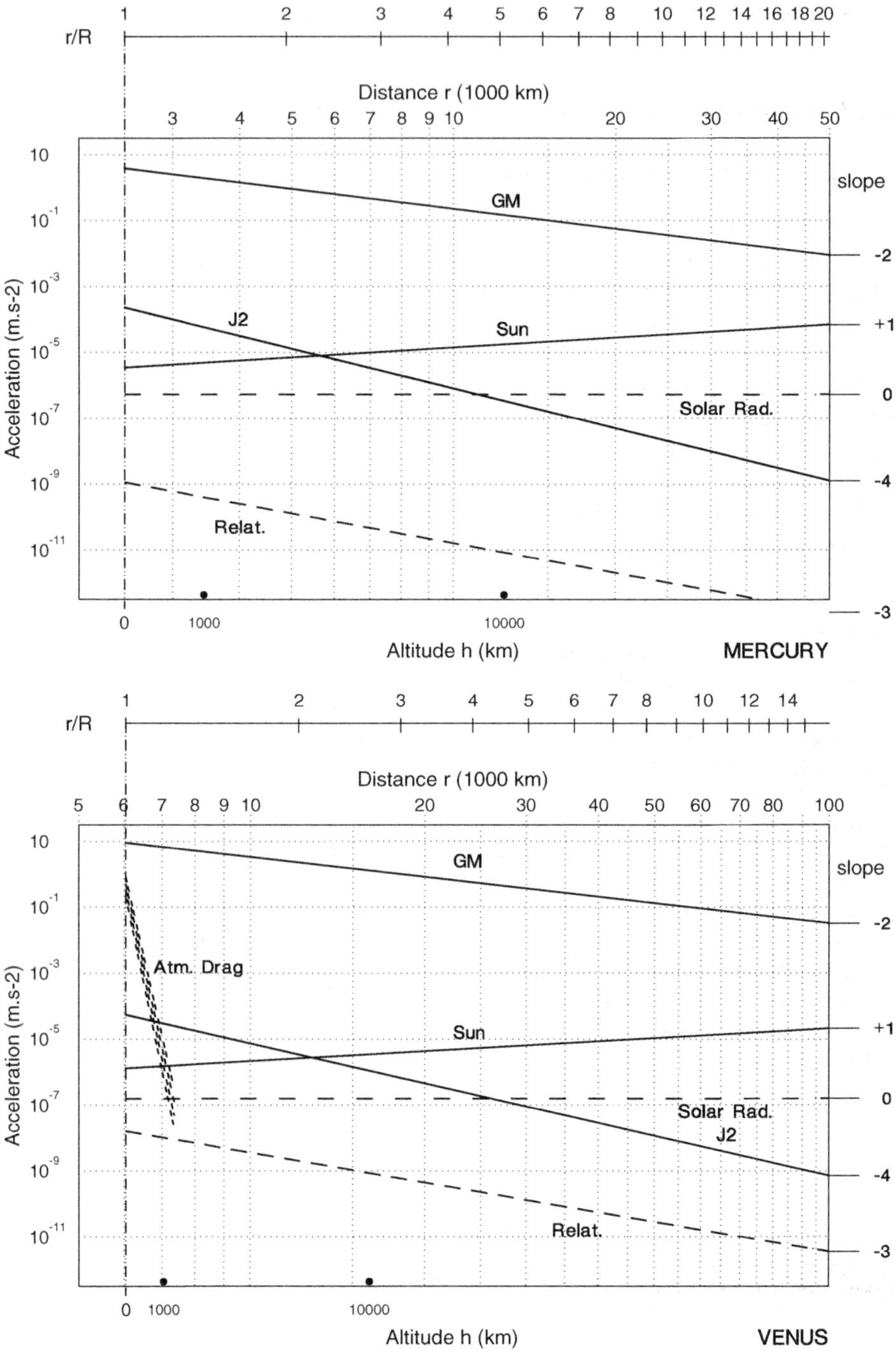

Figure 11.2. Accelerations as a function of the distance r from the satellite to the centre of the planet. Log–log scale. *Upper*: Mercury. *Lower*: Venus

planet in the Solar System, the period $T_{0(h=0)}$ is the shortest. Conversely, the longest period is for a satellite orbiting around Saturn.

Using the mean density d relative to water and expressing the period in minutes, we obtain

$$T_{0(h=0)}\ (\text{min}) \approx 198 d^{-1/2}\ . \tag{11.3}$$

The Keplerian period for semi-major axis a can then be written

$$T_0(\eta)\ (\text{min}) \approx 198\sqrt{\frac{\eta^3}{d}}\ , \tag{11.4}$$

for reduced distance $\eta = a/R$. Figure 11.1 graphs the variation of $T_0/T_{0(h=0)}$ as a function of $\eta = a/R$.

Geographic Maps

Only the telluric planets can be mapped in the sense that we can produce a geographic map of the surface. The mappable area, calculated on the ellipsoid and expressed in millions of km^2, has the following values for each planet: 75 for Mercury, 460 for Venus, 510 for the Earth (140 for the land masses and 370 for the sea bed), and 143 for Mars. For these four planets, this makes a total of 1 188. We also have 18 for Pluto and 2.6 for the largest of the asteroids (and 0.001125 for 433-Eros). We shall discuss maps of natural satellites shortly.

The geography of Mars (volcanoes, impact basins, etc.) was outlined in the last chapter. We shall not dwell further here on the geography of Venus or the other telluric planets.

In this chapter, we shall use the following maps as background for representing ground tracks or orbits:

- for Venus, the topographical map built up from SAR (Synthetic Aperture Radar) data gathered by Magellan,
- for Eros, the topographical map made from altimetry data gathered by NEAR.

In both cases, contours are plotted at 2 km intervals, with the zero altitude contour in bold face. Those at positive altitudes are represented by continuous curves and those at negative altitudes by dashed curves.

For the planets, the zero meridian is chosen arbitrarily (see note on Airy for the Earth and Mars).

11.3 Satellite of Planet in Real Orbit

11.3.1 Perturbative Accelerations

The sphere of influence discussed in Sect. 3.13 informs us of the distance beyond which one can no longer neglect perturbations due to the Sun. Equation (3.103) gives the values of the radius ρ_Σ for all the planets. The results are displayed in Table 11.2b, where they can usefully be compared with the values in Table 11.2c.

The variation of the central acceleration and perturbative accelerations has already been plotted for the Earth in Fig. 3.1, and for Mars in Fig. 10.6. To complete the list of telluric planets, we have plotted the same graphs for Mercury and Venus in Fig. 11.2, using the same notation as in Table 3.3 (in which, of course, we replace the terrestrial acceleration by the acceleration due to the relevant planet).

For these two planets, there is no perturbing acceleration due to the terms J_n, $n \geqslant 3$, since these terms are almost all zero. The very weak acceleration $\gamma_{\mathrm{CCN}.J2}$, due to the term in J_2, is soon supplanted (for $h \sim R$) by the perturbing acceleration γ_{CS} of the solar attraction. For Mercury, the perturbing acceleration γ_{DP} due to solar radiation pressure is still poorly known.

For Venus, atmospheric drag causes an acceleration γ_{DF} which can be very large, and depends on the altitude and shape of the satellite. The solar radiation pressure, which leads to the perturbing acceleration denoted by γ_{DP}, impinges on the satellite both directly and indirectly, by the albedo effect, the mean albedo of Venus being very high (0.76).

11.3.2 Classification of Satellites

Rotational Motion of the Planets

The values given in Table 11.2a to d show that the two planets closest to the Sun have very long periods of rotation J_{sid} : 58.646 day for Mercury[18] (exactly 2/3 of the sidereal period of revolution) and 243.018 day for Venus.[19] For these two planets, the day is longer than the year. This is due to the proximity of the Sun.

Moving further out to the Earth and Mars, this period J_{sid} is about one day, whilst for the giant planets from Jupiter to Neptune, it is of the order of ten hours.

[18] According to astronomers, the proximity of the Sun should have led to a 1:1 resonance phenomenon for Mercury, so that Mercury would always present the same face towards the Sun, like the natural satellites and their planets. This was indeed what was thought up until 1965, when radar measurements from the Earth showed that it had a shorter rotation period, of only 59 days. The Italian astrophysicist Guiseppe Colombo showed that this was a very rare case of a 3:2 resonance, i.e., 3 rotations in 2 revolutions, due to the high eccentricity of Mercury's orbit.

[19] Venus is the brightest celestial body as seen from Earth, apart from the Sun and the Moon. This is partly because it is surrounded by a very thick layer of cloud. The speed of rotation was only measured in 1962, with the advent of the radar. The clouds have a much faster rotational motion, known as superrotation. At an altitude of 70 km, the atmosphere makes one round trip in 4 days, corresponding to winds at 100 m/s, in the direction of rotation of the planet.

Planetosynchronous Satellite

For satellites that are stationary with respect to a planet, we have used (4.55) to calculate the reduced altitude η_{GS} for a Keplerian orbit in terms of J_{sid}. (For the Earth and Mars, the values were calculated for the Keplerian and then the true orbit, in Chaps. 4 and 10, respectively.) They are displayed in Table 11.2c.

The results prompt the following remarks:

- For Mercury and Venus, the values of η_{GS} are so large that this orbit cannot be obtained. The solar perturbative attraction becomes too great for these altitudes and such an orbit would pass way beyond the sphere of influence of the planet.
- For the Earth and Mars, η_{GS} is about 6.
- For the giant planets, η_{GS} is about 2 or 3.
- For Pluto,[20] the position of the stationary satellite is more complex, but fortunately there is no urgency to carry out such a calculation!

Sun-Synchronous Satellite

To satisfy the condition (4.60) for a Sun-synchronous orbit, one must have a large J_2 term if the planet is close to the Sun, since its sidereal period of revolution is short. Conversely, a weak J_2 term is required if the planet is remote from the Sun. Now, as we have already seen, if the planet is close to the Sun, e.g., Mercury, Venus, its rotation about its own axis is blocked and planetary flattening is very low. This in turn implies that J_2 will be low, or almost zero. We conclude that Sun-synchronicity is impossible.

If a planet is far from the Sun, as in the case of the giant planets, its rapid rotation creates a significant flattening effect and J_2 is consequently large. To counterbalance this effect, the orbit must have a very low value of $\cos i$, indeed, practically zero. The orbit is therefore polar, to within a few hundredths of a degree. But we should ask what Sun-synchronicity means for a satellite in orbit around Jupiter, a planet which takes 12 years to go round the Sun, or around Neptune, which takes 165 years.

For the two intermediate planets, the Earth and Mars, this condition can be satisfied. The results for k_h, the constant of Sun-synchronicity, calculated using (4.63), are given in Table 11.2d, where we have also displayed η_m, the value of the maximal reduced distance $\eta_{HS\,max}$, obtained for $i = 180°$.

[20] Pluto is accompanied by Charon, discovered on 2 July 1978 by J.W. Christy. This satellite has such a large relative mass (1/6 of the mass of Pluto) that the Pluton–Charon ensemble can be considered rather as a double planet. The semi-major axis of Charon's circular orbit is $a_P = 19\,460$ km, and it is interesting to note that this value is very close to the value for the planetostationary orbit. We have in fact $a_P/a_{GS} = 1.05$.

Frozen Orbit

In Chap. 7, we saw that a satellite orbit could be frozen by taking advantage of balancing effects between the various variations in the position of the periastron, i.e., secular and long-period variations. The ratio J_3/J_2 arises when we calculate the frozen eccentricity. For a frozen orbit to be useful, the frozen eccentricity e_F, which is of the order of $(1/2)J_3/J_2$, must be less than 0.01, for beyond this, differences in altitude become too great.

The J_3 term is a zonal term (axial symmetry) expressing the effects of asymmetry between the northern and southern hemispheres. For Mercury and Venus, which are practically spherical, J_3 is zero or very low. For the giant planets, J_3 is zero because the plasticity of these planets only generates even zonal coefficients J_{2n} (symmetry relative to the equatorial plane, or north–south symmetry).

It follows that the only two planets that can have a satellite in frozen orbit are the Earth and Mars. As regards Sun-synchronous frozen orbits, recall that the argument of the periastron, related to the sign of J_3, is $\omega_F = 90°$ for the Earth, $\omega_F = 270°$ for Mars.

11.4 Ground Track for a Satellite of a Planet

The ground track of the satellite over several revolutions is characterised by the equatorial shift, which depends for the main part on the angular speed of rotation of the planet. For the Earth or Mars, the equatorial shift is of the order of 25° for a low-orbiting satellite. For the giant planets, it is two to three times greater. But for Mercury and Venus, which rotate very slowly about their axes, the equatorial shift is very slight.

We shall now discuss, with a little more detail, the ground track of satellites in orbit around Mercury, Venus, and the asteroid Eros.

11.4.1 Satellite of Mercury

Mercury has only received the visit of one probe, Mariner-10. However, there are currently three projects under way. The US probe Messenger (Mercury Surface, Space Environment, Geochemistry and Ranging), noting the mythological reference to the main occupation of Mercury or Hermes, was launched in 2004 and will undergo no fewer than six gravity-assist maneuvers,[21] before going into a highly eccentric orbit around the planet in 2011 [h_p = 200 km,

[21] Launch 2004 08 02, Earth flyby (altitude: 2295 km) 2005 07 29, Venus fl. 1 (altitude: 3000 km) 2006 10 23, Venus fl. 2 (altitude: 300 km) 2007 06 04, Mercury fl. 1 (altitude: 200 km) 2008 01 14, Mercury fl. 2 (altitude: 200 km) 2008 10 06, Mercury fl. 3 (altitude: 200 km) 2009 09 29, Mercury Orbit Insertion (MOI) 2011 03 18.

$h_a = 15\,193$ km, $i = 80°$, $T = 720$ min, see Fig. 11.3 (upper)]. The periastron will be located at latitude 60°N to study the Caloris impact basin as closely as possible. The mission in orbit will last for 12 months, i.e., 2 Mercurian solar days.[22]

The Japanese probe Mercury Orbiter and the European probe Bepi-Colombo Mercury Orbiter[23] are due for launch in 2009, or later. The two missions will be combined, although in a manner that has not yet been completely defined. The two orbiters, MPO (Mercury Planetary Orbiter) and MMO (Mercury Magnetospheric Orbiter), should follow eccentric polar orbits ($i = 90°$), with periastron at altitude $h_p = 400$ km. The altitude at apoastron will be $h_a = 1\,500$ km for MPO and $h_a = 12\,000$ km for MMO, in such a way that the period of the second, $T = 560$ min, will be a multiple of the period of the first, $T = 140$ min.

11.4.2 Satellite of Venus

The probe Magellan, named after the sixteenth century Portuguese explorer, orbited Venus from 1990 to 1994, perfectly fulfilling its various missions, in particular the main task of mapping the surface of the planet (achieved to 98%). The result is a topographical map of Venus, shown in Colour Plate XV, in the form of two orthographic equatorial projections. For this map, the Magellan data has been complemented where necessary by data from the probes Venera-15, -16, and Pioneer Venus, as well as by Earth-based radar observations (from the Arecibo observatory).

This radar mapping mission comprised three cycles, each lasting 243 jours, during which the Magellan satellite followed an eccentric polar orbit, shown in Fig. 11.4 (upper), with characteristics:

$$h_p = 289.57 \text{ km}\,, \quad h_a = 8\,458.5 \text{ km}\,, \quad i = 85.7°\,, \quad \omega = 170°\,,$$

$$a = 10\,425.8 \text{ km}\,, \quad e = 0.391\,76\,, \quad T_a = 195.59 \text{ min} = 3.26 \text{ hr}\,.$$

The ground track of this orbit has been shown over one revolution in Fig. 11.5 (upper). The radar mapping only proceeded for 37.2 min during each revolution, from slightly before to slightly after the passage at periastron.

The length of the cycle of 243 days corresponds to J_{sid}, the time the planet takes in a Galilean frame to make a complete round trip relative to the orbit of the satellite, which remains fixed in this frame. Indeed, because Venus is

[22] For a point on Mercury, the day lasts 176 terrestrial days: 88 days of nighttime are followed by 88 days of daytime. This stretch of 176 days corresponds to two revolutions around the Sun. One Mercurian day is exactly equal to two Mercurian years (resonance 3:2).

[23] So named in honour of the Italian mathematician Guiseppe 'Beppi' Colombo (1920–1984), mentioned above.

[MERCURY] Messenger
Orbit - ref.: Mercury

>>>> Time span shown: 4320.0 min = 3.00 days

Equiv. altit. = 7696.5 km a =10136.200 km
Inclination = 80.00 ° e = 0.739577
Period = 719.98 min * rev/day = 2.00
h_a = 15193 km; h_p = 200 km; arg.periapsis: +61.57 °

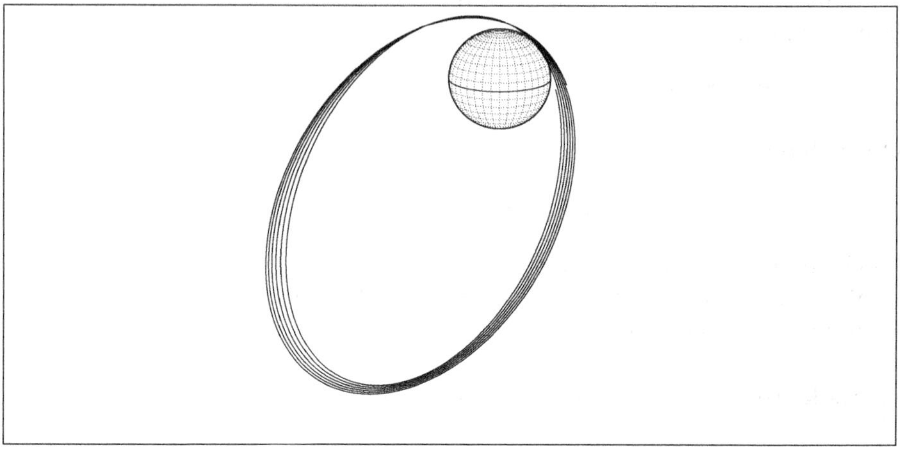

Projection: Orthographic
Property: none
T.:Azimuthal ☿ Graticule: 10°

Map centre: 15.0 ° N; 90.0 °E
Aspect: Oblique
[-90.0 / +75.0 / +0.0] Gr.Mod.: IAU91

Longitude / Initialisation:
A.n.: 160.75 ° - Apo.: 0.00 °

Ιξιων
MC ★ LMD
Ατλας

[VENUS] Venus Express
Orbit - ref.: Venus

>>>> Time span shown: 5760.0 min = 4.00 days

Equiv. altit. = 33125.0 km a =39176.801 km
Inclination = 90.00 ° e = 0.839145
Period = 1424.71 min * rev/day = 1.01
h_a = 66000 km; h_p = 250 km; arg.periapsis: +70.00 °

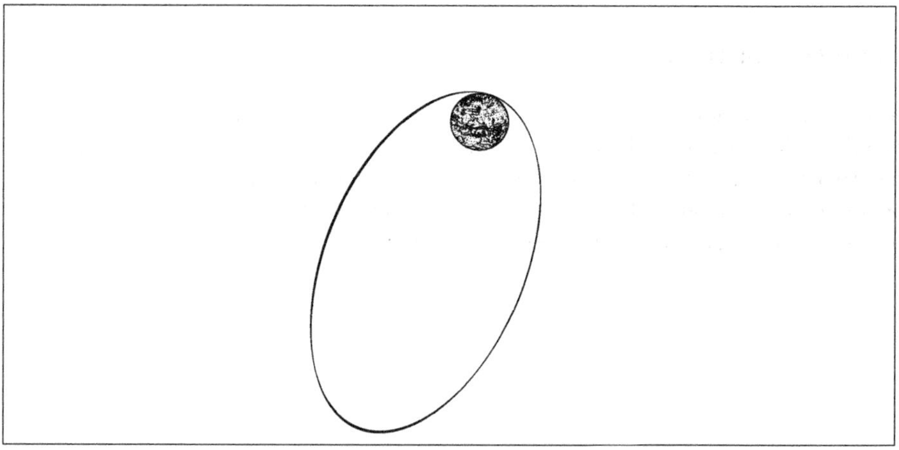

Projection: Orthographic
Property: none
T.:Azimuthal ♀ Graticule: 10°

Map centre: 15.0 ° N; 90.0 °E
Aspect: Oblique
[-90.0 / +75.0 / +0.0] Gr.Mod.: MGNP60

Longitude / Initialisation:
A.n.:-179.28 ° - Apo.: 0.00 °
Magellan Topogr. /h/2km/

Ιξιων
MC ★ LMD
Ατλας

Figure 11.3. Orbit of satellite in a frame moving with the planet. *Upper*: Messenger in orbit around Mercury. *Lower*: Venus Express in orbit around Venus

[VENUS] Magellan
Orbit - ref.: Venus
>>>> Time span shown: 1440.0 min = 1.00 day

Equiv. altit. = 4374.8 km a =10425.835 km
Inclination = 85.70 ° e = 0.391764
Period = 195.59 min * rev/day = 7.36
h_a = 8459 km; h_p = 290 km; arg.periapsis: +170.47 °

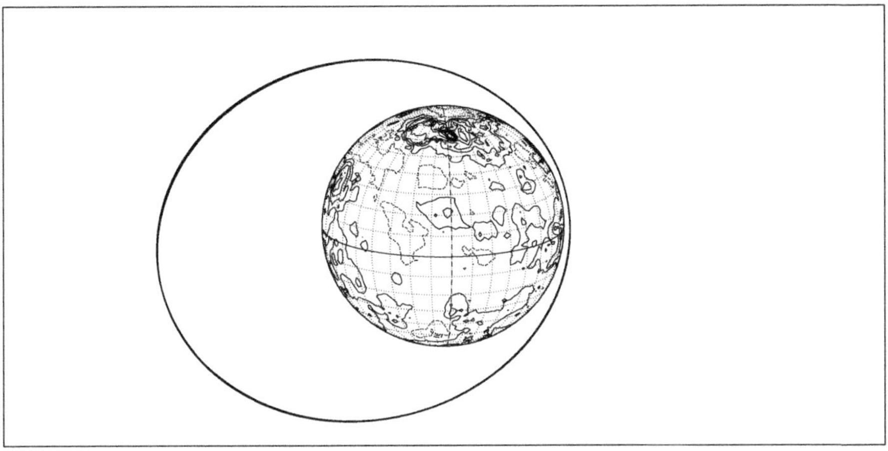

Projection: Orthographic
Property: none
T.:Azimuthal ♀ Graticule: 10°

Map centre: 15.0 ° N; 4.0 ° W
Aspect: Oblique
[-90.0 / +75.0 / +94.0] Gr.Mod.: VGM6A

Longitude / Initialisation:
A.n.: -90.00 ° - Apo.: 89.27 °
Magellan Topogr. /h/2km/

Iξιων
MC ★ LMD
Ατλας

[VENUS] Planet-C 〈 金星プロジエクト 〉
Orbit - ref.: Venus
>>>> Time span shown: 2304.0 min = 1.60 day

Equiv. altit. = 39450.2 km a =45502.000 km
Inclination = 172.00 ° e = 0.860406
Period = 1783.30 min * rev/day = 0.81
h_a = 78600 km; h_p = 300 km; arg.periapsis: +90.00 °

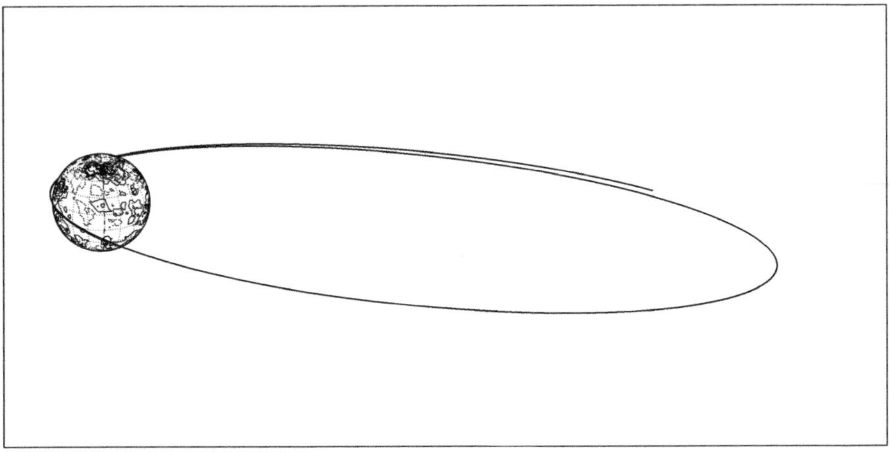

Projection: Orthographic
Property: none
T.:Azimuthal ♀ Graticule: 20°

Map centre: 25.0 ° N; 4.0 ° W
Aspect: Oblique
[-90.0 / +65.0 / +94.0] Gr.Mod.: MGNP60

Longitude / Initialisation:
A.n.: 0.00 ° - Apo.: 89.11 °
Magellan Topogr. /h/2km/

Iξιων
MC ★ LMD
Ατλας

Figure 11.4. Orbit of satellite in a frame moving with the planet. *Upper*: Magellan. *Lower*: Planet-C

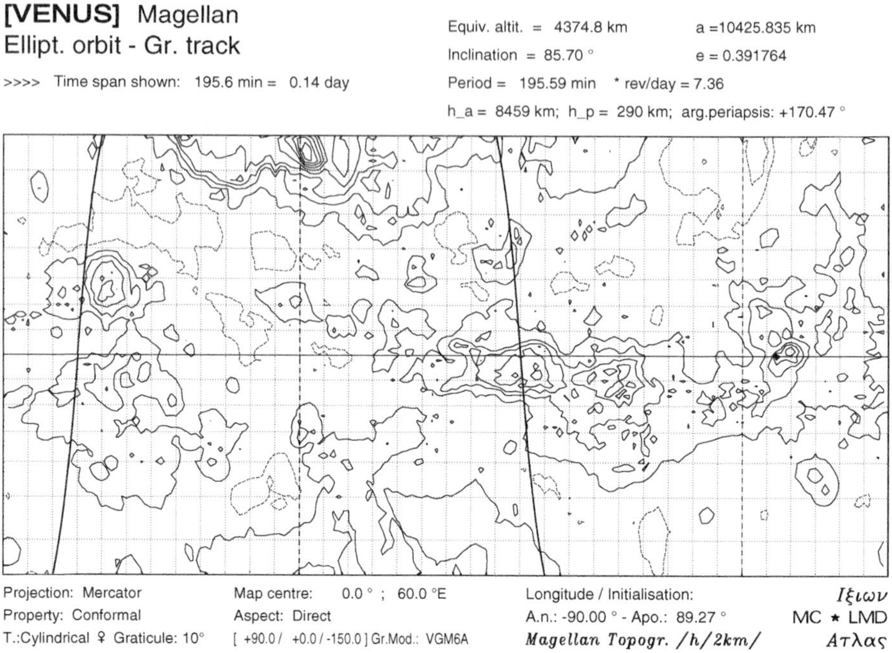

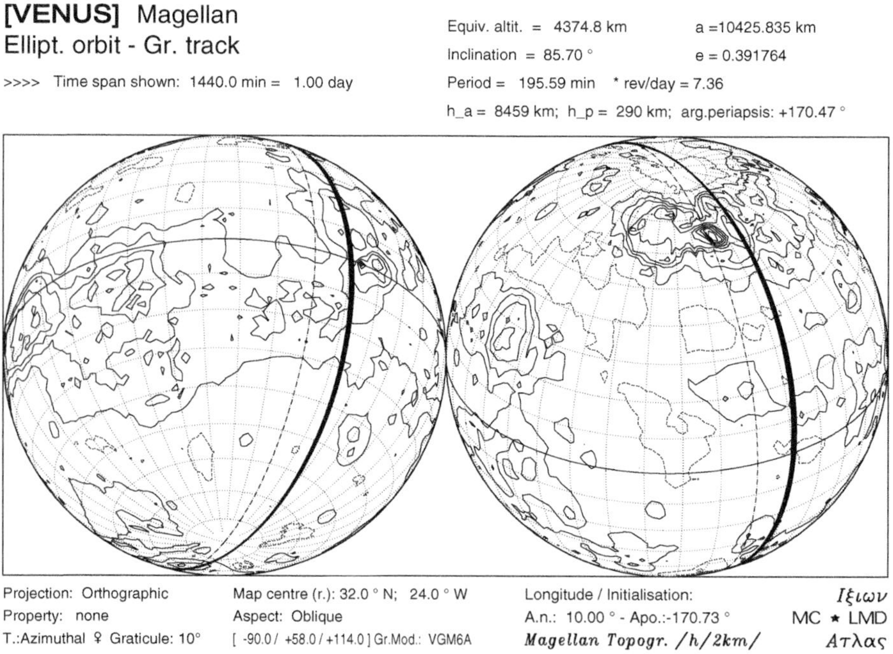

Figure 11.5. Orbital track of the satellite Magellan. *Upper*: over one revolution. *Lower*: over one Earth day

[433-EROS] NEAR/OCM-2
Orbit - Ground track

>>>> Time span shown: 1440.0 min = 1.00 day

Altitude = 196.1 km a = 204.500 km
Inclination = 37.00 °
Period = 14488.19 min * rev/day = 0.10

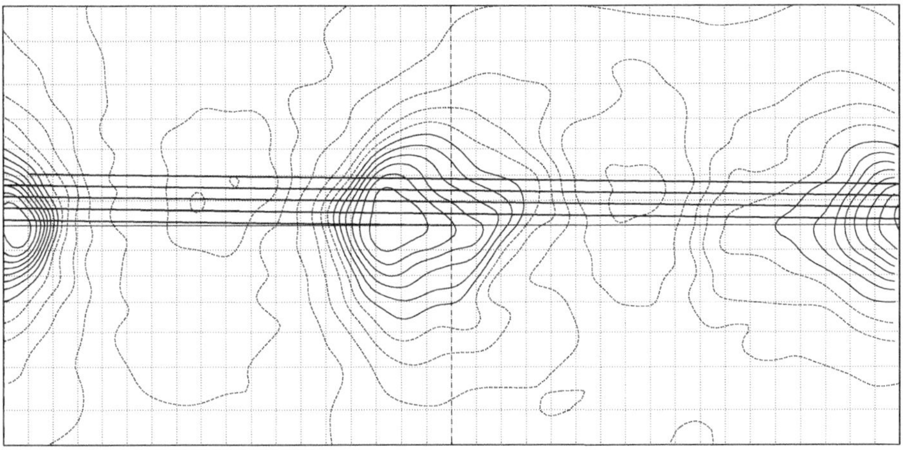

Projection: Mercator Map centre: 0.0 ° ; 0.0 ° Asc. node: 0.00 ° *Iξιων*
Property: Conformal Aspect: Direct App. inclin. = 179.23 ° MC ⋆ LMD
T.:Cylindrical ∗ Graticule: 10° [+0.0 / +0.0 / +0.0] Gr.Mod.: NLR190 ***NEAR Altim. /h/2km/*** *Ατλας*

[433-EROS] NEAR/OCM-6
Orbit - Ground track

>>>> Time span shown: 7.00 days

Altitude = 41.6 km a = 50.000 km
Inclination = 90.00 °
Period = 1761.08 min * rev/day = 0.82

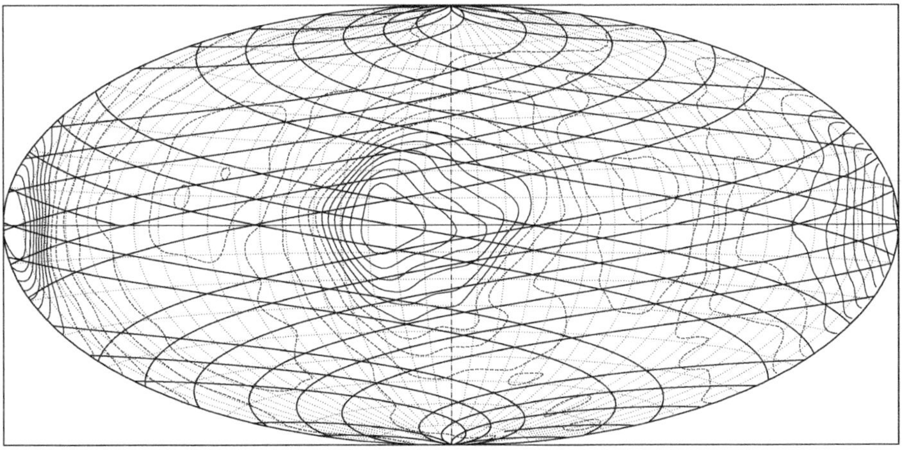

Projection: Hammer-Aitoff Map centre: 0.0 ° ; 0.0 ° Asc. node: -126.00 ° *Iξιων*
Property: Equal area Aspect: Direct App. inclin. = 169.82 ° MC ⋆ LMD
T.:Modif. Azim. ∗ Graticule: 10° [+0.0 / +0.0 / +0.0] Gr.Mod.: NLR190 ***NEAR Altim. /h/2km/*** *Ατλας*

Figure 11.6. Orbital track of the satellite NEAR. *Upper*: over one day. *Lower*: over 7 Earth days an almost perfect sphere ($J_2 = 4.4098 \times 10^{-6}$, a very low value) and the orbit of the satellite is near-polar, the precessional motion is very small. For the orbit defined above, we obtain:

$$\dot{\Omega} = -6 \times 10^{-4}\text{deg/day} \quad \text{and} \quad \dot{\omega} = -4 \times 10^{-3}\text{deg/day} .$$

Following the mapping cycles, the satellite was placed in a circular orbit by air-braking, with $h = 250$ km, to carry out geodetic studies. It was then sacrificed in a final experiment, known as the windmill experiment. For one and a half months, the solar panels were deployed to transform the satellite into a sort of windmill, transmitting back the parameters of the atmosphere which finally consumed it.

The gravitational potential model of degree and order 21, known as JPL-VGM1B (JPL Venus Gravity Model), obtained using Doppler radio tracking data from Pioneer Venus Orbiter, has evolved to degree and order 90 thanks to data gathered by Magellan, to give the model MGNP90 (Magellan plus PVO, 90th degree and order).[24]

The Japanese probe Planet-C (name before launch) should enter a highly eccentric, near-equatorial orbit around Venus. The representation of this orbit in Fig. 11.4 (lower) over slightly more than one revolution shows how little the trajectory shifts from one revolution to the next, in a frame fixed relative to the planet.

The European probe Venus Express, carrying the same type of instruments as Mars Express, is due for launch in November 2005. It will go into a highly eccentric polar orbit with $a = 39\,176$ km, $e = 0.839$ ($h_\text{p} = 250$ km, $h_\text{a} = 66\,000$ km, $T = 1\,425$ min), and periastron at latitude 70°N [see Fig. 11.3 (lower)]. This orbit, almost fixed relative to the planet (as happens for all Venusian satellites), will study the atmosphere, which for its part makes a round trip of the planet very 4 days.

Example 11.1. *Orbital track of the satellite Magellan.*

The ground track of the orbit over one revolution shown in Fig. 11.5 (upper) exhibits no obvious asymmetry, as is the case for eccentric terrestrial orbits. This is due to the fact that the planet is almost motionless (compared with the satellite motion, even at apoastron) in a Galilean frame. The very slow rotation of the planet about its axis gives a very small equatorial shift of about 0.20°, i.e., about 21 km between consecutive ground tracks, as can be seen from Fig. 11.5 (lower), for a time span of

[24] Planetocentric gravitational constant μ, in km^3s^{-2}:

$$\mu = 324\,858.60 \pm 0.05 \quad \text{for JPL-VGM1B (1990)} ,$$

$$\mu = 324\,858.601 \pm 0.014 \quad \text{for MGNP90 (1997)} .$$

Table 11.3. Geodetic and astronomical data for the asteroid 433-Eros. Results from the NEAR mission. Quantities are explained in the caption to Table 11.2. The dimensions of the asteroid are called D_i. The values of the J_2, J_3, and J_4 terms should be multiplied by 10^{-3}. Note the unusual units used here for g_0 and V_e, and the very high values of the terms J_n

$\mu = \mathcal{G}M$	[m^{3}s^{-2}]	4.463×10^5	a_S	[a.u.]	1.458
R	[km]	8.423	N_{sid}	[yr]	1.76
D_1	[km]	34.4	N_{sid}	[day]	643
D_2, D_3	[km]	11.2, 11.2	e	n.d.	0.233
g_0	[mm s^{-2}]	2.1–5.5	i	[deg]	10.8
V_e	[m s^{-1}]	3.1–17.2	J_{sid}	[hr]	5.27026
$T_{0(h=0)}$	[min]	121.18	η_{GS}	n.d.	1.895
d	n.d.	2.67	J_2	10^{-3}	117.4
ρ_Σ	[km]	308	J_3	10^{-3}	3.1
ρ_Σ/R	n.d.	37	J_4	10^{-3}	−37.5

one day, and from Colour Plate XIV, over a time span of six days. In the latter, the Mercator projection used for the map is centered on Maxwell Montes. This great mountain range in Ishtar Terra, the highest on Venus, is 11 km above the mean level of the planet.

11.4.3 Satellite of the Asteroid Eros

The first asteroid discovered was 1-Ceres, by G. Piazzi[25] on 1 January 1801. It is in fact the largest asteroid (R = 454 km). A great many others have since been observed. They are numbered in order of discovery, up to number 7 722 as of 1 January 2000 (see Fig. 3.5). With the sudden increase in automated search programmes, we had reached 253 370 asteroids (85 020 numbered, 168 350 provisional) and 44 moons of asteroids (3 named, 41 provisional) as of 1 September 2004.

On its way to Jupiter, the Galileo probe made the first asteroid flybys as it crossed the Main Belt. These included 951-Gaspra and 243-Ida, with $R \approx$ 20 km. A natural satellite of 243-Ida was even discovered on this occasion, named Dactyl (roughly spherical, with $R \approx 0.7$ km).

[25] Father Giuseppe Piazzi (1746–1826) was an Italian astronomer who made a very accurate catalogue of more than seven thousand stars. While he was doing so, he discovered the first asteroid, which he called Ceres. Having specified the position of the new planet, he was compelled to interrupt observations from his observatory in Palermo due to bad weather. When he resumed, he was unable to relocate his Ceres. It was the young Gauss who found it again, using methods he had just developed and orbital elements provided by Piazzi. In the next seven years, three other asteroids were discovered by H. Olbers. It was Gauss who showed, by determining the orbital elements, that they all belonged to the same belt.

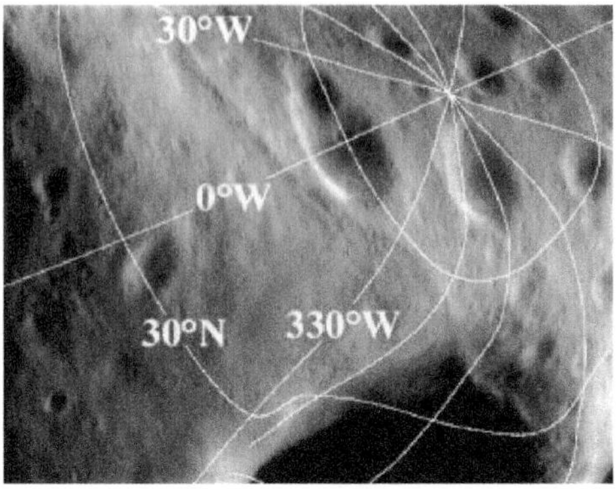

Figure 11.7. North polar region of Eros. Image taken by NEAR–Shoemaker on 31 March 2000 from an orbital altitude of 207 km. The image has been overlain with lines of latitude and longitude. Latitude is measured in degrees from the equator to the pole; longitude is measured in degrees west of a prime meridian. In both cases, the vertex of the angle being measured is the center of Eros. The wandering, curved shapes of the lines are caused by the highly nonspherical and irregular asteroid shape. Credit (legend and photography): NASA and John Hopkins University/Applied Physics Laboratory

The NEAR probe (Near Earth Asteroid Rendezvous), launched on 17 February 1996, observed 253-Mathilde on 27 June 1997 before flirting with Eros on St Valentine's day 2000. The asteroid 433-Eros, made of siliceous rock, has a rather cylindrical shape. From 14 February 2000, the probe, renamed NEAR–Shoemaker, went into orbit to become the first satellite of Eros. From an initially rather eccentric orbit (ellipse with semi-axes $a = 365$ km, $b = 204$ km, $i = 36°$), the satellite gradually moved closer to Eros by means of maneuvers which alternated elliptical and circular orbits. The near-circular ($a = 200$ km), near-polar orbit eventually became rather low ($a = 100$ km) and near-equatorial. On 28 January 2001, the satellite left its approach orbit ($a = 35$ km, $i = 180°$) and landed on the asteroid, transmitting images up until the final impact.

The characteristics of the asteroid are displayed in Table 11.3 (see also Fig. 11.7). The geodetic data come from the NLR190 (NEAR Laser Rangefinder) shape, gravity, and dynamics model. The J_2, J_3, and J_4 terms are given for comparison with other planets.[26]

The Japanese probe Hayabusa (falcon), known as Muses-C before its launch on 9 May 2003, is due to arrive at asteroid 25143-Itokawa in October 2005, from whence it will return a sample to Earth in June 2007.

[26] In fact, these terms refer to the spherical harmonics and are not well-suited to a body which is so far from spherical. In this case, one can decompose the gravitational potential using ellipsoidal harmonics, where the Legendre polynomials are replaced by the Lamé functions of the second kind.

The probe NEAP (Near Earth Asteroid Prospector), with orbiter and lander, should visit asteroid 4660-Nereus. The project was set up by a private company, SpaceDev, which hopes to bring in profits.

With its probe Dawn, NASA plans to visit first the brightest, then the largest of the asteroids, viz., 4-Vesta and 1-Ceres, respectively. Launch is programmed for 27 May 2006, with arrival at Vesta on 30 July 2010, departure from Vesta on 3 July 2111, arrival at Ceres on 20 August 2014, and the end of the Ceres mission on 26 July 2015. The Dawn spacecraft would be the first purely scientific mission to be powered by ion propulsion (xenon thrusters).

Example 11.2. *Orbital track of the satellite NEAR.*

We have represented the ground track of the orbit of the NEAR probe, satellite of 433-Eros. We have considered two types of orbit, known as OCM-2 and OCM-6. (Successive orbits of NEAR are numbered OCM-n, Orbital Correction Maneuver.) In the first case, shown in Fig. 11.6 (upper), the satellite is very high relative to the asteroid, since it takes 10 days (or 46 Eros days) to make one round trip. The ground track thus takes 2.5 days to move from the equator to the maximum latitude. In the second case, shown in Fig. 11.6 (lower), the satellite is on a much lower orbit, although it remains above the planetosynchronous orbit. In this second figure, we have used the cartographic projection chosen by the NEAR science team to represent orbital tracks, i.e., the Hammer–Aitoff projection.

B: Satellite of a Natural Satellite

11.5 Natural Satellites in the Solar System

In previous chapters, we have discussed the three natural satellites of the telluric planets. As far as the giant planets are concerned, the number of known natural satellites was as follows in 2000:

- Jupiter 16: 4 small and very close, then the 4 Galilean moons, then 8 small and distant.
- Saturn 18: 9 small, 8 larger ones including Titan, and the very distant Phoebe.

Since 2000, a great many very small natural satellites have been discovered in orbit around Jupiter and Saturn.

- Uranus 15: 11 small, then the 4 largest, already known in the nineteenth century.
- Neptune 8: 6 small, then Triton and Nereid further out.
- Pluto gravitates in a pair with Charon, which has 1/6 of its mass.

By a small natural satellite, we understand a celestial body with dimensions of the order of a few hundred kilometres or less. Most of these were discovered by Pioneer-11 or the Voyager probes, but the Cassini probe has added a great many others.

These natural satellites have one very important characteristic: they all[27] exhibit synchronous rotation, or 1:1 resonance, i.e., one rotation during one revolution around the planet. They are practically fixed relative to an axis passing through their centre of gravity and the centre of the planet. Like the Moon for the Earth, they always turn the same face towards their planet. This is due to the tidal effect.[28]

Apart from a small number of exceptions,[29] natural satellites have orbits with almost zero eccentricity, lying in the equatorial plane of the associated planet. There is one major exception: the Moon does not gravitate in the equatorial plane of the Earth.[30]

If we exclude the lunar conquest, discussed at the beginning of the chapter in the more general context of space exploration, there is no specific mission to the natural satellites. If we do organise a visit, it is always part of a trip to the associated planetary system as a whole. One exception has been programmed, and this is the mission to Europa, or better still, the JIMO project (Jupiter Icy Moons Orbiter), for 2012. (The spacecraft could orbit Europa for 2 months, then spend 4 months at Ganymede and another 4 months at Callisto.)

---

[27] The only exceptions are two satellites of Saturn, beyond Titan. The most distant, Phoebe, has a retrograde orbit and non-synchronous rotation. The other, Hyperion, trapped between the orbits of Titan and Japet, exhibits chaotic rotation.

[28] The planet exerts a tidal force on the natural satellite, much stronger than that exerted by the natural satellite on the planet. Viscous friction inside the natural satellite with its associated dissipation of energy ends up by slowing down the rotation of the natural satellite. When the rotation becomes synchronous, the natural satellite has been deformed into a shape extended in the direction of the planet. (This deformation of the Moon in the direction of the Earth is very slight, whilst that of Phobos towards Mars is enormous, relative to the size of this moon.) As soon as the natural satellite finds itself in this 1:1 resonance, it remains trapped in that situation by the restoring couple exerted by the planet.

[29] Satellites situated a long way from the attractive planet can have a highly eccentric orbit, like Nereid around Neptune ($e = 0.75$). These may well be former asteroids, captured by the gravitational attraction of the planet.

[30] The motion of the Moon is a very complex problem to analyse. It is a 3-body motion, involving the Moon, the Earth and the Sun ($\mu/\mu_N = 81.300\,59$). The Moon's eccentric orbit makes an angle of $i = 5.1454°$ with the ecliptic. The inclination of the lunar orbit to the equator of the Earth thus varies between $\varepsilon - i = 18.28°$ and $\varepsilon + i = 28.58°$.

11.6 Geodetic and Astronomical Quantities for Natural Satellites

Geodetic and Astronomical Data

We shall be concerned with four natural satellites: the Moon, because it is our own moon, and has been encountered or visited so many times; Europa, because one is led to presume the presence of liquid water; and Titan and Triton, because these satellites carry an atmosphere. Table 11.4 displays geodetic and astronomical data, together with deduced quantities.

Satellite in Keplerian Orbit

Provided its altitude is not too great (in a sense to be defined later), an orbiting satellite (semi-major axis a) around a natural satellite will feel only the attraction of this body. Let μ_{N} be the gravitational constant of this natural satellite, and keep μ for the gravitational constant of the corresponding planet. All the formulas derived for the Keplerian orbit can be applied, replacing μ by μ_{N}, as in (2.16).

The period of the satellite at altitude 0 is given by (2.17) or (11.3). For example, for the Moon, with $d = 3.34$, we obtain for this period $T_{0(h=0)} = 198/\sqrt{3.34} = 108$ min.

The values of $T_{0(h=0)}$ are given in Table 11.4a for various natural satellites. Figure 11.1 graphs the variation of $T_0/T_{0(h=0)}$ as a function of a/R, where R is of course the radius of the natural satellite.

Geographical Maps

Natural satellites of planets can be mapped. The mappable area in millions of km^2 takes the following values for each of the natural satellites considered here: 38 for the Moon, 31 for Europa, 83 for Titan, 23 for Triton. The total area of the natural satellites is 425, including 231 for Jupiter's four Galilean satellites. The geography and geology of the Moon have been the subject of very detailed study since Galileo's first observations of lunar mountains.

In this chapter, we shall use the following maps as background to represent the ground track or orbit:

- For the Moon, the topographical map based on laser altimeter data gathered by Clementine. Contours are plotted in 2 km steps, with the same convention for the curves as we used for planetary maps.
- For Europa, we shall not use a map, which would be difficult to read, but several images compiled by Galileo, showing the rather unusual structure of the ground.

In the case of natural satellites, the zero meridian is not chosen arbitrarily. The origin for longitudes is taken to be the meridian exactly at the centre of the face turned towards the planet.

11.7 Satellite of a Natural Satellite in Real Orbit

11.7.1 Perturbative Accelerations

For a satellite in orbit around a natural satellite, the sphere of influence identifies the region in which the acceleration due to the mother planet is negligible compared to the central acceleration. To evaluate this, we go back to the formulas in Sect. 3.13, replacing μ_S by μ and μ by μ_N, since for the satellite, the planet/Sun system is now replaced by the natural satellite/planet system. Table 11.4 shows the ratios μ/μ_N and the results of the calculation of ρ_Σ. We note that, for Europa, ρ_Σ is very small because its mass is 40 000 times less than the mass of Jupiter.

The variation of the central acceleration and perturbing accelerations with the altitude of the satellite is shown in Fig. 11.8 (upper) for a satellite around the Moon, and in Fig. 11.8 (lower) for a satellite around Europa. The notation for the accelerations is adapted from Table 3.3.

For the central acceleration, (3.22) gives $\gamma_{\mathrm{CCC}}(R) = g(R) = g_0$, which gives 1.62 m s^{-2} for the Moon and 1.31 m s^{-2} for Europa. The main difference between the cases examined up to now, of satellites around planets, is clearly the presence (and importance) of the term γ_{CC1}, the perturbing acceleration due to the central planet.

For a satellite of the Moon close to ground level, $\gamma_{\mathrm{CCN}.J2}$ is greater than γ_{CC1}:

$$\gamma_{\mathrm{CCN}.J2}(R) = 32.8 \times 10^{-5}\ \mathrm{m\,s^{-2}}\,, \quad \gamma_{\mathrm{CC1}}(R) = 2.5 \times 10^{-5}\ \mathrm{m\,s^{-2}}\,.$$

However, above $h \sim 1\,000$ km, γ_{CC1} soon exceeds $\gamma_{\mathrm{CCN}.J2}$.

For a satellite of Europa, γ_{CC1} is always greater than $\gamma_{\mathrm{CCN}.J2}$:

$$\gamma_{\mathrm{CCN}.J2}(R) = 0.8 \times 10^{-3}\ \mathrm{m\,s^{-2}}\,, \quad \gamma_{\mathrm{CC1}}(R) = 1.3 \times 10^{-3}\ \mathrm{m\,s^{-2}}\,.$$

Moreover, this term γ_{CC1} increases with altitude (slope $p = 1$ on a log–log scale) and when the satellite is at an altitude of about 10 000 km, this acceleration, due to Jupiter, is greater than the central acceleration γ_{CCC} due to Europa. In this case, γ_{CC1} can no longer be treated as a perturbation, just as the satellite is no longer a satellite of Europa!

11.7.2 Classification of Satellites

Motion of Natural Satellites

The four natural satellites studied here, like the others, all have synchronous rotation, i.e.,

$$J_{\mathrm{sid}} = N_{\mathrm{sid}}\,,$$

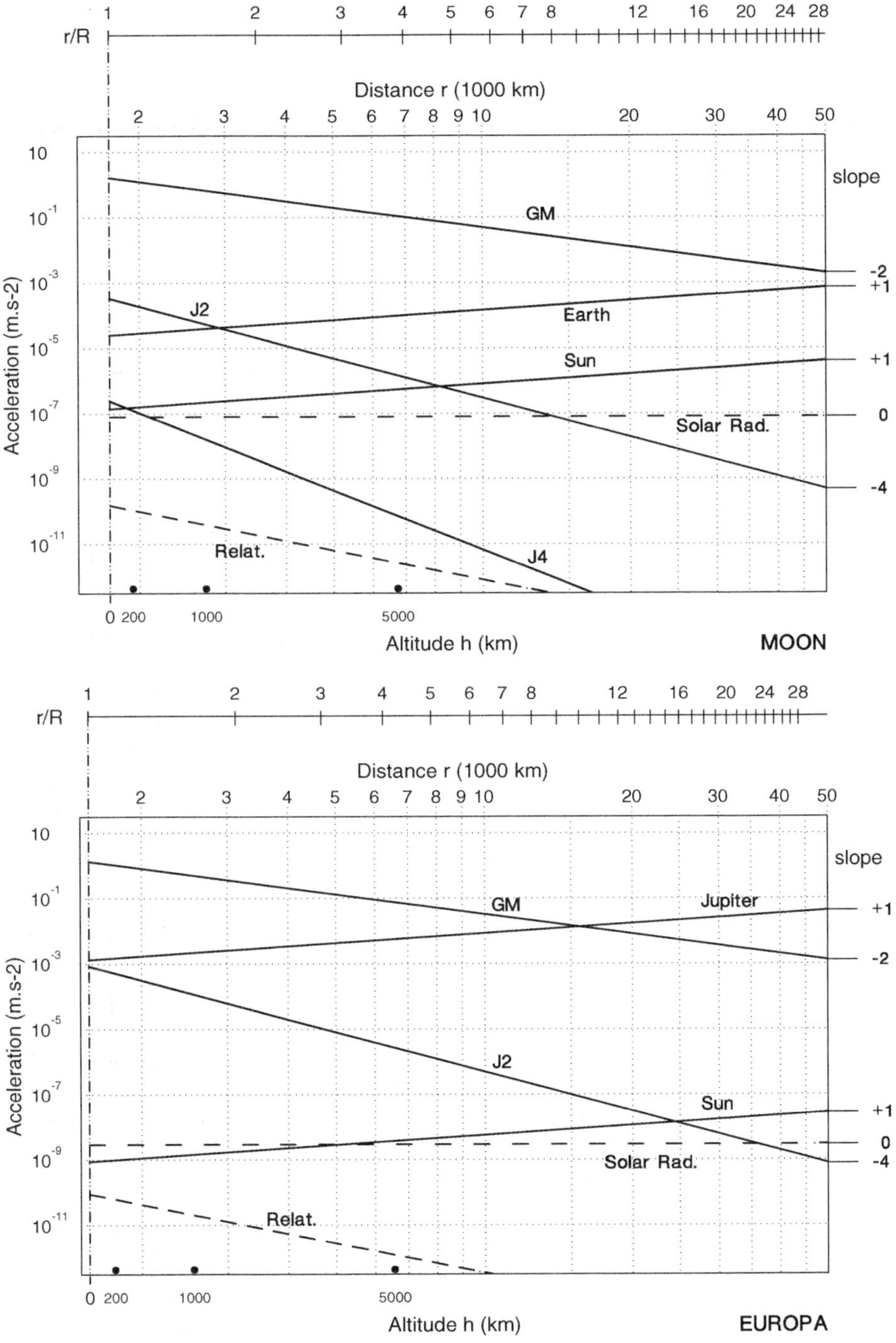

Figure 11.8. Accelerations as a function of the distance r of the satellite from the centre of the natural satellite. Log–log scale. *Upper*: Moon. *Lower*: Europa

Table 11.4. Four natural satellites of planets in the Solar System. Geodetic and astronomical quantities. Data and deduced quantities are the same as those appearing in Table 11.2. Quantities specific to this table: gravitational constant of the natural satellite $\mu_{\rm N}$, semi-major axis $a_{\rm P}$ (planet–natural satellite). The inclination i is taken relative to the plane of the ecliptic (ECL) or the equatorial plane of the planet (EQU). The rotation of the natural satellite is synchronous: $J_{\rm sid} = N_{\rm sid}$. For Triton, retrograde revolution (minus sign)

Satellite	$\mu_{\rm N} = \mathcal{G}M_{\rm N}$ [m^3 s^{-2}]	R [km]	g_0 [m s^{-2}]	$V_{\rm e}$ [km s^{-1}]	$T_{0(h=0)}$ [min]	d n.d.
Moon	4.9028×10^{12}	1737.4	1.62	2.38	108.31	3.34
Europa	3.2014×10^{12}	1561.5	1.31	2.02	114.23	3.01
Titan	8.9782×10^{12}	2575.0	1.35	2.64	144.41	1.88
Triton	1.4279×10^{12}	1352.6	0.78	1.45	137.86	2.06

Satellite	(Satellite) planet/number	$a_{\rm P}$ [km]	$\mu/\mu_{\rm N}$ n.d.	ρ_Σ [km]	ρ_Σ/R n.d.
Moon	Earth I	383 398	81.3	57 433	33.1
Europa	Jupiter II	670 090	39 572.6	8 462	5.4
Titan	Saturn VI	1 221 803	4 225.9	37 709	14.6
Triton	Neptune I	354 759	4 787.8	10 415	7.7

Satellite	$N_{\rm sid} = J_{\rm sid}$ [day]	e n.d.	i [deg]	J_2 10^{-6}
Moon	27.321 662	0.0555	5.15 ECL	203
Europa	3.551 181	0.0090	0.47 EQU	629
Titan	15.945 446	0.0291	0.30 EQU	–
Triton	−5.878 850	0.0000	156.83 EQU	–

where the first term is the sidereal period of rotation of the natural satellite about its own axis and the second is the sidereal period of revolution of the natural satellite about the central planet. For the natural satellites considered here, this period ranges from 27 days for the Moon to less than 4 days for Europa. It should be noted that the revolution (and hence the rotation) of Triton occur in the retrograde direction around Neptune.

Stationary Satellite

We shall study the possibility of placing a satellite in synchronous orbit, and more specifically, in stationary orbit, around a natural satellite. Consider a natural satellite N in orbit (semi-major axis $a_{\rm P}$) around a planet P. If the planet has gravitational constant μ, the natural satellite follows the Keplerian orbit with mean motion $n_{\rm N}$ given by

$$n_{\mathrm{N}} = \sqrt{\frac{\mu}{a_{\mathrm{P}}^3}} \ .$$

Its angular speed of rotation $\dot{\Omega}_{\mathrm{T}}$ about its own axis, relative to a Galilean frame, is thus

$$\dot{\Omega}_{\mathrm{T}} = n_{\mathrm{N}} \ ,$$

since the rotation is synchronous.

Consider a satellite in orbit (semi-major axis a) around a natural satellite N. If the natural satellite has gravitational constant μ_{N}, the satellite follows the Keplerian orbit with mean motion n_0 given by

$$n_0 = \sqrt{\frac{\mu_{\mathrm{N}}}{a^3}} \ .$$

If the motion of the satellite is to be synchronous with that of the natural satellite N, the attractive body, we must satisfy the condition

$$n_0 = \dot{\Omega}_{\mathrm{T}} \ ,$$

and this implies

$$a_{\mathrm{GS}} = a = \left(\frac{\mu_{\mathrm{N}}}{\mu}\right)^{1/3} a_{\mathrm{P}} \ , \tag{11.5}$$

where a_{GS} is the semi-major axis of the stationary orbit.

We now compare this value with ρ_{Σ}, the radius of the sphere of influence. Adapting (3.103) to the present situation, we obtain

$$\rho_{\Sigma} = 2^{-1/5} \left(\frac{\mu_{\mathrm{N}}}{\mu}\right)^{2/5} a_{\mathrm{P}} \ . \tag{11.6}$$

The satellite must stay within the sphere of influence, i.e., it must satisfy the inequality

$$a_{\mathrm{GS}} < \rho_{\Sigma} \ . \tag{11.7}$$

Hence, with the values obtained from (11.5) and (11.6),

$$\left(\frac{\mu_{\mathrm{N}}}{\mu}\right)^{5/3} < \frac{1}{2}\left(\frac{\mu_{\mathrm{N}}}{\mu}\right)^{2} \ .$$

Finally,

$$\mu_{\mathrm{N}} > 8\mu \ . \tag{11.8}$$

This condition is absurd. A natural satellite cannot have greater mass than the central planet. It is thus impossible to obtain a stationary orbit for a satellite of a natural satellite, at least in the sense that we have defined such an orbit here.

If a satellite is stationary with respect to a natural satellite, it is also stationary with respect to the planet, because of the synchronous rotation of the natural satellite. This happens when it occupies one of the five Lagrange points. It then remains fixed relative to the natural satellite–planet system. Only L_4 and L_5 are stable. When the satellite is located at L_4 (or L_5), it forms an equilateral triangle with the planet and the natural satellite. In the case of the Moon and Earth, the satellite is thus located some 380 000 km from the natural satellite. Needless to say, this is not much use for an observation satellite. The mission STARS, since abandoned in this form, was envisaged for the L_5 point of the Earth–Moon system, but the aim was not to observe the Moon.

Sun-Synchronous Satellite

We now study the case of a satellite in Sun-synchronous orbit around a natural satellite. We can calculate the constant of Sun-synchronicity using (4.63). We can also find a relation between k_{hN} and k_{hP}, the constants of Sun-synchronicity for satellites in orbit around a natural satellite N and around the central planet P, respectively. With the corresponding subscripts, we can write

$$k_{\mathrm{hN}} = \frac{3}{2} \frac{T_{\mathrm{sid}}}{T_{0(h=0)\mathrm{N}}} J_{2\mathrm{N}} \,, \qquad k_{\mathrm{hP}} = \frac{3}{2} \frac{T_{\mathrm{sid}}}{T_{0(h=0)\mathrm{P}}} J_{2\mathrm{P}} \,.$$

It is important to note that the sidereal period of revolution T_{sid} is the same in both cases: the natural satellite N takes the same time as the planet P to accomplish one revolution around the Sun. We obtain

$$\frac{k_{\mathrm{hN}}}{k_{\mathrm{hP}}} = \frac{J_{2\mathrm{N}}}{J_{2\mathrm{P}}} \frac{T_{0(h=0)\mathrm{P}}}{T_{0(h=0)\mathrm{N}}} \,. \tag{11.9}$$

Using (11.2) or (11.3) with the mean densities, we then have

$$\frac{k_{\mathrm{hN}}}{k_{\mathrm{hP}}} = \sqrt{\frac{d_{\mathrm{N}}}{d_{\mathrm{P}}}} \frac{J_{2\mathrm{N}}}{J_{2\mathrm{P}}} \,. \tag{11.10}$$

Let us return now to the four natural satellites studied here. For Titan and Triton, we are unable to investigate the question of Sun-synchronicity, because we do not have sufficient knowledge of the J_2 term (and it would not be particularly useful in these cases). For the Moon and Europa, the reader is referred to the two parts of Fig. 11.8 which indicate the various accelerations.

For the Moon, the calculation of the constant of Sun-synchronicity $k_{\mathrm{h}} = k_{\mathrm{hN}}$ using (11.10) yields

$$k_{\mathrm{h}} = 1.4725 \,.$$

This in turn implies a minimal inclination of

$$i_{\mathrm{HS\,min}} = 133^\circ \; .$$

For a satellite in low orbit around the Moon (LLO, Lunar Low Orbiting), the perturbing acceleration γ_{CC1} due to the Earth is less than the perturbing acceleration $\gamma_{\mathrm{CCN}.J2}$ due to flattening. If the LLO satellite is in Sun-synchronous orbit, the effect of the Earth, which is one tenth the effect due to the J_2 term of the Moon, would soon remove the satellite from this Sun-synchronous orbit, in a matter of a few days.

For Europa, the situation is more radical. The term γ_{CC1} is always greater than $\gamma_{\mathrm{CCN}.J2}$, even at zero altitude. The perturbation due to Jupiter's gravity is greater than the one due to the flattening of Europa, whatever the altitude of the satellite. There is therefore no hope of obtaining a Sun-synchronous orbit for a satellite around Europa.

Frozen Orbit

Owing to a lack of accurate enough geodetic data concerning the natural satellites, we can only investigate frozen orbits for satellites around the Moon. In the case of our own natural satellite, the spherical harmonic coefficients of the gravitational potential are well documented. Here are the first few values of $J_n \times 10^6$ in the LPLGM model:

$$J_2 = 203.236\,626$$

$$J_3 = +\,8.475\,906 \qquad J_4 = -\,9.591\,929 \qquad J_5 = +\,0.715\,409$$

$$J_6 = -13.577\,715 \qquad J_7 = -21.774\,733 \qquad J_8 = -\,9.674\,866$$

For a satellite in low near-polar orbit (with frozen perigee $\omega_{\mathrm{F}} = 270^\circ$ since $J_3 > 0$), the frozen eccentricity e_{F} can be approximately calculated using (7.41) to give $e_{\mathrm{F}} \approx 0.02$, which is relatively high for a frozen eccentricity. For other inclinations, e_{F} can take values between 0.01 and 0.001. These calculations are complicated by the presence of the J_7 term, whose value is rather large here.

11.8 Ground Track of a Satellite of a Natural Satellite

11.8.1 Satellite of the Moon

After the conquest of the Moon (1959–1972), discussed at the beginning of the chapter in the more general context of space exploration, further probes and lunar orbiters were sporadic. The probe DSPSE (Deep Space Probe Science Experiment), or Clementine, was launched on 25 January 1994 and flew in lunar orbit for 70 days. It drew up a very accurate topographic map of the Moon, but then failed in its planned encounter with the asteroid

Table 11.5. Measured values of the selenocentric gravitational constant $\mu = \mathcal{G}M$ with estimated error. Historical evolution mentioning the method used and the year

Method	Year	μ [km^{3}s^{-2}]	Error
Laser, LO-4	1980	4 902.799	±0.003
GLGM-1	1993	4 902.802 6	±0.000 1
GLGM-2	1997	4 902.802 9	±0.000 2
LPLGM	1999	4 902.801 06	±0.000 08

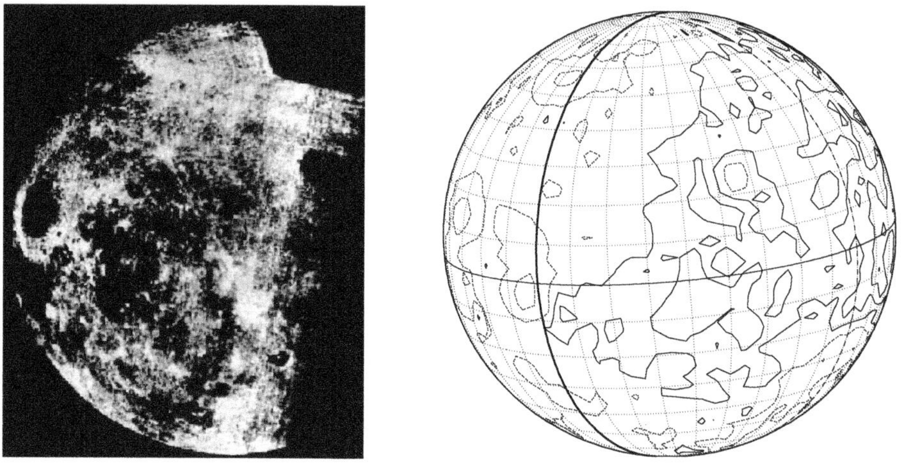

Figure 11.9. *Left*: first image of the hidden face of the Moon. Historic photograph, taken by Luna-3 on 7 October 1959, during a flyby at altitude 66 000 km. Credit: USSR Academy of Sciences. *Right*: Ixion/Atlas reconstitution of the Moon seen by Luna-3 when the image was taken. The meridian (*continuous curve*) demarcates the visible face (to the west, on the left) from the hidden (to the east, on the right)

1620-Geographos. The probe Lunar Prospector, launched on 7 January 1998, went into a near-circular, near-polar orbit (h = 100 km, then h = 40 and h = 30 km). On 31 July 1999, it impacted the Moon near the south pole in a controlled crash to look for evidence of water ice, but none was found.

Lunar gravitational potential models first used laser ranging measurements (LLR, Lunar Laser Ranging) carried out by means of reflectors set up on the Moon, then the satellites Lunar Orbiter-1 to -5, Apollo-15 and -16, Clementine, for the models known as GLGM-1 and 2 (Goddard Lunar Gravity Model). The model known as LPLGM (Lunar Prospector Lunar Gravity Model) also used Lunar Prospector (see Table 11.5).

The Japanese project Selene (Selenological and Engineering Explorer), called Lunar-A before launch, should place an orbiter in circular orbit ($h = 100$ km) for one year.

In the following, we represent several revolutions of the ground track of Clementine and, in memory of the glorious lunar exploration, we give a thought for Luna-3 and Apollo-15. On all maps of the Moon, we show the meridians 90°E and 90°W by a thick continuous line. These symbolically demarcate the visible and hidden faces of the Moon.

Example 11.3. *Discovery of the hidden face of the Moon.*

The Soviet probe Luna-3 was launched on 4 October 1959 and overflew the hidden face of the Moon on 7 October. Unlike its successors, in heliocentric orbit, Luna-3 was in fact a satellite of the Earth on a highly eccentric orbit, with radius at apogee $r_a = 469\,000$ km and period 16.2 days. (However, it also appears under the entry 'space probe' in the index.).

Indeed, it burnt up in the Earth atmosphere in April 1960. The historic photograph of the hidden face is shown on the left in Fig. 11.9. Next to it is a representation of the Moon viewed under the same conditions from a distance of 38 lunar radii. On the left of the photograph and the map is a part of the visible face featuring the dark region of the Mare Crisium (centered on 17.0°N, 59.1°E). The hidden face, incompletely photographed by Luna-3, was soon to be better revealed by Zond-3, and then fully mapped by the US orbiters (Orbiter-3, -4, Explorer-35, Orbiter-5), as they prepared in 1967 the landing areas for the Apollo programme. It was thus discovered that the two faces looked rather different, as can be clearly seen from the map shown in Fig. 11.11 (lower). This difference arises because the lunar crust is thicker on the hidden face than on the visible face, surely a consequence of the tidal effect. The wide basins on the hidden face are not filled with basalts from ancient lava flows, as they are on the visible face. These features have been known as 'seas' since ancient times.

Example 11.4. *Ground track of the satellite Clementine over two days.*

The probe Clementine, in its lunar mapping mission, followed the highly eccentric polar orbit shown in Fig. 11.12 (upper), with pericenter (periselene) at latitude 28°S during the first month and 29°N during the second. Indeed, each measurement cycle lasted one month, the time required by the satellite to observe the whole of the Moon, since this is in fact the time it required to rotate about its own axis in the Galilean frame. The ground track of the orbit is shown in the two parts of Fig. 11.10, over two days, i.e., 9.5 revolutions ($T = 5$ hr). These revolutions correspond to revolutions 103 to 112, on 13 and 14 March 1994 (cycle 1).

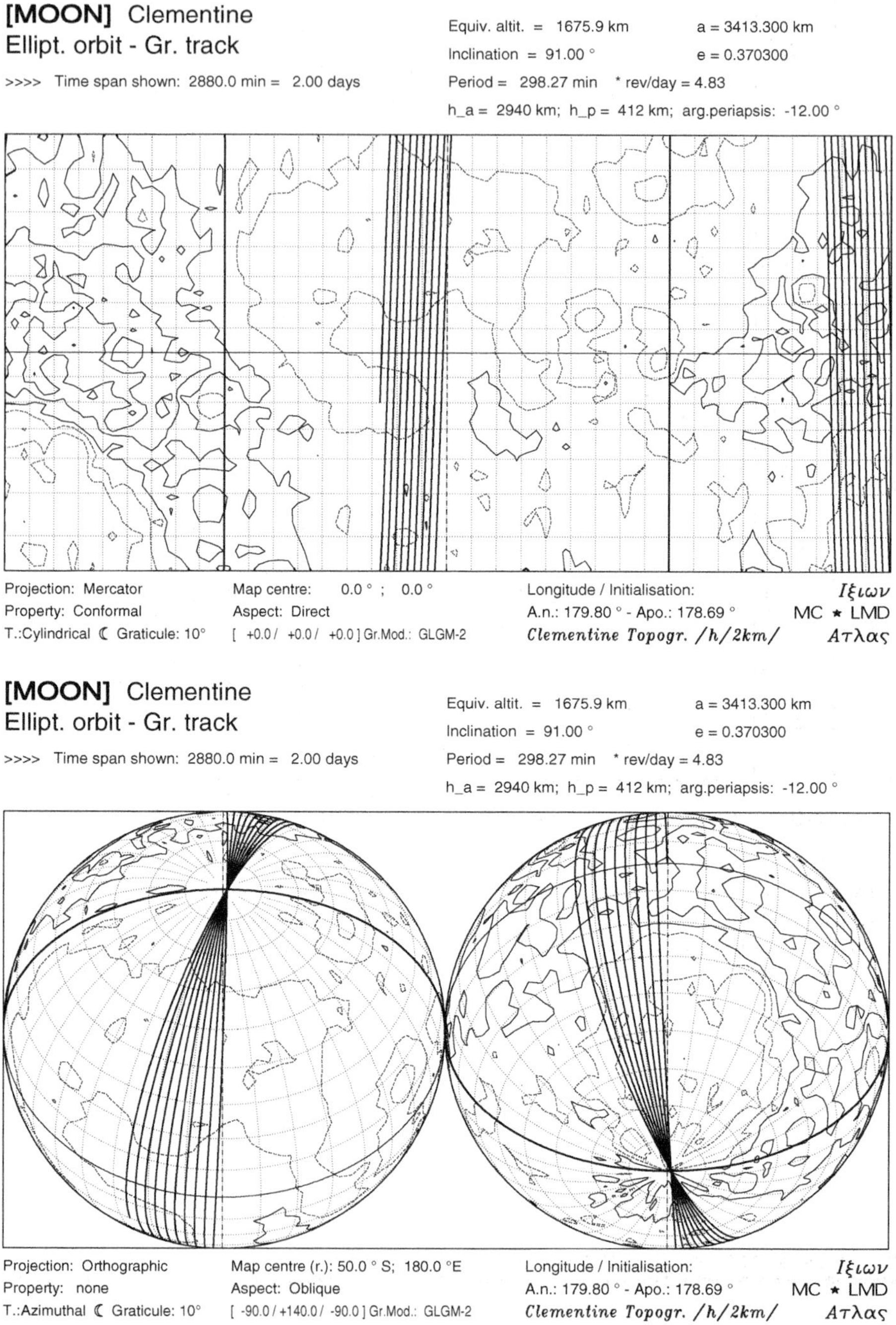

Figure 11.10. Ground track of the satellite Clementine over two days

[MOON] Apollo-15 (Orbiter)
Orbit - ref.: Moon

>>>> Time span shown: 1440.0 min = 1.00 day

Altitude = 113.3 km a = 1850.700 km
Inclination = 154.00 °
Period = 119.00 min * rev/day =12.10
Equat. orbital shift = 30.4 km (1.0 °)

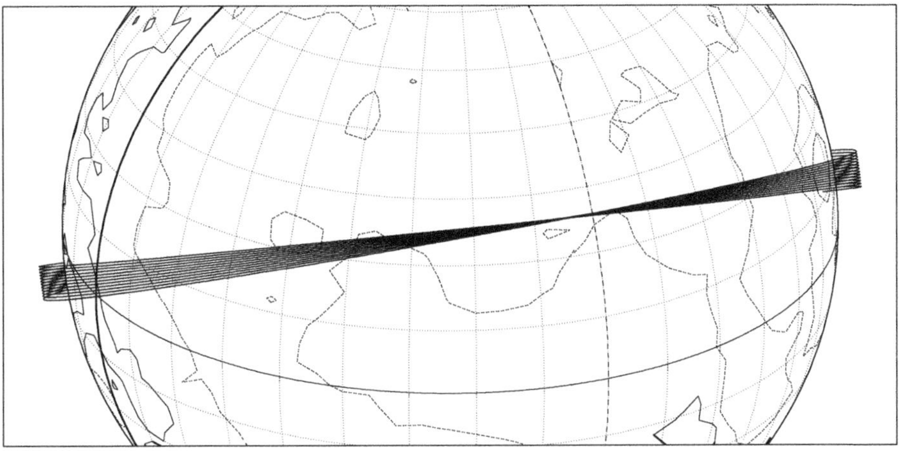

Projection: Orthographic | Map centre: 26.0 ° N; 24.0 ° W | Asc. node: 90.50 ° | Iξιων
Property: none | Aspect: Oblique | MC ★ LMD
T.:Azimuthal ☾ Graticule: 10° | [-90.0 / +64.0 / +114.0] Gr.Mod.: GLGM-2 | *Clementine Topogr. /h/2km/* | Ατλας

[MOON] Apollo-15 (Orbiter)
Orbit - Ground track

>>>> Time span shown: 5443.2 min = 3.78 days

Altitude = 113.3 km a = 1850.700 km
Inclination = 154.00 °
Period = 119.00 min * rev/day =12.10
Equat. orbital shift = 30.4 km (1.0 °)

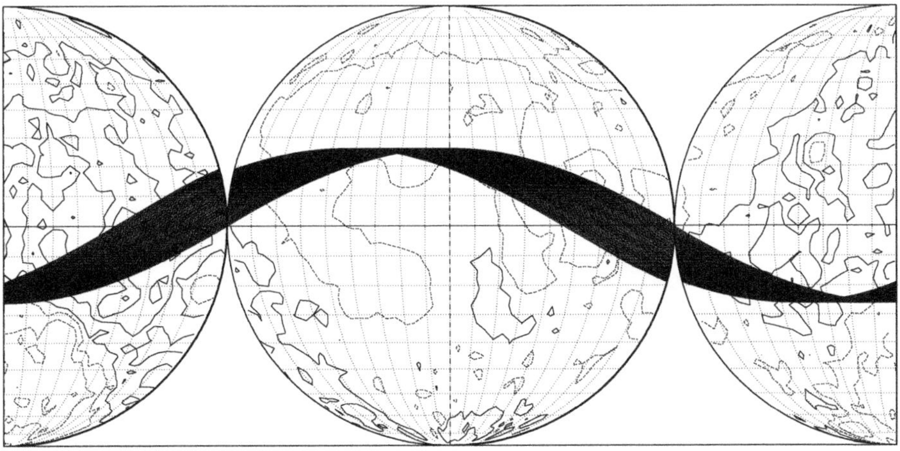

Projection: Mollweide | Map centre: 0.0 ° ; 0.0 ° | Asc. node: 90.50 ° | Iξιων
Property: Equal area | Aspect: Direct [interrupted] | App. inclin. = 154.07 ° | MC ★ LMD
T.:Pseudocyl. ☾ Graticule: 10° | [+0.0 / +0.0 / +0.0] Gr.Mod.: GLGM-2 | *Clementine Topogr. /h/2km/* | Ατλας

Figure 11.11. *Upper*: Orbit of the lunar satellite Apollo-15 (Orbiter) over one day. *Lower*: Ground track over four days (geochemical mapping mission)

[MOON] Clementine
Orbit - ref.: Moon

>>>> Time span shown: 7.00 days

Equiv. altit. = 1675.9 km   a = 3413.300 km
Inclination = 91.00 °   e = 0.370300
Period = 298.27 min   * rev/day = 4.83
h_a = 2940 km; h_p = 412 km; arg.periapsis: -12.00 °

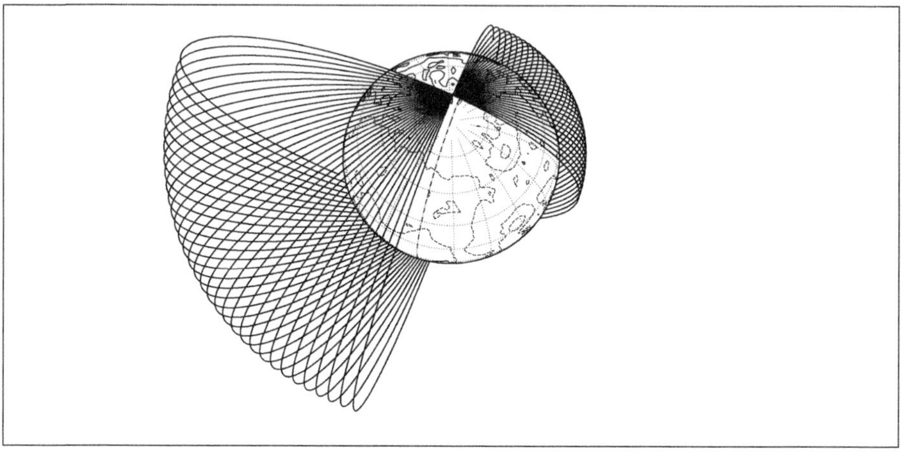

Projection: Orthographic
Property: none
T.:Azimuthal ☾ Graticule: 20°

Map centre: 70.0 ° N; 18.0 °E
Aspect: Oblique
[-90.0 / +20.0 / +72.0] Gr.Mod.: GLGM-2

Longitude / Initialisation:
A.n.: 179.80 ° - Apo.: 178.69 °
Clementine Topogr. /h/2km/

Ιξιων
MC ★ LMD
Ατλας

[EUROPA] Europa Orbiter
Orbit - Ground track

>>>> Time span shown: 720.0 min = 0.50 day
Across track swath

Altitude = 200.0 km   a = 1760.700 km
Inclination = 85.00 °
Period = 136.80 min   * rev/day =10.53
Equat. orbital shift = 263.0 km (9.7 °)
** Half-swath: 35.0° => 144 km [1.0 min]

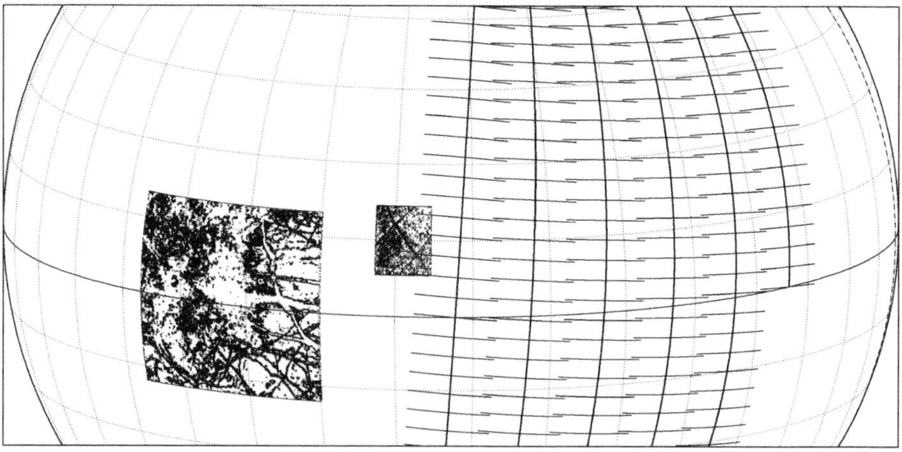

Projection: Orthographic
Property: none
T.:Azimuthal ☾ Graticule: 10°

Map centre: 12.0 ° N; 83.0 ° W
Aspect: Oblique **> zoom : 2.00**
[-90.0 / +78.0 / +173.0] Gr.Mod.: IAU91

Asc. node: -34.00 °
Latit. overlap: 89.7° <--> 90.0°
Galileo Images

Ιξιων
MC ★ LMD
Ατλας

Figure 11.12. *Upper*: Orbit of lunar satellite Clementine over seven days (quarter of a month). *Lower*: Ground track of the Europa Orbiter with swath

[EUROPA] Europa Orbiter
Orbit - Ground track

>>>> Time span shown: 1440.0 min = 1.00 day

Altitude = 200.0 km   a = 1760.700 km
Inclination = 85.00 °
Period = 136.80 min * rev/day =10.53
Equat. orbital shift = 263.0 km (9.7 °)

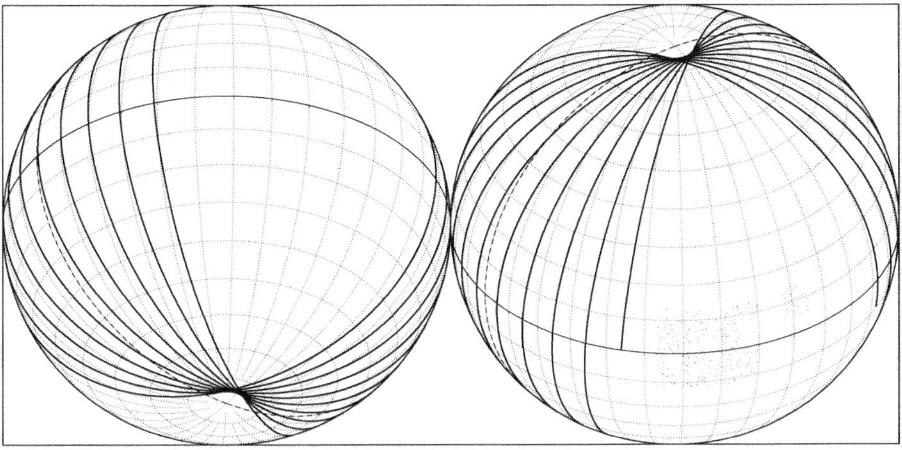

Projection: Orthographic   Map centre (r.): 36.0 ° N; 122.0 ° W   Asc. node: -136.00 °   Ιξιων
Property: none   Aspect: Oblique   MC ⋆ LMD
T.:Azimuthal ☾ Graticule: 10°   [-90.0 / +54.0 / -148.0] Gr.Mod.: IAU91   *Galileo Images*   Ατλας

[TRITON] Triton Orbiter
Orbit - Ground track

>>>> Time span shown: 1440.0 min = 1.00 day

Altitude = 176.6 km   a = 1529.200 km
Inclination = 85.00 °
Period = 165.77 min * rev/day = 8.69
Equat. orbital shift = -166.3 km (-7.0 °)

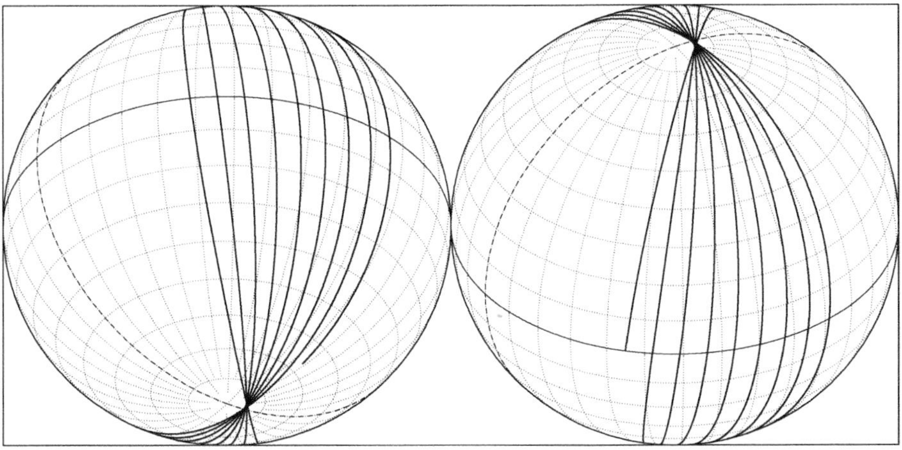

Projection: Orthographic   Map centre (r.): 36.0 ° N; 122.0 ° W   Asc. node: -135.00 °   Ιξιων
Property: none   Aspect: Oblique   MC ⋆ LMD
T.:Azimuthal ☾ Graticule: 10°   [-90.0 / +54.0 / -148.0] Gr.Mod.: IAU91   Ατλας

Figure 11.13. Ground track of a satellite around Europa (*top*) and around Triton (*bottom*)

Note. The very accurate topography established by Clementine revealed a gigantic impact basin called the South Pole–Aitken basin, circular in shape, with diameter 2 500 km. (One diameter extends from the Aitken crater at 16.8°S, 173.4°E, to the South Pole.) It is centered on the point 50°S, 180°E. The basin appears very clearly on the right in Fig. 11.10 (lower), with an orthographic projection centered on the centre of the basin.

Example 11.5. *Ground track of the lunar orbiter Apollo-15 during its geochemical mapping mission.*

In the Apollo human exploration missions of the lunar floor, the capsule remained in lunar orbit with an astronaut aboard [see Fig. 11.11 (upper)]. The two other astronauts left the capsule in the lunar module (LM) and softlanded on the Moon. After a stay lasting one or two days, or six in the case of Apollo-16, they regained the capsule which then left the lunar orbit to return to Earth. During the Apollo-15 mission, the command module carried out a geochemical mapping experiment from its lunar orbit. In fact it measured gamma radiation from the surface, resulting from the natural radioactivity of the crust. The ground track of the orbit during the experiment is shown for a time span of 4 days, between 1 and 4 August 1971, in Fig. 11.11 (lower). The landing site of this mission was located at 26.10°N, 3.65°E, at the limit of the maximum attained latitude. It is clear that, with a lunar orbit, the area scanned over four days not very great.

Note on Map. The Moon is shown in Fig. 11.11 (lower) with an interrupted Mollweide projection, in which the central disk corresponds to the visible face. The altitudes there are lower than on the hidden side.

11.8.2 Satellite of Europa

Jupiter's four large natural satellites, the so-called Galilean moons, each have between half and twice the mass of the Moon. Moving away from the planet, these moons are Io, Europa, Ganymede and Callisto, named after four of Zeus' lovers and companions (spot the boy), whose motions are related by several resonances.[31] Images sent back by Voyager-1 and -2, later refined by those from Galileo, have provided us with a good level of knowledge about these moons.

Europa is covered with water ice (temperature 110 K at the equator and 50 K at the poles). It appears to have an internal heat source, due to tidal effects, and this implies that there may be an ocean of liquid water beneath the frozen surface. Hence the enthusiasm of planetary scientists for further

[31] The mean motions n_i of the first three Galilean moons, $i = 1$ to 3, are related (this is the Laplace resonance):

$$n_1 - 2n_2 = n_2 - 2n_3 = 0.7396°/\text{day} .$$

exploration! The Europa Orbiter mission will determine the thickness of the icy crust. Following complex maneuvers which will take several months, the probe should transit from an orbit around Jupiter to an orbit around Europa which will eventually be circular, low altitude and near-polar.

Example 11.6. *Orbital track of the Europa Orbiter.*

A low-orbiting satellite completes about 10 round trips per (terrestrial) day with an equatorial shift of the order of 10°. In the case of the Europa Orbiter, the orbital characteristics for the scientific mission will be $h = 200$ km, $i = 85°$. Figure 11.12 (lower) shows the ground track of this orbit together with the swath ($f = 35°$) over half a day, whilst Fig. 11.13 (upper) shows the ground track for one day.

11.8.3 Satellite of Titan

The natural satellite Titan is in equatorial orbit around Saturn, with a relatively high eccentricity. It was discovered by Huygens in 1655. In terms of size and mass, it is the second natural satellite of the Solar System, just behind Ganymede. The atmosphere of Titan is 4.5 times as dense as the Earth's atmosphere (at ground level, 1.5 bar with absolute temperature only one third of that on Earth). Photochemical reactions abound due to solar UV radiation, leading to the synthesis of an aerosol layer which masks the surface in the visible. Images taken by Voyager-1 do not show the ground, only the cloud cover. As for Venus, the atmosphere is in superrotation.

The Huygens module of the Cassini probe should make a parachute descent lasting several hours in January 2005. There will not be a satellite around Titan for some time yet.

11.8.4 Satellite of Triton

The natural satellite Triton is in circular equatorial orbit around Neptune, with eccentricity $e = 1.57 \times 10^{-5}$. It was discovered by W. Lassel on 10 October 1846, just two weeks after the discovery of Neptune. As we have already said, the rotation of Triton is synchronous with its revolution, which is retrograde. It is the only large natural satellite to revolve in this direction.

Most of our knowledge of Triton comes from images transmitted by Voyager-2 –see Figure 11.14. As far as it has been observed, these images show a cracked surface, rather like the skin of a melon, and referred to as a cantaloupe terrain. One can make out geysers with plumes rising up to 8 km, although we do not know what they could be ejecting at this temperature. The inclination of Triton with respect to the ecliptic creates a sequence of seasons, as on the Earth and Mars.

A mission to Triton would be a difficult undertaking (it is a long way to Neptune) and it will be some time before we see a satellite in orbit around

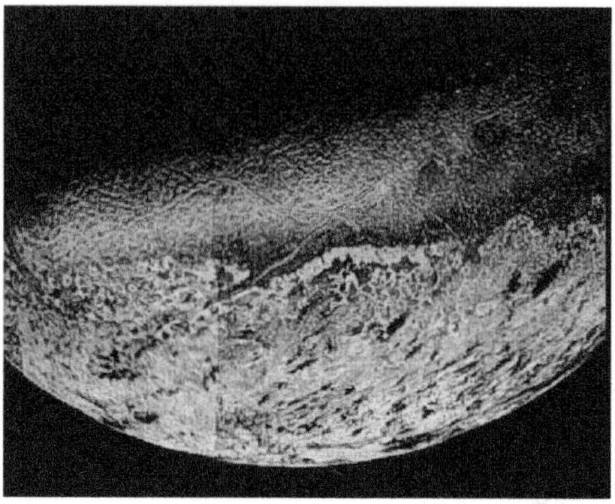

Figure 11.14. Voyager-2 passed by Triton about 5 hours after skimming within 5000 km of the cloudtops of Neptune. Triton is one of the most unusual objects encountered during all the Voyager planetary flybys. This image is a digital mosaic of 12 individual images. The large south polar cap at the bottom of the image is a slowly evaporating later of frozen nitrogen. Voyager data showed that Triton is extremely cold (daytime temperature of 37 K), extremely bright (reflecting nearly 100 % of the sunlight incident upon it) and has a very tenuous atmosphere of nitrogen and methane (with a surface pressure 10 millionths of Earth's atmosphere at sea level). Credit (legend and photography): JPL/NASA

this icy body. However, in the next example, we discuss the ground track of a hypothetical Triton Orbiter!

Example 11.7. *Orbital track of a satellite around Triton.*

We have considered a satellite with the same reduced altitude and the same inclination as the Europa Orbiter. The orbital track has been plotted over one day in Fig. 11.13 (lower). The novelty here is that the shift occurs towards the east. Before bringing this book to a close, we had to find an example of a celestial body in the Solar System which goes round the 'wrong' way!

References

Books are classified in sections according to the main themes covered in this work, and arranged chronologically within each section.

General Mechanics

1. H. Goldstein. *Classical Mechanics*, Addison-Wesley, Cambridge, Mass., 1956
2. E.T. Whittaker & G.N. Watson. *A Course of Modern Analysis*, 4th edn., Cambridge University Press, Cambridge, 1962
3. L. Landau & E. Lifchitz. *Mechanics (Course of Theoretical Physics)*, Vol. 1, Mir, Moscow, 1966, Butterworth–Heinemann 3rd edn., 1976
4. A. Foch. *Mécanique*, Cours de Physique générale de G. Bruhat, Masson, Paris, 1967
5. J. Lelong-Ferrand & J.-M. Arnaudiès. *Cours de Mathématiques*, Vol. 3: *Géométrie et cinématique*, Dunod, Paris, 1977
6. V.I. Arnold. *Mathematical Methods of Classical Mechanics*, Graduate Texts in Mathematics (60), Springer-Verlag, Berlin, 1989
7. J.-P. Pérez. *Mécanique. Points matériels, solides, fluides*, Masson, Paris, 1995

Space Mechanics

8. F. Tisserand. *Traité de mécanique céleste*, Vols. 1, 2, 3, 4, Gauthier-Villars, Paris, 1889, 1891, 1894, 1896
9. F.R. Moulton. *An Introduction to Celestial Mechanics*, The Macmillan Company, New York, 1902 (Dover Publ., New York, 1970)
10. A. Danjon. *Astronomie générale*, J. & R. Sennac, Paris, 1953
11. D. Brouwer & G.M. Clemence. *Methods of Celestial Mechanics*, Academic Press, New York, 1961
12. H.H. Koelle (Ed.). *Handbook of Astronautical Engineering*, McGraw–Hill, New York, 1961
13. D. King-Hele. *Theory of Satellite Orbits in an Atmosphere*, Butterworths Mathematical Texts, Butterworths & Co. Publ., London, 1964
14. P.R. Escobal. *Method of Orbit Determination*, John Wiley & Sons, New York, 1965
15. Y. Hagihara. *Celestial Mechanics*, Vol. 1: *Dynamical Principles and Transformation Theory*; Vol. 2: *Perturbation Theory*, MIT Press, Cambridge, Mass., 1970, 1972
16. B. Morando. *Mouvement d'un satellite artificiel de la Terre*, Gordon & Breach, Paris, 1974

17. A.E. Roy. *Orbital Motion*, Adam Hilger, Bristol, 1982
18. J.-P. Carrou (Ed.). *Mathématiques spatiales*, CNES/Cepadues-Editions, Toulouse, 1984
19. O. Zarrouati. *Trajectoires spatiales*, CNES/Cepadues-Editions, Toulouse, 1987
20. R.H. Battin. *An Introduction to the Mathematics and Methods of Astrodynamics*, AIAA Educations series, Amer. Inst. of Aeronautics and Astronautics, New York, 1987
21. G. Pascoli. *Eléments de mécanique céleste*, Armand Colin, Paris, 1993
22. J.-P. Carrou (Ed.). *Mécanique spatiale*, Vols. 1, 2, CNES/Cepadues-Editions, Toulouse, 1995
23. D. Boccaletti & G. Pucacco. *Theory of Orbits*, Vol. 1: *Integrable systems and non-perturbative methods*; Vol. 2: *Perturbative and Geometrical methods*, Springer-Verlag, Berlin, 2001
24. D. Oliver. *The Shaggy Steeds of Physics: Mathematical Beauty in the Physical World*, 2nd edn., Springer-Verlag, Berlin, 2004

Geodesy

25. W.M. Kaula. *Theory of Satellite Geodesy*, Blaisdell Publ., Waltham, Mass., 1966
26. J.-J. Levallois. *Géodésie générale*, Vols. 1, 2, 3, Eyrolles, Paris, 1969, 1970
27. J.-J. Levallois & J. Kovalevsky. *Géodésie générale*, Vol. 4: *Géodésie spatiale*, Eyrolles, Paris, 1970
28. J.-C. Husson, A. Cazenave, J.-F. Minster (Eds.). *Internal Geophysics and Space*, CNES/Cepadues-Editions, Toulouse, 1985
29. W. Torge. *Geodesy*, Walter de Gruyter, Berlin, 1991
30. L. Lliboutry. *Sciences géométriques et télédétection*, Masson, Paris, 1992
31. G. Seeber. *Satellite Geodesy*, Walter de Gruyter, Berlin, 1993
32. A. Cazenave & K. Feigl. *Formes et mouvements de la Terre. Satellites et géodésie*, Belin/CNRS-Editions, Paris, 1994
33. T.J. Ahrens (Ed.). *Global Earth Physics: A Handbook of Physical Constants*, American Geophysical Union (AGU), Washington, 1995
34. T. Jones. *Splitting the Second: The Story of Atomic Time*, Institute of Physics, Bristol, 2000

History of Celestial Mechanics and Geodesy

35. G. Veis (Ed.). *The Use of Artificial Satellites for Geodesy*, Proceedings of the First International Symposium on the Use of Artificial Satellites for Geodesy, Washington (April 1962), North-Holland, Amsterdam, 1963
36. J.-J. Levallois. *Mesurer la Terre. 300 ans de géodésie française*, Presses de l'Ecole Nationale des Ponts et Chaussées, Paris, 1988
37. J. Dhombres (Ed.). *Leçons de Mathématiques. L'Ecole Normale de l'an III. Edition annotée des cours de Laplace, Lagrange et Monge*, Dunod, Paris, 1992

38. D. King-Hele. *A Tapestry of Orbits*, Cambridge University Press, Cambridge, 1992
39. W. Andrewes (Ed.). *The Quest for Longitude*, Harvard University, Cambridge, Mass., 1993
40. J.L. Greenberg. *The Problem of the Earth's Shape from Newton to Clairaut*, Cambridge University Press, Cambridge, 1995
41. R. D'Hollander. *L'Astrolabe. Histoire, théorie et pratique*, Institut océanographique, Paris, 1999

Mathematics of Cartography

42. A. Germain. *Traité des projections des cartes géographiques*, Arthus Bertrand Editeur, Paris, 1865
43. L. Driencourt & J. Laborde. *Traité des projections cartographiques à l'usage des cartographes et des géodésiens*, Libr. Sc. Hermann, Paris, 1932
44. J.P. Snyder. *Flattening the Earth: Two Thousand Years of Map Projections*, University of Chicago Press, Chicago, 1993
45. J.P. Snyder & P.M. Voxland. *An Album of Map Projections*, U.S. Geological Survey professional paper 1453, Dept. of the Interior, USGPO, Washington, 1994

Satellites and Missions

46. J.R. Wertz & W.J. Larson (Eds.). *Space Mission Analysis and Design*, Space Technology Library, Kluwer Academic Publishers, Dordrecht, 1991
47. R.J. Gurney, J.L. Foster, C.L. Parkinson (Eds.). *Atlas of Satellite Observations Related to Global Change*, Cambridge University Press, Cambridge & New York, 1993
48. S.Q. Kidder & T.H. Vonder Haar. *Satellite Meteorology – An Introduction*, Academic Press, San Diego, 1995
49. M. Courtois (Ed.). *Techniques et technologies des véhicules spatiaux*, Vols. 1, 2, 3, CNES/Cepadues-Editions, Toulouse, 1998
50. O. Montenbruck & E. Gill. *Satellite Orbits: Models, Methods, and Applications*, Springer-Verlag, Berlin, 2000
51. M. Capderou. *Satellites d'observation de la Terre*, Vols. 1, 2, Ecole Polytechnique, Palaiseau, 2000
52. L.-L. Fu & A. Cazenave (Eds.). *Satellite Altimetry and Earth Sciences*, Intern. Geophysics Series, Vol. 69, Academic Press, San Diego, 2001
53. C. Lämmerzahl, C.W. Francis Everitt, F.W. Hehl (Eds). *Gyros, Clocks, Interferometers: Testing Relativistic Gravity in Space*, Lectures Notes in Physics, Vol. 562, Springer-Verlag, Berlin, 2001
54. H.J. Kramer. *Observation of the Earth and its Environment. Survey of Missions and Sensors*, 4th edn., Springer-Verlag, Berlin, 2002
55. P. Fortescue, J. Stark, & G. Swinerd. *Spacecraft Systems Engineering*, 3rd edn., John Wiley & Sons, Chichester, 2003
56. F. Verger, I. Sourbès-Verger, R. Ghirardi. *The Cambridge Encyclopedia of Space – Missions, Applications and Exploration*, Cambridge University Press, Cambridge, 2003

Planets of the Solar System

57. W.M. Kaula. *An Introduction to Planetary Physics. The Terrestrial Planets*, John Wiley & Sons, New York, 1966
58. Z. Kopal. *The Moon in the Post-Apollo Era*, D. Reidel Publishing Company, Dordrecht & Boston, 1974
59. J.-C. Pecker (Ed.). *Astronomie Flammarion*, Vols. 1, 2, Flammarion, Paris, 1985
60. S.K. Atreya, J.B. Pollack, M.S. Matthews (Eds.). *Origin and Evolution of Planetary and Satellite Atmospheres*, The University of Arizona Press, Tucson, 1989
61. H.H. Kieffer, B.M. Jakosky, C.W. Snyder, M.S. Matthews (Eds.). *Mars*, The University of Arizona Press, Tucson, 1992
62. D.P. Cruikshank (Ed.). *Neptune and Triton*, The University of Arizona Press, Tucson, 1995
63. S.W. Bougher, D.M. Hunten, R.J. Phillips (Eds.). *Venus II*, The University of Arizona Press, Tucson, 1997
64. J.-L. Simon, M. Chapront-Touzé, B. Morando, W. Thuillot (Eds.). *Introduction aux éphémérides astronomiques. Supplément explicatif à la Connaissance du Temps*, Bureau des Longitudes, EDP-Sciences, Paris, 1998
65. C.D. Murray, S.F. Dermott. *Solar System Dynamics*, Cambridge University Press, Cambridge, 1999
66. P.R. Weissman, L.-A. McFadden, T.V. Johnson (Eds). *Encyclopedia of the Solar System*, Academic Press, San Diego, 1999
67. D. Fischer. *Mission Jupiter. The Spectacular Journey of the Galileo Spacecraft*, Copernicus Books, New York, 2001
68. R. Greeley, R. Batson. *The Compact NASA Atlas of the Solar System*, Cambridge University Press, Cambridge, 2001
69. A. Brahic. *Planètes et satellites. Cinq leçons d'astronomie*, Vuibert, Paris, 2001
70. F. Forget, F. Costard, P. Lognonné. *La Planète Mars. Histoire d'un autre monde*, Belin, Paris, 2003
71. T. Encrenaz & J.-P. Bibring. *The Solar System*, Springer-Verlag, Berlin, 2004

History of Astronomy

72. A. Beer, P. Beer (Eds). *Kepler – Four Hundred Years*, Pergamon Press, Oxford, 1975
73. J.-R. Roy. *L'Astronomie et son histoire*, Vols. 1, 2, Presses de l'Université du Québec/Editions Masson, Québec & Paris, 1982
74. J. Blamont. *Le Chiffre et le Songe. Histoire politique de la découverte*, Odile Jacob, Paris, 1993
75. M. Lachièze-Rey & J.-P. Luminet. *Figures du ciel. De l'harmonie des sphères à la conquête spatiale*, Seuil/Bibliothèque nationale de France, Paris, 1998

Index

– M –

– P –

– R –

– T –

Ιξίων

The figure from Greek mythology known as Ixion was not, if the truth be told, a particularly savoury one. The King of the Lapiths, he behaved in a decidedly reprehensible manner on the day of his wedding, causing his future father-in-law to fall into a burning pit so that he would not have to pay the dowry.

This act was considered the ultimate crime, for it broke all the rules of hospitality, and indeed, it was reproved by all the gods but one. The only deity who would agree to purify Ixion for the murder was Zeus, a connaisseur when it came to perjury and other misdemeanours. Zeus even felt some compassion for this strong-minded king, inviting him to Olympus and offering him hospitality. As an exceptional sign of friendship, he bade him drink ambrosia, which made him immortal.

Ixion admired Zeus' antics and escapades and, encouraged by the atmosphere of familiarity in the Olympian realm, began to covet Zeus' own wife Hera. But this was where he overstepped the mark! The king of the gods cried out: "A little respect for one's host!" As a punishment, he bound him to a fiery wheel which whirled him forever through the skies.

As he had been made immortal, the poor fellow must still be spinning around up there. One may thus consider Ixion as the first of all artificial satellites, and this is therefore the name we have chosen for our software.

IXION DIXIT

CD Software

The CD software Ixion dixit accompanying this book comprises two parts: Program and Graphs.

Program

The program IxionPC is an interactive orbitography and sampling program. One can select a satellite on the list included, or define one at will. Such a definition can be made directly by feeding in an altitude, inclination, eccentricity, etc., or indirectly, by choosing the period or recurrence cycle.

The program IxionPC establishes the orbit, then provides various points on the ground track. When it is given the characteristics of the swath, it calculates the spatial and temporal sampling. For a given target point, this sampling appears in the form of a table giving the time at which the satellite points at the target and the angles describing the target–satellite geometry. All these calculations can be carried out for the Earth and Mars.

IxionPC is part of the Ixion software, in particular, excluding the graphs part.

Graphs

The Graphs part contains hundreds of charts showing satellite tracks and orbits. This part is not interactive. There is a wide range of orbital parameters in order to provide a varied panorama, reflecting the great diversity of satellite orbits. It also includes a great many sampling tables, most of them taking into account the direction of solar illumination.

In Graphs, sampling is given for Earth- and Mars-orbiting satellites, while the ground track, orbit and swath are also represented for satellites of Mercury, Venus, the asteroid Eros and the natural satellites Europa and the Moon.

How to Use the CD

Insert the CD Ixion dixit in the appropriate drive of your computer. The welcome page will appear with instructions for use of the software.

Colour Plate I. The Earth and the Moon. Image taken by the Galileo probe in December 1992. After three gravity-assist maneuvers (Venus, Earth, Earth), the probe was on its way towards Jupiter to explore the Jovian system (which it did from 1995). On the Earth, one can make out the Pacific and South America, and on the Moon, the Tycho crater. The images of the Earth and the Moon were taken separately and put together with the same colour and albedo scales. Credit: ©NASA/JPL, Galileo Team

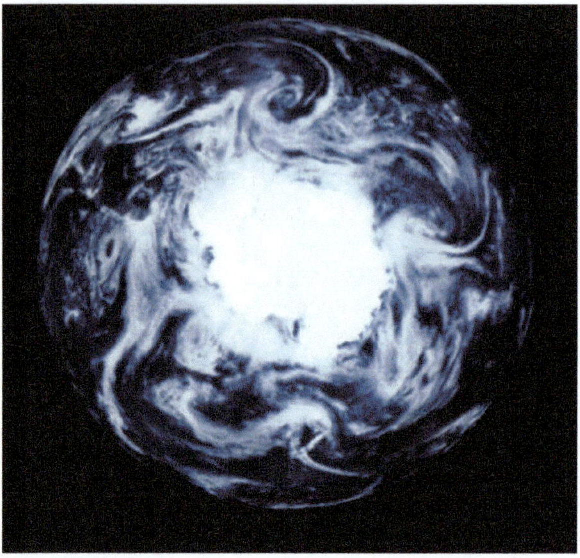

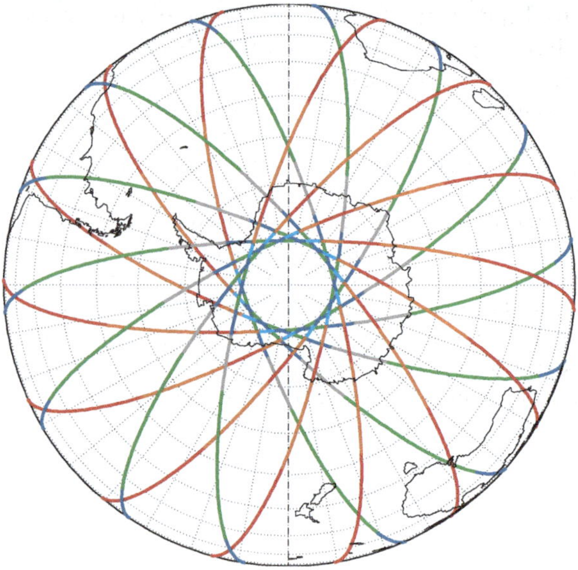

Colour Plate II. *Upper*: rather novel image of the Earth's South Pole taken by Galileo as it headed off for Jupiter in December 1992. Composite image taken over a 24 hr period. Note the characteristic shape of the cloud systems (mid-latitude weather fronts). Credit: ©NASA/JPL, Galileo Team. *Lower*: Ixion/Atlas reconstitution of the map of the Earth seen from the same point, showing the orbital track of an ERS/Envisat-type satellite indicating local time (ascending node 22:15 LMT, *colour scale* as for MetOp-1 in Colour Plate VIII). Using the map, one can make out the dark forms of Australia and South Africa, without cloud cover, on the photograph

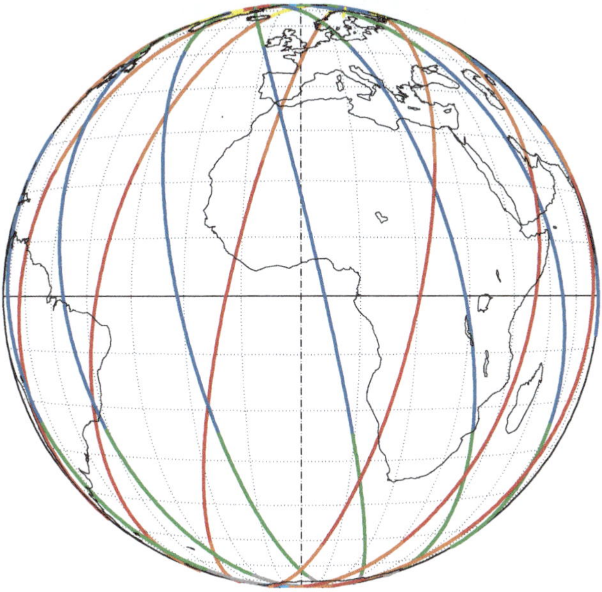

Colour Plate III. *Upper*: classic image of the Earth's disk as viewed by the geostationary satellite METEOSAT-7. Colour image from VIS (visible) channel, 4 September 2001, 12:00 UT. Credit: ©2002 EUMETSAT. *Lower*: Ixion/Atlas reconstitution of the map of the Earth from the same viewing point, showing the orbital track of an ERS/Envisat-type satellite indicating local time (ascending node 22:15 LMT, *colour scale* as for MetOp-1 in Colour Plate VIII)

Colour Plate IV. Brittany, Cornwall, the English Channel and the Iroise sea. Image taken by the MISR instrument (Multi-angle Imaging SpectroRadiometer) aboard Terra. Revolution 7778, 4 June 2001. The turquoise areas off the coast of Brittany reveal the intense efflorescence of phytoplankton (coccolithophores), whose shields made of tiny platelets (of the order of μm) scatter solar light. *At the top* of the picture, the band of cirrus clouds is striped with aircraft trails. Credit: ©NASA/GSFC/LaRC/JPL, MISR Team

Colour Plate V. Fire and smoke across Mato Grosso, Brazil. Image taken by the MODIS instrument (Moderate Resolution Imaging Spectrometer) aboard Terra on 6 September 2004. In central South America, at the border of Mato Grosso state, Brazil (*top*) and Bolivia (*bottom left*) numerous fires choked the skies with smoke on 6 September 2004, when the Terra MODIS instrument captured this image. Actively burning fires detected by MODIS during the overpass are marked in *red*. At the *centre* of the image, a small swath of green hangs down from intact rainforest to the north. Credit: © J. Descloitres, MODIS Land Rapid Response Team, NASA/GSFC

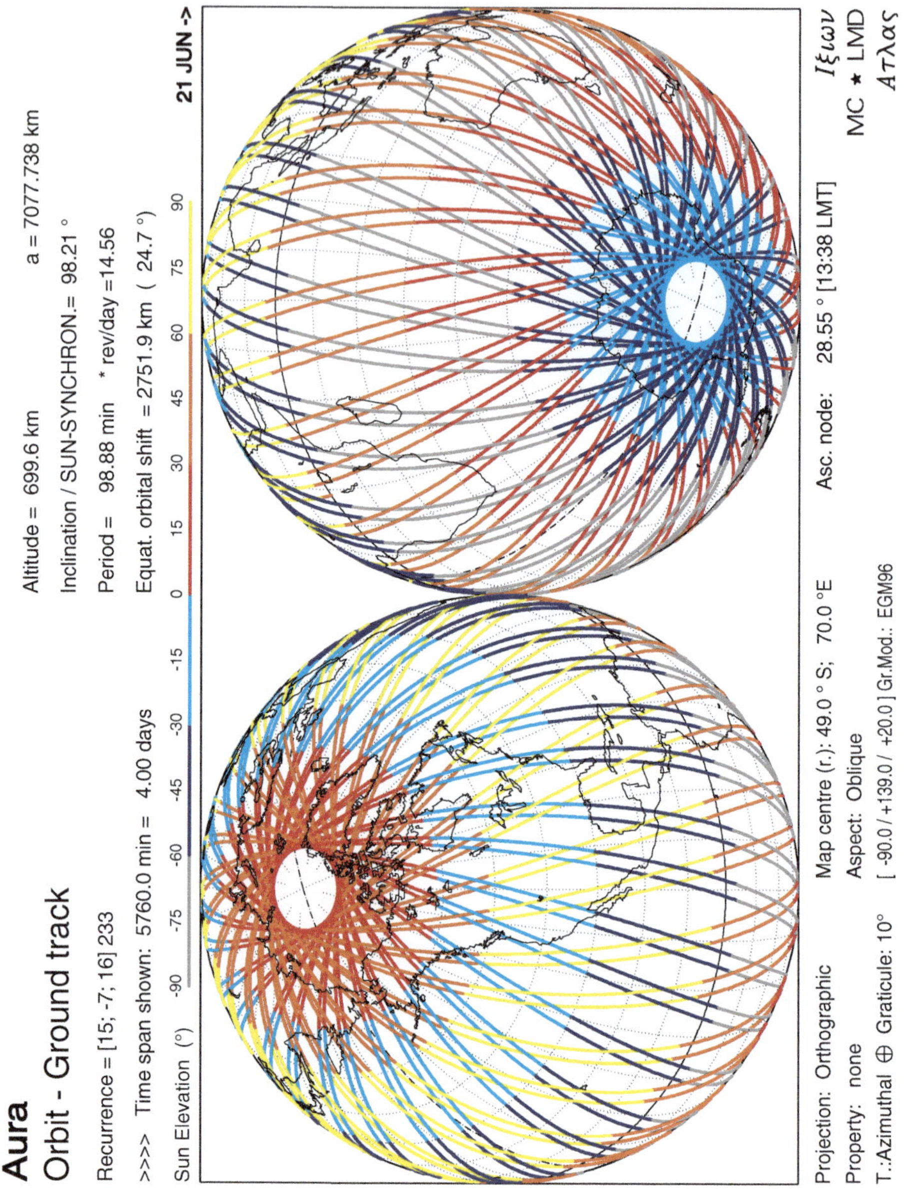

Colour Plate VI. Ground track of the Sun-synchronous satellite Aura, indicating the solar elevation, over four days

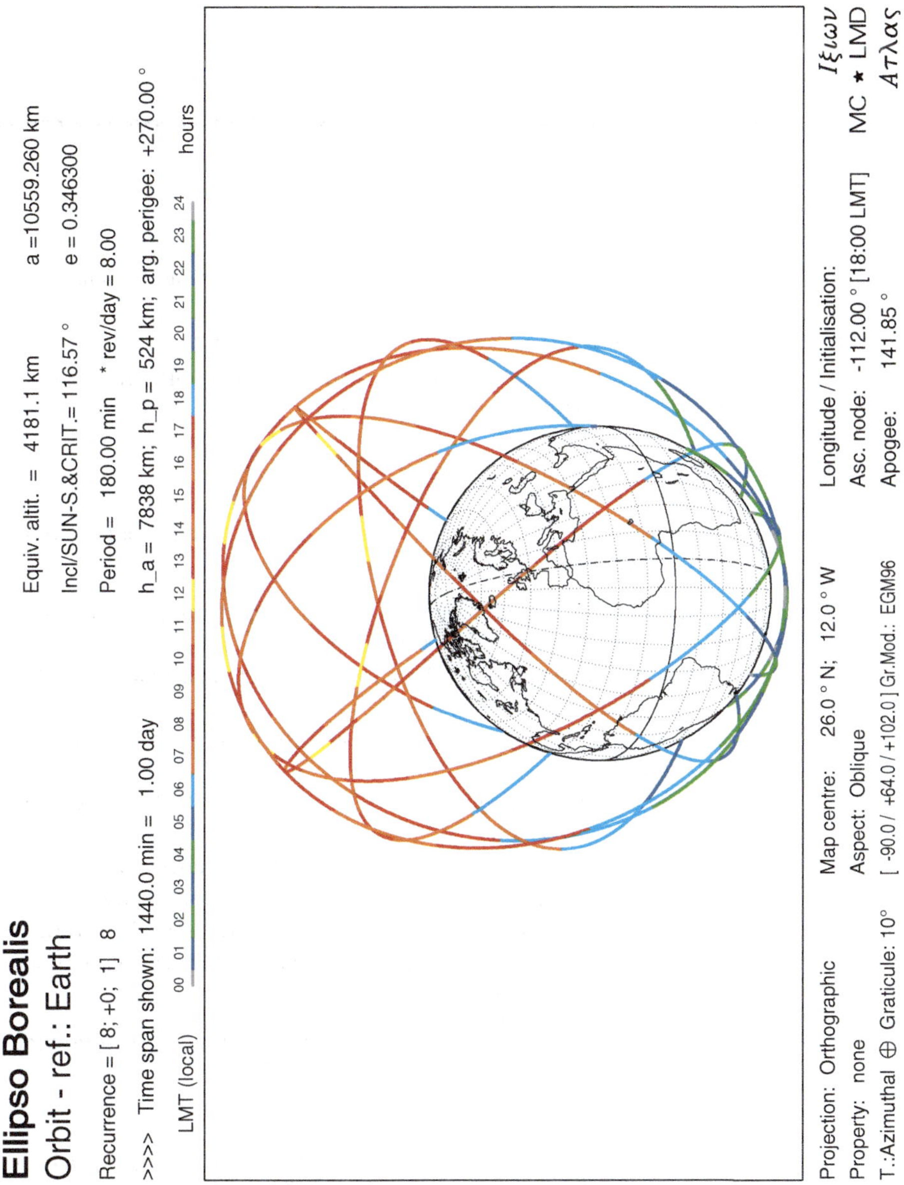

Colour Plate VII. Orbit of the satellite Ellipso Borealis, indicating the local mean time (LMT), over one day. The orbit is Sun-synchronous with critical inclination

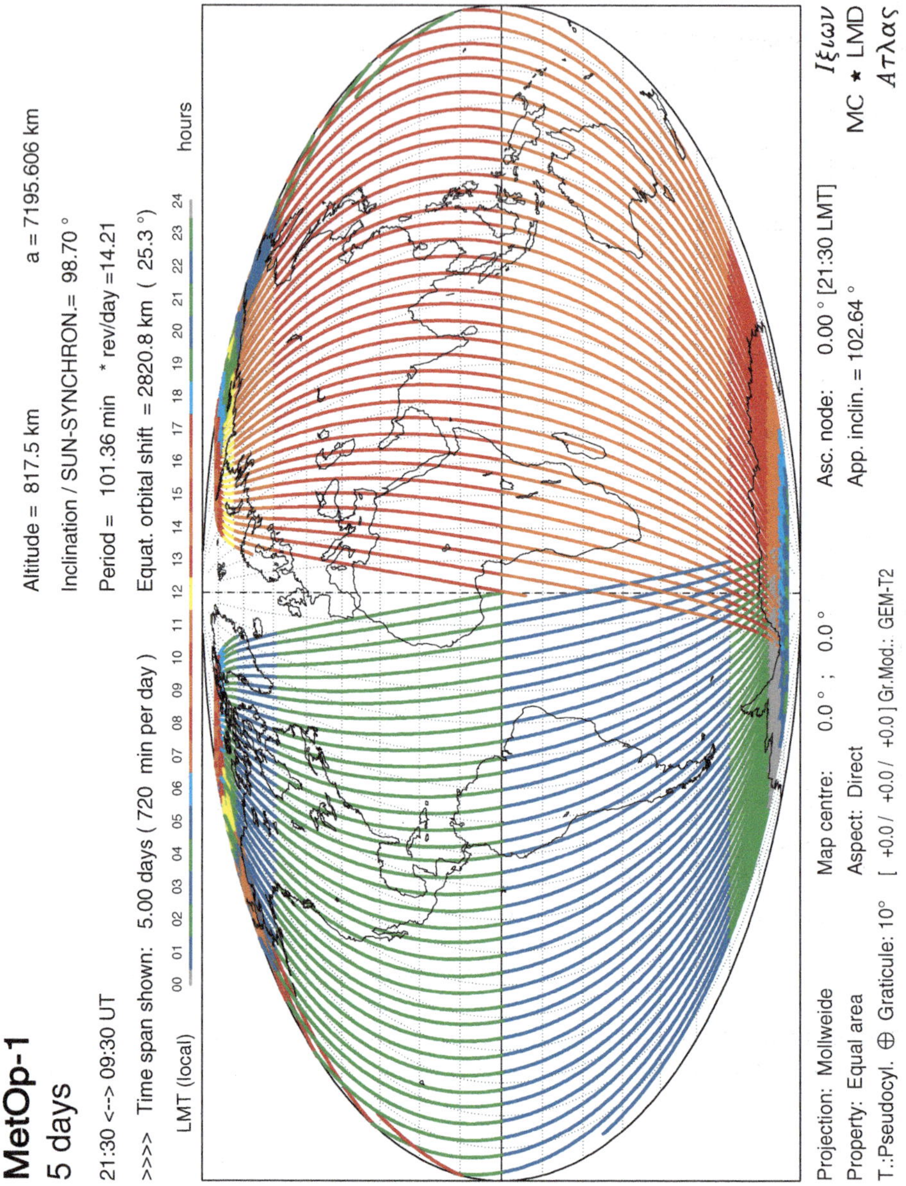

Colour Plate VIII. Ground track of the Sun-synchronous satellite MetOp-1, indicating the local mean time (LMT), over five days, between 21:30 and 09:30 UT each day

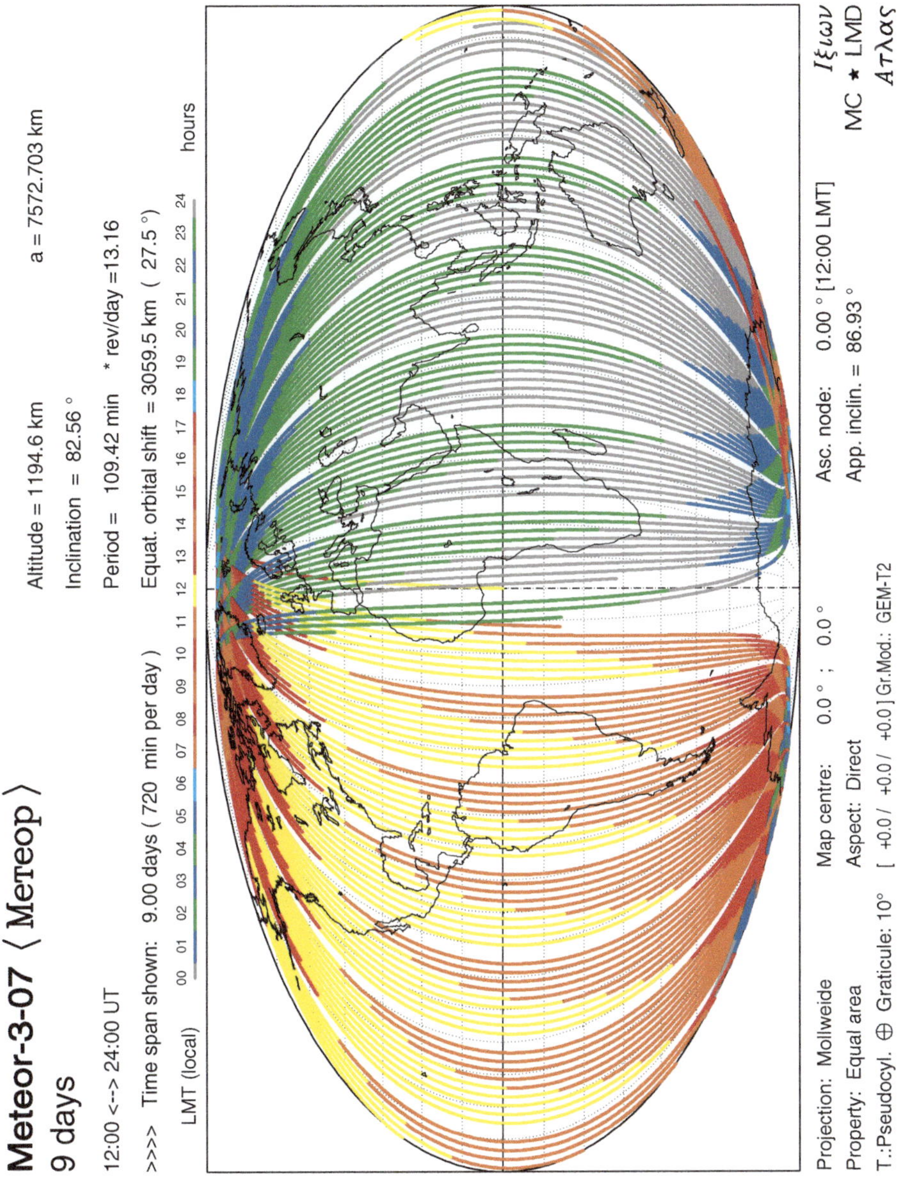

Colour Plate IX. Ground track of the non-Sun-synchronous satellite Meteor-3-07, indicating the local mean time (LMT), over nine days, between 12:00 and 24:00 UT each day

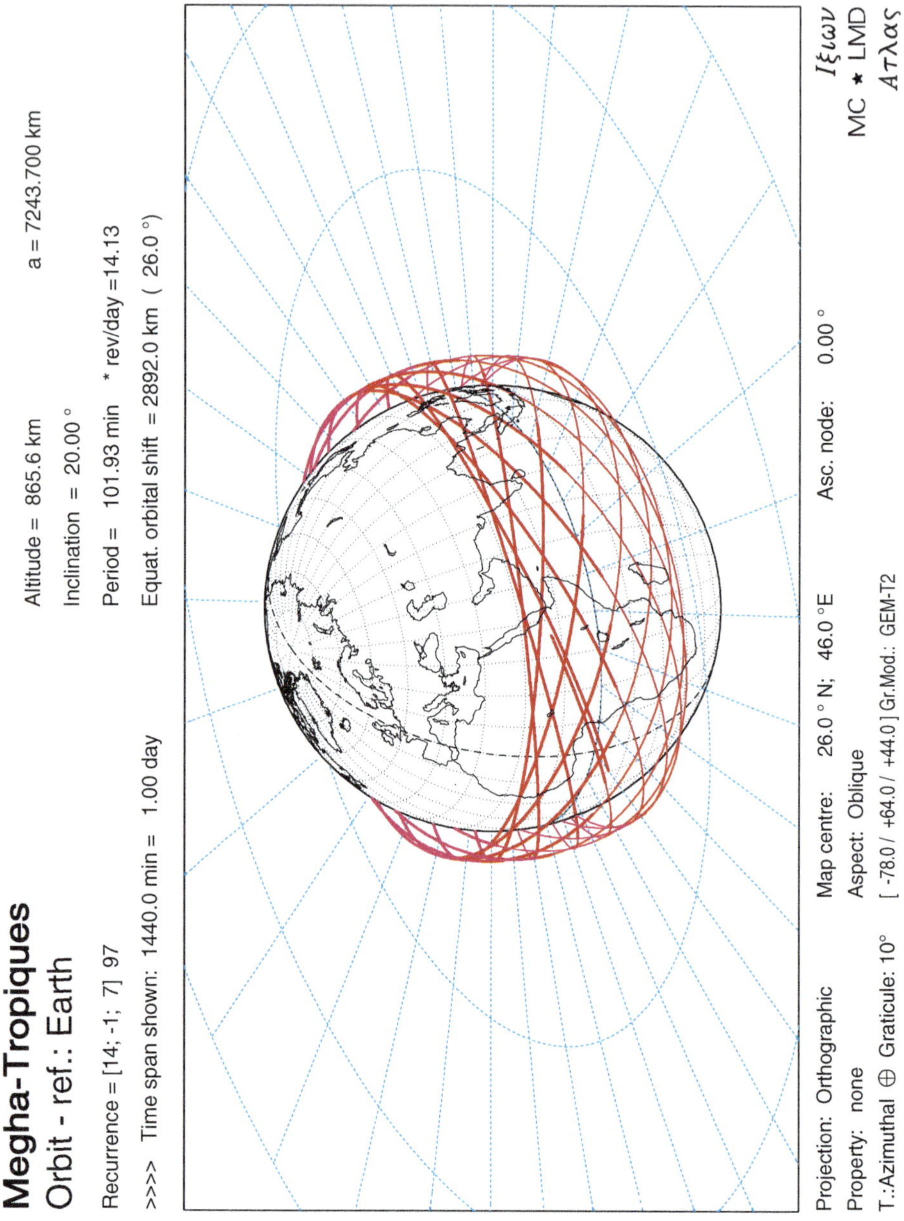

Colour Plate X. Orbit of the satellite Megha-Tropiques relative to the Earth, represented over one day. The Earth's equatorial plane has been shown with a blue grid

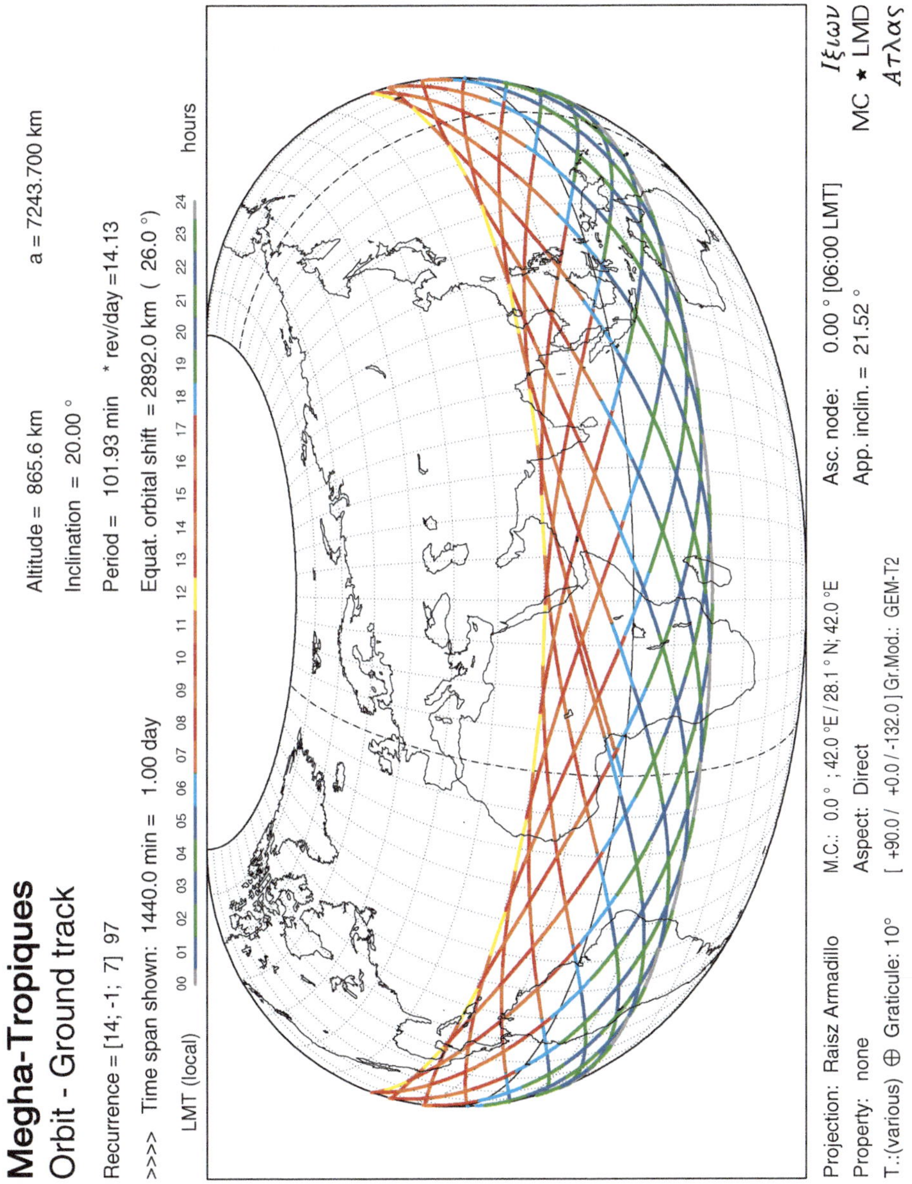

Colour Plate XI. Orbital track of the satellite Megha-Tropiques, indicating the local mean time (LMT), over one day

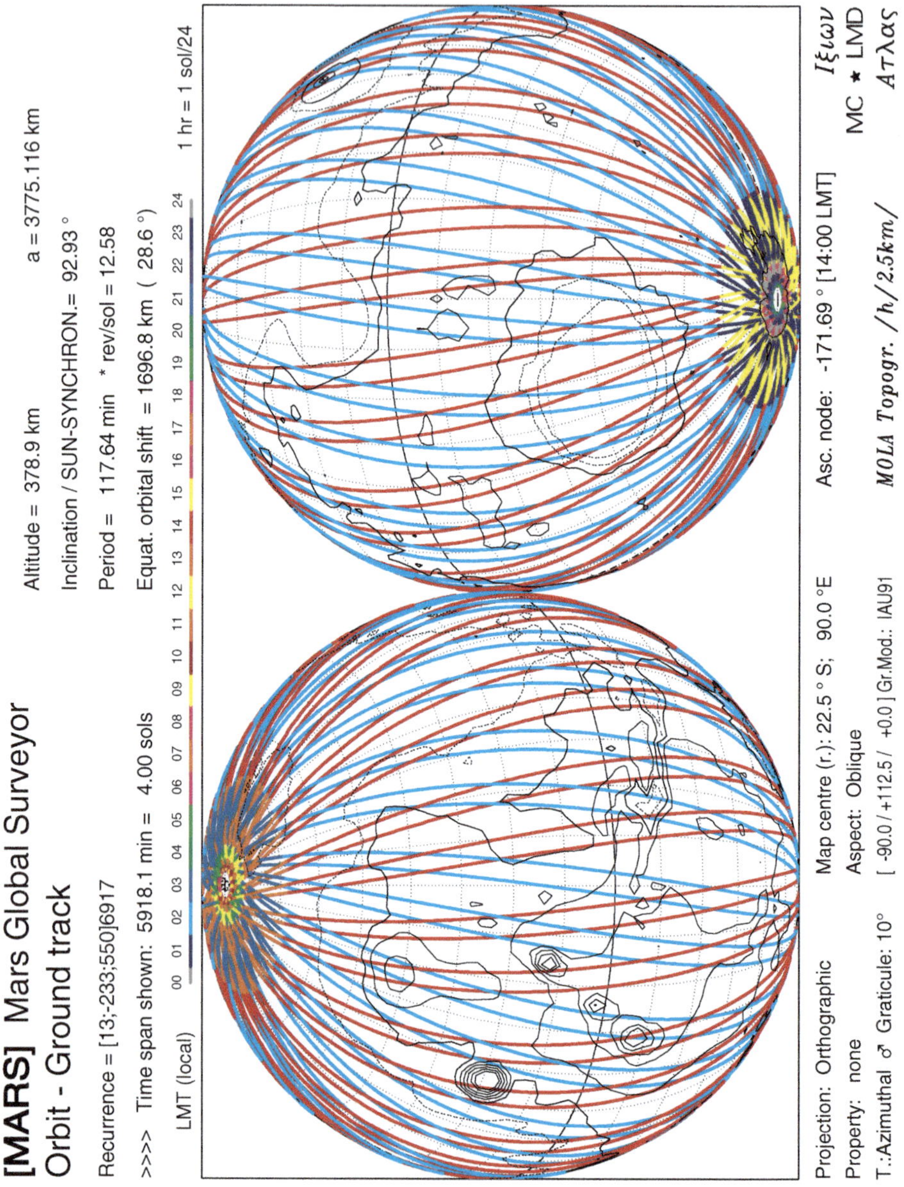

Colour Plate XII. Mars. Ground track of the Sun-synchronous satellite Mars Global Surveyor (MGS), indicating the local mean time (LMT), over four sols (Martian days). The projection used here (oblique orthographic) is the same as the one used for the topography of Mars on the next page

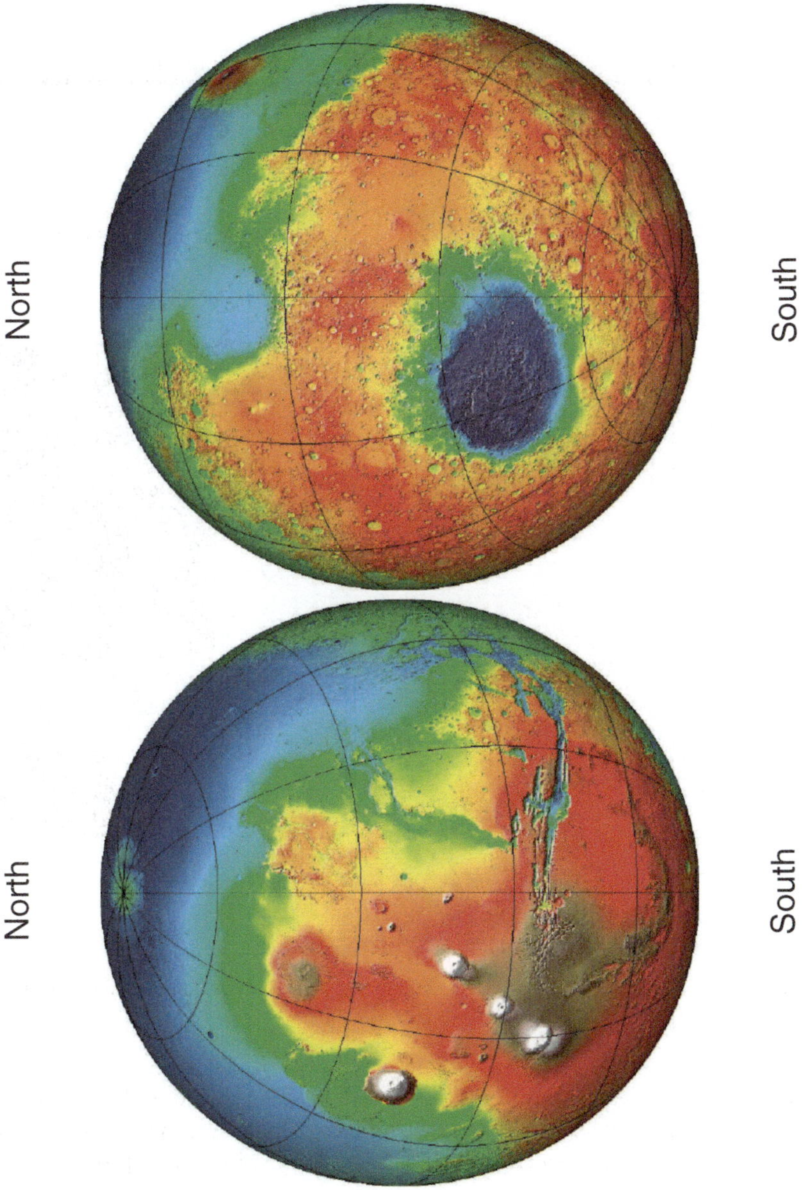

Colour Plate XIII. False colour topographical map of Mars, from data gathered by the MOLA instrument aboard MGS. One can make out Valles Marineris, the volcano Olympus Mons and the Tharsis Rise with its three volcanoes. Almost at the antipodes of the Tharsis Rise is the Hellas impact basin. Credit: ©NASA/JPL, MOLA Team

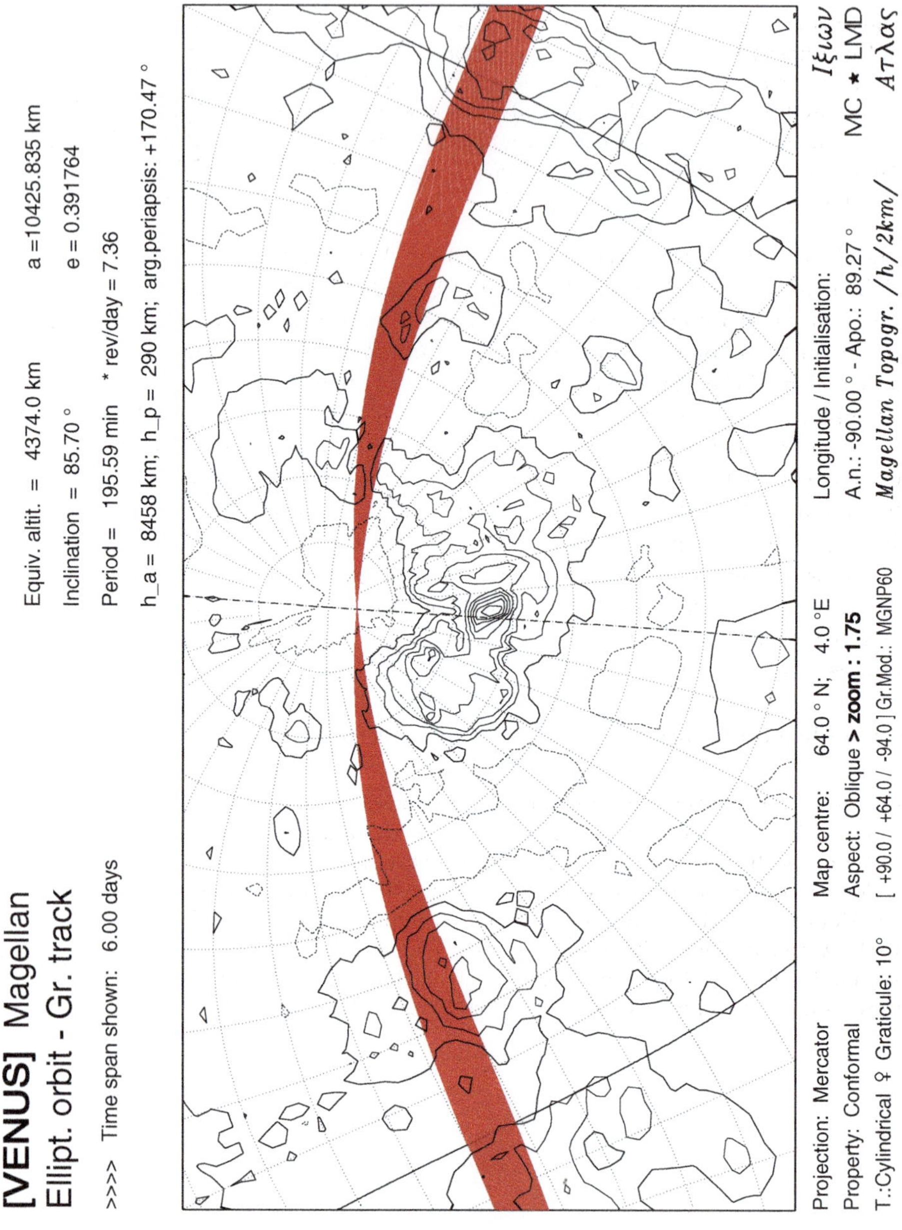

Colour Plate XIV. Venus. Orbital track of the satellite Magellan, over six Earth days. The projection (oblique Mercator) is centered on Maxwell Montes

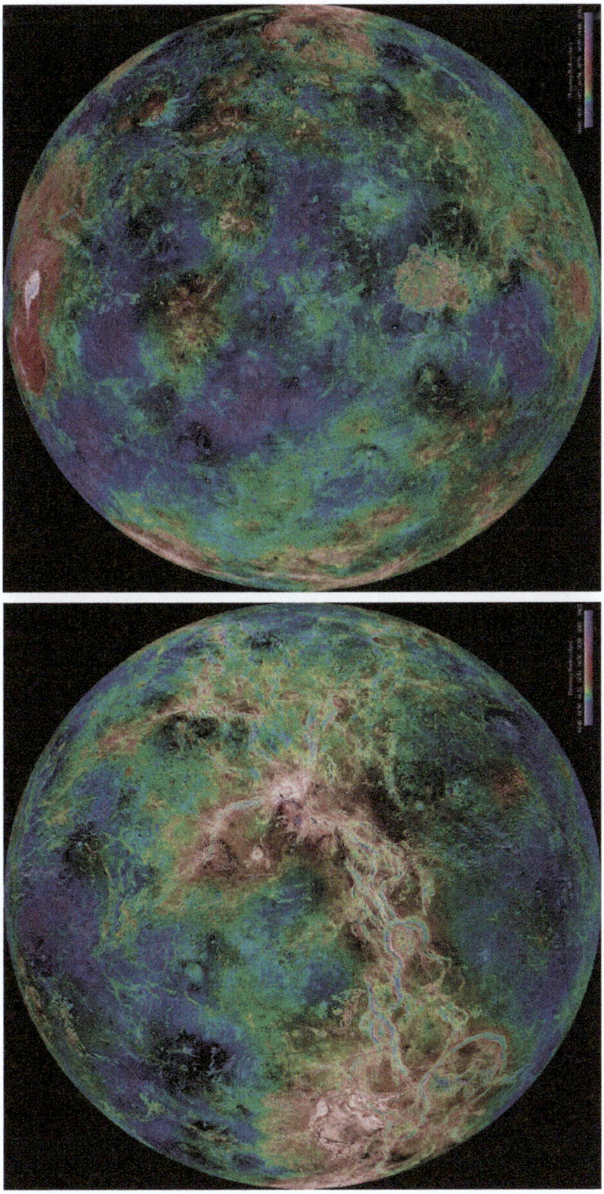

Colour Plate XV. Topographical map of Venus from data collected by the SAR (radar) instrument aboard Magellan. Equatorial orthographic projections centered on longitudes 0° (*top*) and 180° (*bottom*). Maxwell Montes can be made out in the Ishtar Terra region, centered on 64°N, 4°E. Credit: ©NASA/JPL/MIT/USGS, Magellan Team

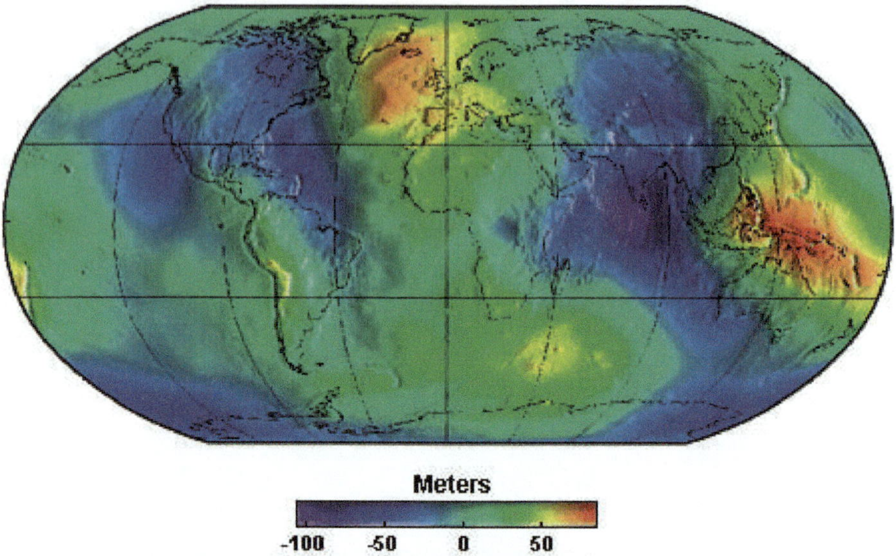

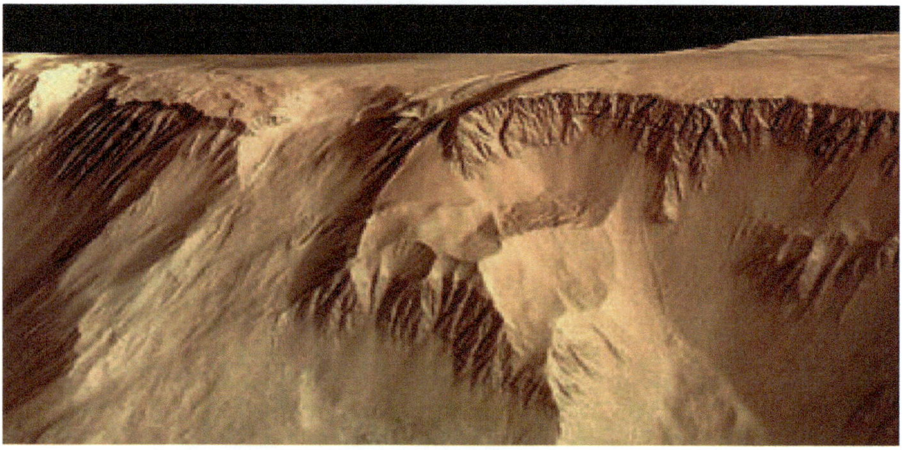

Colour Plate XVI. *Upper*: geopotential obtained using the model EGM96 (NASA GSFC and NIMA Joint Geopotential Model). This shows the difference in metres between the geoid and the terrestrial ellipsoid. Robinson projection. Credit: ©NASA, OSU, NIMA. *Lower*: planet Mars. Close-up perspective view of western flank of Olympus Mons. Images taken by the High Resolution Stereo Camera (HRSC) during Revolution 143 (21 April 2004) from an altitude of 266 km, aboard Mars Express. Colour image created from the nadir and three colour channels. Resolution: 25 m per pixel. Credit: ©ESA, Mars Express, DLR, FUB

Achevé d'imprimer sur les presses de l'Imprimerie BARNÉOUD
B.P. 44 - 53960 BONCHAMP-LÈS-LAVAL
Dépôt légal : Janvier 2005 – N° d'imprimeur : 15584
Imprimé en France

GPSR Compliance

The European Union's (EU) General Product Safety Regulation (GPSR) is a set of rules that requires consumer products to be safe and our obligations to ensure this.

If you have any concerns about our products, you can contact us on ProductSafety@springernature.com

In case Publisher is established outside the EU, the EU authorized representative is:

Springer Nature Customer Service Center GmbH
Europaplatz 3
69115 Heidelberg, Germany

Batch number: 10854370

Printed by Printforce, the Netherlands